BIRKHÄUSER

Operator Theory: Advances and Applications

Vol. 192

Holomorphic Operator Functions of One Variable and Applications

Methods from Complex Analysis in Several Variables

Israel Gohberg
Jürgen Leiterer

Birkhäuser
Basel · Boston · Berlin

Authors:

Israel Gohberg
Department of Mathematical Sciences
Raymond and Beverly Sackler
Faculty of Exact Sciences
Tel Aviv University
69978 Ramat Aviv
Israel
e-mail: gohberg@post.tau.ac.il

Jürgen Leiterer
Institut für Mathematik
Humboldt-Universität zu Berlin
Rudower Chaussee 25
12489 Berlin
Germany
e-mail: leiterer@math.hu-berlin.de

2000 Mathematical Subject Classification: 47-02, 47A56, 47A68, 47B35; 30-02, 30E05; 32L05, 32L10, 32L20

Library of Congress Control Number: 2009927556

Bibliographic information published by Die Deutsche Bibliothek.
Die Deutsche Bibliothek lists this publication in the Deutsche Nationalbibliografie; detailed bibliographic data is available in the Internet at http://dnb.ddb.de

ISBN 978-3-0346-0125-2 Birkhäuser Verlag AG, Basel - Boston - Berlin

Basel · Boston · Berlin
P.O. Box 133, CH-4010 Basel, Switzerland
Part of Springer Science+Business Media
Printed on acid-free paper produced from chlorine-free pulp. TCF∞
Printed in Germany

ISBN 978-3-0346-0125-2

e-ISBN 978-3-0346-0126-9

9 8 7 6 5 4 3 2 1

www.birkhauser.ch

Dedicated to the memory
of our friend, student and colleague

Georg Heinig

Contents

Preface xi

Introduction xiii

Notation xix

1 Elementary properties of holomorphic functions **1**

1.1 Definition and first properties 1
1.2 The maximum principle 3
1.3 Contour integrals 8
1.4 The Cauchy integral theorem 11
1.5 The Cauchy formula 12
1.6 The Hahn-Banach criterion 15
1.7 A criterion for the holomorphy of operator functions 18
1.8 Power series 19
1.9 Laurent series 23
1.10 Isolated singularities 25
1.11 Comments 28

2 Solution of $\overline{\partial} u = f$ and applications **29**

2.1 The Pompeiju formula for solutions of $\overline{\partial} u = f$ on compact sets . . 29
2.2 Runge approximation 37
2.3 Solution of $\overline{\partial} u = f$ on open sets 41
2.4 $\mathcal{O}^E$-cocycles and the Mittag-Leffler theorem 43
2.5 Runge approximation for invertible scalar functions and the Weierstrass product theorem 44
2.6 $\mathcal{O}^E$-cocycles with prescribed zeros and a stronger version of the Mittag-Leffler theorem 52
2.7 Generalization of the Weierstrass product theorem 54
2.8 Comments 58

3 Splitting and factorization with respect to a contour **59**
3.1 Splitting with respect to a contour 59
3.2 Splitting and the Cauchy Integral 61
3.3 Hölder continuous functions split 65
3.4 The splitting behavior of differentiable functions 72
3.5 Approximation of Hölder continuous functions 75
3.6 Example: A non-splitting continuous function 77
3.7 The additive local principle 82
3.8 Factorization of scalar functions with respect to a contour. First remarks 84
3.9 Factorization of Hölder functions 89
3.10 Factorization of Wiener functions 91
3.11 The multiplicative local principle 93
3.12 Comments 95

4 The Rouché theorem for operator functions **97**
4.1 Finite meromorphic Fredholm functions 97
4.2 Invertible finite meromorphic Fredholm functions 101
4.3 Smith factorization 106
4.4 The Rouché theorem 111
4.5 Comments 112

5 Multiplicative cocycles ($\mathcal{O}^G$-cocycles) **113**
5.1 Topological properties of $GL(E)$ 114
5.2 Two factorization lemmas 120
5.3 $\overline{\mathcal{O}}^E$-cocycles 123
5.4 Runge approximation of G-valued functions. First steps 125
5.5 The Cartan lemma 129
5.6 $\mathcal{O}^G$-cocycles. Definitions and statement of the main result 131
5.7 Refinement of the covering 133
5.8 Exhausting by compact sets 137
5.9 Proof of the main theorem in the case of simply connected open sets 141
5.10 Runge approximation of G-valued functions. General case 143
5.11 Proof of the main theorem in the general case 147
5.12 $\mathcal{O}^{\mathcal{G}^\infty(E)}$-cocycles 151
5.13 Weierstrass theorems 153
5.14 Weierstrass theorems for $\mathcal{G}^\infty(E)$ and $\mathcal{G}^\omega(E)$-valued functions . . . 156
5.15 Comments 157

6 Families of subspaces **159**
6.1 The gap metric 159
6.2 Kernel and image of operator functions 172
6.3 Holomorphic sections of continuous families of subspaces 184

6.4 Holomorphic families of subspaces 185
6.5 Example: A holomorphic family of subspaces with jumping isomorphism type 201
6.6 Injective families 203
6.7 Shubin families 204
6.8 Complemented families 206
6.9 Finite dimensional and finite codimensional families 209
6.10 One-sided and generalized invertible holomorphic operator functions 211
6.11 Example: A globally non-trivial complemented holomorphic family of subspaces 214
6.12 Comments 216

7 Plemelj-Muschelishvili factorization 219
7.1 Definitions and first remarks about factorization 220
7.2 The algebra of Wiener functions and other splitting $\mathcal{R}$-algebras . . 222
7.3 Hölder continuous and differentiable functions 230
7.4 Reduction of the factorization problem to functions, holomorphic and invertible on $\mathbb{C}^*$ 237
7.5 Factorization of holomorphic functions close to the unit 240
7.6 Reduction of the factorization problem to polynomials in z and $1/z$ 240
7.7 The finite dimensional case 242
7.8 Factorization of $\mathcal{G}^\infty(E)$-valued functions 245
7.9 The filtration of an opeator function with respect to a contour . . 251
7.10 A general criterion for the existence of factorizations 259
7.11 Comments 267

8 Wiener-Hopf operators, Toeplitz operators and factorization 269
8.1 Holomorphic operator functions 269
8.2 Factorization of $\mathcal{G}^\omega(E)$-valued functions 273
8.3 The space $\mathcal{L}^2(\Gamma, H)$ 276
8.4 Operator functions with values acting in a Hilbert space 287
8.5 Functions close to the unit operator or with positive real part . . . 291
8.6 Block Töplitz operators 297
8.7 The Fourier transform of $\mathcal{L}^1(\mathbb{R}, E)$ 305
8.8 The Fourier isometry $\mathbf{U}$ of $\mathcal{L}^2\big(\mathbb{R}, H\big)$ 313
8.9 The isometry $\mathbf{V}$ from $\mathcal{L}^2(\mathbb{T}, H)$ onto $\mathcal{L}^2(\mathbb{R}, H)$ 317
8.10 The algebra of operator functions $L(H) \oplus \mathcal{L}^1\big(\mathbb{R}, L(H)\big)$ 320
8.11 Factorization with respect to the real line 324
8.12 Wiener-Hopf integral operators in $\mathcal{L}^2\big([0,\infty[, H\big)$ 325
8.13 An example 338
8.14 Comments 340

9 Multiplicative cocycles with restrictions ($\mathcal{F}$-cocycles) 343
9.1 $\mathcal{F}$-cocycles . . . 343
9.2 The main results on cocycles with restrictions. Formulation and reduction to $\mathcal{O}_{D,Z,m}$ . . . 346
9.3 The Cartan lemma with restrictions . . . 347
9.4 Splitting over simply connected open sets after shrinking . . . 352
9.5 Runge approximation on simply connected open sets . . . 354
9.6 Splitting over simply connected open sets without shrinking . . . 358
9.7 Runge approximation. The general case . . . 362
9.8 The Oka-Grauert principle . . . 365
9.9 Comments . . . 367

10 Generalized interpolation problems 369
10.1 Weierstrass theorems . . . 369
10.2 Right- and two-sided Weierstrass theorems . . . 371
10.3 Weierstrass theorems for $\mathcal{G}^\infty(E)$- and $\mathcal{G}^\omega(E)$-valued functions . . . 374
10.4 Holomorphic $\mathcal{G}^\infty(E)$-valued functions with given principal parts of the inverse . . . 377
10.5 Comments . . . 378

11 Holomorphic equivalence, linearization and diagonalization 379
11.1 Introductory remarks . . . 379
11.2 Linearization by extension and equivalence . . . 380
11.3 Local equivalence . . . 386
11.4 A theorem on local and global equivalence . . . 392
11.5 The finite dimensional case . . . 394
11.6 Local and global equivalence for finite meromorphic Fredholm functions . . . 399
11.7 Global diagonalization of finite meromorphic Fredholm functions . 406
11.8 Comments . . . 411

Bibliography 413

Index 419

Preface

This is a book on holomorphic operator functions of a single variable and their applications, which is focussed on the relations between local and global theories. It is based on methods and technics of Complex analysis of scalar and matrix functions of several variables. The applications concern: interpolation, holomorphic families of subspaces and frames, spectral theory of polynomials with operator coefficients, holomorphic equivalence and diagonalization, and Plemelj-Muschelishvili factorization. The book also contains a theory of Wiener-Hopf integral equations with operator-valued kernels and a theory of infinite Töplitz matrices with operator entries.

We started to work on these topics long ago when one of us was a Ph.D. student of the other in Kishinev (now Cisinau) University. Then our main interests were in problems of factorization of operator-valued functions and singular integral operators. Working in this area, we realized from the beginning that different methods and tools from Complex analysis of several variables and their modifications are very useful in obtaining results on factorization for matrix and operator functions. We have in mind different methods and results concerning connections between local and global properties of holomorphic functions. The first period was very fruitful and during it we obtained the basic results presented in this book.

Then World Politics started to interfere in our joint work in the new area. For a long time the authors became separated. One emigrated to Israel, the other was a citizen of East Germany, and the authorities of the second country prevented further meetings and communications of the authors. During that time one of us became more and more involved in Complex analysis of several variables and finally started to work mainly in this area of mathematics. Our initial aims were for a while frozen. Later the political situation in the world changed and after the reunification of Germany the authors with pleasure continued the old projects.

During the time when our projects were frozen, the scientific situation changed considerably. There appeared in the literature new methods, results and applications. In order to cover the old and new material entirely in a modern form and terminology we decided to write this book. As always happens in such cases, during the writing new problems and gaps appear, and the material requires inclusion of additional material with new chapters containing new approaches, new results and plenty of unification and polishing. This work was done by the authors.

We hope the book will be of interest to a number of large groups of experts in pure and applied mathematics as well as for electrical engineers and physicists.

During the work on the book we obtained support of different kinds for our joint activities from the Tel-Aviv University and its School of Mathematical Sciences, the Family of Nathan and Lilly Silver Foundation, the Humboldt Foundation, the Deutsche Forschungsgemeinschaft and the Humboldt University in Berlin and its Institute of Mathematics. We would like to express our sincere gratitude to all these institutions for support and understanding. We would also like to thank the Faculty of Mathematics and Computer Sciences of the Kishinev University and the Institute of Mathematics and Computer Center of the Academy of Sciences of Moldova, where the work on this book was started.

Berlin, Tel-Aviv, November 2008 The authors

Introduction

The book. This book contains a theory and applications of operator-valued holomorphic functions of a single variable. (By *operators* we always mean bounded linear operators between complex Banach spaces.) The applications concern some important problems on factorization, interpolation, diagonalization and others. The book also contains a theory of Wiener-Hopf integral equations with operator-valued kernels and a theory of infinite Töplitz matrices with operator entries.

Our main attention is focussed on the connection between local and global properties of holomorphic operator functions. For this aim, methods from Complex analysis of several variables are used. The exposition of the material appears in style and terms of the latter field.

Multiplicative cocycles. Grauert's theory. The theory of multiplicative cocycles plays a central role in this book. It is a special case of the very deep and powerful theory of cocycles (fiber bundles) on Stein manifolds (any domain in $\mathbb{C}$ is a Stein manifold), which was developed in the 1950s by H. Grauert for cocycles with values in a (finite dimensional) complex Lie group. This theory then was generalized into different interesting directions. In 1968, L. Bungart obtained it for cocycles with values in a *Banach* Lie group, for example, the group of invertible operators in a Banach space.

One of the main statements of Grauert's theory is a principle which is now called the *Oka-Grauert principle*. Non-rigorously, this principle can be stated as follows: *If a holomorphic problem on a Stein manifold has no topological obstructions, then it has a holomorphic solution.* This important principle was first discovered in 1939 by K. Oka in the case of scalar functions.

For domains in the complex plane $\mathbb{C}$, Grauert's theory is much easier but still not simple. It is even not simple for the case of cocycles with values in the group of invertible complex $n \times n$-matrices when no topological obstructions appear.

For operators in infinite dimensional Banach spaces, we meet essential difficulties, which are due to the fact that the group of invertible operators in a Banach space need not be connected. This becomes a topological obstruction if the domain in $\mathbb{C}$ is not simply connected. So, for operator functions, the Oka-Grauert principle is meaningful also for domains in $\mathbb{C}$.

For the problem of Runge approximation, the Oka-Grauert principle claims the following: Runge approximation of a holomorphic invertible operator function by *holomorphic* invertible functions is possible if this is possible by *continuous* invertible functions. From this it follows that such a Runge approximation always holds when the domain is simply connected or the group of invertible operators is connected. The latter is the case for the group of invertible operators in a Hilbert space, and in particular, for the group of invertible complex $n \times n$-matrices.

For simply connected domains, the proof of the Runge approximation theorem for invertible operator functions is not difficult and can be obtained without the theory of cocycles. We show this at the end of Chapter 2. For general domains however, this proof is much more difficult (even in the case of matrix-valued functions) and will be given only in Chapter 5 in the framework of the theory of multiplicative cocycles.

A special type of multiplicative cocycles is given by two open sets D_1 and D_2 in $\mathbb{C}$ and an invertible holomorphic operator function on $D_1 \cap D_2$. For this type, the following is proved:

0.0.1 Theorem. *Let E be a Banach space, let $GL(E)$ be the group of invertible operators in E, let $D_1, D_2 \subseteq \mathbb{C}$ be two open sets, and let $A : D_1 \cap D_2 \to GL(E)$ be holomorphic. Assume that at least one of the following two conditions is satisfied:*

(i) *The union $D_1 \cup D_2$ is simply connected.*

(ii) *All values of A belong to the same connected component of $GL(E)$.*

Then there exist holomorphic operator functions $A_j : D_j \to GL(E)$, $j = 1, 2$, such that

$$A = A_1 A_2^{-1} \qquad \text{on } D_1 \cap D_2 . \tag{0.0.1}$$

If both topological conditions (i) and (ii) in Theorem 0.0.1 are violated, then the assertion of Theorem 0.0.1 is not true. A simple counterexample will be given in Section 5.6.2 for the case when $D_1 \cup D_2$ is an annulus.

The following operator version of the Weierstrass product theorem (on the existence of holomorphic functions with given zeros) is a straightforward consequence of Theorem 0.0.1.

0.0.2 Theorem. *Let E be a Banach space, let $GL(E)$ be the group of invertible operators in E, and let $GL_I(E)$ be the connected component in $GL(E)$ which contains the unit operator I. Let $D \subseteq \mathbb{C}$ be an open set and let Z be a discrete and closed subset of D. Suppose, for each $w \in Z$, a neighborhood $U_w \subseteq D$ of w with $U_w \cap Z = \{w\}$ and a holomorphic operator function $A_w : U_w \setminus \{w\} \to GL(E)$ are given. Further assume that at least one of the following two conditions is fulfilled:*

(i) *The set D is simply connected.*

(ii) *The values of each A_w, $w \in Z$, belong to $GL_I(E)$.*

Then there exist a holomorphic operator function $B : D\backslash Z \to GL(E)$ and a family of holomorphic operator functions $H_w : U_w \to GL(E)$ such that

$$H_w A_w = B \qquad \text{on } U_w \setminus \{w\},, \quad w \in Z\,.$$

The classical Weierstrass product theorem we get for $E = \mathbb{C}$ and $H_w(z) = (z-w)^{\kappa_w}$, $\kappa_w \in \mathbb{N}^*$.

There are also a "right-sided" and a "two-sided" version of Theorem 0.0.2.

Contents. The book consists of an introduction and eleven chapters. Let us now describe in more detail the content of each chapter separately.

The first chapter contains the generalization to functions with values in Banach spaces of the traditional material from Complex analysis of one variable which is usually contained in the beginning of a basic course.

Chapter 2 starts with Pompeiju's integral formula for solutions of the inhomogeneous Cauchy-Riemann equation, the Runge approximation theorem, the Mittag-Leffler theorem, and the Weierstrass product theorem. Then, in Sections 2.6 and 2.7, we present the (less well known) "Anschmiegungsatz" of Mittag-Leffler and a strengthening of the Weierstrass product theorem. In the case of the Weierstrass product theorem and its generalization, in this chapter, we still restrict ourselves to scalar functions. It is one of the main goals of this book, to generalize these results to the case of operator functions, using Grauert's theory of cocycles.

Chapter 3 is dedicated to the splitting problem with respect to a contour for functions with values in a Banach space, as well as to the factorization problem for scalar functions with respect to a contour.

In Chapter 4 we generalize to finite meromorphic Fredholm operator functions the classical Rouché theorem from Complex analysis and the Smith factorization form. The proof is based on the local Smith form.

Chapter 5 is entirely dedicated to the theory of multiplicative cocycles, which were discussed in large before.

Chapter 6 contains a theory of families of subspaces of a Banach space E. First we introduce a complete metric on the set $G(E)$ of closed subspaces of E, the so-called *gap metric*. A **continuous family of subspaces of** E then will be defined as a continuous function with values in $G(E)$, and a **holomorphic** family of subspaces of E will be defined as a continuous family of subspaces which is locally the image of a holomorphic operator function. Vector functions with values in such a family are called **sections** of the family. Note that we do not require that the members of a holomorphic family be complemented in the ambient space. It may even happen they are not pairwise isomorphic. An example is given in Section 6.5.

First we prove the following results: any additive cocycle of holomorphic sections in a holomorphic family of subspaces splits; for any holomorphic operator function A whose image is a holomorphic family of subspaces, and any holomorphic section f of this family, there exists a global holomorphic vector function u that solves the equation $Au = f$; for any holomorphic family of subspaces there exists a global holomorphic operator function with this family as image. Proving this, the main difficulty is the solution of certain local problems (in this generality, published for the first time in this book). In terms of Complex analysis of several variables, the solution of these local problems means that any holomorphic family of subspaces is a so-called *Banach* coherent sheaf (a generalization of the notion of coherent sheaves). After solving this we proceed by standard methods that are well-known in Complex analysis of several variables.

Then we consider holomorphic families of subspaces, which we call **injective** and which have the additional property that, locally, the family can be represented as the image of a holomorphic operator function with zero kernel. We study the problem of a corresponding global representation. Here we need the theory of multiplicative cocycles from Chapter 5. It turns out that this is not always possible, but we have again an Oka-Grauert principle.

Then we study holomorphic families of complemented subspaces (which are injective), where we can prove more precise results than for arbitrary injective families. Again there is an Oka-Grauert principle.

At the end we consider the special case of families of subspaces which are finite dimensional or of finite codimension. Here there are no topological restrictions.

Chapters 7 and 8 are dedicated to factorization of operator functions with respect to a contour and the connection with Wiener-Hopf and Töplitz operators. This type of factorization was in fact considered for the first time in the pioneering works of Plemelj and of Muschelishvili. Because of that we call it **Plemelj-Muschelishvili factorization**. We start with the *local principle*, which quickly follows from the theory of multiplicative cocycles and which allows us to prove theorems on factorization for different classes of operator functions. The local principle reduces the problem to functions which are already holomorphic in a neighborhood of the contour.

For further applications we need a generalization of the theory of multiplicative cocycles. This is the topic of Chapter 9, where we introduce *cocycles with restrictions*. Let us offer an example (which is basic for all cocycles with restrictions). Suppose that in Theorem 0.0.1 an additional set $Z \subseteq D_1 \cup D_2$, discrete and closed in D, and positive integers m_w, $w \in Z$, are given. Assume that the function $A - I$ has a zero of order m_w at each $w \in D_1 \cap D_2 \cap Z$. Then the theory of cocycles with restrictions gives the additional information that the functions A_1 and A_2 in Theorem 0.0.1 can be chosen so that, for all $w \in D_j \cap Z$, $j = 1, 2$, the function $A_j - I$ has a zero of order m_w at w.

In Chapter 10, by means of the theory of cocycles with restrictions, we essentially improve the Weierstrass product Theorem 0.0.2: The functions H_w in this theorem now can be chosen so that, additionally, for each $w \in Z$, the function $H_w - I$ has a zero of an arbitrarily given order m_w at w. This has different consequences that are discussed in this short chapter.

Chapter 11 is dedicated to holomorphic equivalence and its applications to linearization and diagonalization. Let E be a Banach space, let $L(E)$ be the space of bounded linear operators in E, let $GL(E)$ be the group of invertible operators from $L(E)$, let $D \subseteq \mathbb{C}$ be an open set, and let Z be a discrete and closed subset of D. Then two holomorphic operator functions $A, B : D \setminus Z \to L(E)$ are called **(globally) holomorphically equivalent over** D if there exist holomorphic operator functions $S, T : D \to GL(E)$ such that $A = SBT$ on D.

In the first section, results are presented that explain the importance of the notion of holomorphic equivalence in spectral theory of linear operators and holomorphic operator functions. It contains the following two results: 1) For each relatively compact open subset Ω of D, each holomorphic operator function $A : D \to L(E)$, after an appropriate extension, becomes holomorphically equivalent to a function of the form $zI - T$, $z \in \Omega$, where T is a constant operator and I is the identical operator (Theorem 11.2.1). 2) Two operators $T, S \in L(E)$ with the spectra $\sigma(A)$ and $\sigma(B)$ are similar if and only if some extensions of the functions $zI - T$ and $zI - S$ are holomorphically equivalent over some neighborhood of $\sigma(A) \cup \sigma(B)$ (Corollary 11.2.3).

The remainder of this section is devoted to the relation between global and local holomorphic equivalence where two holomorphic operator functions are called **locally holomorphically equivalent** if, for each point, they are holomorphically equivalent over some neighborhood of this point. We prove that two meromorphic operator functions with meromorphic inverse are locally holomorphically equivalent if and only if they are globally holomorphically equivalent (Theorem 11.4.2), and we prove that any finite meromorphic Fredholm operator function is globally holomorphically equivalent to a diagonal function (Theorem 11.7.6). The local fact behind this is the Smith representation of matrices of germs of scalar holomorphic functions.

Acknowledgement. In the beginning of the 1970s, on an invitation of one of us, M.A. Shubin visited Kishinev and gave two talks about applications of Grauert's theory and the theory of coherent analytic sheaves to different results for linear operators. One of the talks was on the local principle for Plemelj-Muschelishvili factorization of matrix functions and the second was about the analysis of holomorphic families of subspaces. These talks had on us an important influence. Very soon after this we came up with a series of papers on operator-valued cocycles in the case of one variable with new direct proofs and also with new results and applications to operator functions. At the end this development led to this book. It is our pleasure to thank M.A. Shubin providing us with the initial input.

Notation

Here we give a list of standard symbols and some remarks concerning the terminology used in this book without further explanation:

- $\mathbb{C}$ is the complex plane, $\mathbb{R}$ is the real axis, $\mathbb{C}^* := \mathbb{C} \setminus \{0\}$, $\mathbb{R}^* := \mathbb{R} \setminus \{0\}$.
- $\mathbb{N}$ is the set of natural numbers (including 0), $\mathbb{N}^* := \mathbb{N} \setminus \{0\}$.
- $\mathbb{Z}$ is the set of entire numbers.

- Banach spaces and Banach algebras are always complex.
- If E, F are Banach spaces, then we denote by $L(E, F)$ the Banach space of bounded linear operators operators from E to F, endowed with the operator norm. We set $L(E) = L(E, E)$, and we denote by $GL(E)$ the group of all invertible operators from $L(E)$. By a *projector* in E we always mean an operator $P \in L(E)$ with $P^2 = P$.
- By an *operator* we always mean a bounded linear operator between two Banach spaces.
- Let E, F be Banach spaces, and let $A \in L(E, F)$. Then we denote by $\operatorname{Im} A$ the image, and by $\operatorname{Ker} A$ the kernel of A. The operator A is called *injective*, if $\operatorname{Ker} A = \{0\}$, and it is called *surjective* if $\operatorname{Im} A = F$.
- The unit operator of a Banach space E will be denoted by I or I_E.

- For $n \in \mathbb{N}^*$ we denote by $L(n, \mathbb{C})$ the algebra of complex $n \times n$ matrices, and by $GL(n, \mathbb{C})$ we denote the group of invertible elements of $L(n, \mathbb{C})$.

- By a *neighborhood* we always mean an *open* neighborhood, if not explicitly stated to be anything else.
- If U is a set in a topological space X, then $\overline{U}$ always denotes the topological closure of U in X (and not the complement).

- By $\mathcal{C}^0$*-functions* or functions *of class* $\mathcal{C}^0$ we mean continuous functions. If Γ is a subset of $\mathbb{C}$ and M is a subset of a Banach space, then we denote by $\mathcal{C}^M(\Gamma)$ or by $(\mathcal{C}^0)^M(\Gamma)$ the set of all continuous functions $f : \Gamma \to M$.

- If $U \subseteq \mathbb{C}$ is an open set, $U \neq \emptyset$, and E is a Banach space, then a function $f : U \to E$ is called C^k or *of class* $\mathcal{C}^k$ on U, $k \in \mathbb{N}^* \cup \{\infty\}$, if it is k times continuously differentiable with respect to the canonical real coordinates of $\mathbb{C}$.

- If $U \subseteq \mathbb{C}$ is an open set, $U \neq \emptyset$, and M is a subset of a Banach space E, then we denote by $(\mathcal{C}^k)^M(U)$ the set of all $\mathbb{C}^k$-functions $f : U \to E$ such that $f(z) \in M$ for all $z \in U$, and by $\mathcal{O}^M(U)$ we denote the set of all holomorphic (Def. 1.1.1) functions $f : U \to E$ such that $f(z) \in M$ for all $z \in U$.

- We set $\mathcal{O}(U) = \mathcal{O}^{\mathbb{C}}(U)$, $\mathcal{C}^k(U) = (\mathcal{C}^k)^{\mathbb{C}}(U)$ and $\mathcal{O}^*(U) = \mathcal{O}^{\mathbb{C}^*}(U)$ for each open $U \subseteq \mathbb{C}$ and $k \in \mathbb{N}$.

- If $K \subseteq \mathbb{C}$ is a (not necessarily open) set of uniqueness for holomorphic functions (for example, the closure of an open set, or an interval) and E is a Banach space, then we speak also about a holomorphic function $f : K \to E$ to say that f is the restriction of an E-valued holomorphic function defined in a neighborhood of K.

- If $D \subseteq \mathbb{C}$ is an open set with piecewise $\mathcal{C}^1$-boundary (Def. 1.4.1), then we denote by ∂D the boundary of D endowed with the orientation defined by D (Sect. 1.4.1), i.e., D lies on the left side of ∂D.

Chapter 1

Elementary properties of holomorphic functions

This chapter is devoted to the basic facts usually contained in a basic course on Complex analysis of one variable. The difference is that we do this for functions with values in a Banach space. Many (not all) of these results will be deduced by the Hahn-Banach theorem from the corresponding scalar fact.

Some care is necessary with respect to the maximum principle. The strong version, that the norm of a non-constant holomorphic function does not admit local maxima, is not true in general. For example, it fails for l^∞ and it is true for Hilbert spaces.

1.1 Definition and first properties

The notion of a holomorphic function with values in a Banach space can be defined as in the scalar case by complex differentiability:

1.1.1 Definition. Let E be Banach space, and let $U \subseteq \mathbb{C}$ be an open set. A function $f : U \to E$ is called **complexly differentiable** or **holomorphic** if, for each $w \in U$,

$$f'(w) := \lim_{z \to w} \frac{f(z) - f(w)}{z - w}$$

exists. Clearly, then also the partial derivatives of f with respect to the canonical real coordinates x, y exist, and the Cauchy-Riemann equation holds:

$$f'(w) = \frac{\partial f}{\partial x}(w) = i\frac{\partial f}{\partial y}(w)\,, \qquad w \in D. \tag{1.1.1}$$

The function $f' : U \to E$, which is then defined, will be called the **complex derivative** or simply the **derivative** of f.

The space of all holomorphic functions from U to E will be denoted by $\mathcal{O}^E(U)$.

1.1.2. From this definition the following facts follow immediately:

- Each holomorphic function with values in a Banach space is continuous.[1]
- If $U \subseteq \mathbb{C}$ is open, E is a Banach space, $f, g \in \mathcal{O}^E(U)$ and $\alpha, \beta \in \mathcal{O}^{\mathbb{C}}(U)$, then $\alpha f + \beta g \in \mathcal{O}^E(U)$, and $(\alpha f + \beta g)' = \alpha' f + \alpha f' + \beta' g + \beta g'$ on U.
- If $U \subseteq \mathbb{C}$ is open, A is a Banach algebra, and $f, g \in \mathcal{O}^A(U)$, then $fg \in \mathcal{O}^A(U)$ and $(fg)' = f'g + fg'$ on U.
- If $U \subseteq \mathbb{C}$ is open, E, F are Banach spaces, $f \in \mathcal{O}^E(U)$, and $A \in \mathcal{O}^{L(E,F)}(U)$, then $Af \in \mathcal{O}^F(U)$ and $(Ag)' = A'g + Ag'$ on U.
- If $U \subseteq \mathbb{C}$ is open, A is a Banach algebra with unit, GA is the group of invertible elements of A, and $f : U \to GA$ is holomorphic, then f^{-1} is holomorphic and $(f^{-1})' = -f^{-1} f' f^{-1}$ on U.
- If $U, V \subseteq \mathbb{C}$ are open, E is a Banach space, and $\alpha : U \to V$, $f : V \to E$ are holomorphic, then the composition $f \circ \alpha$ is holomorphic and
$$(f \circ \alpha)' = \alpha'(f' \circ \alpha) \qquad \text{on } U. \tag{1.1.2}$$
- If $I \subseteq \mathbb{R}$ is an interval, $U \subseteq \mathbb{C}$ is open, E is a Banach space, $\alpha : I \to U$ is differentiable, and $f : U \to E$ is holomorphic, then the composition $f \circ \alpha$ is differentiable on I and
$$(f \circ \alpha)' = \alpha'(f' \circ \alpha) \qquad \text{on } I. \tag{1.1.3}$$

The theorem on uniqueness of holomorphic functions is deeper, but by means of the Hahn-Banach theorem it can be quickly obtained from the scalar fact:

1.1.3 Theorem. *Let $D \subseteq \mathbb{C}$ be a connected open set, and let $z_n \in D$, $n \in \mathbb{N}^*$, be a sequence which converges to a point $z_0 \in D$ such that $z_n \neq z_0$ for all $n \in \mathbb{N}^*$. Further let E be a Banach space, and let $f, g : D \to E$ be two holomorphic functions such that $f(z_n) = g(z_n)$ for all $n \in \mathbb{N}^*$. Then $f \equiv g$ on D.*

Proof. Let E' be the dual of E. Then it follows from the theorem on uniqueness of scalar holomorphic functions that, for all $\Phi \in E'$, $\Phi \circ f \equiv \Phi \circ g$. By the Hahn-Banach theorem this implies that $f \equiv g$. □

The same is true for Liouville's theorem:

1.1.4 Theorem. *Let E be a Banach space, let E' be the dual of E, and let $f : \mathbb{C} \to E$ be a holomorphic function. Suppose, for each $\Phi \in E'$, the function $\Phi \circ f$ is bounded on $\mathbb{C}$ (which is the case, for example, if the function $\|f\|$ is bounded). Then f is constant.*

[1] They are even of class $\mathcal{C}^\infty$, but, as in the scalar case, this can be proved only after the Cauchy formula is obtained.

Proof. It follows from Liouville's theorem for scalar holomorphic functions that $\Phi \circ f$ is constant for all $\Phi \in E'$. By the Hahn-Banach theorem this implies that f is constant. □

1.2 The maximum principle

The following (weak) version of the maximum principle again can be obtained by means of the Hahn-Banach theorem immediately from the maximum principle for scalar holomorphic functions:

1.2.1 Theorem. *Let $D \subseteq \mathbb{C}$ be a bounded open set, let E be a Banach space, and let $f : \overline{D} \to E$ be a continuous function which is holomorphic in D. Denote by ∂D the boundary of D. Then*

$$\max_{z \in \overline{D}} \|f(z)\| = \max_{z \in \partial D} \|f(z)\|. \tag{1.2.1}$$

Proof. Let z_0 be an arbitrary point in $\overline{D}$, and let E' be the dual of E. Then, for each $\Phi \in E'$, the function $\Phi \circ f$ is holomorphic and hence, by the maximum principle for scalar holomorphic functions,

$$\big|(\Phi\big(f(z_0)\big)\big| \leq \max_{z \in \partial D} \big\|(\Phi \circ f)(z)\big\| \leq \|\Phi\| \max_{z \in \partial D} \big\|f(z)\big)\big\|.$$

By the Hahn-Banach theorem, this implies

$$|f(z_0)| \leq \max_{z \in \partial D} \big\|f(z)\big)\big\|.$$

As z_0 was chosen arbitrarily in $\overline{D}$, this implies (1.2.1). □

The strong maximum principle

"If a holomorphic function, defined on a connected open set, admits a local maximum, then it is constant"

is not true for functions with values in an arbitrary Banach space. Indeed, take the space $\mathbb{C}^2$ with the norm $\|(\xi_1, \xi_2)\|_{\max} = \max\{|\xi_1|, |\xi_2|\}$ and consider the holomorphic function $f(z) = (z, 1)$ defined for $|z| < 1$. Clearly f is not constant but $\|f\|_{\max} \equiv 1$.

For functions with values in a Hilbert space we have the strong maximum principle:

1.2.2 Theorem. *Let $D \subseteq \mathbb{C}$ be a connected open set, let H be a Hilbert space, and let $f : D \to H$ be a holomorphic functions such that, for some $z_0 \in D$ and $\varepsilon > 0$,*

$$\|f(z_0)\| \geq \|f(z)\| \qquad \text{for all } |z - z_0| \leq \varepsilon. \tag{1.2.2}$$

Then f is constant.

To prove this we need some facts on scalar holomorphic functions, which are not necessarily contained in a standard course on Complex analysis. We therefore first present these facts with proofs.

1.2.3 Lemma. *Let $D \subseteq \mathbb{C}$ be an open set, and let $\varphi : D \to \mathbb{C}$ be a scalar holomorphic function. Let x, y be the canonical real coordinates on $\mathbb{C}$, and let*

$$\Delta := \frac{\partial^2}{\partial x^2} + \frac{\partial^2}{\partial y^2}$$

be the Laplace operator. Then

$$\Delta|\varphi|^2 = 4|\varphi'|^2. \tag{1.2.3}$$

Proof. We have

$$\frac{\partial^2}{\partial x^2}|\varphi|^2 = \frac{\partial^2}{\partial x^2}(\varphi\overline{\varphi}) = \frac{\partial}{\partial x}\left(\overline{\varphi}\frac{\partial\varphi}{\partial x} + \varphi\frac{\partial\overline{\varphi}}{\partial x}\right) = \frac{\partial\overline{\varphi}}{\partial x}\frac{\partial\varphi}{\partial x} + \overline{\varphi}\frac{\partial^2\varphi}{\partial x^2} + \frac{\partial\varphi}{\partial x}\frac{\partial\overline{\varphi}}{\partial x} + \varphi\frac{\partial^2\overline{\varphi}}{\partial x^2}$$

and, in the same way,

$$\frac{\partial^2}{\partial y^2}|\varphi|^2 = \frac{\partial\overline{\varphi}}{\partial y}\frac{\partial\varphi}{\partial y} + \overline{\varphi}\frac{\partial^2\varphi}{\partial y^2} + \frac{\partial\varphi}{\partial y}\frac{\partial\overline{\varphi}}{\partial y} + \varphi\frac{\partial^2\overline{\varphi}}{\partial y^2}.$$

Since φ is holomorphic and therefore, by the Cauchy-Riemann equation,

$$\varphi' = \frac{\partial\varphi}{\partial x} = i\frac{\partial\varphi}{\partial y} \quad \text{and} \quad \overline{\varphi}' = \frac{\partial\overline{\varphi}}{\partial x} = -i\frac{\partial\overline{\varphi}}{\partial y},$$

this implies

$$\frac{\partial^2}{\partial x^2}|\varphi|^2 = \overline{\varphi}'\varphi' + \overline{\varphi}\frac{\partial^2\varphi}{\partial x^2} + \varphi'\overline{\varphi}' + \varphi\frac{\partial^2\overline{\varphi}}{\partial x^2} = 2|\varphi'|^2 + \overline{\varphi}\frac{\partial^2\varphi}{\partial x^2} + \varphi\frac{\partial^2\overline{\varphi}}{\partial x^2}$$

and, in the same way,

$$\frac{\partial^2}{\partial y^2}|\varphi|^2 = 2|\varphi'|^2 + \overline{\varphi}\frac{\partial^2\varphi}{\partial y^2} + \varphi\frac{\partial^2\overline{\varphi}}{\partial y^2}.$$

Hence

$$\Delta|\varphi|^2 = 4|\varphi'|^2 + \overline{\varphi}\left(\frac{\partial^2\varphi}{\partial x^2} + \frac{\partial^2\varphi}{\partial y^2}\right) + \varphi\left(\frac{\partial^2\overline{\varphi}}{\partial x^2} + \frac{\partial^2\overline{\varphi}}{\partial y^2}\right) = 4|\varphi'|^2 + \overline{\varphi}\Delta\varphi + \varphi\Delta\overline{\varphi}.$$

Since $\Delta\varphi = \Delta\overline{\varphi} = 0$ (real and imaginary part of φ are harmonic), this implies (1.2.3). □

1.2.4 Lemma. *Let $r_0 > 0$, and let φ be a scalar holomorphic function defined on the disc $|z| < r_0$. Set*

$$M(r) = \frac{1}{2\pi}\int_0^{2\pi} \left|\varphi\left(re^{it}\right)\right|^2 dt \qquad \text{for } 0 \le t < r_0. \tag{1.2.4}$$

Then M is of class $\mathcal{C}^\infty$ on $[0, r_0[$, and

$$M'(r) = \frac{2}{\pi r}\int_{|z|<r} |\varphi'|^2 d\lambda \qquad \text{for } 0 < r < r_0. \tag{1.2.5}$$

Here $d\lambda$ is the Lebesgue measure.

Proof. Since the function under the integral in (1.2.4) is of class $\mathbb{C}^\infty$ with respect to t and r, it is clear that M is of class $\mathcal{C}^\infty$, where, by differentiation under the integral sign, we get

$$M'(r) = \frac{1}{2\pi}\int_0^{2\pi} \frac{\partial}{\partial r}|\varphi|^2 dt = \frac{1}{2\pi}\int_0^{2\pi} \left(\frac{\partial|\varphi|^2}{\partial x}\frac{\partial x}{\partial r} + \frac{\partial|\varphi|^2}{\partial y}\frac{\partial y}{\partial r}\right) dt, \tag{1.2.6}$$

where x, y are the canonical real coordinates on $\mathbb{C}$. Since $x(re^{it}) = r\cos t$ and $y(re^{it}) = r\sin t$ and therefore

$$\frac{\partial x}{\partial r} = \cos t \qquad \text{and} \qquad \frac{\partial y}{\partial r} = \sin t,$$

it follows from (1.2.6) that

$$M'(r) = \frac{1}{2\pi}\int_0^{2\pi} \frac{\partial}{\partial r}|\varphi|^2 dt = \frac{1}{2\pi}\int_0^{2\pi} \left(\frac{\partial|\varphi|^2}{\partial x}\cos t + \frac{\partial|\varphi|^2}{\partial y}\sin t\right) dt. \tag{1.2.7}$$

Now we fix $0 < r < r_0$, and we denote by S_r the circle with radius r centered at zero. Let $\tau : S_r \to [0, 2\pi[$ be the function defined by $\tau(re^{it}) = t$, $0 \le t < 2\pi$. Then (1.2.7) can be written

$$M'(r) = \frac{1}{2\pi}\int_{S_r} \left(\frac{\partial|\varphi|^2}{\partial x}(\cos\circ\tau)d\tau + \frac{\partial|\varphi|^2}{\partial y}(\sin\circ\tau)d\tau\right).$$

Since, on S_r, $x = r\cos\circ\tau$, $y = r\sin\circ\tau$ and therefore

$$dx\big|_{S_r} = -r(\sin\circ\tau)d\tau \qquad \text{and} \qquad dy\big|_{S_r} = r(\cos\circ\tau)d\tau,$$

this further implies that

$$M'(r) = \frac{1}{2\pi r}\int_{S_r} \left(\frac{\partial|\varphi|^2}{\partial x}dy - \frac{\partial|\varphi|^2}{\partial y}dx\right).$$

By Stokes' theorem this yields

$$M'(r) = \frac{1}{2\pi r} \int\limits_{|z|<r} \left(\frac{\partial^2 |\varphi|^2}{\partial x^2} dx \wedge dy - \frac{\partial |\varphi|^2}{\partial y} dy \wedge dx \right).$$

Since $dx \wedge dy = -dy \wedge dx = d\lambda$, this implies

$$M'(r) = \frac{1}{2\pi r} \int\limits_{|z|<r} \Delta |\varphi|^2 d\lambda.$$

By Lemma 1.2.3 this means (1.2.5). □

1.2.5 Lemma. *Let $z_0 \in \mathbb{C}$ and $r > 0$. Let φ be a scalar holomorphic function in a neighborhood of the closed disc $|z| \le r$. Then*

$$|\varphi(z_0)|^2 \le \frac{1}{2\pi} \int\limits_0^{2\pi} \left|\varphi\left(z_0 + re^{it}\right)\right|^2 dt \tag{1.2.8}$$

and (1.2.8) *holds with equality, if and only if, φ is constant.*

Proof. We may assume that $z_0 = 0$. It is clear that (1.2.8) holds with equality if φ is constant.

Assume that φ is not constant. Then φ' has not more then a finite number of zeros on $|z| \le r$. Therefore it follows from Lemma 1.2.4 that the function

$$M(r') := \frac{1}{2\pi} \int\limits_0^{2\pi} \left|\varphi\left(z_0 + r'e^{it}\right)\right|^2 dt$$

is strictly monotonicly increasing for $0 \le r' \le r$. Since $M(0) = |\varphi(z_0)|^2$, this implies that

$$|\varphi(z_0)|^2 < \frac{1}{2\pi} \int\limits_0^{2\pi} \left|\varphi\left(z_0 + re^{it}\right)\right|^2 dt.$$ □

Proof of Theorem 1.2.2. Let $\langle \cdot, \cdot \rangle$ be the scalar product of H. Choose an orthonormal basis $\{e_j\}_{j \in I}$ of H. Set

$$f_j(z) = \left\langle f(z), e_j \right\rangle, \qquad z \in D,\ j \in I.$$

Then each f_j is holomorphic and, by (1.2.2),

$$\sum_{j \in I} |f_j(z_0)|^2 = \|f(z_0)\|^2 \ge \|f(z)\|^2 = \sum_{j \in I} |f_j(z)|^2, \quad |z - z_0| \le \varepsilon.$$

It follows that

$$\sum_{j\in I}|f_j(z_0)|^2 = \frac{1}{2\pi}\int_0^{2\pi}\sum_{j\in I}|f_j(z_0)|^2 dt$$
$$\geq \frac{1}{2\pi}\int_0^{2\pi}\sum_{j\in I}\left|f_j\left(z_0+\varepsilon e^{it}\right)\right|^2 dt = \sum_{j\in I}\frac{1}{2\pi}\int_0^{2\pi}\left|f_j\left(z_0+\varepsilon e^{it}\right)\right|^2 dt.$$

Since, on the other hand, by Lemma 1.2.5,

$$\left|f_j(z_0)\right|^2 \leq \frac{1}{2\pi}\int_0^{2\pi}\left|f_j(z_0+\varepsilon e^{it})\right|^2 dt \qquad \text{for all } j\in I, \tag{1.2.9}$$

this implies that

$$|f_j(z_0)|^2 = \frac{1}{2\pi}\int_0^{2\pi}\left|f_j\left(z_0+\varepsilon e^{it}\right)\right|^2 dt \qquad \text{for all } j\in I.$$

Again by Lemma 1.2.5 this means that f_j is constant for all $j \in I$. Hence f is constant. □

Note also the following:

1.2.6 Proposition. *Let $D \subseteq \mathbb{C}$ be an open set, let E be a Banach space, and let $f : D \to E$ be holomorphic. Then $\|f\|$ is subharmonic in D.*[2]

Proof. Let $z_0 \in D$ and $r > 0$ be given such that the closed disc $|z - z_0| \leq r$ is contained in D. Then we have to prove that

$$\|f(z_0)\| \leq \frac{1}{2\pi}\int_0^{2\pi}\left\|f\left(z_0+re^{it}\right)\right\| dt. \tag{1.2.10}$$

Let E' be the dual of E. Then, for each $\Phi \in E'$, $\Phi \circ f$ is holomorphic. Hence, for each $\Phi \in E'$, $\Phi \circ f$ is subharmonic. Hence, for each $\Phi \in E'$,

$$\left\|\Phi\left(f(z_0)\right)\right\| \leq \frac{1}{2\pi}\int_0^{2\pi}\left\|\Phi\left(f\left(z_0+re^{it}\right)\right)\right\| dt \leq \|\Phi\|\frac{1}{2\pi}\int_0^{2\pi}\left\|f\left(z_0+re^{it}\right)\right\| dt.$$

By the Hahn-Banach theorem this implies (1.2.10). □

[2] Recall that a continuous function $\rho : D \to \mathbb{R}$ is called **subharmonic** if, for all $z_0 \in D$ and $r > 0$ such that the closed disc $|z - z_0| \leq r$ is contained in D,

$$\rho(z_0) \leq \frac{1}{2\pi}\int_0^{2\pi}\rho(z_0+re^{it})\,dt.$$

Therefore the maximum principle stated in Theorem 1.2.1 can be viewed also as a consequence of the maximum principle for subharmonic functions.

1.3 Contour integrals

Here we collect a number of definitions for later reference.

1.3.1 Definition ($\mathcal{C}^1$-contours). A set $\Gamma \subseteq \mathbb{C}$ is called a **connected $\mathcal{C}^1$-contour** if there exist real numbers $a < b$ and a $\mathcal{C}^1$-function $\gamma : [a, b] \to \mathbb{C}$ with $\Gamma = \gamma\big([a, b]\big)$ such that:

(i) $\gamma'(t) \neq 0$ for all $a \leq t \leq b$;

(ii) $\gamma(t) \neq \gamma(s)$ for all $a \leq t, s < b$ with $t \neq s$;

(iii) either $\gamma(b) \neq \gamma(t)$ for all $a \leq t < b$
or $\gamma(b) = \gamma(a)$ and $\gamma'(b) = \gamma'(a)$.

Then the function γ is called a **$\mathcal{C}^1$-parametrization** of Γ. If $\gamma(b) = \gamma(a)$, then Γ is called **closed**.

By a (not necessarily connected) **$\mathcal{C}^1$-contour** we mean the union of a finite number of pairwise disjoint connected $\mathcal{C}^1$-contours.

1.3.2 Definition (Piecewise $\mathcal{C}^1$-contours). A set $\Gamma \subseteq \mathbb{C}$ is called a **connected piecewise $\mathcal{C}^1$-contour** in each of the following three cases:

(I) There exist real numbers $a < b$ and a $\mathcal{C}^1$-function $\gamma : [a, b] \to \mathbb{C}$ with $\Gamma = \gamma\big([a, b]\big)$ such that:

   (i) $\gamma'(t) \neq 0$ for all $a \leq t \leq b$;

   (ii) $\gamma(t) \neq \gamma(s)$ for all $a \leq t, s < b$ with $t \neq s$;

   (iii) $\gamma(b) = \gamma(a)$ and $\frac{\gamma'(a)}{\gamma'(b)} \in \mathbb{C} \setminus]-\infty, 0]$.[3]

(II) There exist finitely many real numbers $a = t_1 < \ldots < t_m = b$ and a continuous function $\gamma : [a, b] \to \mathbb{C}$ with $\Gamma = \gamma([a, b])$ such that:

   (i) For each $1 \leq j \leq m - 1$, the function

$$\gamma_j := \gamma\big|_{[t_j, t_{j+1}]}$$

   is of class $\mathcal{C}^1$ on $[t_j, t_{j+1}]$ and $\gamma_j'(t) \neq 0$ for all $t_j \leq t \leq t_{j+1}$.

   (ii) $\frac{\gamma_j'(t_{j+1})}{\gamma_{j+1}'(t_{j+1})} \in \mathbb{C} \setminus]-\infty, 0]$ for $1 \leq j \leq m - 2$.

   (iii) $\gamma(t) \neq \gamma(s)$ for all $a \leq t, s \leq b$ with $t \neq s$.

[3] i.e., either Γ is smooth at $\gamma(b) = \gamma(a)$ or Γ forms a non-zero angle at $\gamma(b) = \gamma(a)$.

(III) If in case (II) condition (iii) is replaced by

(iii′) $\gamma(t) \neq \gamma(s)$ for all $a \leq t, s < b$ with $t \neq s$, $\gamma(b) = \gamma(a)$ and $\frac{\gamma'(a)}{\gamma'(b)} \in \mathbb{C}\backslash]-\infty, 0]$.

The contour Γ is called **closed**, if and only if, $\gamma(b) = \gamma(a)$.

The function γ then is called a **piecewise $\mathcal{C}^1$-parametrization** of Γ.

By a (not necessarily connected) **piecewise $\mathcal{C}^1$-contour** we mean the union of a finite number of pairwise disjoint connected piecewise $\mathcal{C}^1$-contours. Such a contour is called **closed** if each connected component is closed.

1.3.3 Definition (Orientation of a contour). First let $\Gamma \subseteq \mathbb{C}$ be a connected piecewise $\mathcal{C}^1$-contour.

If $\gamma : [a, b] \to \mathbb{C}$ and $\gamma^* : [a^*, b^*] \to \mathbb{C}$ are two piecewise $\mathcal{C}^1$-parametrizations of Γ, then by definition of a piecewise $\mathcal{C}^1$-parametrization, on $[a^*, b^*[$, the function $\gamma^{-1} \circ \gamma^*$ is well defined, and, as it is continuous, it is either strictly monotonicly increasing or strictly monotonicly decreasing. In the first case we call γ and $\gamma*$ **equivalent**.

In this way the set of all piecewise $\mathcal{C}^1$-parametrizations of Γ is divided into two equivalence classes.

We say that Γ is **oriented** if one of these two equivalence classes is chosen. If this is done, then a piecewise $\mathcal{C}^1$-parametrization of Γ is called **positively oriented** if it belongs to the chosen class.

A (not necessarily connected) piecewise $\mathcal{C}^1$-contour Γ is called **oriented** if on each connected component of Γ an orientation is chosen.

1.3.4 Definition. Let Γ be an oriented piecewise $\mathcal{C}^1$-contour, let E be a Banach space, and let $f : \Gamma \to E$ be a continuous function.

If Γ is connected and $\gamma : [a, b] \to \mathbb{C}$ is a positively oriented, piecewise $\mathcal{C}^1$-parametrization of Γ, then it follows from the substitution rule (which follows as in the scalar case from the chain rule) that

$$\int_\Gamma f(z)dz := \int_a^b f(\gamma(z))\gamma'(t)dt \tag{1.3.1}$$

is independent of the choice of γ. Therefore, by (1.3.1) an integral $\int_\Gamma f(z)dz$ is well defined.

If Γ is not connected and $\Gamma_1, \ldots, \Gamma_n$ are the connected components of Γ, then we define

$$\int_\Gamma f(z)dz = \sum_{j=1}^n \int_{\Gamma_j} f(z)dz. \tag{1.3.2}$$

If Γ is a circle of radius r centered at $z_0 \in \mathbb{C}$ and if Γ is oriented by the parametrization $\gamma(t) := z_0 + re^{it}$, $0 \le t \le 2\pi$, then we define

$$\int_{|z-z_0|=r} f(z)dz := \int_\Gamma f(z)dz = ir \int_0^{2\pi} f\big(z_0 + re^{it}\big)e^{it}dt. \tag{1.3.3}$$

1.3.5 Definition. Let Γ be a piecewise $\mathcal{C}^1$-contour.

If Γ is connected and $\gamma : [a,b] \to \mathbb{C}$ is a piecewise $\mathcal{C}^1$-parametrization of Γ, then it follows from the substitution rule (which follows as in the scalar case from the chain rule) that

$$|\Gamma| := \int_a^b |\gamma'(t)|dt \tag{1.3.4}$$

is independent of the choice of γ. Therefore, by (1.3.4) a number $|\Gamma|$ is well defined.

If Γ is not connected and $\Gamma_1, \ldots, \Gamma_n$ are the connected components of Γ, then we define

$$|\Gamma| = \sum_{j=1}^n |\Gamma_j|. \tag{1.3.5}$$

The number $|\Gamma|$ is called the **length** of Γ.

1.3.6 Proposition. *Let Γ be a piecewise $\mathcal{C}^1$-contour, let E be a Banach space, and let $f : \Gamma \to E$ be a continuous function. Then*

$$\left\| \int_\Gamma f(z)dz \right\| \le |\Gamma| \max_{z\in\Gamma} \|f(z)\|. \tag{1.3.6}$$

Proof. We may assume that Γ is connected. Let $\gamma : [a,b] \to \mathbb{C}$ be an oriented, piecewise $\mathcal{C}^1$-parametrization of Γ, and let E' be the dual of E. Then, for each $\Phi \in E'$ with $\|\Phi\| = 1$,

$$\left|\Phi\Big(\int_\Gamma f(z)dz\Big)\right| = \left|\Phi\Big(\int_a^b f\big(\gamma(t)\big)\gamma'(t)dt\Big)\right| = \left|\int_a^b \Phi\Big(f\big(\gamma(t)\big)\Big)\gamma'(t)dt\right|$$

$$\le \int_a^b \left|\Phi\Big(f\big(\gamma(t)\big)\Big)\gamma'(t)\right|dt \le \max_{a\le t\le b}\left|f\big(\gamma(t)\big)\right| \int_a^b |\gamma'(t)|dt = |\Gamma| \max_{z\in\Gamma}\|f(z)\|.$$

By the Hahn-Banach theorem, this implies (1.3.6). □

1.4 The Cauchy integral theorem

1.4.1. Let $D \subseteq \mathbb{C}$ be an open set. We shall say that D has a **piecewise $\mathcal{C}^1$-boundary** if the boundary of D (in $\mathbb{C}$) is a closed piecewise $\mathcal{C}^1$-contour Γ (Def. 1.3.2) such that each point of Γ is also a boundary point of $\mathbb{C} \setminus \overline{D}$.

Let $D \subseteq \mathbb{C}$ be an open set with **piecewise $\mathcal{C}^1$-boundary**. Then different orientations of Γ are possible (more than two if Γ is not connected). One of these orientations is of particular interest: The **orientation defined by** D: This is the orientation of Γ such that D is "on the left" of Γ with respect to this orientation.[4] In this case, we also say that Γ **is oriented by** D or **oriented as the boundary of** D.

By ∂D we denote the boundary of D if it is oriented by D.

1.4.2 Theorem (Cauchy integral theorem). *Let $D \subseteq \mathbb{C}$ be a bounded open set with piecewise $\mathcal{C}^1$-boundary, let E be a Banach space, and let $f : \overline{D} \to E$ be a continuous function which is holomorphic in D. Then*

$$\int_\Gamma f(z)dz = 0. \tag{1.4.1}$$

Proof. Let E' be the dual of E. Then, for each $\Phi \in E'$, $\Phi \circ f$ is a scalar function which is continuous on $\overline{D}$ and holomorphic in D. Hence, by the Cauchy integral theorem for scalar functions,

$$\Phi\Big(\int_{\partial D} f(z)dz\Big) = \int_{\partial D} \Phi\big(f(z)\big)dz = 0 \quad \text{for all } \Phi \in E'. \tag{1.4.2}$$

By the Hahn-Banach theorem this implies (1.4.1). □

1.4.3. Recall that an open set $D \subseteq \mathbb{C}$ is called **simply connected** if it is connected and, for any continuous function $\gamma : [0,1] \to D$ with $\gamma(0) = \gamma(1)$, there exists a continuous function

$$H : [0,1] \times [0,1] \longrightarrow D$$

such that $H(0,t) = H(1,t)$ for all $0 \le t \le 1$, $H(\cdot, 0) = \gamma$ for all $0 \le s \le 1$ and $H(\cdot, 1)$ is a constant.

We need also the following homotopy version of the Cauchy integral theorem for functions with values in a Banach space.

[4] A possible formal definition: We say that D is **on the left** of Γ if the following condition is satisfied: If $\gamma : [a,b] \to \mathbb{C}$ is a positively oriented piecewise $\mathcal{C}^1$-parametrization of one of the connected components of Γ, then $\varepsilon i\gamma'(t) \in D$ for each $a \le t \le b$ such that $\gamma(t)$ is a smooth point of Γ and any sufficiently small $\varepsilon > 0$.

1.4.4 Theorem. *Let $D \subseteq \mathbb{C}$ be a simply connected open set, let E be a Banach space, let $f : D \to E$ be holomorphic, and let $\gamma_1, \gamma_2 : [0,1] \to D$ be two piecewise $\mathcal{C}^1$-functions with $\gamma_1(0) = \gamma_2(0)$ and $\gamma_1(1) = \gamma_2(1)$.*[5] *Then*

$$\int_0^1 f\big(\gamma_1(z)\big)\gamma_1'(z)dz = \int_0^1 f\big(\gamma_1(z)\big)\gamma_2'(z)dz. \tag{1.4.3}$$

Proof. Let E' be the dual of E. Then, by the homotopy version of the Cauchy integral theorem for scalar functions, for each $\Phi \in E'$,

$$\begin{aligned}\Phi\bigg(\int_0^1 f\big(\gamma_1(z)\big)\gamma_1'(z)dz\bigg) &= \int_0^1 \Phi\Big(f\big(\gamma_1(z)\big)\gamma_1'(z)dz \\ &= \int_0^1 \Phi\Big(f\big(\gamma_1(z)\big)\gamma_2'(z)\Big)dz = \Phi\bigg(\int_0^1 f\big(\gamma_2(z)\gamma_2'(z)dz\bigg).\end{aligned}$$

By the Hahn-Banach theorem this implies (1.4.3). □

1.5 The Cauchy formula

1.5.1 Theorem (Cauchy formula). *Let $D \subseteq \mathbb{C}$ be a bounded open set with piecewise $\mathcal{C}^1$-boundary, let E be a Banach space, and let $f : \overline{D} \to E$ be a continuous function which is holomorphic in D. Then*

$$f(w) = \frac{1}{2\pi i}\int_{\partial D} \frac{f(z)}{z-w}dz, \qquad w \in D. \tag{1.5.1}$$

Proof. Let $w \in D$ be given. Let E' be the dual of E. Then, for each $\Phi \in E'$, $\Phi \circ f$ is a scalar function which is continuous on $\overline{D}$ and holomorphic in D. Hence, by the Cauchy formula for scalar functions,

$$\Phi\big(f(w)\big) = \frac{1}{2\pi i}\int_{\partial D} \frac{\Phi\big(f(z)\big)}{z-w}dz = \Phi\bigg(\frac{1}{2\pi i}\int_{\partial D}\frac{f(z)}{z-w}dz\bigg) \quad \text{for all } \Phi \in E'.$$

By the Hahn-Banach theorem this implies (1.5.1). □

1.5.2 Lemma. *Let $\Gamma \subseteq \mathbb{C}$ be an oriented, piecewise $\mathcal{C}^1$-contour (not necessarily closed), let E be a Banach space, and let $f : \Gamma \to E$ be continuous. Let $n \in \mathbb{N}^*$ and set*

$$F(z) = \int_\Gamma \frac{f(\zeta)}{(\zeta - z)^n}d\zeta\,, \qquad z \in \mathbb{C} \setminus \Gamma. \tag{1.5.2}$$

---

[5] Here we do not assume that the images $\gamma_1\big([0,1]\big)$ and $\gamma_2\big([0,1]\big)$ are piecewise $\mathcal{C}^1$-contours in the sense of Definition 1.3.2.

Then F is holomorphic on $\mathbb{C}\setminus\Gamma$ and

$$F'(z) = n\int_\Gamma \frac{f(\zeta)}{(\zeta-z)^{n+1}}d\zeta\,, \qquad z\in\mathbb{C}\setminus\Gamma. \tag{1.5.3}$$

Moreover

$$\lim_{|z|\to\infty}\|F(z)\| = 0. \tag{1.5.4}$$

Proof. Using the estimate from Proposition 1.3.6, we get

$$\|F(z)\| \le |\Gamma|\max_{\zeta\in\Gamma}\frac{\|f(\zeta)\|}{|\zeta-z|^n}.$$

As Γ is compact and therefore

$$\lim_{|z|\to\infty}\max_{\zeta\in\Gamma}\frac{1}{|\zeta-z|^n} = 0,$$

this implies (1.5.4).

It remains to prove that F is holomorphic on $\mathbb{C}\setminus\Gamma$. Let $w\in\mathbb{C}\setminus\Gamma$ be given. We must prove that

$$\lim_{z\to w}\int_\Gamma\left(\frac{\frac{1}{(\zeta-z)^n}-\frac{1}{(\zeta-w)^n}}{z-w}-\frac{n}{(\zeta-w)^{n+1}}\right)f(\zeta)d\zeta = 0. \tag{1.5.5}$$

We have

$$\begin{aligned}
&\frac{\frac{1}{(\zeta-z)^n}-\frac{1}{(\zeta-w)^n}}{z-w}-\frac{n}{(\zeta-w)^{n+1}}\\
&=\frac{(\zeta-w)^{n+1}-(\zeta-z)^n(\zeta-w)-n(\zeta-z)^n(z-w)}{(z-w)(\zeta-z)^n(\zeta-w)^{n+1}}\\
&=\frac{(\zeta-z+z-w)^{n+1}-(\zeta-z)^n(\zeta-z+z-w)-n(\zeta-z)^n(z-w)}{(z-w)(\zeta-z)^n(\zeta-w)^{n+1}}\\
&=\frac{\sum_{k=0}^{n+1}\binom{n+1}{k}(\zeta-z)^{n+1-k}(z-w)^k-(\zeta-z)^{n+1}-(n+1)(\zeta-z)^n(z-w)}{(z-w)(\zeta-z)^n(\zeta-w)^{n+1}}\\
&=\frac{\sum_{k=2}^{n+1}\binom{n+1}{k}(\zeta-z)^{n+1-k}(z-w)^k}{(z-w)(\zeta-z)^n(\zeta-w)^{n+1}}\\
&=\frac{\sum_{k=2}^{n+1}\binom{n+1}{k}(\zeta-z)^{n+1-k}(z-w)^{k-2}}{(\zeta-z)^n(\zeta-w)^{n+1}}(z-w).
\end{aligned}$$

If $\varepsilon>0$ is chosen so small that the disc $|z-w|\le\varepsilon$ is contained in $\mathbb{C}\setminus\Gamma$, this implies that, for some constant $C<\infty$,

$$\left|\frac{\frac{1}{(\zeta-z)^n}-\frac{1}{(\zeta-w)^n}}{z-w}-\frac{n}{(\zeta-w)^{n+1}}\right|\le C|z-w| \qquad \text{if } |z-w|<\varepsilon \text{ and } \zeta\in\Gamma.$$

By the estimate from Proposition 1.3.6 this further implies

$$\left\| \int_\Gamma \left(\frac{\frac{1}{(\zeta-z)^n} - \frac{1}{(\zeta-w)^n}}{z-w} - \frac{n}{(\zeta-w)^{n+1}} \right) f(\zeta)d\zeta \right\| \le |\Gamma| \max_{\zeta\in\Gamma} \|f(\zeta)\| C|z-w|$$

for $|z-w|<\varepsilon$, which proves (1.5.5). □

In view of this lemma, the Cauchy formula immediately implies:

1.5.3 Corollary. *Any holomorphic function with values in a Banach space is infinitely times complexly differentiable (Def. 1.1.1). In particular, it is of class $\mathcal{C}^\infty$.*

Moreover, if D and f are as in Theorem 1.5.1 and if we denote by $f^{(n)}$ the n-th complex derivative of f in D, then

$$f^{(n)}(w) = \frac{n!}{2\pi i} \int_{\partial D} \frac{f(z)}{(z-w)^{n+1}} dz, \qquad w \in D. \tag{1.5.6}$$

1.5.4 Theorem. *Let $D \subseteq \mathbb{C}$ be an open set, and let M be a subset of D such that there exists a piecewise $\mathcal{C}^1$-contour in $\mathbb{C}$ with $M \subseteq \Gamma$. Further, let E be a Banach space, and let $f : D \to E$ be a continuous function which is holomorphic on $D\setminus M$. Then f is holomorphic on D.*

Proof. Let w be an arbitrary point in M. Set

$$\Delta_\varepsilon = \left\{ z \in \mathbb{C} \,\middle|\, |z-w| < \varepsilon \right\} \qquad \text{for } \varepsilon > 0.$$

It is sufficient to prove that, for sufficiently small $\varepsilon > 0$, f is holomorphic on Δ_ε.

By hypothesis there exists a piecewise $\mathcal{C}^1$-contour Γ in $\mathbb{C}$ with $M \subseteq \Gamma$. Enlarging Γ if necessary, we may achieve that w is an "inner point" of Γ, i.e., that, for some $\varepsilon_0 > 0$, the disc Δ_{ε_0} is divided by Γ into two connected open sets $\Delta^+_{\varepsilon_0}$ and $\Delta^-_{\varepsilon_0}$. Moreover, we can choose $0 < \varepsilon < \varepsilon_0$ so small that the closed disc $\overline{\Delta}_\varepsilon$ is contained in D, and the open sets

$$\Delta^+_\varepsilon := \Delta_\varepsilon \cap \Delta^+_{\varepsilon_0} \qquad \text{and} \qquad \Delta^-_\varepsilon := \Delta_\varepsilon \cap \Delta^-_{\varepsilon_0}$$

have piecewise $\mathcal{C}^1$-boundaries $\partial\Delta^+_\varepsilon$ and $\partial\Delta^-_\varepsilon$. Then, by Cauchy's formula and by Cauchy's integral theorem,

$$\frac{1}{2\pi i} \int_{\partial\Delta^+_\varepsilon} \frac{f(\zeta)}{\zeta - z} d\zeta = \begin{cases} f(z) & \text{if } z \in \Delta^+_\varepsilon, \\ 0 & \text{if } z \in \Delta^-_\varepsilon \end{cases}$$

and

$$\frac{1}{2\pi i} \int_{\partial\Delta^-_\varepsilon} \frac{f(\zeta)}{\zeta - z} d\zeta = \begin{cases} f(z) & \text{if } z \in \Delta^-_\varepsilon, \\ 0 & \text{if } z \in \Delta^+_\varepsilon. \end{cases}$$

Since the orientation of $\Gamma \cap \Delta_\varepsilon$ as a part of $\partial\Delta_\varepsilon^+$ is different from its orientation as a part of $\partial\Delta_\varepsilon^-$, this implies that

$$\frac{1}{2\pi i} \int\limits_{|\zeta - w| = \varepsilon} \frac{f(\zeta)}{\zeta - z} d\zeta = f(z) \qquad \text{if } z \in \Delta_\varepsilon \setminus \Gamma.$$

Since f is continuous on Δ_ε, this further implies that

$$\frac{1}{2\pi i} \int\limits_{|\zeta - w| = \varepsilon} \frac{f(\zeta)}{\zeta - z} d\zeta = f(z) \qquad \text{for all } z \in \Delta_\varepsilon.$$

In view of Lemma 1.5.2 this proves that f is holomorphic on Δ_ε. □

1.6 The Hahn-Banach criterion

1.6.1 Theorem (Hahn-Banach criterion). *Let $D \subseteq \mathbb{C}$ be an open set, let E be a Banach space, let E' be the dual of E, and let $f : D \to E$ be a function. Then the following two conditions are equivalent:*

(i) *The function f is holomorphic on D.*

(ii) *For each $\Phi \in E'$, the scalar function $\Phi \circ f$ is holomorphic on D.*

Proof. (i) $\Rightarrow$ (ii) is obvious.

Suppose (ii) is satisfied.

We first prove that then f is continuous. Consider an arbitrary point $z_0 \in D$ and choose $r > 0$ so small that the disc $|z - z_0| \le r$ is contained in D. Since, for each $\Phi \in E'$, $\Phi \circ f$ is holomorphic on D and, hence, $\Phi \circ f$ is bounded on $|z - z_0| \le r$, the set

$$\Big\{ \Phi\big(f(z)\big) \;\Big|\; |z - z_0| \le r \Big\}$$

is bounded for each $\Phi \in E'$. Since weakly bounded sets are strongly bounded, it follows that the set $\{f(z) \,|\, |z - z_0| \le r\}$ is bounded in E, i.e., we have a constant $C < \infty$ such that

$$\|f(z)\| \le C \qquad \text{for all } |z - z_0| \le r. \tag{1.6.1}$$

Now let z_n, $n \in \mathbb{N}^*$, be a sequence which converges to z_0 such that $|z_n - z_0| < r$ for all $n \in \mathbb{N}^*$. From the Cauchy formula, the estimate from Proposition 1.3.6 and estimate (1.6.1) then it follows that, for all $\Phi \in E'$,

$$\Big|\Phi\big(f(z_0) - f(z_n)\big)\Big| = \frac{1}{2\pi} \Bigg| \int\limits_{|z - z_0| = r} \Phi\big(f(z)\big) \left(\frac{1}{z - z_0} - \frac{1}{z - z_n} \right) dz \Bigg|$$

$$\le r \max_{|z - z_0| = r} \Big|\Phi\big(f(z)\big)\Big| \left| \frac{1}{z - z_0} - \frac{1}{z - z_n} \right| \le r\|\Phi\|C \max_{|z - z_0| = r} \left| \frac{1}{z - z_0} - \frac{1}{z - z_n} \right|.$$

By the Hahn-Banach theorem this implies that

$$\|f(z_0) - f(z_n)\| \leq rC \max_{|z-z_0|=r} \left| \frac{1}{z-z_0} - \frac{1}{z-z_n} \right|.$$

Hence $\lim_{n\to\infty} f(z_n) = f(z_0)$.

To prove that f is holomorphic, we again consider an arbitrary point $z_0 \in D$ and choose $r > 0$ so small that the disc $|z - z_0| \leq r$ is contained in D. As f is continuous, by Lemma 1.5.2 the function

$$F(z) := \frac{1}{2\pi i} \int\limits_{|\zeta - z_0|=r} \frac{f(\zeta)}{\zeta - z} d\zeta\,, \qquad |z - z_0| < r,$$

is holomorphic. Therefore it remains to prove that $f(z) = F(z)$ for $|z - z_0| < r$. Let such z be given. Since, for each $\Phi \in E'$, the function $\Phi \circ f$ is holomorphic, it follows from the Cauchy formula and then from the definition of F that

$$\Phi\big(f(z)\big) = \frac{1}{2\pi i} \int\limits_{|\zeta - z_0|=r} \frac{\Phi\big(f(\zeta)\big)}{\zeta - z} d\zeta = \Phi\left(\frac{1}{2\pi i} \int\limits_{|\zeta - z_0|=r} \frac{f(\zeta)}{\zeta - z} d\zeta \right) = \Phi\big(F(z)\big)$$

for all $\Phi \in E'$. By the Hahn-Banach theorem this implies that $f(z) = F(z)$. □

We conclude this section with some applications of Theorem 1.6.1.

1.6.2 Theorem. *Let $D \subseteq \mathbb{C}$ be a simply connected open set (Section 1.4.3), let E be a Banach space, and let $f : D \to E$ be holomorphic. Then there exists a holomorphic function $F : D \to E$ such that $F' = f$.*

Proof. Fix $z_0 \in D$. Then, for each $z \in D$, we choose a $\mathcal{C}^1$-function $\gamma : [0, 1] \to D$ with $\gamma(0) = z_0$ and $\gamma(1) = z$ and define

$$F(z) = \int_0^1 f\big(\gamma(\zeta)\big)\gamma'(\zeta) d\zeta.$$

As D is connected, such a function γ always exists, and, as D is even simply connected, by the homotopy version of the Cauchy integral Theorem 1.4.4, this definition does not depend on the choice of γ. So F is well defined. It remains to prove that F is holomorphic and $F' = f$.

Let E' be the dual of E. Then, by the corresponding scalar fact, for each $\Phi \in E'$, the function $\Phi \circ F$ is holomorphic and $(\Phi \circ F)' = \Phi \circ f$. By the Hahn-Banach criterion, Theorem 1.6.1, this means that F itself is holomorphic, and thus this implies that $F' = f$. □

1.6.3 Lemma. *Let $D \subseteq \mathbb{C}$ be an open set, let E be a Banach space, and let $f : D \to E$ be a function. Let E' be the dual of E, and suppose there exists a sequence* *of holomorphic functions $f_n : D \to E$, $n \in \mathbb{N}$, such that, for all $\Phi \in E'$ and each compact $K \subseteq D$,*

$$\lim_{n\to\infty} \max_{z\in K} \left\| \Phi\Big(f(z)\Big) - \Phi\Big(f_n(z)\Big) \right\| = 0. \tag{1.6.2}$$

Then f is holomorphic on D.

Proof. From (1.6.2) it follows that $\Phi \circ f$ is holomorphic for each $\Phi \in E'$. Hence, by Theorem 1.6.1, f is holomorphic. □

1.6.4 Theorem. *Let $D \subseteq \mathbb{C}$ be an open set, let E be a Banach space, and let $f_n : D \to E$, $n \in \mathbb{N}$, be a sequence of holomorphic functions, which converges, uniformly on each compact subset of D, to some function $f : D \to E$. Then f is holomorphic on D, and the sequence of complex derivatives f_n' converges, uniformly on each compact subset of D, to f'.*

Proof. It follows immediately from Lemma 1.6.3 that f is holomorphic. It remains to prove that f_n' converges to f', uniformly on each compact subset of D. Let $z_0 \in D$ and $r > 0$ be given such that the closed disc $|z - z_0| \le r$ is contained in D. It is sufficient to prove that

$$\lim_{n\to\infty} \max_{|z-z_0|\le r} \|f'(z) - f_n'(z)\| = 0. \tag{1.6.3}$$

Choose r' with $r < r'$ such that also the closed disc $|z - z_0| \le r'$ is contained in D. Then, by the Cauchy formula for the complex derivative (1.5.6),

$$f'(z) - f_n'(z) = \frac{1}{2\pi i} \int\limits_{|\zeta - z_0| = r'} \frac{f(\zeta) - f_n(\zeta)}{(\zeta - z)^2} d\zeta$$

for $|z - z_0| \le r$. By the estimate from Proposition 1.3.6 this implies that

$$\|f'(z) - f_n'(z)\| \le r' \max_{|\zeta - z_0| = r'} \frac{\|f(\zeta) - f_n(\zeta)\|}{|\zeta - z|^2} \le \frac{r'}{(r' - r)^2} \max_{|\zeta - z_0| = r'} \|f(\zeta) - f_n(\zeta)\|$$

for $|z - z_0| \le r$. As f_n converges to f, uniformly on $|z - z_0| = r'$, this implies 15.3.08. □

We already observed that also the holomorphic functions with values in a Banach space satisfy the Cauchy-Riemann equation (1.1.1). If we additionally assume that the function is continuously differentiable (with respect to the canonical real coordinates), then, as in the scalar case, the Cauchy-Riemann equation is also sufficient for holomorphy:

1.6.5 Theorem (Cauchy-Riemann criterion). *Let $D \subseteq \mathbb{C}$ be an open set, let E be a Banach space, and let $f : D \to E$ be a $\mathcal{C}^1$-function. Then f is holomorphic, if and only if,*

$$\frac{\partial f}{\partial x} = i \frac{\partial f}{\partial y} \qquad \text{on } D. \tag{1.6.4}$$

Proof. We already observed that (1.6.4) is necessary for the holomorphy of f.

Now we assume that (1.6.4) is satisfied. Let E' be the dual of E. Since f is of class $\mathcal{C}^1$, then it follows from (1.6.4) that, for each $\Phi \in E'$,

$$\frac{\partial(\Phi \circ f)}{\partial x} = i\frac{\partial(\Phi \circ f)}{\partial y} \qquad \text{on } D.$$

Hence, $\Phi \circ f$ is holomorphic for all $\Phi \in E'$, and it follows from the Hahn-Banach criterion, Theorem 1.6.1, that f is holomorphic. □

1.7 A criterion for the holomorphy of operator functions

1.7.1 Theorem. *Let $D \subseteq \mathbb{C}$ be an open set, let E, F be Banach spaces, let $A : D \to L(E, F)$ be a holomorphic operator function. Then the following two conditions are equivalent:*

(i) *A is holomorphic on D.*

(ii) *For each vector $x \in E$, the vector function Ax is holomorphic on D.*

Proof. (i) $\Rightarrow$ (ii) is obvious.

Assume (ii) is satisfied.

We first prove that then A is continuous. Consider an arbitrary point $z_0 \in D$ and choose $r > 0$ so small that the disc $|z - z_0| \leq r$ is contained in D. Since, for each $x \in E$, Ax is holomorphic on D and, hence, Ax is bounded on $|z - z_0| \leq r$, the set

$$\left\{ A(z)x \;\middle|\; |z - z_0| \leq r \right\}$$

is bounded for each $x \in E$. By the Banach-Steinhaus theorem, it follows that the set

$$\left\{ A(z) \;\middle|\; |z - z_0| \leq r \right\}$$

is bounded in $L(E, F)$, i.e., we have a constant $C < \infty$ such that

$$\|A(z)\| \leq C \qquad \text{for all } |z - z_0| \leq r. \tag{1.7.1}$$

Now let z_n, $n \in \mathbb{N}^*$, be a sequence which converges to z_0 such that $|z_n - z_0| < r$ for all $n \in \mathbb{N}^*$. From the Cauchy formula, the estimate from Proposition 1.3.6 and estimate (1.7.1) then it follows that, for all $x \in E$,

$$\begin{aligned}\big\|A(z_0)x - A(z_n)x\big\| &= \frac{1}{2\pi}\left\| \int\limits_{|z-z_0|=r} A(z)x\left(\frac{1}{z - z_0} - \frac{1}{z - z_n}\right) dz \right\| \\ &\leq rC\|x\| \max_{|z-z_0|=r}\left|\frac{1}{z - z_0} - \frac{1}{z - z_n}\right|.\end{aligned}$$

Hence

$$\|A(z_0) - A(z_n)\| \le rC \max_{|z-z_0|=r} \left| \frac{1}{z-z_0} - \frac{1}{z-z_n} \right|,$$

which further implies that $\lim_{n\to\infty} A(z_n) = A(z_0)$.

To prove that A is holomorphic, we again consider an arbitrary point $z_0 \in D$ and choose $r > 0$ so small that the disc $|z - z_0| \le r$ is contained in D. As A is continuous, by Lemma 1.5.2 the function

$$F(z) := \frac{1}{2\pi i} \int\limits_{|\zeta - z_0|=r} \frac{A(\zeta)}{\zeta - z} d\zeta\,, \qquad |z - z_0| < r,$$

is holomorphic. Therefore it remains to prove that $A(z) = F(z)$ for $|z - z_0| < r$. Let such z be given. Since, for each $x \in E$, the function Ax is holomorphic, it follows from the Cauchy formula and then from the definition of F that

$$A(z)x = \frac{1}{2\pi i} \int\limits_{|\zeta - z_0|=r} \frac{A(\zeta)x}{\zeta - z} d\zeta = \left(\frac{1}{2\pi i} \int\limits_{|\zeta - z_0|=r} \frac{A(\zeta)}{\zeta - z} d\zeta \right) x = F(z)x$$

for all $x \in E$. Hence $A(z) = F(z)$. □

1.8 Power series

1.8.1. Let E be a Banach space, and let

$$\sum_{n=0}^{\infty} a_n (z - z_0)^n\,, \qquad z_0 \in \mathbb{C}, \tag{1.8.1}$$

be a power series with coefficients $a_n \in E$. The series (1.8.1) is called **convergent** in a point $\zeta \in \mathbb{C}$ if the series of vectors

$$\sum_{n=0}^{\infty} a_n (\zeta - z_0)^n$$

converges in E (with respect to the norm), and it is called **absolutely convergent** in $\zeta \in \mathbb{C}$ if

$$\sum_{n=0}^{\infty} \|a_n\| |\zeta - z_0|^n < \infty.$$

1.8.2 Theorem (Abel's lemma). *Let E be a Banach space, and let*

$$\sum_{n=0}^{\infty} (z - z_0)^n a_n\,, \qquad z_0 \in \mathbb{C}, \tag{1.8.2}$$

be a power series with coefficients $a_n \in E$. *Set*

$$\rho = \frac{1}{\limsup\limits_{n\to\infty} \sqrt[n]{\|a_n\|}}. \tag{1.8.3}$$

If $|\zeta - z_0| > \rho$, *then the series* (1.8.2) *does not converge in* ζ. *If* $|\zeta - z_0| < \rho$, *then the series* (1.8.2) *converges absolutely in* ζ. *Moreover, then, for all* $r < \rho$,

$$\sum_{n=0}^{\infty} \max_{|\zeta - z_0| \le r} |\zeta - \zeta_0|^n \|a_n\| < \infty. \tag{1.8.4}$$

Proof. First let $|\zeta - z_0| > \rho$. This means that

$$|\zeta - z_0| \limsup_{n\to\infty} \|a_n\|^{1/n} > 1,$$

i.e.,

$$\limsup_{n\to\infty} \|a_n\| |\zeta - z_0|^n > 1.$$

Hence the sequence $a_n(\zeta - z_0)^n$ does not converge to zero.

Now let $r < \rho$ be given. Then

$$r \limsup_{n\to\infty} \|a_n\|^{1/n} < 1.$$

Choose q with

$$r \limsup_{n\to\infty} \sqrt[n]{\|a_n\|} < q < 1.$$

Then we can find $n_0 \in \mathbb{N}$ such that

$$r \sqrt[n]{\|a_n\|} \le q \qquad \text{for all } n \ge n_0.$$

It follows that

$$\sum_{n=0}^{\infty} \max_{|\zeta - z_0| \le r} \|a_n\| |\zeta - \zeta_0|^n \le \sum_{n=0}^{\infty} q^n < \infty. \qquad \square$$

1.8.3. The number $\rho \in [0, \infty]$ defined by (1.8.3) is called the **radius of convergence** of the power series in (1.8.1).

1.8.4 Theorem. *Let* E *be a Banach space, let*

$$\sum_{n=0}^{\infty} (z - z_0)^n a_n\,, \qquad z_0 \in \mathbb{C}, \tag{1.8.5}$$

be a power series with coefficients $a_n \in E$, *and let* ρ *be the radius of convergence of it. Then:*

(i) *The power series*

$$\sum_{n=1}^{\infty} n(z-z_0)^{n-1}a_n \tag{1.8.6}$$

also has the radius of convergence ρ.

(ii) *The function defined by*

$$f(\zeta) = \sum_{n=0}^{\infty} (\zeta - z_0)^n a_n\,, \qquad |\zeta - z_0| < \rho, \tag{1.8.7}$$

is holomorphic on the open disc $|z - z_0| < \rho$, *and*

$$f'(\zeta) = \sum_{n=1}^{\infty} n(\zeta - z_0)^{n-1} a_n\,, \qquad |\zeta - z_0| < \rho. \tag{1.8.8}$$

(iii) *If* f *is the holomorphic function defined by* (1.8.6), *then*

$$a_n = \frac{f^{(n)}(z_0)}{n!}\,, \qquad n \in \mathbb{N}. \tag{1.8.9}$$

Proof. Part (i): By definition (1.8.3) of the radius of convergence, assertion (i) is equivalent to the equality

$$\frac{1}{\limsup\limits_{n\to\infty} \sqrt[n]{\|a_n\|}} = \frac{1}{\limsup\limits_{n\to\infty} \sqrt[n-1]{n\|a_n\|}}.$$

But the latter relation follows (for example) from the fact that the *scalar* power series

$$\sum_{n=0}^{\infty} \|a_n\|(z-z_0)^n \qquad \text{and} \qquad \sum_{n=1}^{\infty} n\|a_n\|(z-z_0)^{n-1}$$

have the same radius of convergence.

Part (ii): From (1.8.4) it follows that the sequence of partial sums

$$\sum_{n=0}^{N} a_n(z-z_0)^n\,, \qquad N \in \mathbb{N},$$

converges to f, uniformly on each compact subset of the disc $|z - z_0| < \rho$. By Theorem 1.6.4 this implies that f is holomorphic and that the sequence of partial sums

$$\sum_{n=1}^{N} na_n(z-z_0)^{n-1}\,, \qquad N \in \mathbb{N},$$

converge to f', uniformly on each compact subset of the disc $|z - z_0| < \rho$.

Part (iii): This follows, as in the scalar case, by repeated application of (i) and (ii). □

1.8.5 Theorem. *Let $D \subseteq \mathbb{C}$ be an open set, let E be a Banach space, and let $f : D \to E$ be a holomorphic function. Let $z_0 \in D$ and let $r > 0$ such that the open disc $|z - z_0| < r$ is contained in D. Then there exists a uniquely determined power series*

$$\sum_{n=0}^{\infty} (z - z_0)^n f_n$$

with coefficients $f_n \in E$ with radius of convergence such that

$$f(\zeta) = \sum_{n=0}^{\infty} (\zeta - z_0)^n f_n \tag{1.8.10}$$

for all ζ in a neighborhood of z_0. Then

$$f_n = \frac{f^{(n)}(z_0)}{n!} = \frac{1}{2\pi i} \int\limits_{|z-z_0|=r'} \frac{f(z)}{(z - z_0)^{n+1}} dz \qquad \text{for all } 0 < r' < r. \tag{1.8.11}$$

Moreover, then $\rho \geq r$ and (1.8.10) holds for all ζ wtih $|\zeta - z_0| < r$.

Proof. The statement on uniqueness and the first equality in (1.8.10) follows from part (iii) of Theorem 1.8.4. To prove the remaining statements, we define vectors by

$$f_n = \frac{1}{2\pi i} \int\limits_{|z-z_0|=r'} \frac{f(z)}{(z - z_0)^{n+1}} dz\,, \qquad n \in \mathbb{N}\,, \tag{1.8.12}$$

where $0 < r' < r$. (By the Cauchy integral theorem this is independent of the choice of r'.) It remains to prove that, for $|\zeta - z_0| < r$, the series

$$\sum_{n_0}^{\infty} (\zeta - z_0)^n f_n$$

converges to $f(\zeta)$. Let such ζ be given. Then, as in the scalar case, we choose r' with $|\zeta - z_0| < r' < r$ and obtain by means of the Cauchy formula

$$\begin{aligned}
f(\zeta) &= \frac{1}{2\pi i} \int\limits_{|z-z_0|=r'} \frac{f(z)}{z - \zeta} dz = \frac{1}{2\pi i} \int\limits_{|z-z_0|=r'} \frac{1}{1 - \frac{\zeta - z_0}{z - z_0}} \frac{f(z)}{z - z_0} dz \\
&= \frac{1}{2\pi i} \int\limits_{|z-z_0|=r'} \sum_{n=0}^{\infty} \left(\frac{\zeta - z_0}{z - z_0} \right)^n \frac{f(z)}{z - z_0} dz \\
&= \sum_{n=0}^{\infty} (\zeta - z_0)^n \left(\frac{1}{2\pi i} \int\limits_{|z-z_0|=r} \frac{f(z)}{(z - z_0)^{n+1}} dz \right) = \sum_{n=0}^{\infty} (\zeta - z_0)^n f_n.
\end{aligned}$$

□

1.9 Laurent series

1.9.1 Theorem. *Let $z_0 \in \mathbb{C}$, let $0 \le r < R \le \infty$, let E be a Banach space, and let f be an E-valued function defined and holomorphic in $r < |z - z_0| < R$. Then there exists a uniquely determined Laurent series*

$$\sum_{n=-\infty}^{\infty} (z - z_0)^n f_n \tag{1.9.1}$$

with coefficients $f_n \in E$ such that

$$\sum_{n=-\infty}^{\infty} \max_{r' \le |\zeta - z_0| \le R'} |\zeta - z_0|^n \|f_n\| < \infty \qquad \text{for } r < r' < R' < R \tag{1.9.2}$$

and

$$f(\zeta) = \sum_{n=-\infty}^{\infty} (\zeta - z_0)^n f_n \qquad \text{for } r < |\zeta - z_0| < R. \tag{1.9.3}$$

Moreover, then

$$f_n = \frac{1}{2\pi i} \int\limits_{|z-z_0|=\rho} \frac{f(z)}{(z - z_0)^{n+1}} dz \qquad \text{for } r < \rho < R. \tag{1.9.4}$$

Proof. Uniqueness and formula (1.9.4)*:* Suppose we have a Laurent series with (1.9.2) and (1.9.3). Then, for each $r < \rho < R$,

$$\frac{1}{2\pi i} \int\limits_{|z-z_0|=\rho} \frac{f(z)}{(z - z_0)^{n+1}} dz = \sum_{k=-\infty}^{\infty} \frac{f_k}{2\pi i} \int\limits_{|z-z_0|=\rho} \frac{(z - z_0)^k}{(z - z_0)^{n+1}} dz = f_n.$$

Existence: We define a function f_+ on the disc $|z - z_0| < R$ as follows: If a point z with $|z - z_0| < R$ is given, then we choose a number ρ with $|z - z_0| < \rho < R$ and set

$$f_+(z) = \frac{1}{2\pi i} \int\limits_{|\zeta-z_0|=\rho} \frac{f(\zeta)}{\zeta - z} d\zeta. \tag{1.9.5}$$

By the Cauchy integral theorem, this definition is independent of the choice of ρ, and, by Lemma 1.5.2, f_+ is holomorphic on the disc $|z - z_0| < R$. Furthermore, we define a function f_- on $|z - z_0| > r$ as follows: If a point z with $|z - z_0| > r$ is given, then we choose a number ρ with $|z - z_0| > \rho > r$ and set

$$f_-(z) = -\frac{1}{2\pi i} \int\limits_{|\zeta-z_0|=\rho} \frac{f(\zeta)}{\zeta - z} d\zeta. \tag{1.9.6}$$

Again, by the Cauchy integral theorem, this definition is independent of the choice of ρ, and, by Lemma 1.5.2, f_- is holomorphic on $|z - z_0| > r$ and

$$\lim_{|z|\to\infty} f_-(z) = 0. \tag{1.9.7}$$

Then

$$f(z) = f_+(z) + f_-(z) \qquad \text{for } r < |z - z_0| < R. \tag{1.9.8}$$

Indeed, let z with $r < |z - z_0| < R$ be given. Then we choose ρ_+ and ρ_- with $r < \rho_- < |z - z_0| < \rho_+ < R$ and obtain, by the Cauchy-Integral formula,

$$f(z) = \frac{1}{2\pi i} \int\limits_{|\zeta - z_0| = \rho_+} \frac{f(\zeta)}{(\zeta - z)} d\zeta - \frac{1}{2\pi i} \int\limits_{|\zeta - z_0| = \rho_-} \frac{f(\zeta)}{(\zeta - z)} d\zeta = f_+(z) + f_-(z).$$

As f_+ is holomorphic on the disc $|z - z_0| < R$, it can be represented by a power series (Theorem 1.8.5):

$$f_+(z) = \sum_{n=0}^{\infty} (z - z_0)^n f_n, \qquad |z - z_0| < R, \tag{1.9.9}$$

where, by Abel's lemma (Theorem 1.8.2),

$$\sum_{n=0}^{\infty} \max_{|z - z_0| \le R'} |z - z_0|^n \|f_n\| < \infty \qquad \text{for } R' < R. \tag{1.9.10}$$

Now we consider the function

$$F(z) := f_-\left(z_0 + \frac{1}{z}\right),$$

which is defined and holomorphic for $0 < |z| < 1/r$. As $\lim_{|z|\to\infty} f_-(z) = 0$, this function extends continuously to 0 with $F(0) := 0$. By Theorem 1.5.4, this extended F is holomorphic on the whole disc $|z| < 1/r$. Therefore also F can be represented by a power series

$$F(z) = \sum_{n=1}^{\infty} z^n F_n, \qquad |z| < \frac{1}{r}, \tag{1.9.11}$$

where

$$\sum_{n=1}^{\infty} \max_{|z| \le 1/r'} |z|^n \|F_n\| < \infty, \qquad r' > r. \tag{1.9.12}$$

Set $f_n = F_{-n}$ for $n \le -1$. Then it follows from (1.9.11) and (1.9.12) that

$$f_-(z) = F\left(\frac{1}{z - z_0}\right) = \sum_{n=1}^{\infty} \left(\frac{1}{z - z_0}\right)^n F_n = \sum_{n=-\infty}^{-1} (z - z_0)^n f_n, \quad |z - z_0| > r,$$

and

$$\sum_{n=-\infty}^{-1} \max_{|z-z_0|\geq r'} |z-z_0|^n \|f_n\| = \sum_{n=1}^{\infty} \max_{1/|z-z_0|\leq 1/r'} \left|\frac{1}{z-z_0}\right|^n \|F_n\| < \infty\,, \qquad r' > r.$$

Together with (1.9.8), (1.9.9) and (1.9.10) this shows that the Laurent series

$$\sum_{n=-\infty}^{\infty} (z-z_0)^n f_n$$

has the required properties. □

1.10 Isolated singularities

1.10.1. Let $D \subseteq \mathbb{C}$ be an open set, let E be a Banach space, and let f be an E-valued holomorphic function with the domain of definition D. Then, as in the scalar case, a point $z_0 \in \mathbb{C}$ is called an **isolated singularity** of f, if $\{z_0\} \cup D$ is open and $z_0 \notin D$.

1.10.2. Let $z_0 \in \mathbb{C}$, let U be a neighborhood of z_0, let E be a Banach space, and let $f : U \setminus \{z_0\} \to E$ be a holomorphic function. (In other words, in the sense of Section 1.10.1, we assume that z_0 is an isolated singularity of some holomorphic function f.) Further let $\varepsilon > 0$ be the maximal radius such that the punctured disc $0 < |z - z_0| < \varepsilon$ is still contained in U.

By Theorem 1.9.1 then there exists a uniquely determined Laurent series

$$\sum_{n=-\infty}^{\infty} (z-z_0)^n f_n$$

which converges in $0 < |z - z_0| < \varepsilon$ such that

$$f(\zeta) = \sum_{n=-\infty}^{\infty} (\zeta - z_0)^n f_n \qquad \text{for all } 0 < |\zeta - z_0| < \varepsilon. \tag{1.10.1}$$

This Laurent series will be called the **Laurent series of f at z_0**, the formula (1.10.1) will be called the **Laurent expansion of f at z_0**, and the vector f_{-1} will be called the **residuum** of f at z_0. As in the scalar case one sees that, for each $\varepsilon > 0$ such that the punctured disc $0 < |z - z_0| \leq \varepsilon$ is contained in U,

$$f_{-1} = \frac{1}{2\pi i} \int_{|z-z_0|=\varepsilon} f(z)dz. \tag{1.10.2}$$

The isolated singularity z_0 will be called a **removable singularity of f** if $f_n = 0$ for all negative integers n. If, moreover, $f_0 = 0$, then it will be called a **zero of f**.

If z_0 is a zero of f and $f \not\equiv 0$ in a neighborhood of z_0, then (by uniqueness of the Laurent expansion) there exist positive integers n with $f_n \neq 0$ – the smallest of them will be called the **order of the zero** z_0.

The isolated singularity z_0 will be called a **pole** of f if there exists a negative integer p such that $f_p \neq 0$ and $f_n = 0$ for all integers $n \leq p - 1$. The integer p then is called the **order of the pole** z_0.

If z_0 is not a removable singularity of F and not a pole of f, then z_0 is called an **essential singularity of** f.

1.10.3 Theorem (Riemann's theorem on removable singularities). *Let E be a Banach space, and let z_0 be an isolated singularity of an E-valued holomorphic function f defined in a deleted neighborhood of z_0. If f is bounded, then z_0 is removable as a singularity of f.*

Proof. Let

$$f(z) = \sum_{n=-\infty}^{\infty} f_n (z - z_0)^n$$

be the Laurent series of f at z_0. We have to prove that $f_n = 0$ for $n \leq -1$. Let $n \leq -1$ be given. Choose $r > 0$ sufficiently small. Then

$$C := \sup_{0<|z-z_0|<r} |f(z)| < \infty,$$

and, by (1.9.4),

$$f_n = \frac{1}{2\pi i} \int\limits_{|z-z_0|=\varepsilon} \frac{f(z)}{(z-z_0)^{n+1}}\, dz \qquad \text{for } 0 < \varepsilon < r.$$

In view of the estimate given in Proposition 1.3.6, this implies

$$|f_n| \leq \varepsilon \max_{|z-z_0|=\varepsilon} |f(z)| \frac{1}{\varepsilon^{n+1}} \leq \frac{C}{\varepsilon^n} \qquad \text{for } 0 < \varepsilon < r.$$

As $n \leq -1$, this further implies that $f_n = 0$. □

1.10.4 Theorem. *Let E be a Banach space, and let z_0 be an isolated singularity of an E-valued holomorphic function f defined in a deleted neighborhood of z_0. Then the following two conditions are equivalent.*

(i) *There exist constants $C < \infty$ and $c > 0$ such that, for some $\varepsilon > 0$,*

$$c|z-z_0|^N \leq \|f(z)\| \leq C|z-z_0|^N \qquad \text{for } 0 < |z-z_0| < \varepsilon. \tag{1.10.3}$$

(ii) *The Laurent series of f at z_0 is of the form*

$$f(z) = \sum_{n=N}^{\infty} f_n (z-z_0)^n \qquad \text{with } f_N \neq 0. \tag{1.10.4}$$

Proof. First assume that (i) is satisfied. Then

$$g(z) := \frac{f(z)}{(z - z_0)^N}$$

is a holomorphic function in a deleted neighborhood of z_0 which is bounded from above and below. This implies, by Riemann's Theorem 1.10.3, that g extends holomorphically to z_0, where $g(z_0) \neq 0$. Let

$$g(z) = \sum_{n=0}^{\infty} (z - z_0)^n g_n$$

be the potential series of g at z_0. Then

$$f(z) = (z - z_0)^N g(z) = \sum_{n=0}^{\infty} (z - z_0)^{n+N} g_n = \sum_{n=N}^{\infty} (z - z_0)^n g_{n-N}$$

is the Laurent expansion of f. As $g_0 = g(z_0) \neq 0$, it is of the form (1.10.4).

Now we assume that condition (ii) is satisfied. Then

$$g(z) := \frac{f(z)}{(z - z_0)^N} = \sum_{n=N}^{\infty} (z - z_0)^{n-N} f_n = \sum_{n=0}^{\infty} (z - z_0)^n f_{n+N}$$

is holomorphic in a neighborhood of z_0, where $g(z_0) = f_N \neq 0$. Choose $\varepsilon > 0$ so small that

$$\frac{\|f_N\|}{2} \leq \|g(z)\| \leq 2\|f_N\| \qquad \text{for } \|z - z_0\| < \varepsilon.$$

As $f(z) = (z - z_0)^N g(z)$, then (1.10.4) holds with $C = 2|f_N|$ and $c = |f_N|/2$. □

1.10.5 Theorem (Residue theorem). *Let $D \subseteq \mathbb{C}$ be an open set with piecewise C^1-boundary ∂D, let $z_1, \ldots, z_n$ be a finite number of points in D, let E be a Banach space, and let $f : D \setminus \{z_1, \ldots, z_n\} \to E$ be a continuous function which is holomorphic in D. If we denote by $res_{z_j} f$ the residuum of f at z_j, then*

$$\sum_{j=1}^{n} res_{z_j} f = \frac{1}{2\pi i} \int_{\partial D} f(z) dz. \tag{1.10.5}$$

Proof. Choose $\varepsilon > 0$ so small that the closed discs $|z - z_j| \leq \varepsilon$, $1 \leq j \leq n$, are pairwise disjoint and contained in D. Then by the Cauchy integral theorem

$$\int_{\partial D} f(z) dz = \sum_{j=1}^{n} \int_{|z - z_j| = \varepsilon} f(z) dz.$$

By (1.10.2) this implies (1.10.5). □

1.10.6. Let $D \subseteq \mathbb{C}$ be an open set, and let E be a Banach space. If we say that f is an E-valued **holomorphic function with isolated singularities** on D, then we mean that there is a set $Z \subseteq D$, which is discrete and closed in D (i.e., without accumulation points in D), such that f is an E-valued holomorphic function on $D \setminus Z$. In this case we say also that $f : D \to E$ **is holomorphic with isolated singularities**. The points from Z (which then are isolated singularities of f in the sense of Section 1.10.1) are called the **singular points** of f, and the points from $D \setminus Z$ are called the **regular points** of f.

If all singular points of f are either removable or poles, then f is called **meromorphic** on D.

1.11 Comments

Except for Theorem 1.2.2 about the strong maximum principle in Hilbert spaces (which is possibly new), the results of this chapter are coa.

Chapter 2

Solution of $\overline{\partial} u = f$ and applications

In complex analysis of several variables, the inhomogeneous Cauchy-Riemann equation is an important tool. For results in Complex analysis of one variable this equation is also important, but it is missing in many standard books.

For the aim of the present book, the inhomogeneous Cauchy-Riemann equation is basic. Therefore we dedicate the present chapter to it and its applications. We give these results with full proofs, not using the corresponding scalar fact, even if it would be possible to deduce a result by the Hahn-Banach theorem from the corresponding scalar fact.

Moreover, here we present also some results, which are specific for *scalar* functions, with full proofs, as these results are difficult to find in the literature.

2.1 The Pompeiju formula for solutions of $\overline{\partial} u = f$ on compact sets

In this section, E is a Banach space.

Let $D \subseteq \mathbb{C}$ be an open set, and let $u : D \to E$ be a C^1-function. Then we use the abbreviation

$$\overline{\partial} u = \frac{\partial u}{\partial \overline{z}} = \frac{1}{2}\left(\frac{\partial u}{\partial x} + i\frac{\partial u}{\partial y}\right)$$

where x, y are the canonical real coordinates on $\mathbb{C}$ and $z = x + iy$. The function $\overline{\partial} u$ is called the **Cauchy-Riemann derivative** of u. By the Cauchy-Riemann criterion (Theorem 1.6.5), u is holomorphic if and only if $\overline{\partial} u = 0$. This implies that, for all holomorphic functions $h : D \to \mathbb{C}$,

$$\overline{\partial}(hu) = h\overline{\partial} u. \tag{2.1.1}$$

Moreover, if $\varphi\colon D \to \mathbb{C}$ is a C^1-function with compact support, then partial integration gives

$$\int_D \varphi(\overline{\partial}u)\,d\lambda = -\int_D (\overline{\partial}\varphi)u\,d\lambda$$

where $d\lambda$ is the Lebesgue measure.

This can be used to define $\overline{\partial}u$ in the sense of distributions if u is not of class C^1. In this book, we are interested only in the following special case:

2.1.1 Definition. Let $D \subseteq \mathbb{C}$ be an open set, and let $u : D \to E$ be a continuous function. We say that u has a *continuous* Cauchy-Riemann derivative if there exists a continuous function $v : U \to E$ such that

$$\int_D \varphi v\,d\lambda = -\int_D (\overline{\partial}\varphi)u\,d\lambda \tag{2.1.2}$$

for all C^∞-functions $\varphi\colon D \to \mathbb{C}$ with compact support. It is clear that this function v then is uniquely determined. We call it the **Cauchy-Riemann derivative** of u and denote it by $\overline{\partial}u$.

Instead of "u has a continuous Cauchy-Riemann derivative" we say also "$\overline{\partial}u$ is continuous". Moreover, if u and f are two continuous functions, then writing "$\overline{\partial}u = f$" we mean that "u has a continuous Cauchy-Riemann derivative and $\overline{\partial}u = f$".

2.1.2 Proposition. *Let $D \subseteq \mathbb{C}$ be an open set, let $u : D \to E$ be a continuous function such that $\overline{\partial}u$ is continuous on D, and let $\psi : D \to \mathbb{C}$ be a C^1-function. Then also $\overline{\partial}(\psi u)$ is continuous on D, and*

$$\overline{\partial}(\psi u) = (\overline{\partial}\psi)u + \psi\overline{\partial}u.$$

Proof. For all C^∞-functions $\varphi\colon D \to \mathbb{C}$ with compact support, we have

$$\begin{aligned}
\int_D \varphi\big((\overline{\partial}\psi)u + \psi\overline{\partial}u\big)\,d\lambda &= \int_D \varphi(\overline{\partial}\psi)u\,d\lambda + \int_D \varphi\psi\overline{\partial}u\,d\lambda \\
&= \int_D \varphi(\overline{\partial}\psi)u\,d\lambda - \int_D \overline{\partial}(\varphi\psi)u\,d\lambda \\
&= \int_D \varphi(\overline{\partial}\psi)u\,d\lambda - \int_D (\overline{\partial}\varphi)\psi u\,d\lambda - \int_D \varphi(\overline{\partial}\psi)u\,d\lambda \\
&= -\int_D (\overline{\partial}\varphi)\psi u\,d\lambda.
\end{aligned}$$

□

2.1.3 Lemma. *Let $D \subseteq \mathbb{C}$ be an open set, let $u\colon D \to E$ be a continuous function such that also $\overline{\partial}u$ is continuous on D, and let $K \subseteq D$ be compact. Then there* *exists a sequence* $(u_n)_{n\in\mathbb{N}}$ *of* C^∞*-functions* $u_n : D \to E$ *such that, uniformly on* K*, both*

$$\lim_{n\to\infty} u_n = u \quad \textit{and} \quad \lim_{n\to\infty} \overline{\partial} u_n = \overline{\partial} u.$$

Proof. Choose an open neighborhood V of K which is relatively compact in D. Further, let φ be a real non-negative C^∞-function on $\mathbb{C}$ with

$$\int_{\mathbb{C}} \varphi\, d\lambda = 1 \qquad \text{and} \qquad \varphi(z) = 0 \quad \text{if} \quad |z| > 1.$$

Now let a continuous map $u : D \to E$ be given such that $\overline{\partial} u$ is also continuous on D. Then we set

$$u_\varepsilon(z) = \varepsilon^2 \int_V \varphi\Big(\frac{z-\zeta}{\varepsilon}\Big) u(\zeta)\, d\lambda(\zeta), \qquad z \in \mathbb{C}\,, \;\; \varepsilon > 0. \tag{2.1.3}$$

Since

$$\varepsilon^2 \int_{\mathbb{C}} \varphi\Big(\frac{z-\zeta}{\varepsilon}\Big)\, d\lambda(\zeta) = 1 \quad \text{for all } z \in \mathbb{C} \quad \text{and} \quad \varphi\Big(\frac{z-\zeta}{\varepsilon}\Big) = 0 \quad \text{if} \quad |z-\zeta| > \varepsilon,$$

then, for sufficiently small $\varepsilon > 0$ and all $z \in K$,

$$\big|u_\varepsilon(z) - u(z)\big| \le \varepsilon^2 \int_V \varphi\Big(\frac{z-\zeta}{\varepsilon}\Big) \big|u(\zeta) - u(z)\big|\, d\lambda(\zeta) \le \max_{\zeta\in\overline{V}, |z-\zeta|\le\varepsilon} \big|u(\zeta) - u(z)\big|.$$

Since $\overline{V}$ is compact, it follows that $\lim_{\varepsilon\to 0} u_\varepsilon = u$, uniformly on K. Since also $\overline{\partial} u$ is continuous on D, then, in the same way, we get $\lim_{\varepsilon\to 0}(\overline{\partial} u)_\varepsilon = \overline{\partial} u$, uniformly on K. This completes the proof, because $\overline{\partial} u_\varepsilon = (\overline{\partial} u)_\varepsilon$. Indeed, differentiating under the integral in (2.1.3) we obtain

$$(\overline{\partial} u_\varepsilon)(z) = \varepsilon^2 \int_V \overline{\partial}_z \varphi\Big(\frac{z-\zeta}{\varepsilon}\Big) u(\zeta)\, d\lambda(\zeta) = -\varepsilon^2 \int_V \overline{\partial}_\zeta \varphi\Big(\frac{z-\zeta}{\varepsilon}\Big) u(\zeta)\, d\lambda(\zeta),$$

which implies, by (2.1.2),

$$(\overline{\partial} u_\varepsilon)(z) = \varepsilon^2 \int_V \varphi\Big(\frac{\zeta - z}{\varepsilon}\Big) (\overline{\partial} u)(\zeta)\, d\lambda(\zeta) = (\overline{\partial} u)_\varepsilon(z). \qquad \square$$

2.1.4 Theorem (Cauchy formula for continuous functions). *Let* D *be a bounded open set with piecewise* C^1*-boundary* ∂D*, and let* $u : \overline{D} \to E$ *be a continuous function such that also* $\overline{\partial} u$ *is continuous on* D*. Moreover we assume that* $\overline{\partial} u$ *admits a continuous extension to* $\overline{D}$*. Then*

$$u(z) = \frac{1}{2\pi i} \int\limits_{\partial D} \frac{u(\zeta)}{\zeta - z} d\zeta - \frac{1}{\pi} \int\limits_{D} \frac{\overline{\partial} u(z)}{\zeta - z} d\lambda(\zeta), \qquad z \in D. \tag{2.1.4}$$

Proof. First consider the case when u is of class C^1 in a neighborhood of $\overline{D}$. Let $z \in D$. Set $D_\varepsilon = \{\zeta \in D \mid |\zeta - z| > \varepsilon\}$ for sufficiently small $\varepsilon > 0$. Then, by Stokes' formula, with $\zeta = x + iy$,

$$\int_{\partial D_\varepsilon} \frac{u(\zeta)}{\zeta - z}\, d\zeta = \int_{\partial D_\varepsilon} \frac{u(\zeta)}{\zeta - z}\, dx \,+\, i \int_{\partial D_\varepsilon} \frac{u(\zeta)}{\zeta - z}\, dy$$

$$= -\int_{D_\varepsilon} \frac{\partial}{\partial y} \frac{u(\zeta)}{\zeta - z} d\lambda(\zeta) + i \int_{D_\varepsilon} \frac{\partial}{\partial x} \frac{u(\zeta)}{\zeta - z} d\lambda(\zeta) = +2i \int_{D_\varepsilon} \frac{(\overline{\partial} u)(\zeta)}{\zeta - z} d\lambda(\zeta).$$

Since

$$\lim_{\varepsilon \searrow 0} \int_{\partial D_\varepsilon} \frac{u(\zeta)}{\zeta - z} d\zeta = \int_{\partial D} \frac{u(\zeta)}{\zeta - z} d\zeta - \lim_{\varepsilon \searrow 0} \int_{|\zeta - z| = \varepsilon} \frac{u(\zeta)}{\zeta - z} d\zeta = \int_{\partial D} \frac{u(\zeta)}{\zeta - z} d\zeta - 2\pi i\, u(z),$$

this implies (2.1.4).

Now we consider the case when u admits a continuous extension $\widetilde{u}$ to a neighborhood of $\overline{D}$ such that $\overline{\partial}\widetilde{u}$ is also continuous in this neighborhood. Then by Lemma 2.1.3 we can find a sequence $(u_n)_{n \in \mathbb{N}}$ of C^∞-maps $u_n \colon \mathbb{C} \to E$ such that, uniformly on $\overline{D}$, both

$$\lim_{n \to \infty} u_n = u \quad \text{and} \quad \lim_{n \to \infty} \overline{\partial} u_n = \overline{\partial} u.$$

Since the required equation(2.1.4) is already proved for each u_n, passing to the limit, it follows for u.

Finally, consider the general case. Then we take a sequence of bounded open sets $(D_n)_{n \in \mathbb{N}}$ with piecewise $\mathcal{C}^1$-boundary such that $\overline{D}_n \subseteq D$ and

$$\lim_{n \to \infty} \int_{\partial D_n} \frac{u(\zeta)}{\zeta - z} d\zeta = \int_{\partial D} \frac{u(\zeta)}{\zeta - z} d\zeta \quad , \quad \lim_{n \to \infty} \int_{D_n} \frac{\overline{\partial} u(\zeta)}{\zeta - z} d\lambda(\zeta) = \int_D \frac{\overline{\partial} u(\zeta)}{\zeta - z} d\lambda(\zeta).$$

Since the required equation(2.1.4) is already proved for each $u\big|_{D_n}$, passing to the limit, it follows for u. □

2.1.5 Theorem. *Let $D \subseteq \mathbb{C}$ be an open set, and let $u : D \to E$ be a continuous function such that $\overline{\partial} u = 0$ on D. Then u is holomorphic.*

Proof. Since the assertion is local, we may assume that D is a disc and u is defined and continuous in a neighborhood of $\overline{D}$ and that $\overline{\partial} u = 0$ in this neighborhood. Then the second integral in the Cauchy formula (2.1.4) vanishes, i.e.,

$$u(z) = \frac{1}{2\pi i} \int_{\partial D} \frac{u(\zeta)}{\zeta - z} d\zeta\,, \qquad z \in D.$$

Now the assertion follows by differentiation under the integral. □

2.1.6 Definition. Let $0 < \alpha < 1$. First let $M \subseteq \mathbb{C}$ be an arbitrary set. For any function $f : M \to E$, we set

$$\|f\|_{M,0} = \sup_{z \in M} \|f(z)\|$$

and

$$\|f\|_{M,\alpha} = \|f\|_{M,0} + \sup_{z,w \in M, z \neq w} \frac{\|f(z) - f(w)\|}{|z-w|^\alpha}.$$

We write $\|f\|_0$ instead of $\|f\|_{M,0}$ and $\|f\|_\alpha$ instead of $\|f\|_{M,\alpha}$ if it is clear which set M we mean.

A function $f : M \to E$ is called (locally) **Hölder continuous with exponent** α if, for each point $z_0 \in M$, there exists a neighborhood U of z_0 such that

$$\|f\|_{U \cap M,\alpha} < \infty.$$

The space of all such functions will be denoted by $(C^\alpha)^E(M)$. Instead of *Hölder continuous with exponent* α we say also **Hölder-α continuous** or **of class $\mathcal{C}^\alpha$** .

A function $f : M \to E$ will be called **Hölder continuous** if there exists $0 < \alpha < 1$ such that f is Hölder continuous with exponent α.

Sometimes, for practical reasons, continuous functions will be called **Hölder continuous functions with exponent** 0 (although they are not Hölder continuous).

Now let $D \subseteq \mathbb{C}$ be an open set, and $k \in \mathbb{N}^* \cup \{\infty\}$. Recall that $(\mathcal{C}^k)^E(D)$ denotes the space of all E-valued $\mathcal{C}^k$ functions on D.

A function $f : D \to E$ is called **of class $\mathcal{C}^{k+\alpha}$** or simply $\mathcal{C}^{k+\alpha}$ if f is $\mathcal{C}^k$ on D and, moreover, the partial derivatives of order k of f are of class $\mathcal{C}^\alpha$ on D. The space of all such functions will be denoted by $(\mathcal{C}^{k+\alpha})^E(D)$.

2.1.7 Definition (Pompeiju operator Π_D). Let $D \subseteq \mathbb{C}$ be a bounded open set. Then, for any bounded continuous function $f \colon D \to E$ and each $z \in \mathbb{C}$, we define

$$(\Pi_D f)(z) = -\frac{1}{\pi} \int_D \frac{f(\zeta)}{\zeta - z}\, d\lambda(\zeta).$$

The operator Π_D will be called the **Pompeiju operator** on D.

2.1.8 Lemma. *Let $d < \infty$. Then there exists $C < \infty$ such that*

$$\int_{|\zeta|<d} \left| \frac{1}{\zeta - z} - \frac{1}{\zeta - w} \right| d\lambda(\zeta) \leq C|z-w| \left| \log|z-w| \right| \tag{2.1.5}$$

for all $z, w \in \mathbb{C}$.

Proof. Let $z, w \in \mathbb{C}$ be given. Since, for $z = w$, the left-hand side of (2.1.5) is zero, we may assume that $z \neq w$. We have

$$\int\limits_{|\zeta - z| < \frac{1}{2}|z-w|} \frac{d\lambda(\zeta)}{|\zeta - z|} = \int\limits_{0}^{\frac{1}{2}|z-w|} \int\limits_{|\zeta - z| = r} \frac{|d\zeta|}{r} dr = \pi |z - w| . \tag{2.1.6}$$

Moreover, if $|\zeta - z| \leq \frac{1}{2}|z - w|$, then $|\zeta - w| > \frac{1}{2}|z - w|$. Therefore

$$\int\limits_{|\zeta - z| < \frac{1}{2}|z-w|} \frac{d\lambda(\zeta)}{|\zeta - w|} \leq \frac{2}{|z - w|} \int\limits_{|\zeta - z| < \frac{1}{2}|z-w|} d\lambda(\zeta) = \frac{\pi}{2}|z - w| .$$

Together with (2.1.5) this yields

$$\int\limits_{|\zeta - z| < \frac{1}{2}|z-w|} \left| \frac{1}{\zeta - z} - \frac{1}{\zeta - w} \right| d\lambda(\zeta) \leq \frac{3\pi}{2}|z - w| .$$

Hence

$$\int\limits_{\min(|\zeta - z|, |\zeta - w|) < \frac{1}{2}|z-w|} \left| \frac{1}{\zeta - z} - \frac{1}{\zeta - w} \right| d\lambda(\zeta) \leq 3\pi |z - w| . \tag{2.1.7}$$

It remains to estimate

$$I(z, w) := \int\limits_{\substack{|\zeta| < d \\ \min(|\zeta - z|, |\zeta - w|) > \frac{1}{2}|z-w|}} \left| \frac{1}{\zeta - z} - \frac{1}{\zeta - w} \right| d\lambda(\zeta) .$$

If $|\zeta - w| \geq \frac{1}{2}|z - w|$, then

$$|\zeta - z| \leq |\zeta - w| + |z - w| \leq |\zeta - w| + 2|\zeta - w| = 3|\zeta - w|$$

and therefore

$$\left| \frac{1}{\zeta - z} - \frac{1}{\zeta - w} \right| = \frac{|z - w|}{|\zeta - z||\zeta - w|} \leq \frac{1}{3} \frac{|z - w|}{|\zeta - z|^2}.$$

Hence

$$I(z, w) \leq \frac{|z - w|}{3} \int\limits_{\substack{|\zeta| < d \\ |\zeta - z| > \frac{1}{2}|z-w|}} \frac{d\lambda(\zeta)}{|\zeta - z|^2} \leq \frac{|z - w|}{3} \int\limits_{2d > |\zeta - z| > \frac{1}{2}|z-w|} \frac{d\lambda(\zeta)}{|\zeta - z|^2}$$

$$= \frac{|z - w|}{3} \int\limits_{\frac{1}{2}|z-w|}^{2d} \int\limits_{|\zeta - z| = r} \frac{|d\zeta|}{r^2} dr = \frac{2\pi}{3}|z - w| \int\limits_{\frac{1}{2}|z-w|}^{2d} \frac{dr}{r} .$$

Since

$$\int_{\frac{1}{2}|z-w|}^{2d} \frac{dr}{r} = \log(2d) - \log\Big(\frac{1}{2}|z-w|\Big) = \log(4d) - \log|z-w|$$
$$\leq \log(4d) + \big|\log|z-w|\big|,$$

this yields

$$I(z,w) \leq \frac{2\pi\log(4d)}{3}|z-w| + \frac{2\pi}{3}|z-w|\big|\log|z-w|\big|.$$

Together with (2.1.7) this implies that there is a constant $C < \infty$ (depending on d) satisfying (2.1.5) $\square$

2.1.9 Theorem. *Let $D \subseteq \mathbb{C}$ be a bounded open set, let Π_D be the Pompeiju operator (Def. 2.1.7), and let $f : D \to E$ be a bounded continuous function.*

(i) *Then there exists a constant $C < \infty$ such that*

$$\Big\|(\Pi_D f)(z) - (\Pi_D f)(w)\Big\| \leq C|z-w|\Big|\log|z-w|\Big| \tag{2.1.8}$$

for all $z, w \in \mathbb{C}$. In particular,

$$\Pi_D f\Big|_{\overline{D}} \in \bigcap_{0<\alpha<1} (C^\alpha)^E(\overline{D}). \tag{2.1.9}$$

(ii) *If, moreover, f is of class $\mathcal{C}^k$ on D, $k \in \mathbb{N}^*$, then*

$$\Pi_D f\Big|_{D} \in \bigcap_{0<\alpha<1} (C^{k+\alpha})^E(D). \tag{2.1.10}$$

Proof. Part (i) follows immediately from Lemma 2.1.8.

We prove part (ii). Let $k \in \mathbb{N}$ be given. Since the statement is local, it is sufficient to prove that, for each $\eta \in D$, there exists a neighborhood $U \subseteq D$ of η such that

$$\Pi_D f\Big|_{U} \in \bigcap_{0<\alpha<1} (C^{k+\alpha})^E(U). \tag{2.1.11}$$

Let $\eta \in D$ be given. Choose a $\mathcal{C}^\infty$-function χ with compact support $\operatorname{supp}\chi \subseteq D$ and $\chi \equiv 1$ in a neighborhood $U \subseteq D$ of η. We have

$$\Pi_D f = \Pi_D(\chi f) + \Pi_D\big((1-\chi)f\big).$$

Since $(1-\chi)f \equiv 0$ in U, by differentiation under the integral, we see that $\Pi_D\big((1-\chi)f\big)$ is $\mathcal{C}^\infty$ in U. Therefore it remains to prove that $\Pi_D(\chi f)$ is $\mathcal{C}^{k+\alpha}$ in U.

As χ has compact support, the substitution $\zeta \to \zeta + z$ gives

$$\big(\Pi_D(\chi f)\big)(z) = -\frac{1}{\pi}\int_D \frac{(\chi f)(\zeta)}{\zeta - z}\, d\lambda(\zeta) = -\frac{1}{\pi}\int_{\mathbb{C}} \frac{(\chi f)(\zeta + z)}{\zeta} d\lambda(\zeta)$$

for all $z \in U$. Again by differentiation under the integral, we see that $\Pi(\chi f)$ is of class $\mathcal{C}^k$ on U, where, for all $\nu, \mu \in \mathbb{N}$ with $0 \le \nu + \mu \le k$,

$$\frac{\partial^{\nu+\mu}\big(\Pi_D(\chi f)\big)}{\partial x^\nu \partial y^\mu}(z) = -\frac{1}{\pi}\int_{\mathbb{C}} \frac{\partial^{\nu+\mu}(\chi f)}{\partial x^\nu \partial y^\mu}(\zeta + z)\frac{d\lambda(\zeta)}{\zeta}\,, \qquad z \in U\,. \tag{2.1.12}$$

So it is proved that $\Pi_D(f)$ is of class $\mathcal{C}^k$ on U.

It remains to prove that the derivatives of order k of $\Pi_D(\chi f)$ belong to $\bigcap_{0<\alpha<1}(C^\alpha)^E(U)$. Let $\nu, \mu \in \mathbb{N}$ with $\nu + \mu = k$ be given. By the substitution $\zeta \to \zeta - z$, from (2.1.12) we get

$$\frac{\partial^{\nu+\mu}\big(\Pi_D(\chi f)\big)}{\partial x^\nu \partial y^\mu}(z) = -\frac{1}{\pi}\int_{\mathbb{C}} \frac{\partial^{\nu+\mu}(\chi f)}{\partial x^\nu \partial y^\mu}(\zeta)\frac{d\lambda(\zeta)}{\zeta - z}\,, \qquad z \in U\,. \tag{2.1.13}$$

Since $\operatorname{supp}\chi$ is a compact subset of D,

$$C := \max_{\zeta \in \operatorname{supp}\chi}\left\| \frac{\partial^{\nu+\mu}(\chi f)}{\partial x^\nu \partial y^\mu}(\zeta)\right\| < \infty,$$

and, by (2.1.13),

$$\left\| \frac{\partial^{\nu+\mu}\big(\Pi_D(\chi f)\big)}{\partial x^\nu \partial y^\mu}(z) - \frac{\partial^{\nu+\mu}\big(\Pi_D(\chi f)\big)}{\partial x^\nu \partial y^\mu}(w)\right\| \le C\int\limits_D \left|\frac{1}{\zeta - z} - \frac{1}{\zeta - w}\right| d\lambda(\zeta)$$

for all $z, w \in U_\eta$. Now it follows again from Lemma 2.1.8 that

$$\frac{\partial^{\nu+\mu}\big(\Pi_D(\chi f)\big)}{\partial x^\nu \partial y^\mu}$$

belongs to $\bigcap_{0<\alpha<1}(C^\alpha)^E(U)$. □

2.1.10 Theorem. *Let $D \subseteq \mathbb{C}$ be a bounded open set, and let $f : D \to E$ be continuous and bounded. Then $\Pi_D f$ (Def. 2.1.7) has a continuous Cauchy-Riemann derivative on D and*

$$\overline{\partial}\Pi_D f = f \qquad \text{on } D. \tag{2.1.14}$$

2.1.11 Remark. It is easy to see that the assertion of Theorem 2.1.10 is true for any open set D and any continuous f such the integral

$$\int_D \frac{f(\zeta)}{\zeta - z} d\lambda(\zeta)$$

converges (without the hypotheses on boundedness of D and f). We will not use this generalization. Instead, in Section 2.3,we will use an approximation argument to deduce from Theorem 2.1.10 the solvability of $\overline{\partial}u = f$ for arbitrary continuous functions f on arbitrary open sets. For that we need the Runge approximation theorem presented in Section 2.2. First, in the next section, we give a first application of Theorem 2.1.10 and Theorem 2.1.9.

Proof of Theorem 2.1.10. By Definition 2.1.1 we have to prove that

$$\int_D \varphi f \, d\lambda = -\int_D (\overline{\partial}\varphi)(\Pi_D f)\, d\lambda$$

for any C^∞-function $\varphi\colon D \to \mathbb{C}$ with compact support in D. Let such φ be given. Then, by the Cauchy formula (Theorem 2.1.4), $\varphi = \Pi_D(\overline{\partial}\varphi)$. Interchanging the order of integration, this yields the required relation:

$$\begin{aligned}\int_D \varphi f\, d\lambda = \int_D \Pi_D(\overline{\partial}\varphi) f\, d\lambda &= \int_D \left(-\frac{1}{\pi}\int_D \frac{(\overline{\partial}\varphi)(\zeta)}{\zeta - z}\, d\lambda(\zeta)\right) f(z)\, d\lambda(z)\\ &= \int_D (\overline{\partial}\varphi)(\zeta)\left(\frac{1}{\pi}\int_D \frac{f(z)}{z-\zeta}\, d\lambda(z)\right) d\lambda(\zeta) = -\int_D (\overline{\partial}\varphi)(\Pi_D f)\, d\lambda.\end{aligned}$$

□

2.1.12 Theorem (Regularity of $\overline{\partial}$). *Let $D \subseteq \mathbb{C}$ an open set, let $f : D \to E$ be of class $\mathcal{C}^k$ on D, $k \in \mathbb{N} \cup \{\infty\}$, and let $u : D \to E$ be a continuous function with continuous Cauchy-Riemann derivative such that*

$$\overline{\partial}u = f \qquad \text{on } D.$$

Then

$$u \in \bigcap_{0<\alpha<1} \mathcal{C}^{k+\alpha}(D).$$

Proof. Since the assertion is local, we may assume that both D and f are bounded. Then it follows from theorems 2.1.10 and 2.1.9 that there exists

$$v \in \bigcap_{0<\alpha<1} \mathcal{C}^{k+\alpha}(D) \qquad \text{with} \qquad \overline{\partial}v = f \qquad \text{on } D,$$

namely $v = \Pi_D f$. Then $\overline{\partial}(u - v) = f - f = 0$ on D. By Theorem 2.1.5 this means that $u - v$ is holomorphic. As holomorphic functions are of class $\mathcal{C}^\infty$ and $v \in \bigcap_{0<\alpha<1} \mathcal{C}^{k+\alpha}(D)$, it follows that $u \in \bigcap_{0<\alpha<1} \mathcal{C}^{k+\alpha}(D)$. □

2.2 Runge approximation

In this section E is a Banach space.

2.2.1 Theorem (Mergelyan approximation). *Let $D \subseteq \mathbb{C}$ be a bounded open set with piecewise C^1-boundary, and let $f : \overline{D} \to E$ be a continuous function which is holomorphic in D. Then, for each $\varepsilon > 0$, there exists a neighborhood U of $\overline{D}$ and a function $\widetilde{f} \in \mathcal{O}^E(U)$ such that*

$$\|f(z) - \widetilde{f}(z)\| < \varepsilon \qquad \text{for all } z \in \overline{D}. \tag{2.2.1}$$

Proof. Take a finite number of real non-negative C^∞ functions $\chi_1, \dots, \chi_n$ on $\mathbb{C}$ with sufficiently small supports $\operatorname{supp}\chi_j$ (how small, we say below) such that $\sum_{j=1}^n \chi_j = 1$ in some neighborhood of ∂D. Set

$$f_j(z) = \frac{1}{2\pi i}\int_{\partial D} \frac{\chi_j(\zeta)f(\zeta)}{\zeta - z}d\zeta, \quad z \in \mathbb{C} \setminus (\operatorname{supp}\chi_j \cap \partial D), \quad 1 \le j \le n.$$

Differentiation under the sign of integration shows that each f_j is holomorphic on $\mathbb{C} \setminus (\operatorname{supp}\chi_j \cap \partial D)$. From Cauchy's formula we get

$$\sum_{j=1}^n f_j(z) = \frac{1}{2\pi i}\int_{\partial D} \frac{\sum_j^n \chi_j(\zeta)f(\zeta)}{\zeta - z}d\zeta = \frac{1}{2\pi i}\int_{\partial D} \frac{f(\zeta)}{\zeta - z}d\zeta = f(z) \tag{2.2.2}$$

for all $z \in D$. By Proposition 2.1.2, $\overline{\partial}(\chi_j f) = (\overline{\partial}\chi_j)f$ is continuous on $\overline{D}$. Therefore, by the Cauchy formula for continuous functions (Theorem 2.1.4) and by Definition 2.1.7 of Π_D,

$$\chi_j f = f_j + \Pi_D\big(\overline{\partial}(\chi_j f)\big) \qquad \text{on } D.$$

Since $\chi_j f$ is continuous on $\overline{D}$ and, by Theorem 2.1.9, also $\Pi_D\big(\overline{\partial}(\chi_j f)\big)$ is continuous on $\overline{D}$, this implies that each f_j admits a continuous extension from D to $\operatorname{supp}\chi_j \cap \partial D$, which we denote by $f_j^{\overline{D}}$. Then it follows from (2.2.2) that

$$f(z) = f_1^{\overline{D}}(z) + \ldots + f_n^{\overline{D}}(z) \qquad \text{for all } z \in \overline{D}.$$

Since ∂D is piecewise C^1 and each $f_j^{\overline{D}}$ extends to a holomorphic function outside $\operatorname{supp}\chi_j \cap \partial D$, now we can choose the supports $\operatorname{supp}\chi_j$ so small that, by small shifts, for each j, we can find a neighborhood U_j of $\overline{D}$ and a function $\widetilde{f}_j \in \mathcal{O}^E(U_j)$ such that

$$\|f_j^{\overline{D}}(z) - \widetilde{f}_j(z)\| < \frac{\varepsilon}{n} \qquad \text{for } z \in \overline{D}.$$

Setting $U = U_1 \cap \ldots \cap U_n$ and $\widetilde{f} = \widetilde{f}_1 + \ldots + \widetilde{f}_n$, we complete the proof. □

2.2.2 Theorem (Runge approximation). *Let $D \subseteq \mathbb{C}$ be a bounded open set (possibly not connected) with piecewise C^1-boundary, and let $f : \overline{D} \to E$ be a continuous function which is holomorphic in D.*

(i) *If $\mathbb{C}\setminus\overline{D}$ is connected, then f can be approximated uniformly on $\overline{D}$ by E-valued polynomials.*

(ii) *If $\mathbb{C} \setminus \overline{D}$ is not connected and $U_1, \ldots, U_N$ are the bounded connected components of $\mathbb{C} \setminus \overline{D}$, then for any choice of points $p_1 \in U_1, \ldots, p_N \in U_N$, the function f can be approximated uniformly on $\overline{D}$ by E-valued rational functions which are holomorphic on $\mathbb{C} \setminus \{p_1, \ldots, p_N\}$.*

Proof. Denote by U_∞ the unbounded connected component of $\mathbb{C} \setminus \overline{D}$ and set $R = \max_{z \in \overline{D}} |z|$. We now proceed in 4 steps:

Step 1. First consider the case when f is holomorphic in some neighborhood of the disc $|z| \leq R$. Then the potential series of f at zero gives a uniform approximation of f on this disc by E-valued polynomials. Since $\overline{D}$ is contained in this disc, this completes the proof in this case.

Step 2. Now we consider the case when f is of the form

$$f(z) = \frac{b}{z - \xi} \qquad \text{with } b \in E \text{ and } \xi \in U_\infty. \tag{2.2.3}$$

Then we choose a continuous curve $\gamma\colon [0,1] \to \mathbb{C} \setminus \overline{D}$ with $\gamma(0) = \xi$ and $|\gamma(1)| > R$ and take $0 = t_0 < t_1 < \ldots < t_n = 1$ such that

$$|t_{j-1} - t_j| < \min_{z \in \overline{D},\, 0 \leq t \leq 1} |z - \gamma(t)| \qquad \text{for } 1 \leq j \leq n.$$

Then, for $1 \leq j \leq N$,

$$\begin{aligned} \frac{b}{z - \gamma(t_j)} &= \frac{b}{z - \gamma(t_{j-1})} \cdot \frac{1}{1 - \frac{\gamma(t_j) - \gamma(t_{j-1})}{z - \gamma(t_{j-1})}} \\ &= \frac{b}{z - \gamma(t_{j-1})} \sum_{k=0}^{\infty} \left(\frac{\gamma(t_j) - \gamma(t_{j-1})}{z - \gamma(t_{j-1})} \right)^k \end{aligned}$$

where the series converges uniformly in $z \in \overline{D}$. Therefore, for each j, the function $b/(z - \gamma(t_j))$ can be approximated uniformly on $\overline{D}$ by E-valued polynomials in $1/(z - \gamma(t_{j-1}))$. It follows that the map

$$\frac{b}{z - \xi} = \frac{b}{z - \gamma(t_0)}$$

can be approximated uniformly on $\overline{D}$ by E-valued polynomials in

$$\frac{1}{z - \gamma(t_n)} = \frac{1}{z - \gamma(1)}.$$

Since $\gamma(1) > R$ and therefore, as we saw in step 1,

$$\frac{1}{z - \gamma(1)}$$

can be approximated uniformly on $\overline{D}$ by polynomials, this completes the proof for functions of the form (2.2.3).

Step 3. Here we consider the case when $\mathbb{C} \setminus \overline{D}$ is not connected, $U_1, \dots, U_N$ are the bounded connected components of $\mathbb{C} \setminus \overline{D}$, $p_1 \in U_1, \dots, p_N \in U_N$ are the chosen points and, for some $1 \le j \le N$, f is of the form

$$f(z) = \frac{b}{z - \xi} \qquad \text{with } b \in E \text{ and } \xi \in U_j. \tag{2.2.4}$$

Then we choose a continuous curve $\gamma : [0,1] \to U_j$ with $\gamma(0) = \xi$ and $\gamma(1) = p_j$, and, in the same way as in step 2, we see that

$$\frac{b}{z - \xi}$$

can be approximated uniformly on $\overline{D}$ by E-valued polynomials in

$$\frac{1}{z - p_j},$$

which completes the proof for functions of the form (2.2.4).

Step 4. Consider the general case. By Theorem 2.2.1 we may assume that f is defined and holomorphic in some neighborhood V of $\overline{D}$. Take a bounded open set G with $\mathcal{C}^1$-boundary such that $\overline{D} \subseteq G$ and $\overline{G} \subseteq V$. Then, by Cauchy's formula,

$$f(z) = \frac{1}{2\pi i} \int_{\partial G} \frac{f(\zeta)}{\zeta - z} d\zeta, \qquad z \in \overline{D}.$$

If $\mathbb{C} \setminus \overline{D}$ is connected, then passing to Riemann sums, this shows that f can be approximated uniformly on $\overline{D}$ by linear combinations of functions of the form

$$\frac{f(\xi)}{z - \xi}$$

with $\xi \in \partial G \subseteq U_\infty$. Since, as we saw in step 2, such functions can be approximated uniformly on $\overline{D}$ by E-valued polynomials, this completes the proof if $\mathbb{C} \setminus \overline{D}$ is connected. If $\mathbb{C} \setminus \overline{D}$ is not connected, $U_1, \dots, U_N$ are the bounded connected components of $\mathbb{C} \setminus \overline{D}$ and $p_1 \in U_1, \dots, p_N \in U_N$ are the chosen points, then the same argument with Riemann sums yields that f can be approximated uniformly on $\overline{D}$ by linear combinations of functions of the form

$$\frac{f(\xi)}{z - \xi}$$

with $\xi \in \partial G \subseteq U_\infty \cup U_1 \cup \ldots \cup U_N$. Since, as we saw in steps 2 and 3, all such functions can be approximated uniformly on $\overline{D}$ by E-valued rational functions from $\mathcal{O}(\mathbb{C} \setminus \{p_1, \ldots p_N\}, E)$, this completes the proof also if $\mathbb{C} \setminus \overline{D}$ is not connected. $\square$

2.2.3 Remark. In the approximation Theorems 2.2.1 and 2.2.2, the hypothesis that ∂D is piecewise $\mathcal{C}^1$ can be essentially weakened. But then the proof of Theorem 2.2.1 (which is used in the proof of Theorem 2.2.2) becomes more difficult. Let us mention also the following approximation theorem without any hypothesis on the smoothness of the boundary:

2.2.4 Corollary (to Theorem 2.2.2). *Let $D \subseteq \mathbb{C}$ be a bounded open set which consists of a finite number of connected components $D_1, \dots, D_N$, each of which is star shaped, such that $\overline{D}_j \cap \overline{D}_k = \emptyset$ for $j \neq k$, and let $f : \overline{D} \to E$ be a continuous function which is holomorphic in D. Then f can be approximated uniformly on $\overline{D}$ by E-valued polynomials.*

Proof. Since each D_j is star shaped, we have points $z_j \in D_j$ with $z_j + t(z - z_j) \in D_j$ for all $z \in D_j$ and (consequently) $z_j + t(z - z_j) \in \overline{D}_j$ for all $z \in \overline{D}_j$. Then, for each $\varepsilon > 0$,

$$f_\varepsilon(z) := f\big(z_j + (1-\varepsilon)(z - z_j)\big)\,, \qquad z \in \overline{D}_j\,,\ 0 \le t \le 1,$$

is holomorphic in a neighborhood of $\overline{D}$ and $\lim\limits_{\varepsilon \to 0} \max\limits_{z \in \overline{D}} \|f(z) - f_\varepsilon(z)\| = 0$. Now we can continue as in the proof of Theorem 2.2.2 or we can apply Theorem 2.2.2 to a slightly larger open set with $\mathbb{C}^1$-boundary. □

2.3 Solution of $\overline{\partial}u = f$ on open sets

In this section E is a Banach space.

By Theorem 2.1.10, for each bounded open set $D \subseteq \mathbb{C}$ and each bounded continuous function $f :\to E$, there exists a continuous function $u : D \to E$ with $\overline{\partial}u = f$. As a first important consequence of the Runge approximation Theorem 2.2.2 now we get the following stronger result:

2.3.1 Theorem. *Let $D \subseteq \mathbb{C}$ be an arbitrary open set, and let $f : D \to E$ be an arbitrary continuous function.*

(i) *Then there exists a continuous function $u : D \to E$, which has a continuous Cauchy-Riemann derivative on D (Def. 2.1.1), such that*

$$\overline{\partial}u = f \qquad \text{on } D.$$

(ii) *Any such function u is automatically of class $\mathcal{C}^\alpha$ on D, for all $0 < \alpha < 1$. Moreover, if $f \in (\mathcal{C}^k)^E(D)$, $k \in \mathbb{N}^*$, then any such function u is automatically of class $(\mathcal{C}^{k+\alpha})^E$ on D, for all $0 < \alpha < 1$.*

Proof. Part (ii) follows from Theorem 2.1.12. We prove part (i). Take a sequence $(D_n)_{n \in \mathbb{N}}$ of open sets $D_n \subseteq D$ such that:

- For all $n \in \mathbb{N}$, D_n is bounded, D_n has $\mathcal{C}^1$ boundary, $\overline{D}_n \subseteq D_{n+1}$, and each bounded connected component of $\mathbb{C} \setminus D_n$ (if there is any) contains at least one point of $\mathbb{C} \setminus D$.
- $\bigcup_{n=1}^{\infty} D_n = D$.

Then, by Theorem 2.1.10, there is a sequence $(\widetilde{u}_n)_{n\in\mathbb{N}}$ of continuous functions $\widetilde{u}_n \colon D_n \to E$ with continuous Cauchy-Riemann derivatives such that $\overline{\partial}\widetilde{u}_n = f\big|_{D_n}$.

Now we construct inductively a sequence $(u_n)_{n\in\mathbb{N}}$ of continuous functions $u_n : D_n \to E$ such that also

$$\overline{\partial}u_n = f\big|_{D_n} \tag{2.3.1}$$

for all $n \in \mathbb{N}$, and moreover

$$\max_{z\in\overline{D}_{n-2}} \big|u_n(z) - u_{n-1}(z)\big| \le \frac{1}{2^n} \tag{2.3.2}$$

if $n \geq 2$.

Beginning of the induction: $u_0 := \widetilde{u}_0$, $u_1 := \widetilde{u}_1$.

Hypothesis of induction: Assume, for some $k \in \mathbb{N}^*$, we already have continuous functions $u_n : D_n \to E$, $0 \le n \le k$, such that (2.3.1) holds for all $0 \le n \le k$, and (2.3.2) holds if $2 \le n \le k$.

Step of induction: Since $\overline{\partial}u_k = f = \widetilde{u}_{k+1}$ on D_k, the difference $u_k - \widetilde{u}_{k+1}$ is holomorphic on the neighborhood D_k of $\overline{D}_{k-1}$. As each bounded connected component of $\mathbb{C} \setminus \overline{D}_{k-1}$ contains at least one point of $\mathbb{C} \setminus D$, we can apply the Runge approximation Theorem 2.2.2 and obtain a holomorphic function $h : D \to E$ with

$$\max_{z\in\overline{D}_{k-1}} \big|u_k(z) - \widetilde{u}_{k+1}(z) - h(\zeta)\big| \le \frac{1}{2^{k+1}}.$$

Setting $u_{k+1} := \widetilde{u}_{k+1} + h\big|_{D_{k+1}}$, we obtain a continuous function $u_{k+1} : D_{k+1} \to E$ such that (2.3.1) and (2.3.2) holds also for $n = k + 1$.

The sequence $(u_n)_{n\in\mathbb{N}}$ is constructed.

By (2.3.2), there is a well-defined continuous function $u : D \to E$ such that, for each $k \in \mathbb{N}$, the sequence $(u_n)_{n>k}$ converges to u, uniformly on D_k. It remains to prove that $\overline{\partial}u = f$ on D. It is sufficient to show this on each D_k. Let $k \in \mathbb{N}$ be given. Then, uniformly on D_k,

$$u - u_k = \lim_{k\le n\to\infty} \left(u_n - u_k\right).$$

Since the functions $u_n - u_k$, $k \le n$, are holomorphic on D_k, and the uniform limits of holomorphic functions are holomorphic, it follows that $u - u_k$ is holomorphic on D_k. Hence $\overline{\partial}u = \overline{\partial}u_k = f$ on D_k. □

2.4 $\mathcal{O}^E$-cocycles and the Mittag-Leffler theorem

In this section E is a Banach space.

Let $U \subseteq \mathbb{C}$ be an open set. Recall that, in this book, for $U \neq \emptyset$, we denote by $\mathcal{O}^E(U)$ the space of E-valued holomorphic functions on U. For $U = \emptyset$, we set $\mathcal{O}^E(U) = \{0\}$, where 0 is the zero vector of E.

2.4.1 Definition. Let $D \subseteq \mathbb{C}$ be an open set, and let $\mathcal{U} = \{U_j\}_{j\in I}$ be an open covering of D. Denote by $C^1(\mathcal{U}, \mathcal{O}^E)$ the space of families $f = \{f_{jk}\}_{j,k\in I}$ with $f_{jk} \in \mathcal{O}^E(U_j \cap U_k)$. A family $f = \{f_{jk}\}_{j,k\in I} \in C^1(\mathcal{U}, \mathcal{O}^E)$ will be called an **$(\mathcal{U}, \mathcal{O}^E)$-cocycle** if, for all $j, k, l \in I$ with $U_j \cap U_k \cap U_l \neq \emptyset$,

$$f_{jk} + f_{kl} = f_{jl} \qquad \text{on } U_j \cap U_k \cap U_l. \tag{2.4.1}$$

Note that then, in particular,

$$f_{jk} = -f_{kj} \qquad \text{on } U_j \cap U_k \qquad \text{and} \qquad f_{jj} = 0 \qquad \text{on } U_j. \tag{2.4.2}$$

The space of all $(\mathcal{U}, \mathcal{O}^E)$-cocycles will be denoted by $Z^1(\mathcal{U}, \mathcal{O}^E)$. If the covering $\mathcal{U}$ is not specified, then we speak also about **$\mathcal{O}^E$-cocycles** over D.

We call such cocycles also **additive** to point out the difference from the *multiplicative* cocycles, which we introduce in Section 5.6. Due to P. Cousin the elements of $Z^1(\mathcal{U}, \mathcal{O}^E)$ are also called **additive Cousin problems**. To call the elements of $Z^1(\mathcal{U}, \mathcal{O}^E)$ *problems* is due to the fact that cocycles were first studied in Complex analysis of several variables, where the elements of $Z^1(\mathcal{U}, \mathcal{O}^E)$ give rise to problems which not always have solutions. In the case of a single variable however, these problems always can be solved. This is the statement of the following theorem.

2.4.2 Theorem. *Let $D \subseteq \mathbb{C}$ be an open set, let $\mathcal{U} = \{U_j\}_{j\in I}$ be an open covering of D, and let $f \in Z^1(\mathcal{U}, \mathcal{O}^E)$. Then there exists a family $\{h_j\}_{j\in I}$ of functions $h_j \in \mathcal{O}^E(U_j)$ with*

$$f_{jk} = h_j - h_k \qquad on U_j \cap U_k \tag{2.4.3}$$

for all $j, k \in I$ with $U_j \cap U_k \neq \emptyset$.

Proof. Take a $\mathcal{C}^\infty$-partition of unity $\{\chi_j\}_{j\in I}$ subordinated to the covering $\{U_j\}_{j\in I}$ and define $\mathcal{C}^\infty$-maps $\varphi_j : U_j \to E$ setting

$$\varphi_j = -\sum_{\nu\in I} \chi_\nu f_{\nu j}.$$

Then, by (2.4.2),

$$\varphi_j - \varphi_k = -\sum_{\nu\in I} \chi_\nu (f_{\nu j} - f_{\nu k}) = \sum_{\nu\in I} \chi_\nu (f_{j\nu} + f_{\nu k}) = \sum_{\nu\in I} \chi_\nu f_{jk} = f_{jk}.$$

Since the functions f_{jk} are holomorphic, i.e., $\overline{\partial} f_{jk} = 0$, it follows that

$$\overline{\partial}\varphi_j = \overline{\partial}\varphi_k \qquad \text{on } U_j \cap U_k.$$

Hence there is a well-defined $\mathcal{C}^\infty$-function $\varphi : D \to E$ with $\varphi = \overline{\partial}\varphi_j$ on U_j. Now, by Theorem 2.3.1, we can solve the equation $\varphi = \overline{\partial}u$ with some $\mathcal{C}^\infty$-function $u : D \to E$. Setting $h_j = c_j - u$ on U_j, we complete the proof. □

For many purposes the special case of Theorem 2.4.2 is sufficient when the covering consists only of two sets:

2.4.3 Corollary (to Theorem 2.4.2). *Let $D_1, D_2 \subseteq \mathbb{C}$ be two open sets with $D_1 \cap D_2 \neq \emptyset$. Then, for each holomorphic function $f : D_1 \cap D_2 \to E$, there exist holomorphic functions $f_j : D_j \to E$, $j = 1, 2$, such that*

$$f = f_1 + f_2 \qquad \text{on } D_1 \cap D_2.$$

An example of such an application is the Mittag-Leffler theorem:

2.4.4 Theorem (Mittag-Leffler theorem). *Let $D \subseteq \mathbb{C}$ be an open set, let Z be a discrete and closed subset of D, and assume that, for each $w \in Z$, a holomorphic function $f_w : \mathbb{C} \setminus \{w\} \to E$ of the form*

$$f_w(z) = \sum_{n=-\infty}^{-1} (z - w)^n f_{wn}$$

is given. Then there exists a holomorphic function $f : D \setminus Z \to E$ such that, for each $w \in Z$, f_w is the principal part of the Laurent expansion of f at w.

Proof. Since Z is discrete and closed in D, we can find a family $\{U_w\}_w \in Z$ of open subsets of D such that, U_w is a neighborhood of w and $U_w \cap U_v = \emptyset$ if $w \neq v$. Set $D_1 = \bigcup_{w \in Z} U_w$ and $D_2 = D \setminus Z$. Then $D_2 \cap D_2$ is the disjoint union of the punctured sets $U_w \setminus \{w\}$. Therefore, setting

$$g = f_w \qquad \text{on } U_w \setminus \{w\},$$

we obtain a holomorphic function $g : D_1 \cap D_2 \to E$. Now from Corollary 2.4.3 we get holomorphic functions $h : D_1 \to E$ and $f : D_2 \to E$ such that $g = f + h$ on $D_1 \cap D_2$. Then, for all $w \in Z$,

$$f - f_w = g - h - f_w = -h \qquad \text{on } U_w \setminus \{w\},$$

and therefore f_w is the principal part of the Laurent expansion of f at w. □

2.5 Runge approximation for invertible scalar functions and the Weierstrass product theorem

Recall that, in this book, for a non-empty open set $U \subseteq \mathbb{C}$, we denote by $\mathcal{O}^*(U)$ the multiplicative group of holomorphic functions $f : U \to \mathbb{C}^*$, where $\mathbb{C}^* := \mathbb{C} \setminus \{0\}$.

2.5.1. Recall that an open set $D \subseteq \mathbb{C}$, which is already connected, is simply connected (Section 1.4.3) if and only if $\mathbb{C} \setminus D$ is connected (see, for example, Theorem 13.11 in [Ru]). Together with the Riemann mapping theorem this implies the following facts, which will be used throughout this book without further reference:

(i) For a connected, bounded open set $D \subseteq \mathbb{C}$ with $\mathcal{C}^1$-boundary the following are equivalent:

- D is simply connected.
- $\mathbb{C} \setminus \overline{D}$ is connected.
- The boundary of D is connected.
- The boundary of D is homeomorphic to the unit circle.

(ii) Let $D \subseteq \mathbb{C}$ be a bounded, connected open set with piecewise $\mathcal{C}^1$-boundary, which is not simply connected, let $U_0, U_1, \ldots, U_m$ be the connected components of $\mathbb{C} \setminus \overline{D}$, where U_0 is the unbounded connected component of $\mathbb{C} \setminus \overline{D}$, and let Γ_j be the boundary of U_j, $0 \leq j \leq m$. Then $U_1, \ldots, U_m$ and $D \cup \overline{U}_1 \cup \ldots \cup \overline{U}_m$ are simply connected, and the contours $\Gamma_0, \Gamma_1, \ldots, \Gamma_m$ are homeomorphic to the unit circle.

2.5.2 Lemma. *Let $D \subseteq \mathbb{C}$ be a connected bounded open set with piecewise $\mathcal{C}^1$-boundary. Suppose D is not simply connected. Let $\Gamma_0, \Gamma_1, \ldots, \Gamma_m$ be the connected components of the boundary of D where Γ_0 is the boundary of the unbounded component of $\mathbb{C} \setminus \overline{D}$. Suppose, for each $0 \leq j \leq m$, a smooth*[1] *point a_j of Γ_j is chosen. Let U be a neighborhood of $\overline{D}$. Then there exist simply connected open sets $U_0, \ldots, U_m$ such that:*

- *The sets $U_1, \ldots, U_m$ are pairwise disjoint,*
- $\overline{D} \subseteq U_0 \cup \ldots \cup U_m \subseteq U$,

and, for all $1 \leq j \leq m$,

- $a_j \in U_j$,
- $\Gamma_j \cap U_0 = \Gamma_j \setminus \{a_j\}$,
- *$U_j \cap U_0$ consists of precisely two connected components.*

Proof. Since Γ_0 and Γ_1 are parts of the piecewise $\mathcal{C}^1$-boundary of D and D is connected, first we can find a contour γ_1, diffeomorphic to the closed interval $[0, 1]$, which starts at a_1, transversally to Γ_1, which ends at some smooth point $b_1 \in \Gamma_0$, transversally to Γ_0, and which lies, except for these two points, in D.

Then $\Gamma_1 \setminus \gamma_1 = \Gamma_1 \setminus \{a_1\}$ is still connected (as Γ_1 is homeomorphic to the circle). Since D is connected, this easily implies that $D \setminus \gamma_1$ is still connected.

[1] It is not important that a_j is a *smooth* point of Γ_j, but this simplifies the arguments in the proof.

Next, since also Γ_2 is piecewise $\mathcal{C}^1$, since γ_1 meets Γ_0 and Γ_1 transversally and $D \setminus \gamma_1$ is connected, we can find a contour γ_2, which is diffeomorphic to $[0,1]$, which starts at a_2, transversally to Γ_2, which ends at some smooth point $b_2 \in \Gamma_0$, transversally to Γ_0, and which lies, except for these two points, in $D \setminus \gamma_1$.

Proceeding in this way, we get pairwise disjoint contours $\gamma_1, \dots, \gamma_m$ and smooth points $b_1, \dots, b_m$ of Γ_0 such that $D \setminus (\gamma_1 \cup \dots \cup \gamma_m)$ is connected and, for each $1 \le j \le m$:

- γ_j is diffeormorphic to $[0,1]$;
- γ_j meets Γ_j transversally at a_j;
- γ_j meets Γ_0 transversally at b_j
- $\gamma_j \setminus \{a_j, b_j\} \subseteq D$.

Then $\Gamma_0 \cup \Gamma_1 \cup \dots \Gamma_m \cup \gamma_1 \cup \dots \cup \gamma_m$ is connected. Since this is the boundary of $\mathbb{C} \setminus \big(D \setminus (\gamma_1 \cup \dots \cup \gamma_m)\big)$, it follows that $\mathbb{C} \setminus \big(D \setminus (\gamma_1 \cup \dots \cup \gamma_m)\big)$ is connected. Hence $D \setminus (\gamma_1 \cup \dots \cup \gamma_m)$ is simply connected.

Since each γ_j is diffeormorphic to $[0,1]$ and meets $\Gamma_0 \cup \Gamma_1 \cup \dots \Gamma_m$ transversally, now we can find a neighborhood V of $\overline{D}$ such that, for $1 \le j \le m$, there exists a closed contour γ_j' in V which is diffeomorphic to the open interval $]0,1[$ and such that $\gamma_j \subseteq \gamma_j'$. Since $D \setminus (\gamma_1 \cup \dots \cup \gamma_m)$ is simply connected, by shrinking V, we may achieve that also

$$U_0 := V \setminus (\gamma_1 \cup \dots \cup \gamma_m)$$

is simply connected. Moreover, we can achieve that V has $\mathbb{C}^1$-boundary which is met transversally by $\gamma_1', \dots, \gamma_m'$. Choose $\varepsilon > 0$ sufficiently small and set

$$U_j = \left\{ z \in W \;\middle|\; \min_{w \in \gamma_j'} |z - w| < \varepsilon \right\}.$$

Since γ_j' is diffeomorphic to $[0,1]$, then it is clear that U_j is simply connected and $U_j \setminus \gamma_j'$ is the union of two simply connected open sets V_j and V_j. □

2.5.3 Definition. Let $D \subseteq \mathbb{C}$ be a bounded, connected open set (possibly, not simply connected) with piecewise $\mathcal{C}^1$ boundary. Let Γ be the union of some of the connected components of the boundary of D. Assume Γ is oriented (not necessarily by D). Let $f : \Gamma \to \mathbb{C}^*$ be a holomorphic function[2]. Then we define

$$\operatorname{ind}_\Gamma f := \frac{1}{2\pi i} \int\limits_\Gamma \frac{f'(z)}{f(z)}\, dz\,. \tag{2.5.1}$$

[2]By this we mean the following: If $K \subseteq \mathbb{C}$ is a set of uniqueness for holomorphic functions (as, for example Γ, or the closure of an open set), then we say that $f : K \to E$ **is a holomorphic function** if $f = \widetilde{f}|_\Gamma$, where $\widetilde{f}$ is a holomorphic function defined in some neighborhood of K. By f' then we mean the function $\widetilde{f}'|_\Gamma$.

We recall, with proofs, some well known facts about this index.

2.5.4 Proposition. *Let $D \subseteq \mathbb{C}$ be a bounded, connected open set with piecewise $\mathcal{C}^1$ boundary ∂D.*

(i) *Let Γ be the union of some of the connected components of ∂D, oriented somehow. Let $f, g : \Gamma \to \mathbb{C}^*$ be two holomorphic functions. Then*

$$\operatorname{ind}_\Gamma(fg) = \operatorname{ind}_\Gamma f + \operatorname{ind}_\Gamma g\,. \tag{2.5.2}$$

(ii) *Let $f : \partial D \to \mathbb{C}^*$ be a holomorphic function which admits a meromorphic extension to D, let N be the number of zeros of f in D, counted according to their multiplicities, and let P be the number of poles of f in D, also counted according to their multiplicities. If ∂D is oriented by D, then*

$$\operatorname{ind}_{\partial D} f = N - P\,. \tag{2.5.3}$$

(iii) *Let $f : \partial D \to \mathbb{C}^*$ be a holomorphic function which admits a meromorphic extension to $(\mathbb{C}\setminus D)\cup\{\infty\}$, let N be the number of zeros of f in $(\mathbb{C}\setminus D)\cup\{\infty\}$, counted according to their multiplicities, and let P be the number of poles of f in $(\mathbb{C}\setminus D)\cup\{\infty\}$, also counted according to their multiplicities. If ∂D is oriented by D, then*

$$\operatorname{ind}_{\partial D} f = P - N\,. \tag{2.5.4}$$

(iv) *Let Γ be the union of some of the connected components of ∂D, oriented somehow. Then, for any holomorphic functions $f : \partial D \to \mathbb{C}^*$, the index $\operatorname{ind}_\gamma f$ is an integer.*

Proof. (i)

$$\begin{aligned}\operatorname{ind}_\Gamma(fg) &= \frac{1}{2\pi i}\int\limits_\Gamma \frac{(fg)'(z)}{(fg)(z)}\,dz = \frac{1}{2\pi i}\int\limits_\Gamma \frac{(f'g+fg')(z)}{(fg)(z)}\,dz \\ &= \frac{1}{2\pi i}\int\limits_\Gamma \frac{f'(z)}{f(z)}\,dz + \frac{1}{2\pi i}\int\limits_\Gamma \frac{g'(z)}{g(z)}\,dz = \operatorname{ind}_\Gamma f + \operatorname{ind}_\Gamma g\,.\end{aligned}$$

(ii) By Cauchy's theorem we may pass to small circles surrounding the zeros and poles of f. Therefore, we may assume that, for some $w \in D$, the function f is of the form $f(z) = (z-w)^\kappa g(z)$, where $\kappa \in \mathbb{Z}$ and g is holomorphic and invertible on $\overline{D}$. Then $\kappa = N - P$ and

$$\frac{f'(z)}{f(z)} = \frac{\kappa}{z-w} + \frac{g'(z)}{g(z)}\,, \qquad z \in \partial D,$$

where g'/g is holomorphic on $\overline{D}$. Hence

$$\operatorname{ind}_{\partial D} f = \frac{\kappa}{2\pi i}\int\limits_{\partial} D\frac{dz}{z-w} = \kappa = N - P.$$

(iii) Set
$$D^* = \left\{ z \in \mathbb{C} \,\middle|\, \frac{1}{z} \in (\mathbb{C} \setminus \overline{D}) \cup \{\infty\} \right\},$$
and let ∂D^* be the boundary of D^*, oriented by D^*. Let $-\partial D$ be the boundary of D oriented by $\mathbb{C}\setminus\overline{D}$. Since $1/z$ maps a neighborhood of $\mathbb{C}\setminus D$ biholomorphically to a neighborhood of $\overline{D}^* \setminus \{0\}$ and since biholomorphic maps respect the orientation, $-\partial D$ is mapped by $1/z$ to ∂D^*. Therefore
$$\operatorname{ind}_{\partial D} f = -\frac{1}{2\pi i} \int\limits_{-\partial D} \frac{f'(z)}{f(z)}\, dz = \frac{1}{2\pi i} \int\limits_{\partial D^*} \frac{f'(1/z)}{f(1/z)} \frac{1}{z^2}\, dz.$$
Setting
$$F(z) = f\left(\frac{1}{z}\right), \qquad z \in \partial D^*,$$
this implies that
$$\operatorname{ind}_{\partial D} f = -\frac{1}{2\pi i} \int\limits_{\partial D^*} \frac{F'(z)}{F(z)} dz = -\operatorname{ind}_{\partial D^*} F = P - N,$$
where the last equality follows from part (ii) of the proposition, as $F(z) = f(1/z)$ admits a meromorphic extension to $\overline{D}^*$, where N is the number of zeros of F and P is the number of poles of F, both counted with multiplicities.

(iv) We may assume that D is simply connected, and Γ is the boundary of D (Section 2.5.1). Let a holomorphic function $f : \Gamma \to \mathbb{C}^*$ be given. Choose a neighborhood U of Γ such that U has a $\mathcal{C}^1$-boundary, $\mathbb{C} \setminus \overline{U}$ consists of not more than two connected components, and f is defined, holomorphic and invertible on $\overline{U}$. By the Runge approximation Theorem 2.2.2, we can find a sequence ρ_n of rational functions which converges to f uniformly in $\overline{U}$. We may assume that $\rho_n \neq 0$ on $\overline{U}$. Then the functions ρ_n'/ρ_n converge to f'/f uniformly on Γ. Hence
$$\operatorname{ind}_\Gamma f = \lim_{n\to\infty} \operatorname{ind}_\Gamma \rho_n \,.$$
This implies that $\operatorname{ind}_\Gamma f$ is an integer, as, by part (ii) of the proposition, each $\operatorname{ind}_\Gamma \rho_n$ is an integer. □

Recall that (see, e.g., Theorem 13.11 in [Ru]), for each simply connected open set $D \subseteq \mathbb{C}$ and any $f \in \mathcal{O}^*(D)$, there exists $g \in \mathcal{O}(D)$ with $e^g = f$. In the case of connected open sets, which are not simply connected, this is not always true. There are topological obstructions, described in the following theorem.

2.5.5 Theorem. *Suppose D is connected but not simply connected. Then for any holomorphic function $f : \overline{D} \to \mathbb{C}^*$ the following are equivalent:*

(i) *There exists a holomorphic function $g : \overline{D} \to \mathbb{C}$ with $e^g = f$ on $\overline{D}$.*

(ii) *For any connected component* Γ *of the boundary of* D,

$$\operatorname{ind}_\Gamma f = 0.$$

Proof. Let $\Gamma_0, \ldots, \Gamma_m$ be the connected components of the boundary of D such that Γ_0 is the boundary of the unbounded component of $\mathbb{C} \setminus \overline{D}$.

If condition (i) is fulfilled, it is clear that

$$\operatorname{ind}_{\Gamma_j} f = \frac{1}{2\pi i} \int\limits_{\Gamma_j} \frac{f'(z)}{f(z)}\, dz = \frac{1}{2\pi i} \int\limits_{\Gamma_j} g'(z)\, dz = 0\,, \qquad 0 \le j \le m.$$

Now assume that (ii) is satisfied.

For $1 \le j \le m$, we fix a smooth point a_j of Γ_j. Let $U_0, U_1, \ldots, U_m$ be the sets from Lemma 2.5.2. Since the sets U_j are simply connected, then we have holomorphic functions $g_j : U_j \to \mathbb{C}$ such that

$$e^{g_j} = f\big|_{U_j}\,, \qquad 0 \le j \le m\,. \tag{2.5.5}$$

It follows that

$$e^{g_j - g_0} = 1 \quad \text{on} \quad U_0 \cap U_j\,, \qquad 1 \le j \le m\,. \tag{2.5.6}$$

Since $a_j \in U_j$ and $\Gamma_j \cap U_0 = \Gamma_j \setminus \{a_j\}$, we can choose two different points $b_j, c_j \in \Gamma_j \cap U_j \cap U_0$, so that one of the two closed connected contours Γ_j' and Γ_j'' with the boundary points b_j and c_j, into which Γ_j is divided by b_j and c_j, is contained in U_j and the other one is contained in U_0, say $\Gamma_j' \subseteq U_j$ and $\Gamma_j'' \subseteq U_0$. Moreover, by changing the notation if necessary, we may assume that, with respect to the orientation of Γ_j, the point b_j is the starting point of Γ_j' and the endpoint of Γ_j'', and c_j is the starting point of Γ_j'' and the end point of Γ_j'. Then

$$\int\limits_{\Gamma_j'} g_j'(z) dz = g_j(c_j) - g_j(b_j) \quad \text{and} \quad \int\limits_{\Gamma_j''} g_0'(z) dz = g_0(b_j) - g_0(c_j)\,. \tag{2.5.7}$$

Let C_j, B_j be the two connected components of $U_j \cap U_0$. By changing the notation if necessary, we may assume that $c_j \in C_j$ and $b_j \in B_j$. Then it follows from (2.5.6) that, for some integers k_j, n_j

$$g_j - g_0 \equiv k_j 2\pi i \quad \text{on } C_j \qquad \text{and} \qquad g_j - g_0 \equiv n_j 2\pi i \quad \text{on } B_j\,. \tag{2.5.8}$$

Now, from condition (ii) we get

$$0 = \frac{1}{2\pi i} \int\limits_{\Gamma_j} \frac{f'(z)}{f(z)}\, dz = \frac{1}{2\pi i} \int\limits_{\Gamma_j'} \frac{f'(z)}{f(z)}\, dz + \frac{1}{2\pi i} \int\limits_{\Gamma_j''} \frac{f'(z)}{f(z)}\, dz\,, \qquad 1 \le j \le m\,.$$

Since $e^{g_j} = f\big|_{U_j}$ for all $0 \le j \le m$, this implies that

$$0 = \frac{1}{2\pi i}\int\limits_{\Gamma_j'} g_j'(z)\,dz + \frac{1}{2\pi i}\int\limits_{\Gamma_j''} g_0'(z)\,dz\,, \qquad 1 \le j \le m\,.$$

Together with (2.5.7) and (2.5.8) this gives

$$0 = g_j(c_j) - g_j(b_j) + g_0(b_j) - g_0(c_j) = (k_j - n_j)2\pi i\,, \qquad 1 \le j \le m\,.$$

Hence $k_j = n_j$ and therefore, again by (2.5.8),

$$g_j \equiv g_0 \qquad \text{on } C_j \cup B_j = U_j \cap U_0\,, \qquad 1 \le j \le m\,.$$

Since $\overline{D} \subseteq U_0 \cup U_1 \cup \ldots \cup U_m$ and the sets $U_1, \ldots, U_m$ are pairwise disjoint, it follows that there is a global holomorphic function g on $\overline{D}$ with $g = g_j$ on U_j, $0 \le j \le m$, such that (by (2.5.5) $e^g = f$ on $\overline{D}$. □

2.5.6 Theorem (Runge approximation for invertible functions). *Let $D \subseteq \mathbb{C}$ be a bounded open set with piecewise $\mathcal{C}^1$-boundary ∂D, and let $f : \overline{D} \to \mathbb{C}^*$ be a continuous function which is holomorphic in D.*

(i) *If $\mathbb{C}\setminus\overline{D}$ is connected, then f can be approximated uniformly on $\overline{D}$ by functions from $\mathcal{O}^*(\mathbb{C})$.*

(ii) *If $\mathbb{C} \setminus \overline{D}$ is not connected and $U_1, \ldots, U_m$ are the bounded connected components of $\mathbb{C} \setminus \overline{D}$, then for any choice of points $p_1 \in U_1, \ldots, p_m \in U_m$, f can be approximated uniformly on $\overline{D}$ by functions from $\mathcal{O}^*\big(\mathbb{C} \setminus \{p_1, \ldots, p_m\}\big)$.*

Proof. By the Mergelyan approximation Theorem 2.2.1, we may assume that f is holomorphic in some neighborhood of $\overline{D}$. Let $D_1, \ldots, D_k$ be the connected components of D, and let $U_k^1, \ldots, U_k^{m_k}$ be the bounded connected components of $\mathbb{C} \setminus \overline{D}_k$ (if there are any), and let Γ_k^j be the boundary of U_k^j endowed with the orientation defined by D, i.e., $\Gamma_k^j = -\partial U_k^j$. Set (cf. Def. 2.5.3)

$$\kappa_k^j = \operatorname{ind}_{\Gamma_k^j} f.$$

By hypothesis, there are points $p_j^k \in U_j^k \cap \{p_1, \ldots, p_m\}$. Then, by Proposition 2.5.4 ,

$$\operatorname{ind}_{\Gamma_k^j}\Big((z - p_j^k)^{\kappa_k^j} f(z)\Big) = 0\,.$$

Therefore, by Theorem 2.5.5, there exists a holomorphic function $g : \overline{D} \to \mathbb{C}$ with

$$e^{g(z)} = (z - p_j^k)^{\kappa_k^j} f(z) \qquad \text{for all } z \in \overline{D}\,.$$

By the Runge approximation Theorem 2.2.2, the function g can be apporoximated uniformly on $\overline{D}$ by functions from $\mathcal{O}\big(\mathbb{C} \setminus \{p_1, \ldots, p_m\}\big)$. Hence the function

$$f(z) = \frac{e^{g(z)}}{(z - p_j^k)^{\kappa_k^j}}$$

can be approximated uniformly on $\overline{D}$ by functions of the form

$$e^{g_k(z)} = \frac{e^{h(z)}}{(z-p_j^k)^{\kappa_k^j}}$$

with $h \in \mathcal{O}(\mathbb{C} \setminus \{p_1, \ldots, p_m\})$, which belong to $\mathcal{O}^*(\mathbb{C} \setminus \{p_1, \ldots, p_m\})$. □

2.5.7 Theorem (Weierstrass product theorem). *Let $D \subseteq \mathbb{C}$ be an open set, let Z be a discrete and closed subset of D, and assume that, for each $w \in Z$, a number $m_w \in \mathbb{N}^*$ is given. Then there exists a holomorphic function $f : D \to \mathbb{C}$ such that $f(z) \neq 0$ for $z \in D \setminus Z$ and, for each $w \in Z$, f has a zero precisely of order m_w.*

Proof. Choose a sequence of open sets $D_n \subseteq D$, $n \in \mathbb{N}$, such that, for all $n \in \mathbb{N}$:

- D_n has piecewise $\mathcal{C}^1$-boundary ∂D_n.
- $\overline{D}_n \subseteq D_{n+1}$.
- $\bigcup_{n=0}^{\infty} D_n = D$.
- Each bounded, connected component of $\mathbb{C} \setminus D_n$ contains at least one point of $\mathbb{C} \setminus D$.

Next we inductively construct a sequence $(f_n)_{n\in\mathbb{N}}$ of functions $f_n \in \mathcal{O}(D)$ such that, for all $n \in \mathbb{N}$:

(a) If $w \in Z \cap \overline{D}_n$, then f_n has a zero at w precisely of order m_w.

(b) If $z \in D \setminus (Z \cap \overline{D}_n)$, then $f_n(z) \neq 0$.

(c) If $n \geq 1$, then we have $\varepsilon_n \in \mathcal{O}(D)$ such that

$$f_n = (1 + \varepsilon_n) f_{n-1} \tag{2.5.9}$$

$$|\varepsilon_n(z)| < 2^{-n}\Big(1 + \max_{\zeta \in \overline{D}_{n-1}} |f_{n-1}(\zeta)|\Big) \qquad \text{for all } z \in \overline{D}_{n-1}\,. \tag{2.5.10}$$

We start with $f_0(z) := (z-w_1)^{m_{w_1}} \ldots (z-w_s)^{m_{w_s}}$ where $\{w_1, \ldots, w_s\} := Z \cap \overline{D}_0$. Now we assume that functions $f_0, \ldots, f_{k-1} \in \mathcal{O}(D)$ are already constructed such that (a), (b), (c) hold for $0 \leq n \leq k-1$. Then we set $u(z) = (z-w_1)^{m_{w_1}} \ldots (z-w_r)^{m_{w_r}}$ where $\{w_1, \ldots, w_r\} := Z \cap \overline{D}_{k+1}$. Then $u f_{k-1}^{-1} \neq 0$ on $\overline{D}_{k-1}$. Hence, by the Runge approximation Theorem 2.5.6, we can find $g \in \mathcal{O}^*(D)$ such that the function $\varepsilon_k := g u f_{k-1}^{-1} - 1$ satisfies (2.5.6) for $n = k$. It remains to set $f_k = (1 + \varepsilon_k) f_{k-1}$. The sequence $(f_n)_{n\in\mathbb{N}}$ is constructed.

Now from property (c) of the sequence $(f_n)_{n\in\mathbb{N}}$ it follows that $|f_{n+1} - f_n| < 2^{-n-1}$ on $\overline{D}_n$. Hence this sequence converges uniformly on the compact subsets of D. Set $f := \lim_{n\to\infty} f_n$. It remains to prove that f has precisely the prescribed zeros. It is sufficient to check this over each $\overline{D}_n$. Let $n \in \mathbb{N}$ be given. By (2.5.9) and (2.5.10), then for all $m > n$ and $z \in \overline{D}_n$,

$$f_m(z) = f_n(z) \prod_{j=n+1}^{m} \big(1 + \varepsilon_j(z)\big) = f_n(z) \exp\Bigg(\sum_{j=n+1}^{m} \log\big(1 + \varepsilon_j(z)\big)\Bigg)$$

and hence,

$$f\Big|_{\overline{D}_n} = f_n\Big|_{\overline{D}_n} \exp\Bigg(\sum_{j=n+1}^{\infty} \log(1 + \varepsilon_j)\Big|_{\overline{D}_n}\Bigg).$$

Since, over $\overline{D}_n$, the function f_n has precisely the prescribed zeros, this implies that, over $\overline{D}_n$, the function f has precisely the prescribed zeros. □

2.6 $\mathcal{O}^E$-cocycles with prescribed zeros and a stronger version of the Mittag-Leffler theorem

In this section E is a Banach space.

The Weierstrass product Theorem 2.5.7 makes it possible to improve the results of Section 2.4. Now we can consider additive Cousin problems (see Def. 2.4.1) with *prescribed zeros* and solve them with the same prescibed zeros. This is the topic of the present section.

2.6.1 Definition. By a **data of zeros** we mean a pair (Z, m) where $Z \subseteq \mathbb{C}$ and $m = \{m_w\}_{w \in z}$ is a family of integers $m_w \geq 0$.

Let such a data be given.

Then, for an open set $U \subseteq \mathbb{C}$, we denote by $\mathcal{O}^E_{Z,m}(U)$ the space of functions $f \in \mathcal{O}^E(U)$ such that, for each $w \in Z \cap U$, f has a zero of order $\geq m_w$ at w. The functions from $\mathcal{O}^E_{Z,m}(U)$ will be called $\mathcal{O}^E_{Z,m}$**-functions on** U.

For $E = \mathbb{C}$ we write also $\mathcal{O}_{Z,m}(U)$ instead of $\mathcal{O}^E_{Z,m}(U)$.

Now let $D \subseteq \mathbb{C}$ be an open set, and let $\mathcal{U} = \{U_j\}_{j \in I}$ be an open covering of D. Then we denote by $C^1(\mathcal{U}, \mathcal{O}^E_{Z,m})$ the space of all $f \in C^1(\mathcal{U}, \mathcal{O}^E)$ (see Def. 2.4.1) such that $f_{jk} \in \mathcal{O}^E_{Z,m}(U_j \cap U_k)$ for all $j, k \in I$. Further we set (see again Def. 2.4.1)

$$Z^1(\mathcal{U}, \mathcal{O}^E_{Z,m}) = Z^1(\mathcal{U}, \mathcal{O}^E) \cap C^1(\mathcal{U}, \mathcal{O}^E_{Z,m}).$$

The elements of $Z^1(\mathcal{U}, \mathcal{O}^E_{Z.m})$ will be called $(\mathcal{U}, \mathcal{O}^E_{Z,m})$**-cocycles**. If the covering $\mathcal{U}$ is not specified, then we speak also about $\mathcal{O}^E_{Z,m}$**-cocycles over** D.

There is the following improvement of Theorem 2.4.2:

2.6.2 Theorem. *Let $D \subseteq \mathbb{C}$ be an open set, and let (Z, m) be a data of zeros such that $Z \cap D$ is discrete and closed in D. Let $\mathcal{U} = \{U_j\}_{j \in I}$ be an open covering* *of D, and let $f \in Z^1(\mathcal{U}, \mathcal{O}^E_{Z,m})$. Then there exists a family $\{h_j\}_{j\in I}$ of functions $h_j \in \mathcal{O}^E_{Z,m}(U_j)$ such that, for all $j,k \in I$ with $U_j \cap U_k \neq \emptyset$,*

$$f_{jk} = h_j - h_k \qquad on \quad U_j \cap U_k\,, \quad j,k \in I. \tag{2.6.1}$$

Proof. By the Weierstrass product Theorem 2.5.7, there exists a holomorphic function $\varphi : D \to \mathbb{C}$ such that $\varphi(z) \neq 0$ for $z \in D \setminus Z$ and, for each $w \in Z$, φ has a zero precisely of order m_w. Setting $\widetilde{f}_{jk} = f_{jk}/\varphi$, $j,k \in I$, then we obtain an $(\mathcal{U}, \mathcal{O}^E)$-cocycle, and from Theorem 2.4.2 we obtain a family $\{\widetilde{h}_j\}_{j\in I}$ of functions $\widetilde{h}_j \in \mathcal{O}^E_{Z,m}(U_j)$ such that, for all $j,k \in I$ with $U_j \cap U_k \neq \emptyset$,

$$\widetilde{f}_{jk} = \widetilde{h}_j - \widetilde{h}_k \qquad on \quad U_j \cap U_k\,, \quad j,k \in I.$$

It remains to set $h_j = \varphi \widetilde{h}_j$. □

We point out again the special case of coverings by two sets:

2.6.3 Corollary (to Theorem 2.4.2). *Let $D \subseteq \mathbb{C}$ be an open set, and let (Z,m) be a data of zeros such that $Z \cap D$ is discrete and closed in D. Let $D_1, D_2 \subseteq D$ be two open subsets with $D = D_1 \cup D_2$ and $D_1 \cap D_2 \neq \emptyset$, and let $f \in \mathcal{O}^E_{Z,m}(D_1 \cap D_2)$. Then there exist $f_j \in \mathcal{O}^E_{Z,m}(D_j)$, $j = 1,2$, such that*

$$f = f_1 + f_2 \qquad on\ D_1 \cap D_2.$$

By means of this corollary, now we obtain the following version of the Mittag-Leffler theorem, which is stronger than Theorem 2.4.4, but also due to Mittag-Leffler (see the historical remarks on page 116 of [Re][3])

2.6.4 Theorem (Mittag-Leffler theorem). *Let $D \subseteq \mathbb{C}$ be an open set, let Z be a discrete and closed subset of D, and assume that, for each $w \in Z$, a number $m_w \in \mathbb{N}$ and a holomorphic function $f_w : \mathbb{C} \setminus \{w\} \to E$ of the form*

$$f_w(z) = \sum_{n=-\infty}^{m_w} (z-w)^n f_{wn}$$

are given. Then there exists a holomorphic function $f : D \setminus Z \to E$ such that, for each $w \in Z$, f_w is the first part of the Laurent expansion of f, i.e., the difference $f - f_w$ has a zero of order $\geq m_w$ at w.

Proof. Since Z is discrete and closed in D, we can find a family $\{U_w\}_w \in Z$ of open subsets of D such that, U_w is a neighborhood of w and $U_w \cap U_v = \emptyset$ if $w \neq v$.

[3]This is the only book where we found the scalar case of Theorem 2.6.4. There, this theorem is called *Anschmiegungssatz von Mittag-Leffler.* We do not know whether there is a corresponding commonly used name in English. Therefore, we call it just *Mittag-Leffler theorem.*

Set $D_1 = \bigcup_{w \in Z} U_w$ and $D_2 = D \setminus Z$. Then $D_2 \cap D_2$ is the disjoint union of the punctured sets $U_w \setminus \{w\}$. Therefore, setting

$$g = f_w \qquad \text{on } U_w \setminus \{w\},$$

we obtain a holomorphic function $g : D_1 \cap D_2 \to E$. Since $D_1 \cap D_2 \cap Z = \emptyset$, the function g can be interpreted as a function from $\mathcal{O}^E_{Z,m}(D_1 \cap D_2)$. Then from Corollary 2.6.3 we get functions $h \in \mathcal{O}^E_{Z,m}(D_1)$ and $f \in \mathcal{O}^E(D_2)$ such that $g = f + h$ on $D_1 \cap D_2$. It follows that

$$f - f_w = g - h - f_w = -h \qquad \text{on } U_w \setminus \{w\}.$$

Since, for all $w \in Z$, the function h has a zero of order $\geq m_w$ at w, this completes the proof. □

2.7 Generalization of the Weierstrass product theorem

Recall that the theory of cocycles with prescribed zeros, which was developed in the preceding Section 2.6, is based on the Weierstrass product Theorem 2.5.7. In turn, this theory allows us to prove the following generalization of the Weierstrass product theorem:

2.7.1 Theorem. *Let $D \subseteq \mathbb{C}$ be an open set, and let Z be a discrete and closed subset of D. Suppose, for each $w \in Z$, a neighborhood $U_w \subseteq D$ of w and a holomorphic function $f_w \in \mathcal{O}^*(U_w \setminus \{w\})$ are given such that $U_w \cap Z = \{w\}$, $w \in \mathbb{Z}$.*

(i) *Then there exists a family of holomorphic functions $h_w \in \mathcal{O}^*(U_w)$, $w \in Z$, and a holomorphic function $f \in \mathcal{O}^*(D \setminus Z)$ such that*

$$h_w f_w = f \qquad \text{on } U_w \setminus \{w\} \qquad \text{for all } w \in Z\,. \tag{2.7.1}$$

(ii) *Moreover, for any given family of numbers $m_w \in \mathbb{N}^*$, $w \in Z$, we can achieve that, for each $w \in Z$, the function $h_w - 1$ has a zero of order $\geq m_w$.*

The claim of the classical Weierstrass product Theorem 2.5.7 is the special case of part (i) of Theorem 2.7.1 obtained by setting $f_w(z) = (z - w)^{m_w}$.

Proof of Theorem 2.7.1. Let some family $m = \{m_w\}_{w \in Z}$ of numbers $m_w \in \mathbb{N}^*$ be given. Choose small discs V_w around w such that $\overline{V}_w \subseteq U_w$, $w \in Z$, and $\overline{V}_w \cap \overline{V}_v = \emptyset$ for all $w, v \in Z$ with $w \neq v$. Then it is sufficient to find holomorphic functions $h_w \in \mathcal{O}^*(V_w)$, $w \in Z$, and a holomorphic function $f \in \mathcal{O}^*(D \setminus Z)$ such that

$$h_w f_w = f \qquad \text{on } V_w \setminus \{w\} \qquad \text{for all } w \in Z\,. \tag{2.7.2}$$

Indeed, since f is holomorphic and invertible on $D \setminus Z$ and f_w is holomorphic and invertible on $U_w \setminus \{w\}$, then it follows from (2.7.2) that each h_w admits a holomorphic extension to U_w, which is invertible on $U_w \setminus \{w\}$ and satisfies (2.7.1).

Set $\kappa_w = \operatorname{ind}_{\partial V_w} f_w$ for $w \in Z$ (cf. Def. 2.5.3). By the Weierstrass product Theorem 2.5.7, we can find a holomorphic function $\varphi : D \setminus Z \to \mathbb{C}^*$ such that, for each $w \in Z$, φ has a zero precisely of order κ_w at w. Set

$$\widetilde{f}_w = \varphi^{-1} f_w \qquad \text{on } V_w \setminus \{w\},\ \ w \in Z.$$

Then, by Proposition 2.5.4,

$$\operatorname{ind}_{\partial V_w} \widetilde{f}_w = \operatorname{ind}_{\partial V_w} \varphi^{-1} + \operatorname{ind}_{\partial V_w} f_w = 0.$$

Hence, by Theorem 2.5.5, for each $w \in Z$, we can find a neighborhood Θ_w of ∂V_w and a holomorphic function $g_w : \Theta_w \to \mathbb{C}$ with

$$e^{g_w} = \widetilde{f}_w \qquad \text{on } \Theta_w\,.$$

Since the sets $\overline{V}_w$ are pairwise disjoint, we may assume that also the sets Θ_w are pairwise disjoint. Moreover, we may assume that $w \notin \Theta_w$. Set

$$D_1 = \bigcup_{w \in Z} (V_w \cup \Theta_w) \quad \text{and} \quad D_2 = \Big(D \setminus \bigcup_{w \in Z} V_w\Big) \cup \bigcup_{w \in Z} \Theta_w.$$

Then

$$D_1 \cap D_2 = \bigcup_{w \in Z} \Theta_w$$

and, since the sets Θ_w are pairwise disjoint, setting

$$g = g_w \qquad \text{on } \Theta_w,$$

we obtain a holomorphic function $g : D_1 \cap D_2 \to \mathbb{C}$. Since $Z \cap D_1 \cap D_2 = \emptyset$, this function can be interpreted as a function from $\mathcal{O}_{Z,m}(D_1 \cap D_2)$ (see Def. 2.6.1). Then from Corollary 2.6.3 we get functions $g_j \in \mathcal{O}_{Z,m}(D_j)$, $j = 1, 2$, with $g = g_1 + g_2$ on $D_1 \cap D_2$. Set

$$f = \varphi e^{g_2} \text{ on } D_2 \qquad \text{and} \qquad h_w = e^{-g_1} \text{ on } V_w.$$

Since, for each $w \in Z$, the function g_1 has a zero of order $\geq m_w$ at w, then, for each $w \in Z$, also the function $h_w - 1 = e^{-g_1} - 1$ has a zero of order $\geq m_w$ at w, and we have

$$h_w f_w = e^{-g_1} \varphi \widetilde{f}_w = e^{-g_1} \varphi e^{g_w} = \varphi e^{g_2} = f \quad \text{on } D_2 \cap V_w = \Theta_w$$

on Θ_w. It remains to observe that from this relation it follows that f admits a holomorphic and invertible extension to $D \setminus Z$, and this extension satisfies (2.7.2). □

We point out also the following generalization of the Weierstrass product theorem, which follows from Theorem 2.7.1:

2.7.2 Theorem. *Let $D \subseteq \mathbb{C}$ be an open set, let Z be a discrete and closed subset of D, and assume that, for each $w \in Z$, a number $m_w \in \mathbb{N}$ and a holomorphic function $f_w \in \mathcal{O}^*(U_w \setminus \{w\})$ are given such that $U_w \cap Z = \{w\}$, $w \in \mathbb{Z}$, and f_w has a pole or a removable singularity at w. Let*

$$f_w(z) = \sum_{n=n_w}^{\infty} f_{w,n}(z-w)^n$$

be the Laurent expansion of f_w at $z \in Z$. Then, for each given family $k_w \in \mathbb{N}^*$, $z \in Z$, there exists a holomorphic function $f \in \mathcal{O}^*(D \setminus Z)$ such that, for each $w \in Z$, the Laurent expansion of f at w is of the form

$$f(z) = \sum_{n=n_w}^{k_w} f_{w,n}(z-w)^n + \sum_{n=k_w+1}^{\infty} f_n(z-w)^n .$$

The claim of the classical Weierstrass product Theorem 2.5.7 is the special case obtained by setting $f_w(z) = (z-w)^{m_w}$ and $k_w = m_w$.

Proof of Theorem 2.7.2. By Theorem 2.7.1 we can find a family of holomorphic functions $h_w \in \mathcal{O}^*(U_w)$, $w \in Z$, and a holomorphic function $f \in \mathcal{O}^*(D \setminus Z)$ such that, for all $w \in Z$,

$$h_w f_w = f \qquad \text{on } U_w \setminus \{w\} \tag{2.7.3}$$

and, moreover, the Laurent expansion of h_w at w is of the form

$$h_w(z) = 1 + \sum_{n=k_w-n_w+1}^{\infty} h_{w,n}(z-w)^n .$$

Hence, for the Laurent expansion of f at w, we get

$$\begin{aligned} f(z) &= \sum_{n=n_w}^{\infty} f_{w,n}(z-w)^n + \sum_{n=k_w-n_w+1}^{\infty} h_{w,n}(z-w)^n \sum_{n=n_w}^{\infty} f_{w,n}(z-w)^n \\ &= \sum_{n=n_w}^{\infty} f_{w,n}(z-w)^n + h_{k_w-n_w+1} f_{w,k_w+1}(z-w)^{k_w+1} + \dots \,. \end{aligned}$$

□

We conclude this section with a discussion of the relation between the Mittag-Leffler Theorem 2.6.4 and the generalized Weierstrass product theorems 2.7.1 and 2.7.2.

There are two differences. The first difference is a disadvantage of the Mittag-Leffler therem: Even if the given local functions f_w are different from zero in a punctured neighborhood of w, the global function f given by the Mittag-Leffler Theorem 2.6.4 can have zeros outside Z, whereas the global function f given by the generalized Weierstrass product theorems 2.7.1 and 2.7.2 is different from zero everywhere on $D \setminus Z$.

The second difference is a disadvantage of the generalized Weierstrass theorems: In the Mittag-Leffler Theorem 2.6.4, we do not require that the given functions f_w are different from zero in some punctured neighborhood of w. It is even allowed that there is an infinite number of zeros in each punctured neighborhood of w (in the case of an essential singularity this is possible). Moreover, the Mittag-Leffler Theorem 2.6.4 preserves arbitrary starting pieces of the given Laurent expansions, whereas for the generalized Weierstrass Theorem 2.7.2, this is only true for poles and removable singularities.

We consider the simplest non-trivial example:

2.7.3 Example. Let $D \subseteq \mathbb{C}$ be a connected open set with $Z := \{0,1\} \subseteq D$, and let

$$f_0(z) := \exp\left(\frac{1}{z}\right) \qquad \text{and} \qquad f_1(z) := \frac{1}{z-1}\,.$$

Then there is no holomorphic function $f : D \setminus \{0,1\} \to \mathbb{C}^*$ which has, at the same time, the following two properties:

a) (as claimed in Theorem 2.7.1) There are neighborhoods $U_j \subseteq D$ of j and functions $h_j \in \mathcal{O}^*(U_j)$, $j = 0,1$, such that

$$f(z) = e^{1/z} h_0(z) \qquad \text{for } z \in U_0 \setminus \{0\}\,,$$

$$f(z) = \frac{h_1(z)}{z-1} \qquad \text{for } z \in U_1 \setminus \{1\}\,.$$

b) (as claimed in the Mittag-Leffler Theorem 2.6.4) $f(z) - e^{1/z}$ is holomorphic at 0, and $f(z) - \frac{1}{z-1}$ is holomorphic at 1.

Indeed, assume there exists such a function f. Since $e^{1/z}$ is holomorphic at 1, whereas f has the pole $\frac{1}{z-1}$ there, then $f(z) \not\equiv e^{1/z}$. Since $f(z) = e^{1/z} h_0(z)$, it follows that $h_0(z) \not\equiv 1$. Choose a neighborhood $V_0 \subseteq U_0$ of 0 so small that $h_0(z) - 1 \neq 0$ for $z \in V_0 \setminus \{0\}$. Then

$$e^{1/z} = \frac{h_0(z)e^{1/z} - e^{1/z}}{h_0(z) - 1} = \frac{f(z) - e^{1/z}}{h_0(z) - 1} \qquad \text{for } z \in V_0 \setminus \{0\}.$$

Since $f(z) - e^{1/z}$ and $h_0(z) - 1$ are holomorphic at 0, this implies that the singularity of $e^{1/z}$ at 0 is not essential, which is not true.

Moreover, if in this example, f is a function with the properties claimed in the Mittag-Leffler Theorem 2.6.4, then this function has zeros in any punctured neighborhood of 0 – in distinction to $e^{1/z}$. This is due to the following proposition which is a consequence of Picard's theorem:

2.7.4 Proposition. *Let $w \in \mathbb{C}$, let $W \subseteq \mathbb{C}$ be a connected neighborhood of w, and let $f_w : W \setminus \{w\} \to \mathbb{C}^*$ be a holomorphic function which has an essential singularity at w. Further let $f : W \setminus \{w\} \to \mathbb{C}$ be a holomorphic function such that $f - f_w$ has a removable singularity at w. Then either $f \equiv f_w$ on W or f has zeros in each punctured neighborhood of w.*

Proof. Assume $f \not\equiv f_w$ on W.

Since $f_w \neq 0$ on $W \setminus \{w\}$ and f_w has an essential singularity at w, $1/f_w$ is a well-defined holomorphic function on $W \setminus \{w\}$ which also has an essential singularity at w. Since $f - f_w$ is holomorphic and not identically zero on W, it follows that also

$$g := \frac{f_w - f}{f_w}$$

has an essential singularity at w. Since $f \not\equiv f_w$ on W, there is a punctured neighborhood of w where g has no zeros. By Picard's theorem this implies that $1 - g$ has zeros in any punctured neighborhood of w. Hence

$$\frac{f}{f_w} = 1 - \frac{f_w - f}{f_w} = 1 - g$$

has zeros in any punctured neighborhood of w. Since $f_w \neq 0$ on $W \setminus \{w\}$, this implies that f has zeros in any punctured neighborhood of w. □

2.8 Comments

The results of Sections 2.1–2.4 are well-known in the case of scalar functions. The proofs given here are straightforward generalizations of the proofs in the scalar case. In a large part, we follow the presentation of the corresponding scalar results in the first chapter of Hörmanders book [Ho].

The theorem on Runge approximation for invertible functions (Section 2.5) has also been well known for a long time, even in the case of several complex variables, but in the literature we did not find a direct proof for it in the case of one variable. (For several variables this result is widely published, but much more difficult.)

The material of the last two sections, in this form, probably appears here for the first time.

Chapter 3

Splitting and factorization with respect to a contour

This chapter contains mostly well-known material presented in a form needed for some of the further chapters. This material can not always be found concentrated in one place with complete proofs. The main theme in this chapter is to study continuous functions on a closed contour which admit an additive splitting as a sum or a product (with additional properties) of two functions; one continuous and analytic inside relative to the contour and the second outside. Not all continuous functions admit a splitting. We give here complete descriptions when continuous functions admit such representations and an example when this does not happen. We prove that functions from the algebras of Hölder, differentiable, and Wiener functions admit additive and multiplicative splittings inside these algebras under natural conditions. A local principle is also deduced.

3.1 Splitting with respect to a contour

In this section, E is a Banach space and $D_+ \subseteq \mathbb{C}$ is a bounded open set with piecewise $\mathcal{C}^1$-boundary Γ (Section 1.4.1), and $D_- := \mathbb{C} \setminus \overline{D}_+$.

3.1.1. Let $U \subseteq \mathbb{C}$ be an open set such that $\mathbb{C} \setminus U$ is bounded. We say that a function $f : U \cup \{\infty\} \to E$ is **holomorphic, continuous** etc. if f is holomorphic, continuous etc. on U and $f(1/z)$ is holomorphic, continuous etc. on

$$\left\{ z \in \mathbb{C} \;\middle|\; \frac{1}{z} \in U \right\} \cup \{0\}.$$

3.1.2 Definition. Let $f : \Gamma \to E$ be a continuous function. We say that f **splits (additively) with respect to** Γ if there exist functions $f_- : \overline{D}_- \cup \{\infty\} \to E$ and

$f_+ : \overline{D}_+ \to E$, where f_- is continuous on $\overline{D}_-$ and holomorphic in $D_- \cup \{\infty\}$ and f_+ is continuous on $\overline{D}_+$ and holomorphic in D_+, such that

$$f = f_+ + f_- \qquad \text{on } \Gamma . \tag{3.1.1}$$

The pair (f_+, f_-) or the expression $f = f_+ + f_-$ then will be called a **splitting of f with respect to Γ**. To underline the difference from Definition 3.7.1 below, in this case we say also that f **globally** splits, and (f_+, f_-) is a **global** splitting of f with respect to Γ.

3.1.3 Proposition. *If (f_+, f_-) is a splitting of a continuous function $f : \Gamma \to E$ with respect to Γ, then by adding a constant we can always achieve that $f_-(\infty) = 0$. With this additional property, the splitting (f_+, f_-) is uniquely determined by f.*

Proof. Indeed, let $(\widetilde{f}_+, \widetilde{f}_-)$ be a second splitting of f with $\widetilde{f}_-(\infty) = 0$. Then

$$f_+ - \widetilde{f}_+ = f_- - \widetilde{f}_- \qquad \text{on } \Gamma ,$$

and it follows from Theorem 1.5.4 that the function defined by

$$h = \begin{cases} f_+ - \widetilde{f}_+ & \text{on } \overline{D}_+ , \\ f_- - \widetilde{f}_- & \text{on } \overline{D}_- \cup \{\infty\} , \end{cases}$$

is a well-defined holomorphic function on $\mathbb{C} \cup \{\infty\}$ which vanishes at ∞. Hence, by Liouville's theorem, this function identically vanishes, i.e., $f_\pm = \widetilde{f}_\pm$. □

3.1.4. This uniquely determined splitting with $f_-(\infty) = 0$ will be referred to as the **splitting with respect to a contour vanishing at infinity**.

Not every continuous function $f : \Gamma \to E$ splits with respect to Γ. In Section 3.6 we give an example.

There are different additional conditions which ensure the existence of a splitting. For example, the class of Wiener functions:

3.1.5 Definition. Let $\mathbb{T} = \{z \in \mathbb{C} \mid |z| = 1\}$ be the unit circle. We denote by $W(E)$ the space of functions $f : \mathbb{T} \longrightarrow E$ of the form

$$f(z) = \sum_{n=-\infty}^{\infty} z^n f_n \qquad \text{with } \|f\|_W := \sum_{n=-\infty}^{\infty} \|f_n\| < \infty . \tag{3.1.2}$$

The functions in $W(E)$ are called **Wiener functions** with values in E.

3.1.6. Each E-valued Wiener function splits with respect to the unit circle. Indeed, if f is such a function and written in the form (3.1.2), then the pair (f_+, f_-) defined by

$$f_+(z) := \sum_{n=0}^{\infty} z^n f_n , \qquad |z| \le 1 ,$$

and

$$f_-(z) := \sum_{n=-\infty}^{-1} z^n f_n\,, \qquad |z| \geq 1\,,$$

is a splitting of f with respect to the unit circle, where

$$\|f_+\|_W \leq \|f\|_W \qquad \text{and} \qquad \|f_+\|_W \leq \|f\|_W. \tag{3.1.3}$$

3.2 Splitting and the Cauchy Integral

As in the previous section, E is a Banach space, $D_+ \subseteq \mathbb{C}$ is a bounded open set with piecewise $\mathcal{C}^1$-boundary Γ oriented by D_+ (Section 1.4.1), and $D_- = \mathbb{C} \setminus \overline{D}_+$.

3.2.1. Let $f : \Gamma \to E$ be a continuous function.

Then we set

$$\widehat{f}(z) = \frac{1}{2\pi i} \int_\Gamma \frac{f(\zeta)}{\zeta - z} d\zeta \qquad \text{for } z \in \mathbb{C} \setminus \Gamma. \tag{3.2.1}$$

This function $\widehat{f}$ will be called the **Cauchy integral** with respect to Γ of f.

Recall that, by Lemma 1.5.2, $\widehat{f}$ is holomorphic on $\mathbb{C} \setminus \Gamma$, where the complex derivative is given by

$$\widehat{f}'(z) = \frac{1}{2\pi i} \int_\Gamma \frac{f(\zeta)}{(\zeta - z)^2} d\zeta\,, \qquad z \in \mathbb{C} \setminus \Gamma. \tag{3.2.2}$$

Moreover, by this lemma,

$$\lim_{|z| \to \infty} \widehat{f}(z) = 0.$$

By Riemann's theorem on removable singularities 1.10.3, this implies that $\widehat{f}$ is holomorphic on $\big(\mathbb{C} \cup \{\infty\}\big) \setminus \Gamma$, where $\widehat{f}(\infty) = 0$ (in the sense as defined in Section 3.1.1).

3.2.2 Theorem. *Let $f : \Gamma \to E$ be continuous, and let*

$$\widehat{f}_+ := \widehat{f}\Big|_{D_+} \quad \textit{and} \quad \widehat{f}_- := \widehat{f}\Big|_{D_- \cup \{\infty\}}$$

be the two parts of the Cauchy integral (3.2.1) *of f. Then the following two conditions are equivalent:*

(i) *The function f splits with respect to Γ.*

(ii) *The function $\widehat{f}_+$ admits a continuous extension to $\overline{D}_+$, and $\widehat{f}_-$ admits a continuous extension to $\overline{D}_-$.*

In that case $f = \widehat{f}_+ - \widehat{f}_-$ on Γ, i.e., $(\widehat{f}_+, -\widehat{f}_-)$ is the splitting of f which vanishes at infinity (*Section* 3.2.1).

Proof. (i)⇒(ii): Let $f = f_+ + f_-$ be the splitting of f with $f_-(\infty) = 0$. As f_- is holomorphic at ∞ and $f_-(\infty) = 0$, then we have (Theorem 1.10.4)

$$f_-(\zeta) = O\left(\frac{1}{|\zeta|}\right) \qquad \text{for } |\zeta| \to \infty.$$

By Cauchy's integral theorem and the estimate established in Proposition 1.3.6, this implies that, for every fixed $z \in \mathbb{C} \setminus \Gamma$,

$$\lim_{R\to\infty} \int_{|\zeta|=R} \frac{f_-(\zeta)}{\zeta - z} d\zeta = 0. \tag{3.2.3}$$

We have to prove that $f_+ = \widehat{f}_+$ on D_+ and $f_- = \widehat{f}_-$ on D_-.

First let $z \in D_+$. If $0 < R < \infty$ is so large that $\overline{D}_+$ is contained in the disc $|\zeta| < R$, then, by Cauchy's integral theorem,

$$\int_\Gamma \frac{f_-(\zeta)}{\zeta - z} d\zeta = \int_{|\zeta|=R} \frac{f_-(\zeta)}{\zeta - z} d\zeta.$$

By (3.2.3) this implies that

$$\int_\Gamma \frac{f_-(\zeta)}{\zeta - z} d\zeta = 0.$$

As $f = f_+ + f_-$ on Γ and, by Cauchy's integral formula,

$$\frac{1}{2\pi i} \int_\Gamma \frac{f_+(\zeta)}{\zeta - z} d\zeta = f_+(z),$$

this further implies that

$$\widehat{f}_+(z) = \frac{1}{2\pi i} \int_\Gamma \frac{f(\zeta)}{\zeta - z} d\zeta = f_+(z).$$

Now let $z \in D_-$ be given. Then, by Cauchy's integral theorem,

$$\int_\Gamma \frac{f_+(\zeta)}{\zeta - z} d\zeta = 0. \tag{3.2.4}$$

First assume that z belongs to a bounded component of D_-. Let Γ_0 be the part of the boundary of Γ which is the boundary of this component (endowed with the orientation of Γ). Then, again by Cauchy's integral theorem,

$$\int_{\Gamma\setminus\Gamma_0} \frac{f_-(\zeta)}{\zeta - z} d\zeta = 0 \tag{3.2.5}$$

and, by Cauchy's integral formula,

$$\frac{1}{2\pi i} \int_{\Gamma_0} \frac{f_-(\zeta)}{\zeta - z} d\zeta = -f_-(z). \tag{3.2.6}$$

From (3.2.4)–(3.2.6) and the definition of $\widehat{f}_-$ we obtain

$$\widehat{f}_-(z) = \frac{1}{2\pi i}\int_\Gamma \frac{f(\zeta)}{\zeta - z}d\zeta = \frac{1}{2\pi i}\int_\Gamma \frac{f_+(\zeta)+f_-(\zeta)}{\zeta - z}d\zeta = -f_-(z).$$

Finally we consider the case when z belongs to the unbounded component of D_-. Let Γ_∞ be the part of Γ which is the boundary of this component. Then, by Cauchy's integral theorem,

$$\int_{\Gamma\setminus\Gamma_\infty} \frac{f(\zeta)}{\zeta - z}d\zeta = 0. \tag{3.2.7}$$

By Cauchy's integral formula, for all sufficiently large $R < \infty$,

$$f_-(z) = -\int_{\Gamma_\infty} \frac{f_-(\zeta)}{\zeta - z}d\zeta + \int_{|\zeta|=R} \frac{f_-(\zeta)}{\zeta - z}d\zeta.$$

Together with (3.2.3) and (3.2.7) this implies

$$f_-(z) = -\int_{\Gamma_\infty} \frac{f_-(\zeta)}{\zeta - z}d\zeta = -\int_\Gamma \frac{f_-(\zeta)}{\zeta - z}d\zeta.$$

By (3.2.4) and the definition of $\widehat{f}_-$, this further implies that

$$f_-(z) = -\int_\Gamma \frac{f_+(\zeta)+f_-(\zeta)}{\zeta - z}d\zeta = -\int_\Gamma \frac{f(\zeta)}{\zeta - z}d\zeta = -\widehat{f}_-(z).$$

(ii)⇒(i): Let $z_0 \in \Gamma$ and $\varepsilon > 0$ be given. We have to prove that

$$\left|\widehat{f}_+(z_0) - \widehat{f}_-(z_0) - f(z_0)\right| < \varepsilon, \tag{3.2.8}$$

where $\widehat{f}_+$ denotes the continuous extension of $\widehat{f}$ from D_+ to Γ, and $\widehat{f}_-$ denotes the continuous extension of $\widehat{f}$ from D_- to Γ. Let Γ_0 be the connected component of Γ with $z_0 \in \Gamma_0$ (endowed with the orientation of Γ), and let $\gamma : [a,b] \to \Gamma_0$ be a piecewise $\mathcal{C}^1$-parametrization of Γ (Def. 1.3.2). Since f, f_+ and f_- are continuous on Γ, we may assume that $z_0 = \gamma(t_0)$, where $a < t_0 < b$ is a smooth point of γ and $\gamma'(t_0) \neq 0$. Then we can find $\delta > 0$, $0 < c < 1$ and sequences $z_n^+ \in D_+$, $z_n^- \in D_-$, $n \in \mathbb{N}^*$, such that

$$|\gamma(t) - z_0| \geq c|t - t_0| \qquad \text{for } |t - t_0| < \delta, \tag{3.2.9}$$

$$\left|f(\gamma(t)) - f(z_0)\right| < \frac{c^2\,\varepsilon}{32 \max\limits_{|t-t_0|\leq\delta} |\gamma'(t)|} \qquad \text{for } |t - t_0| < \delta, \tag{3.2.10}$$

$$\lim_{n\to\infty} z_n^+ = \lim_{n\to\infty} z_n^- = z_0, \tag{3.2.11}$$

and, for all $n \in \mathbb{N}^*$,

$$\left|\gamma(t) - z_n^+\right|, \left|\gamma(t) - z_n^-\right| \geq \frac{1}{2}\left|\gamma(t) - z_0\right| + \frac{1}{4}\left|z_n^+ - z_n^-\right| \quad \text{if } |t - t_0| \leq \delta. \tag{3.2.12}$$

Since $0 < c < 1$, it follows from (3.2.9) and (3.2.12) that, for all $n \in \mathbb{N}^*$,

$$\left|\gamma(t) - z_n^+\right|, \left|\gamma(t) - z_n^-\right| \geq \frac{c}{4}\Big(\left|t - t_0\right| + \left|z_n^+ - z_n^-\right|\Big) \quad \text{if } |t - t_0| \leq \delta. \tag{3.2.13}$$

Set

$$\widehat{f}_+^{\Gamma_0}(z_n^+) = \frac{1}{2\pi i}\int_{\Gamma_0} \frac{f(\zeta)}{\zeta - z_n^+} d\zeta \quad \text{and} \quad \widehat{f}_-^{\Gamma_0}(z_n^-) = \frac{1}{2\pi i}\int_{\Gamma_0} \frac{f(\zeta)}{\zeta - z_n^-} d\zeta.$$

Then it follows from (3.2.11) that

$$\lim_{n\to\infty}\left(\widehat{f}_+(z_n^+) - \widehat{f}_-(z_n^-) - \left(\widehat{f}_+^{\Gamma_0}(z_n^+) - \widehat{f}_-^{\Gamma_0}(z_n^-)\right)\right)$$
$$\lim_{n\to\infty} \frac{1}{2\pi i}\int\limits_{\Gamma\setminus\Gamma_0} f(\zeta)\left(\frac{1}{\zeta - z_n^+} - \frac{1}{\zeta - z_n^-}\right) d\zeta = 0.$$

To prove (3.2.8), it is therefore sufficient to find n_ε such that

$$\left|\widehat{f}_+^{\Gamma_0}(z_n^+) - \widehat{f}_-^{\Gamma_0}(z_n^-) - f(z_0)\right| < \varepsilon \qquad \text{for } n \geq n_\varepsilon. \tag{3.2.14}$$

If $-\Gamma_0$ is the boundary of the unbounded component of D_- and n is sufficiently large, then, by Cauchy's integral formula and Cauchy's integral theorem,

$$\frac{1}{2\pi i}\int\limits_{\Gamma_0} \frac{f(z_0)}{\zeta - z_n^+} d\zeta = f(z_0) \quad \text{and} \quad \frac{1}{2\pi i}\int\limits_{\Gamma_0} \frac{f(z_0)}{\zeta - z_n^-} d\zeta = 0.$$

If $-\Gamma_0$ is the boundary of one of the bounded components of D_-, then for all sufficiently large n,

$$\frac{1}{2\pi i}\int\limits_{\Gamma_0} \frac{f(z_0)}{\zeta - z_n^+} d\zeta = 0 \quad \text{and} \quad \frac{1}{2\pi i}\int\limits_{\Gamma_0} \frac{f(z_0)}{\zeta - z_n^-} d\zeta = -f(z_0).$$

In both cases, for all sufficiently large n,

$$\frac{1}{2\pi i}\int\limits_{\Gamma_0} f(z_0)\left(\frac{1}{\zeta - z_n^+} - \frac{1}{\zeta - z_n^-}\right) d\zeta = f(z_0),$$

and therefore

$$\widehat{f}_+^{\Gamma_0}(z_n^+) - \widehat{f}_-^{\Gamma_0}(z_n^-) - f(z_0) = \frac{1}{2\pi i}\int_{\Gamma_0}\Big(f(\zeta) - f(z_0)\Big)\Big(\frac{1}{\zeta - z_n^+} - \frac{1}{\zeta - z_n^-}\Big)d\zeta$$

$$= \frac{1}{2\pi i}\int_a^b \Big(f\big(\gamma(t)\big) - f(z_0)\Big)\Big(\frac{1}{\gamma(t) - z_n^+} - \frac{1}{\gamma(t) - z_n^-}\Big)\gamma'(t)\,dt$$

$$= \frac{1}{2\pi i}\Big(I_1(n) + I_2(n)\Big),$$

where

$$I_1(n) := \int_{|t-t_0|\ge\delta} \Big(f\big(\gamma(t)\big) - f(z_0)\Big)\Big(\frac{1}{\gamma(t) - z_n^+} - \frac{1}{\gamma(t) - z_n^-}\Big)\gamma'(t)\,dt,$$

$$I_2(n) := \int_{|t-t_0|\le\delta} \Big(f\big(\gamma(t)\big) - f(z_0)\Big)\Big(\frac{1}{\gamma(t) - z_n^+} - \frac{1}{\gamma(t) - z_n^-}\Big)\gamma'(t)\,dt.$$

From (3.2.11) it follows that $\lim_{n\to\infty} I_1(n) = 0$. Therefore, to prove (3.2.14), it is sufficient to prove that

$$\big\|I_2(n)\big\| < \varepsilon \qquad \text{for all } n \in \mathbb{N}^*.$$

Let $n \in \mathbb{N}^*$ be given. Then it follows from (3.2.10) that

$$\big\|I_2(n)\big\| \le \frac{c^2\varepsilon}{32}\int_{|t-t_0|\le\delta}\left|\frac{1}{\gamma(t) - z_n^+} - \frac{1}{\gamma(t) - z_n^-}\right|dt$$

$$= \frac{c^2\varepsilon}{32}\int_{|t-t_0|\le\delta}\frac{|z_n^+ - z_n^-|}{|\gamma(t) - z_n^+|\,|\gamma(t) - z_n^-|}\,dt.$$

In view of (3.2.13), this further implies that

$$\big\|I_2(n)\big\| \le \frac{\varepsilon}{2}\big|z_n^+ - z_n^-\big|\int_{|t-t_0|\le\delta}\frac{dt}{\big(|t-t_0| + |z_n^+ - z_n^-|\big)^2}$$

$$= \varepsilon\big|z_n^+ - z_n^-\big|\int_0^\delta \frac{ds}{\big(s + |z_n^+ - z_n^-|\big)^2} < \varepsilon.$$

□

3.3 Hölder continuous functions split

As in the previous two sections, E is a Banach space, $D_+ \subseteq \mathbb{C}$ is a bounded open set with piecewise $\mathcal{C}^1$-boundary Γ oriented by D_+ (Section 1.4.1), and $D_- = \mathbb{C}\setminus\overline{D}_+$.

Here we prove that each Hölder continuous function $f : \Gamma \to E$ splits with respect to Γ.

3.3.1 Lemma. *Let $0 < \alpha < 1$. Then there exists a constant $C < \infty$ such that, for each Hölder-α continuous function $f : \Gamma \to E$ (Def. 2.1.6) the Cauchy-integral $\widehat{f}$, for all $z \in \mathbb{C} \setminus \Gamma$, satisfies the estimate*

$$\|\widehat{f}'(z)\| \le C\|f\|_{\Gamma,\alpha}\Big(\operatorname{dist}(z,\Gamma)\Big)^{\alpha-1} \tag{3.3.1}$$

where $\operatorname{dist}(z,\Gamma) := \min_{\zeta\in\Gamma}|\zeta - z|$, *and* $\|f\|_{\Gamma,\alpha}$ *is the Hölder norm introduced in Definition* 2.1.6.

Proof. Let a Hölder-α continuous function $f : \Gamma \to E$ be given.

For $z \in \mathbb{C}$ and $\varepsilon > 0$, we denote by $\Delta_\varepsilon(z)$ the open disc with radius ε centered at z.

It follows from (3.2.2) (and the estimate from Proposition 1.3.6) that, for each $z \in \mathbb{C}$ with $\operatorname{dist}(z,\Gamma) \ge 1$,

$$\|\widehat{f}'(z)\| \le \frac{|\Gamma|}{2\pi}\max_{\zeta\in\Gamma}\|f(\zeta)\|\Big(\operatorname{dist}(z,\Gamma)\Big)^{-2} \le \frac{|\Gamma|}{2\pi}\|f\|_{\Gamma,\alpha}\Big(\operatorname{dist}(z,\Gamma)\Big)^{\alpha-1}.$$

Moreover, it is clear that, for each neighborhood U of Γ, there exists a constant $C_U < \infty$ such that, for all $z \in \mathbb{C} \setminus U$ with $\operatorname{dist}(z,\Gamma) \le 1$,

$$\|\widehat{f}'(z)\| \le C_U\|f\|_{\Gamma,\alpha}\Big(\operatorname{dist}(z,\Gamma)\Big)^{\alpha-1}.$$

Therefore, it is sufficient to prove that, for each point $z_0 \in \Gamma$ there exists $\varepsilon_0 > 0$ such that (3.3.1) holds for all $z \in \Delta_{\varepsilon_0}(z_0) \setminus \Gamma$ with a constant $C < \infty$ which is independent of f.

Let $z_0 \in \Gamma$ be given, and let Γ_0 be the connected component of Γ with $z_0 \in \Gamma_0$. Choose a piecewise $\mathcal{C}^1$-parametrization $\gamma : [-3,3] \to \Gamma_0$ of Γ_0 with $\gamma(0) = z_0$ (Def. 1.3.2). Then it follows from the properties of γ listed in Definition 1.3.2 that there exist constants $0 < c_1 < C_1 < \infty$ such that

$$c_1|t-s| \le |\gamma(t)-\gamma(s)| \le C_1|t-s| \qquad \text{for all } -2 \le s,t \le 2. \tag{3.3.2}$$

Choose $0 < \varepsilon_0 < 1$ so small that

$$\varepsilon_0 < 2c_1 \tag{3.3.3}$$

and

$$\overline{\Delta}_{3\varepsilon_0}(z_0) \cap \big(\Gamma \setminus \gamma([-1,1])\big) = \emptyset. \tag{3.3.4}$$

Now let $z \in \Delta_{\varepsilon_0}(z_0)$ be given. Set $\varepsilon = \operatorname{dist}(z,\Gamma)$. As $z_0 \in \Gamma$, then $\varepsilon \le \varepsilon_0$. Take a point $z' \in \Gamma$ with $|z-z'| = \operatorname{dist}(z,\Gamma) = \varepsilon$. Since $z \in \Delta_{\varepsilon_0}(z_0)$ and $\varepsilon \le \varepsilon_0$, then $z' \in \Delta_{2\varepsilon_0}(z_0)$. Since $z' \in \Gamma \cap \Delta_{2\varepsilon_0}(z_0)$, it follows from (3.3.4) that $z' \in \gamma([-1,1])$.

Let $-1 \le t' \le 1$ be the parameter with $\gamma(t') = z'$ (which is uniquely determined by (3.3.2)).

As Γ is closed and therefore

$$\int_\Gamma \frac{f(z')}{(\zeta - z)^2} d\zeta = f(z') \int_\Gamma \frac{d\zeta}{(\zeta - z)^2} = 0,$$

it follows (see (3.2.2)) that

$$\widehat{f}'(z) = \frac{1}{2\pi i} \int_\Gamma \frac{f(\zeta) - f(z')}{(\zeta - z)^2} d\zeta. \tag{3.3.5}$$

Since $-1 \le t' \le 1$ and, by (3.3.3), $2\varepsilon/c_1 \le 2\varepsilon_0/c_1 < 1$, we have

$$-2 \le t' - \frac{2\varepsilon}{c_1} < t' + \frac{2\varepsilon}{c_1} < 2.$$

Therefore (3.3.5) can be written

$$\widehat{f}'(z) = I_1 + I_2 + I_3, \tag{3.3.6}$$

where

$$I_1 = \frac{1}{2\pi i} \int_{\Gamma \setminus \gamma([-2,2])} \frac{f(\zeta) - f(z')}{(\zeta - z)^2} d\zeta,$$

$$I_2 = \frac{1}{2\pi i} \int_{t'-2\varepsilon/c_1}^{t'+2\varepsilon/c_1} \frac{f\big(\gamma(t)\big) - f(z')}{(\gamma(t) - z)^2} \gamma'(t) dt,$$

$$I_3 = \frac{1}{2\pi i} \int_{-2}^{t'-2\varepsilon/c_1} \frac{f\big(\gamma(t)\big) - f(z')}{(\gamma(t) - z)^2} \gamma'(t) dt + \frac{1}{2\pi i} \int_{t'+2\varepsilon/c_1}^{2} \frac{f\big(\gamma(t)\big) - f(z')}{(\gamma(t) - z)^2} \gamma'(t) dt.$$

Since $|z - z_0| \le \varepsilon_0$ and, by (3.3.4), $|\zeta - z_0| \ge 3\varepsilon_0$ for $\zeta \in \Gamma \setminus \gamma([-2,2])$, we have

$$|\zeta - z| \ge 2\varepsilon_0 \qquad \text{for } \zeta \in \Gamma \setminus \gamma([-2,2]).$$

This implies (by Proposition 1.3.6)

$$\|I_1\| \le \frac{|\Gamma|}{4\pi\varepsilon_0^2} \max_{\zeta \in \Gamma} \|f(\zeta)\| \le \frac{|\Gamma|}{4\pi\varepsilon_0^2} \|f\|_{\Gamma,\alpha} \le \frac{|\Gamma|}{4\pi\varepsilon_0^2} \|f\|_{\Gamma,\alpha} \varepsilon^{\alpha-1}, \tag{3.3.7}$$

where the last inequality follows from $0 < \varepsilon < 1$.

Set

$$C_2 = \max_{-2 \le t \le 2} |\gamma'(t)|.$$

Since $|\gamma(t) - z| \geq \operatorname{dist}(z, \Gamma) = |z - z'| = \varepsilon$ for all $-3 \leq t \leq 3$, then it follows that

$$\|I_2\| \leq \frac{C_2}{2\pi\varepsilon^2} \int_{t'-2\varepsilon/c_1}^{t'+2\varepsilon/c_1} \big\|f(\gamma(t)) - f(z')\big\| dt. \tag{3.3.8}$$

As

$$\|f(\gamma(t)) - f(z')\| \leq \|f\|_{\Gamma,\alpha} |\gamma(t) - z'|^\alpha$$

and $z' = \gamma(t')$, it follows from (3.3.2) that

$$\|f(\gamma(t)) - f(z')\| \leq \|f\|_{\Gamma,\alpha} C_1 |t - t'|^\alpha \qquad \text{for all } -3 \leq t \leq 3.$$

Together with (3.3.8) this gives

$$\begin{aligned} \|I_2\| \leq \frac{C_2 \|f\|_{\Gamma,\alpha} C_1}{2\pi\varepsilon^2} \int_{t'-2\varepsilon/c_1}^{t'+2\varepsilon/c_1} |t - t'|^\alpha dt &\leq \frac{C_2 \|f\|_{\Gamma,\alpha} C_1}{2\pi\varepsilon^2} \frac{2}{1+\alpha} \left(\frac{2\varepsilon}{c_1}\right)^{1+\alpha} \\ &\leq \frac{C_2 C_1 2^{2+\alpha}}{2\pi(1+\alpha) c_1^{1+\alpha}} \|f\|_{\Gamma,\alpha} \varepsilon^{\alpha-1} \leq \frac{C_2 C_1 2}{c_1^2} \|f\|_{\Gamma,\alpha} \varepsilon^{\alpha-1}. \end{aligned} \tag{3.3.9}$$

For $|t - t'| \geq 2\varepsilon/c_1$, we have by (3.3.2)

$$\frac{1}{2} |\gamma(t) - \gamma(t')| \geq \frac{1}{2} c_1 |t - t'| \geq \varepsilon.$$

As $|\gamma(t') - z| = |z' - z| = \varepsilon$, this implies that, for $|t - t'| \geq 2\varepsilon/c_1$,

$$|\gamma(t) - z| \geq |\gamma(t) - \gamma(t')| - |\gamma(t') - z| = |\gamma(t) - \gamma(t')| - \varepsilon \geq \frac{1}{2} |\gamma(t) - \gamma(t')|.$$

Hence

$$I_3 \leq \int_{-2}^{t'-2\varepsilon/c_1} \frac{\big\|f(\gamma(t)) - f(z')\big\|}{|\gamma(t) - \gamma(t')|^2} \gamma'(t)\, dt + \int_{t'+2\varepsilon/c_1}^{2} \frac{\big\|f(\gamma(t)) - f(z')\big\|}{|\gamma(t) - \gamma(t')|^2} \gamma'(t)\, dt.$$

Taking into account that

$$\|f(\gamma(t)) - f(z')\| = \|f(\gamma(t)) - f(\gamma(t'))\| \leq \|f\|_{\Gamma,\alpha} |\gamma(t) - \gamma(t')|^\alpha$$

and the definition of C_2, this implies that

$$I_3 \leq \|f\|_{\Gamma,\alpha} C_2 \Bigg(\int_{-2}^{t'-2\varepsilon/c_1} |\gamma(t) - \gamma(t')|^{\alpha-2} dt + \int_{t'+2\varepsilon/c_1}^{2} |\gamma(t) - \gamma(t')|^{\alpha-2} dt \Bigg).$$

By (3.3.2) this further implies that

$$\begin{aligned} I_3 \leq &\|f\|_{\Gamma,\alpha} C_2 C_1 \Big(\int_{-2}^{t'-2\varepsilon/c_1} |t-t'|^{\alpha-2} dt + \int_{t'+2\varepsilon/c_1}^{2} |t-t'|^{\alpha-2} dt \Big) \\ \leq &2\|f\|_{\Gamma,\alpha} C_2 C_1 \int_{t'+2\varepsilon/c_1}^{\infty} |t-t'|^{\alpha-2} dt = 8\|f\|_{\Gamma,\alpha} C_2 C_1 \frac{1}{1-\alpha} \Big(\frac{2\varepsilon}{c_1}\Big)^{\alpha-1} \\ \leq &\frac{2C_2C_1c_1^{1-\alpha}}{1-\alpha} \|f\|_{\Gamma,\alpha} \varepsilon^{1-\alpha}. \end{aligned} \tag{3.3.10}$$

From (3.3.6), (3.3.7), (3.3.9) and (3.3.10) follows (3.3.1) if

$$C := \frac{|\Gamma|}{4\pi\varepsilon_0^2} + \frac{C_2 C_1 2}{c_1^2} + \frac{2C_2 C_1 c_1^{1-\alpha}}{1-\alpha}.$$

This completes the proof of Lemma 3.3.1. □

Now we can prove that any Hölder continuous function $f : \Gamma \to E$ splits with respect to Γ. Moreover, we prove the following stronger result:

3.3.2 Theorem. *Let $0 < \alpha < 1$, and let $f : \Gamma \to E$ be Hölder continuous with exponent α. Set $f_+ = \widehat{f}\big|_{D_+}$ and $f_- = -\widehat{f}\big|_{D_-}$ where $\widehat{f}$ is the Cauchy integral of f (Section 3.2.1). Then:*

(i) *f_+ admits a Hölder-α continuous extension to $\overline{D}_+$, and f_- admits a Hölder-α continuous extension to $\overline{D}_- \cup \{\infty\}$.*

(ii) *If we denote these extensions also by f_+ and f_-, then*

$$f = f_+ + f_- \qquad \text{on } \Gamma. \tag{3.3.11}$$

Proof. By Theorem 3.2.2 we only have to prove part (i). For that it is sufficient to find a neighborhood Θ of Γ and a constant $C < \infty$ such that

$$\|f_+(z_1) - f_+(z_2)\| \leq C\|f\|_{\Gamma,\alpha}|z_1 - z_2|^\alpha \tag{3.3.12}$$

for all $z_1, z_2 \in \Theta \cap D_+$ and

$$\|f_-(z_1) - f_-(z_2)\| \leq C\|f\|_{\Gamma,\alpha}|z_1 - z_2|^\alpha \tag{3.3.13}$$

for all $z_1, z_2 \in \Theta \cap D_-$.

The proofs of these two estimates are analogous, and we may restrict ourselves to (3.3.12). Let z_0 be an arbitrary point in Γ. It is sufficient to find constants $\varepsilon_0 > 0$ and $C < \infty$ such that (3.3.12) holds true for all $z_1, z_2 \in \Delta_{\varepsilon_0}(z_0) \cap D_+$, where we again denote by $\Delta_\varepsilon(z)$ the open disc with radius ε centered at z.

By Lemma 3.3.1, there is a constant $C_0 < \infty$ with

$$\|f'_+(z)\| \leq C_0 \|f\|_{\Gamma,\alpha} \Big(\operatorname{dist}(z,\Gamma)\Big)^{\alpha-1}, \qquad z \in D_+. \tag{3.3.14}$$

Since Γ is piecewise $\mathbb{C}^1$, we can find constants $0 < \varepsilon_1$, $c < 1$ and a complex number v with $|v| = 1$ such that, for each $z \in D_+ \cap \Delta_{\varepsilon_0}(z_0)$,

$$\Delta_{ct}(z + tv) \subseteq D_+ \qquad \text{for all } 0 \leq t \leq \varepsilon_1. \tag{3.3.15}$$

Set

$$\varepsilon_0 = c\varepsilon_1/4 \qquad \text{and} \qquad C = \frac{C_0 2^{\alpha+1}}{c\alpha} + C_0.$$

We claim that, with this choice of ε_0 and C, estimate (3.3.12) holds true for all $z_1, z_2 \in \Delta_{\varepsilon_0}(z_0) \cap D_+$.

Let $z_1, z_2 \in \Delta_{\varepsilon_0}(z_0) \cap D_+$ be given. Set $\varepsilon = |z_1 - z_2|$. Then $2\varepsilon/c \leq 4\varepsilon_0/c = \varepsilon_1$ and it follows from (3.3.15) that

$$z_1 + tv \in D_+ \quad \text{and} \quad z_2 + tv \in D_+ \qquad \text{for all } 0 \leq t \leq \frac{2\varepsilon}{c},$$

and

$$(1-t)\left(z_1 + \frac{2\varepsilon}{c}v\right) + t\left(z_2 + \frac{2\varepsilon}{c}v\right) \in D_+ \qquad \text{for all } 0 \leq t \leq \frac{2\varepsilon}{c}.$$

Therefore

$$\begin{aligned}
\big\|f_+(z_1) - f_+(z_2)\big\| &\leq \Big\|f_+(z_1) - f_+\Big(z_1 + \frac{2\varepsilon}{c}v\Big)\Big\| + \Big\|f_+(z_2) - f_+\Big(z_2 + \frac{2\varepsilon}{c}v\Big)\Big\| \\
&\quad + \left\|f_+\left(z_1 + \frac{2\varepsilon}{c}v\right) - f_+\left(z_2 + \frac{2\varepsilon}{c}v\right)\right\| \\
&= \left\|\int_0^{2\varepsilon/c} f'_+\Big(z_1 + tv\Big)v\,dt\right\| + \left\|\int_0^{2\varepsilon/c} f'_+\Big(z_2 + tv\Big)v\,dt\right\| \\
&\quad + \left\|\int_0^1 f'_+\Big((1-t)\Big(z_1 + \frac{2\varepsilon}{c}v\Big) + t\Big(z_2 + \frac{2\varepsilon}{c}v\Big)\Big)(z_2 - z_1)\,dt\right\|.
\end{aligned}$$

Since $|v| = 1$ and $|z_2 - z_1| = \varepsilon$, this implies that

$$\begin{aligned}
\|f_+(z_1) - f_+(z_2)\| &\leq \int_0^{2\varepsilon/c} \Big\|f'_+\Big(z_1 + tv\Big)\Big\|\,dt + \int_0^{2\varepsilon/c} \Big\|f'_+\Big(z_2 + tv\Big)\Big\|\,dt \\
&\quad + \varepsilon \int_0^1 \Big\|f'_+\Big((1-t)\Big(z_1 + \frac{2\varepsilon}{c}v\Big) + t\Big(z_2 + \frac{2\varepsilon}{c}v\Big)\Big)\Big\|dt.
\end{aligned} \tag{3.3.16}$$

It follows from (3.3.14) and (3.3.15) that

$$\Big\|f_+'\Big(z_j+tv\Big)\Big\| \le C_0\|f\|_{\Gamma,\alpha}(ct)^{\alpha-1} \qquad \text{for } 0\le t\le \frac{2\varepsilon}{c} \text{ and } j=1,2.$$

Hence

$$\int_0^{2\varepsilon/c} \Big\|f_+'\Big(z_j+tv\Big)\Big\|\,dt \le C_0c^{\alpha-1}\|f\|_{\Gamma,\alpha}\int_0^{2\varepsilon/c} t^{\alpha-1}dt = \frac{C_0 2^\alpha}{c\alpha}\|f\|_{\Gamma,\alpha}\varepsilon^\alpha \tag{3.3.17}$$

for $j=1,2$. Moreover it follows from (3.3.15) that

$$\Delta_{2\varepsilon}\Big(z_2+\frac{2\varepsilon}{c}v\Big)\subseteq D_+.$$

As

$$\Big\|\Big(z_1+\frac{2\varepsilon}{c}v\Big)-\Big(z_2+\frac{2\varepsilon}{c}v\Big)\Big\| = |z_1-z_2| = \varepsilon$$

and therefore

$$(1-t)\Big(z_1+\frac{2\varepsilon}{c}v\Big)+t\Big(z_2+\frac{2\varepsilon}{c}v\Big)\in\overline{\Delta}_\varepsilon\Big(z_2+\frac{2\varepsilon}{c}v\Big), \qquad 0\le t\le 1,$$

this yields

$$(1-t)\Big(z_1+\frac{2\varepsilon}{c}v\Big)+t\Big(z_2+\frac{2\varepsilon}{c}v\Big)\in D_+, \qquad 0\le t\le 1,$$

and

$$\operatorname{dist}\bigg((1-t)\Big(z_1+\frac{2\varepsilon}{c}v\Big)+t\Big(z_2+\frac{2\varepsilon}{c}v\Big),\,\Gamma\bigg)\ge\varepsilon, \qquad 0\le t\le 1.$$

Together with (3.3.14) this implies that

$$\varepsilon\int_0^1 \bigg\|f_+'\bigg((1-t)\Big(z_1+\frac{2\varepsilon}{c}v\Big)+t\Big(z_2+\frac{2\varepsilon}{c}v\Big)\bigg)\bigg\|dt \le C_0\|f\|_{\Gamma,\alpha}\varepsilon^\alpha. \tag{3.3.18}$$

Estimate (3.3.12) now follows from (3.3.16), (3.3.17) and (3.3.18). □

3.3.3 Corollary. *Let $P\subseteq\mathbb{C}$ be a set such that, in each connected component of D_+ and in each connected component of D_-, lies at least one point from P. Then any continuous function $f:\Gamma\to E$ can be approximated uniformly on Γ by holomorphic functions defined in $\mathbb{C}\setminus P$.*

Proof. Since any continuous function on Γ can be approximated uniformly on Γ by $\mathcal{C}^\infty$-functions, we may assume that f is of class $\mathbb{C}^\infty$. Then, by Theorem 3.3.2, there exists a global splitting (f_+,f_-) of f with respect to Γ. It remains to apply the Runge approximation Theorem 2.2.2 to f_+ and f_-. □

3.3.4. From the proof of Theorem 3.3.2 it is clear that the constant C in (3.3.12) and (3.3.13) is independent of f. Note also the following corollary of Theorem 3.3.2: If $f_+ : \overline{D}_+ \to E$ ($f : \overline{D}_-\{\infty\} \to E$) is a continuous function which is Hölder-α continuous on Γ and holomorphic in D_+ ($D_- \cup \{\infty\}$), then this function is automatically also Hölder-α continuous on $\overline{D}_+$ ($D_- \cup \{\infty\}$).

3.4 The splitting behavior of differentiable functions

3.4.1 Definition ($\mathcal{C}^k$-contours). Let $k \in \mathbb{N}^*$.

A set $\Gamma \subseteq \mathbb{C}$ is called a **closed connected $\mathcal{C}^k$-contour** if there exist real numbers $a < b$ and a $\mathcal{C}^k$-function $\gamma : [a,b] \to \mathbb{C}$ with $\Gamma = \gamma\big([a,b]\big)$ such that

(i) $\gamma'(t) \neq 0$ for $a \leq t \leq b$;

(ii) $\gamma(t) \neq \gamma(s)$ for $a \leq t, s < b$ with $t \neq s$;

(iii) $\gamma^{(n)}(b) = \gamma^{(n)}(a)$ for $0 \leq n \leq k$.

The function γ then is called a **$\mathcal{C}^k$-parametrization** of Γ.

By a (not necessarily connected) **closed $\mathcal{C}^k$-contour** we mean the union of a finite number of closed connected $\mathcal{C}^k$-contours.

We shall say that an open set $D \subseteq \mathbb{C}$ has a **$\mathcal{C}^k$-boundary** if the boundary of D (in $\mathbb{C}$) is a closed $\mathcal{C}^k$-contour and each point of this boundary is also a boundary point of $\mathbb{C} \setminus \overline{D}$.

3.4.2. Let $\Gamma \subseteq \mathbb{C}$ be a closed $\mathcal{C}^k$-contour, $k \in \mathbb{N}^*$. Then, by the inverse function theorem, for each $z_0 \in \Gamma$, there is a neighborhood U of z_0 and a $\mathcal{C}^k$-diffeomorphism Φ from U onto an open set $V \subseteq \mathbb{C}$ such that

$$\Phi(U \cap \Gamma) = \Big\{ z \in V \,\Big|\, \operatorname{Im} z = 0 \Big\}.$$

3.4.3 Definition. Let $\Gamma \subseteq \mathbb{C}$ be a closed $\mathcal{C}^k$-contour, $k \in \mathbb{N}^*$, let E be a Banach space, and let $f : \Gamma \to E$ be a function. It follows from the observation in the preceding Section 3.4.2 that then the following two conditions are equivalent:

(a) If $\gamma : [a,b] \to \Gamma_0$ is a $\mathcal{C}^k$-parametrization of a connected component Γ_0 of Γ, then the composition $f \circ \gamma$ is k times continuously differentiable on $[a,b]$.

(b) The function f admits an extension to a neighborhood of Γ which is of class $\mathcal{C}^k$.

If these two equivalent conditions are satisfied, then f is called **of class $\mathcal{C}^k$** on Γ.

If f is of class $\mathcal{C}^1$ on Γ, then, for each $\zeta \in \Gamma$, we define

$$f'(\zeta) = \frac{\big(f \circ \gamma\big)'\big(\gamma^{-1}(\zeta)\big)}{\gamma'\big(\gamma^{-1}(\zeta)\big)}, \tag{3.4.1}$$

where $\gamma : [a,b] \to \mathbb{C}$ is an arbitrary $\mathcal{C}^k$-parametrization of the connected component of Γ which contains the point ζ. It follows from the chain rule that this definition does not depend on the choice of γ. The continuous function $f' : \Gamma \to E$ defined in this way will be called the **derivative of f with respect to** Γ. Note that if f is holomorphic in a neighborhood of Γ, then this is the restriction to Γ of the complex derivative of f.

We write also $f^{(1)}$ for f' and $f^{(0)}$ for f.

If f is of class $\mathcal{C}^k$ on Γ, then we define $f^{(2)}, \ldots, f^{(k)}$, by setting

$$f^{(n)} = \left(f^{(n-1)}\right)' \qquad \text{for } 2 \le n \le k.$$

The function f is called **of class $\mathcal{C}^{k+\alpha}$ on** Γ, $0 < \alpha < 1$ if f is of class $\mathcal{C}^k$ on Γ, and $f^{(k)}$ is Hölder continuous with exponent α on Γ.

3.4.4 Definition. Let $D_+ \subseteq \mathbb{C}$ be a bounded open set with $\mathcal{C}^k$-boundary Γ, $k \in \mathbb{N}^*$, let $D_- := \mathbb{C} \setminus \overline{D}_+$, let E be a Banach space, and let $0 < \alpha < 1$.

If $f : D_- \to E$ is a holomorphic function, then we say that f is of **class $\mathcal{C}^{k+\alpha}$ on $\overline{D}_-$** if all complex derivatives $f^{(n)}$, $0 \le n \le k$, (which are well defined on D_-) admit continuous extensions to $\overline{D}_-$, where $f^{(k)}$ is Hölder continuous with exponent α on $\overline{D}_-$. If this is the case, then these extensions will be denoted also by $f^{(n)}$. Note that then $f^{(n)}\big|_\Gamma$ is of class $\mathcal{C}^{k-n}$ on Γ (in the sense of the preceding definition) and $\left(f^{(n)}\big|_\Gamma\right)' = f^{(n+1)}\big|_\Gamma$ for $0 \le n \le k-1$.

Correspondingly we define what it means that a holomorphic function $f : D_+ \to E$ is of **class $\mathcal{C}^{k+\alpha}$ on $\overline{D}_+$**.

It is the aim of the present section to prove the following

3.4.5 Theorem. *Let $D_+ \subseteq \mathbb{C}$ be a bounded open set with $\mathcal{C}^k$-boundary Γ, $k \in \mathbb{N}^*$, let $D_- := \mathbb{C} \setminus \overline{D}_+$, let E be a Banach space, and let $f : \Gamma \to E$ be of class $\mathcal{C}^{k+\alpha}$ on Γ, $0 < \alpha < 1$. Then f splits with respect to Γ (Def. 3.1.2).*

Moreover, if $f = f_+ + f_-$ is an arbitrary splitting of f with respect to Γ, then f_+ is of class $\mathcal{C}^{k+\alpha}$ on $\overline{D}_+$, and f_- is of class $\mathcal{C}^{k+\alpha}$ on $\overline{D}_-$.

That then f splits with respect to Γ, we know already from Theorem 3.3.2. To prove the additional assertion, we begin with the following

3.4.6 Lemma. *Let Γ be an oriented closed $\mathcal{C}^k$-contour. Let E be a Banach space, let $f : \Gamma \to E$ be of class $\mathcal{C}^k$, and let*

$$\widehat{f}(z) := \frac{1}{2\pi i}\int_\Gamma \frac{f(\zeta)}{\zeta - z}d\zeta, \qquad z \in \mathbb{C} \setminus \Gamma,$$

be the Cauchy integral of f with respect to Γ (which is a holomorphic function, by Lemma 1.5.2). Then, for $0 \le n \le k$,

$$\left(\widehat{f}\,\right)^{(n)}(z) = \frac{1}{2\pi i}\int_\Gamma \frac{f^{(n)}(\zeta)}{\zeta - z}d\zeta \qquad \textit{for all } z \in \mathbb{C} \setminus \Gamma. \tag{3.4.2}$$

Proof. For the proof we may assume that Γ is connected. Then we have a positively oriented $\mathcal{C}^k$-parametrization $\gamma : [a, b] \to \Gamma$ of Γ.

For $n = 0$, (3.4.2) holds by definition. Assume (3.4.2) is already proved for some n with $0 \le n \le k - 1$, and let $z \in \mathbb{C} \setminus \Gamma$ be given. Then it follows from Lemma 1.5.2 that

$$\begin{aligned}(\widehat{f}\,)^{(n+1)}(z) &= \frac{1}{2\pi i} \int_\Gamma \frac{f^{(n)}(\zeta)}{(\zeta - z)^2} d\zeta = \frac{1}{2\pi i} \int_a^b \frac{f^{(n)}(\gamma(t))}{(\gamma(t) - z)^2} \gamma'(t) dt \\ &= -\frac{1}{2\pi i} \int_a^b (f^{(n)} \circ \gamma)(t) \, \frac{d}{dt} \frac{1}{\gamma(t) - z} \, dt.\end{aligned}$$

As $\gamma(a) = \gamma(b)$, integrating by parts, this implies

$$\begin{aligned}(\widehat{f}\,)^{(n+1)}(z) &= \frac{1}{2\pi i} \int\limits_a^b \frac{(f^{(n)} \circ \gamma)'(t)}{\gamma(t) - z} \, dt = \frac{1}{2\pi i} \int\limits_a^b \frac{\frac{(f^{(n)} \circ \gamma)'(t)}{\gamma'(t)}}{\gamma(t) - z} \, \gamma'(t) \, dt \\ &= \frac{1}{2\pi i} \int\limits_\Gamma \frac{\frac{(f^{(n)} \circ \gamma)'(\gamma^{-1}(\zeta))}{\gamma'(\gamma^{-1}(\zeta))}}{\zeta - z} \, d\zeta = \frac{1}{2\pi i} \int\limits_\Gamma \frac{f^{(n+1)}(\zeta)}{\zeta - z} d\zeta.\end{aligned}$$

□

Proof of Theorem 3.4.5. Let

$$(\widehat{f})_+(z) := \frac{1}{2\pi i} \int_\Gamma \frac{f(\zeta)}{\zeta - z} d\zeta, \ z \in D_+, \quad (\widehat{f})_-(z) := \frac{1}{2\pi i} \int_\Gamma \frac{f(\zeta)}{\zeta - z} d\zeta, \ z \in D_-,$$

be the two parts of the Cauchy integral of f. By Theorem 3.2.2, we only have to prove that $(\widehat{f})_\pm$ is of class $\mathcal{C}^{k+\alpha}$ on $\overline{D}_\pm$. Since the functions $(\widehat{f})_\pm$ are holomorphic in $D_\pm$, this means that each of the functions $(\widehat{f})_\pm^{(n)}$, $1 \le n \le k$, (here $(\widehat{f})_\pm^{(n)}$ denotes the n-th complex derivative of $(\widehat{f})_\pm$ on $D_\pm$) is Hölder-α continuous on $\overline{D}_\pm$.

Let $f^{(n)}$, $1 \le n \le k$, be the n-th derivative of f with respect to Γ (Def. 3.4.3), and let

$$(\widehat{f^{(n)}})_+(z) := \frac{1}{2\pi i} \int_\Gamma \frac{f(\zeta)}{\zeta - z} d\zeta, \ z \in D_+, \quad (\widehat{f^{(n)}})_-(z) := \frac{1}{2\pi i} \int_\Gamma \frac{f(\zeta)}{\zeta - z} d\zeta, \ z \in D_-,$$

be the two parts of the Cauchy integral of $f^{(n)}$. Since, by hypothesis, the functions $f^{(n)}$, $1 \le n \le k$, are Hölder-α continuous on Γ, it follows from Theorem 3.3.2 that each of the functions $\big(\widehat{f^{(n)}}\big)_\pm$, $1 \le n \le k$, admits a Hölder-α continuous extension to $\overline{D}_\pm$.

As, by Lemma 3.4.6, $(\widehat{f})_\pm^{(n)}(z) = \big(\widehat{f^{(n)}}\big)_\pm(z)$, $z \in D_\pm$, $1 \le n \le k$, this completes the proof. □

3.5 Approximation of Hölder continuous functions

In this section, $D_+ \subseteq \mathbb{C}$ is a bounded open set with piecewise $\mathcal{C}^1$-boundary Γ, $D_- := \mathbb{C} \setminus \overline{D}_+$, and E is a Banach space.

For $0 < \alpha < 1$, the space of scalar Hölder-α continuous functions on Γ is not separable. Therefore, it is impossible to approximate such functions in the norm $\|\cdot\|_{\Gamma,\alpha}$ (Def. 2.1.6) by functions which are holomorphic in a neighborhood of Γ. However, for $0 \le \beta < \alpha$, this is possible with respect to the norm $\|\cdot\|_{\Gamma,\beta}$. In the present section we prove this. Moreover, if Γ is a $\mathcal{C}^k$-contour, $k \in \mathbb{N}^*$, then we obtain the corresponding fact for functions which are of class $\mathcal{C}^{k+\alpha}$ on Γ (Def. 3.4.3). We prove:

3.5.1 Theorem. *Let $0 \le \beta < \alpha < 1$, $k \in \mathbb{N}$ and let $f : \Gamma \to E$ be a function.*

If $k = 0$, then we assume that f is Hölder continuous with exponent α.

If $k \ge 1$, then we assume that Γ is of class $\mathbb{C}^k$ and f is of class $\mathcal{C}^{k+\alpha}$ on Γ (Def. 3.4.3).

Then, for each $\varepsilon > 0$, there exist a neighborhood U of Γ and a holomorphic function $\widetilde{f} : U \to E$ such that, for all $0 \le n \le k$, (see Def. 2.1.6 for $\|\cdot\|_{\Gamma,\beta}$)

$$\|f^{(n)} - \widetilde{f}^{(n)}\|_{\Gamma,\beta} < \varepsilon. \tag{3.5.1}$$

Here $f^{(n)}$ denotes the n-th complex derivative of f with respect to Γ (Def. 3.4.3) and $\widetilde{f}^{(n)}$ is the n-th complex derivative of $\widetilde{f}$ (as a holomorphic function on U).

In the proof of this theorem we use the following simple lemma:

3.5.2 Lemma. *Let $\omega \subseteq K \subseteq \mathbb{C}$ be two compact sets such that, for some vector $\theta \in \mathbb{C}$, $|\theta| = 1$, and some $\varepsilon_0 > 0$,*

$$\omega \subseteq K + \varepsilon\theta \qquad \text{if } 0 \le \varepsilon \le \varepsilon_0.$$

Let $0 \le \beta < \alpha < 1$, and let $f : K \to E$ be Hölder continuous with exponent α. Set

$$f_\varepsilon(z) = f(z - \varepsilon\theta) \qquad \text{for } z \in K + \varepsilon\theta \text{ and } 0 \le \varepsilon \le \varepsilon_0.$$

Then

$$\lim_{\varepsilon \to 0} \|f - f_\varepsilon\|_{\omega,\beta} = 0.$$

For the definition of $\|\cdot\|_{\omega,\beta}$, see Def. 2.1.6.

Proof. Let $0 \le \varepsilon < \varepsilon_0$ be given. As $|\theta| = 1$ and $z, z - \varepsilon\theta \in K$ for $z \in \omega$, then

$$\|f - f_\varepsilon\|_{\omega,0} = \max_{z\in\omega} \|f(z) - f(z - \varepsilon\theta)\| \le \|f\|_{K,\alpha}\, \varepsilon^\alpha.$$

Moreover, if $z, w \in \omega$ with $\varepsilon \le |z - w|$, then

$$\begin{aligned}\big\|f(z) - f_\varepsilon(z) - \big(f(w) - f_\varepsilon(w)\big)\big\| &\le \big\|f(z) - f(z - \varepsilon\theta)\big\| + \big\|f(w) - f(w - \varepsilon\theta)\big\| \\ &\le 2\|f\|_{K,\alpha}\, \varepsilon^\alpha \le 2\|f\|_{K,\alpha}\, \varepsilon^{\alpha-\beta}|z - w|^\beta,\end{aligned}$$

and if $z, w \in \omega$ with $\varepsilon \geq |z-w|$, then also

$$\begin{aligned}\big\|f(z) - f_\varepsilon(z) - \big(f(w) - f_\varepsilon(w)\big)\big\| &\leq \big\|f(z) - f(w)\big\| + \big\|f(w - \varepsilon\theta_-) - f(z - \varepsilon\theta_-)\big\| \\ &\leq 2\|f\|_{K,\alpha}|z-w|^\alpha \leq 2\|f\|_{K,\alpha}\, \varepsilon^{\alpha-\beta}|z-w|^\beta.\end{aligned}$$

Hence

$$\|f - f_\varepsilon\|_{\omega,\beta} \leq \|f\|_{K,\alpha}\big(\varepsilon^\alpha + 2\varepsilon^{\alpha-\beta}\big).$$ □

Proof of Theorem 3.5.1. By Theorem 3.3.2, $f = f_+ + f_-$ on Γ, where

$$f_+(z) = \frac{1}{2\pi i}\int_\Gamma \frac{f(\zeta)}{\zeta - z}d\zeta\ , z \in D_+, \quad f_-(z) = -\frac{1}{2\pi i}\int_\Gamma \frac{f(\zeta)}{\zeta - z}d\zeta\ , z \in D_-, \tag{3.5.2}$$

and the functions $f_\pm$ are Hölder-α continuous on $\overline{D}_\pm$. If $k \geq 1$ (and hence, by hypotheses of the theorem, Γ is of class $\mathcal{C}^k$), then, by Theorem 3.4.5, these functions are even of class $\mathcal{C}^{k+\alpha}$.

It is now sufficient to approximate each of the functions f_+ and f_- separately. Since the proofs are the same, we restrict ourselves to the function f_+. Note that the following arguments are parallel to the proof of the Mergelyan approximation Theorem 2.2.1.

Take a finite number of real non-negative C^∞ functions $\chi_1, \dots, \chi_N$ on $\mathbb{C}$ with sufficiently small supports $\operatorname{supp}\chi_j$ (how small, we say below) such that $\sum_{j=1}^N \chi_j = 1$ in some neighborhood of ∂D. Set

$$f_{+,j}(z) = \frac{1}{2\pi i}\int_\Gamma \frac{\chi_j(\zeta)f(\zeta)}{\zeta - z}d\zeta, \quad z \in \mathbb{C}\setminus(\operatorname{supp}\chi_j \cap \Gamma), \quad 1 \leq j \leq N.$$

Again by theorems 3.3.2 and 3.4.5, the functions $f_{+,j}$ admit extensions of class $\mathcal{C}^\alpha$ from D_+ to $\overline{D}_+$, which are even of class $\mathcal{C}^{k+\alpha}$ if $k \geq 1$. We denote these extensions by $f_{+,j}^{\overline{D}_+}$. Then it follows from the first equation in (3.5.2) that

$$f(z) = f_{+,1}^{\overline{D}_+}(z) + \ldots + f_{+,n}^{\overline{D}_+}(z) \qquad \text{for all } z \in \overline{D}_+.$$

Since Γ is piecewise C^1 and each $f_{+j}^{\overline{D}_+}$ extends to a holomorphic function outside $\operatorname{supp}\chi_j \cap \Gamma$, now we can choose the supports $\operatorname{supp}\chi_j$ so small that we can apply Lemma 3.5.2 to this situation: By small shifts, for each j, we can find a bounded neighborhood U_j of Γ and a function $\widetilde{f}_j \in \mathcal{O}^E(U_j)$ such that

$$\big\|\big(f_{+,j}^{\overline{D}_+}\big)^{(n)} - \widetilde{f}_j^{(n)}\big\|_{U_j,\beta} < \frac{\varepsilon}{N} \quad \text{for } 0 \leq n \leq k.$$

Setting $U = U_1 \cap \ldots \cap U_N$ and $\widetilde{f} = \widetilde{f}_1 + \ldots + \widetilde{f}_N$, we complete the proof. □

3.6 Example: A non-splitting continuous function

Let $\mathbb{T}$ be the unit circle in the complex plane. In this section, we construct a continuous function $\Omega : \mathbb{T} \to \mathbb{C}$ which does not split with respect to $\mathbb{T}$. We begin with the following lemma of Abel: [1]

3.6.1 Lemma. *Let $[a, b]$ be a real interval, and let $\varphi_n : [a, b] \to \mathbb{C}$, $n \in \mathbb{N}$, be a sequence of continuous functions such that*

$$C := \sup_{a \le x \le b, n \in \mathbb{N}} \left| \sum_{k=0}^{n} \varphi_k(x) \right| < \infty. \tag{3.6.1}$$

Further, let α_n, $n \in \mathbb{N}$, be a monotonicly decreasing sequence of real numbers tending to zero. Then the series of functions

$$\sum_{k=0}^{\infty} \alpha_k \varphi_k \tag{3.6.2}$$

converges uniformly on $[a, b]$ to a continuous function $S : [a, b] \to \mathbb{C}$ such that

$$\max_{a \le x \le b} |S(x)| \le C\alpha_0. \tag{3.6.3}$$

Proof. Set

$$S_n = \sum_{k=0}^{n} \alpha_k \varphi_k \qquad \text{and} \qquad \Phi_n = \sum_{k=0}^{n} \varphi_k.$$

Then $\varphi_k = \Phi_k - \Phi_{k-1}$ for $k \ge 1$ and therefore

$$\begin{aligned} S_n = \alpha_0 \varphi_0 + \sum_{k=1}^{n} \alpha_k (\Phi_k - \Phi_{k-1}) &= \sum_{k=0}^{n} \alpha_k \Phi_k - \sum_{k=1}^{n} \alpha_k \Phi_{k-1} \\ &= \sum_{k=0}^{n} \alpha_k \Phi_k - \sum_{k=0}^{n-1} \alpha_{k+1} \Phi_k = \alpha_n \Phi_n + \sum_{k=0}^{n-1} (\alpha_k - \alpha_{k+1}) \Phi_k. \end{aligned} \tag{3.6.4}$$

Note that by (3.6.1)

$$\max_{a \le x \le b, n \in \mathbb{N}} |\Phi_n(x)| \le C. \tag{3.6.5}$$

As $\lim_{n\to\infty} \alpha_n = 0$, this implies that

$$\lim_{n\to\infty} \max_{a \le x \le b} |\alpha_n \Phi_n(x)| = 0. \tag{3.6.6}$$

[1] Our source for this section is the book [Bar]. There this lemma is called *Abel's lemma*, in distinction to the present book, where by *Abel's lemma* we mean Theorem 1.8.2.

Since $\alpha_k - \alpha_{k+1} \geq 0$, (3.6.5) moreover yields

$$\sum_{k=0}^{n-1} \max_{a\leq x\leq b} \Big|(\alpha_k - \alpha_{k+1})\Phi_k(x)\Big| \leq C \sum_{k=0}^{n-1}(\alpha_k - \alpha_{k+1}) = C(\alpha_0 - \alpha_n).$$

Using again that $\lim_{n\to\infty} \alpha_n = 0$, this implies that the series

$$\sum_{k=0}^{\infty}(\alpha_k - \alpha_{k+1})\Phi_k$$

converges uniformly on $[a, b]$ to a continuous function $S : [a, b] \to \mathbb{C}$ satisfying (3.6.3). Together with (3.6.4) and (3.6.2) this further implies that the series (3.6.2) converges uniformly on $[a, b]$ to S. □

Further we need the following lemma:

3.6.2 Lemma. *Set $D_n(x) = \sum_{k=1}^{n} \sin(kx)$ for $x \in \mathbb{R}$ and $n \in \mathbb{N}^*$. Then*

$$|D_n(x)| \leq \frac{4}{|x|} \qquad \text{for all } 0 < |x| \leq \pi \text{ and } n \in \mathbb{N}^*. \tag{3.6.7}$$

Proof. Recall that $\cos(s + t) = \cos t \cos s - \sin t \sin s$, $\cos(s - t) = \cos t \cos s + \sin t \sin s$ and therefore

$$2 \sin t \sin s = \cos(s - t) - \cos(s + t)$$

for all $s, t \in \mathbb{R}$. Hence

$$\begin{aligned} 2\sin\frac{x}{2} D_n(x) &= \sum_{k=1}^{n} 2\sin\frac{x}{2}\sin(kx) = \sum_{k=1}^{n}\left(\cos\left(kx - \frac{x}{2}\right) - \cos\left(kx + \frac{x}{2}\right)\right) \\ &= \sum_{k=0}^{n-1}\cos\left(kx + \frac{x}{2}\right) - \sum_{k=1}^{n}\cos\left(kx + \frac{x}{2}\right) = \cos\frac{x}{2} - \cos\left(nx + \frac{x}{2}\right) \end{aligned}$$

and

$$D_n(x) = \frac{\cos\frac{x}{2} - \cos\left(nx + \frac{x}{2}\right)}{2\sin\frac{x}{2}}$$

for $0 < |x| < 2\pi$. As $|\sin t| \geq |t|/2$ for $-\pi/2 < t < \pi/2$, this implies (3.6.7). □

3.6.3 Theorem. (i) *The series*

$$\sum_{k=2}^{\infty}\frac{z^k - z^{-k}}{k\log k}, \qquad z \in \mathbb{T}, \tag{3.6.8}$$

converges uniformly on $\mathbb{T}$ to a continuous function $\Omega : \mathbb{T} \to \mathbb{C}$.

(ii) *The function* Ω *from part* (i) *does not split with respect to* $\mathbb{T}$ (*Def.* 3.1.2).

Proof. We first prove part (i). Since, for $z = e^{ix}$, $-\pi \le x \le \pi$,

$$z^k - z^{-k} = e^{ikx} - e^{-ikx} = 2i\sin(kx),$$

this is equivalent to the assertion that the series

$$\sum_{k=2}^{\infty} \frac{\sin(kx)}{k\log k}, \qquad -\pi \le x \le \pi, \tag{3.6.9}$$

converges uniformly on $[0, 2\pi]$ to a continuous function $\omega : [-\pi, \pi] \to \mathbb{C}$. Recall that by Lemma 3.6.2,

$$|D_k(x)| \le \frac{4}{|x|} \qquad \text{for all } 0 < |x| \le \pi \text{ and } k \ge 2, \tag{3.6.10}$$

where $D_k(x) := \sum_{\nu=1}^{k} \sin(\nu x)$. By Lemma 3.6.1 this implies that, for each $\delta > 0$, the series (3.6.8) converges uniformly on $[-\pi, -\delta] \cup [\delta, \pi]$. In particular, this series converges for each fixed $x \in [-\pi, \pi]$ (for $x = 0$ this is trivial, as $\sin 0 = 0$), and we can define

$$\omega(x) = \sum_{k=2}^{\infty} \frac{\sin(nx)}{n\log n} \qquad \text{for all } -\pi \le x \le \pi.$$

Let

$$r_n(x) := \omega(x) - \sum_{k=2}^{n} \frac{\sin(kx)}{k\log k} = \sum_{k=n+1}^{\infty} \frac{\sin(kx)}{k\log k} \qquad \text{for } -\pi \le x \le \pi \text{ and } n \ge 2.$$

We have to prove that $\lim_{n\to\infty} r_n = 0$ uniformly on $[-\pi, \pi]$. For that, it is sufficient to prove that

$$|r_n(x)| \le \frac{16}{\log n} \qquad \text{for } |x| \le \pi \text{ and } n \ge 2. \tag{3.6.11}$$

For $x = 0$ this is trivial. Let $n \ge 2$ and $x \ne 0$ with $|x| \le \pi$ be given. We distinguish two cases:

First Case: $|x| \ge \frac{1}{n}$. Then

$$\begin{aligned}
r_n(x) &= \sum_{k=n+1}^{\infty} \frac{1}{k\log k}\big(D_k(x) - D_{k-1}(x)\big) \\
&= \sum_{k=n+1}^{\infty} \frac{1}{k\log k} D_k(x) - \sum_{k=n+1}^{\infty} \frac{1}{k\log k} D_{k-1}(x) \\
&= \sum_{k=n+1}^{\infty} \frac{1}{k\log k} D_k(x) - \sum_{k=n}^{\infty} \frac{1}{(k+1)\log(k+1)} D_k(x) \\
&= \frac{1}{(n+1)\log(n+1)} D_{n+1}(x) + \sum_{k=n}^{\infty} \left(\frac{1}{k\log k} - \frac{1}{(k+1)\log(k+1)}\right) D_k(x).
\end{aligned}$$

By (3.6.10) this implies that

$$|r_n(x)| \le \frac{4}{|x|(n+1)\log(n+1)} + \frac{4}{|x|}\sum_{k=n}^{\infty}\left|\frac{1}{k\log k} - \frac{1}{(k+1)\log(k+1)}\right|.$$

As

$$\frac{1}{k\log k} > \frac{1}{(k+1)\log(k+1)} \qquad \text{for } k \ge 2$$

and (therefore)

$$\sum_{k=n}^{\infty}\left|\frac{1}{k\log k} - \frac{1}{(k+1)\log(k+1)}\right| = \frac{1}{n\log n},$$

it follows that

$$|r_n(x)| \le \frac{8}{|x| n\log n}.$$

Since $|x| \ge \frac{1}{n}$, this implies (3.6.11),

Second case: $|x| < \frac{1}{n}$. Let $N \ge n$ be the number in $\mathbb{N}$ with

$$\frac{1}{N} > |x| \ge \frac{1}{N+1}.$$

Since $|\sin(kx)| \le k|x|$, then

$$\sum_{k=n+1}^{N}\frac{\sin(kx)}{k\log k} \le \sum_{k=n+1}^{N}\frac{|x|}{\log k} \le (N-n)\frac{|x|}{\log(n+1)}.$$

As $|x| < 1/N$ and $\log(n+1) > \log n$, this yields

$$\sum_{k=n+1}^{N}\frac{\sin(kx)}{k\log k} \le \sum_{k=n+1}^{N}\frac{|x|}{\log k} \le \frac{1}{\log n}. \tag{3.6.12}$$

Moreover, as in the first case, using (3.6.10), we get

$$\begin{aligned}
&\left|\sum_{k=N+1}^{\infty}\frac{\sin(kx)}{k\log k}\right| \\
&\le \frac{1}{(N+1)\log(N+1)}|D_{N+1}(x)| + \sum_{k=N}^{\infty}\left(\frac{1}{k\log k} - \frac{1}{(k+1)\log(k+1)}\right)|D_k(x)| \\
&\le \frac{4}{|x|(N+1)\log(N+1)} + \frac{4}{|x|}\frac{1}{N\log N}.
\end{aligned}$$

As $|x| \ge 1/(N+1)$ and $N \ge n$, this implies

$$\left|\sum_{k=N+1}^{\infty}\frac{\sin(kx)}{k\log k}\right| \le \frac{4}{\log(N+1)} + \frac{4(N+1)}{N\log N} \le \frac{12}{\log n}.$$

Together with (3.6.12) this implies (3.6.11).

Now we prove part (ii). Set

$$D_+ = \left\{ z \in \mathbb{C} \,\middle|\, |z| < 1 \right\} \qquad \text{and} \qquad D_- = \left\{ z \in \mathbb{C} \,\middle|\, |z| > 1 \right\}.$$

Assume Ω splits with respect to $\mathbb{T}$, i.e., $\Omega = \Omega_+ + \Omega_-$ where $\Omega_+ : \overline{D}_+ \to \mathbb{C}$ and $\Omega_- : \overline{D}_- \cup \{\infty\} \to \mathbb{C}$ are continuous functions which are holomorphic in D_+ and $D_- \cup \{\infty\}$, respectively. After adding a constant we may assume that $\Omega_-(\infty) = 0$. By the Cauchy formula, we have

$$\Omega_+(z) = \frac{1}{2\pi i} \int_{\mathbb{T}} \frac{\Omega_+(\zeta)}{\zeta - z} d\zeta \qquad \text{for } z \in D_+. \tag{3.6.13}$$

By the Cauchy integral theorem, for all $1 < R < \infty$ and $z \in D_+$, we have

$$\left| \frac{1}{2\pi i} \int_{\mathbb{T}} \frac{\Omega_-(\zeta)}{\zeta - z} d\zeta \right| = \left| \frac{1}{2\pi i} \int_{|\zeta| = R} \frac{\Omega_-(\zeta)}{\zeta - z} d\zeta \right| \le \frac{1}{(R-1)\pi} \max_{|\zeta|=R} |\Omega_-(\zeta)|.$$

As $\Omega_-(\infty) = 0$, this implies that

$$\frac{1}{2\pi i} \int_{\mathbb{T}} \frac{\Omega_-(\zeta)}{\zeta - z} d\zeta = 0 \qquad \text{for all } z \in D_+.$$

Together with (3.6.13), this gives

$$\frac{1}{2\pi i} \int_{\mathbb{T}} \frac{\Omega(\zeta)}{\zeta - z} d\zeta = \Omega_+(z) \qquad \text{for all } z \in D_+.$$

As, by part (i) of this theorem, the series (3.6.8) converges uniformly on $\mathbb{T}$ to Ω, this further implies that

$$\Omega_+(z) = \lim_{n \to \infty} \frac{1}{2\pi i} \int_{\mathbb{T}} \left(\sum_{k=2}^{n} \frac{1}{k \log k} \frac{\zeta^k - \zeta^{-k}}{\zeta - z} \right) d\zeta \qquad \text{for } z \in D_+.$$

Taking into account that

$$\frac{1}{2\pi i} \int_{\mathbb{T}} \frac{\zeta^{-k}}{\zeta - z} d\zeta = 0 \qquad \text{and} \qquad \frac{1}{2\pi i} \int_{\mathbb{T}} \frac{\zeta^{k}}{\zeta - z} d\zeta = z^k \qquad \text{for } k \ge 1 \text{ and } z \in D_+,$$

it follows that

$$\Omega_+(z) = \sum_{k=2}^{\infty} \frac{z^k}{k \log k}, \qquad z \in D_+.$$

Since $\sum_{k=2}^{\infty} 1/(k \log k) = \infty$, from this we get

$$\lim_{\varepsilon \downarrow 0} \Omega_+(1 - \varepsilon) = \infty,$$

which is a contradiction to the assumption that Ω_+ is continuous on $\overline{D}_+$. □

3.7 The additive local principle

In this section, E is a Banach space and $D_+ \subseteq \mathbb{C}$ is a bounded open set with piecewise $\mathcal{C}^1$-boundary Γ oriented by D_+ (Section 1.4.1), and $D_- = \mathbb{C} \setminus \overline{D}_+$.

3.7.1 Definition. Let $U \subseteq \mathbb{C}$ be an open set with $U \cap \Gamma \neq \emptyset$, and let $f : \Gamma \cap U \to E$ be a continuous function. We say that f **splits (additively) over** U **with respect to** Γ if there exists a pair of E-valued functions (f_+, f_-), where f_+ is continuous on $U \cap \overline{D}_+$ and holomorphic in $U \cap D_+$, and f_- is continuous on $U \cap \overline{D}_-$ and holomorphic in $U \cap D_-$, such that

$$f = f_+ + f_- \qquad \text{on } U \cap \Gamma . \tag{3.7.1}$$

Then the pair (f_+, f_-) or the representation $f = f_+ + f_-$ will be called a **splitting with respect to** Γ **over** U.

Let $f : \Gamma \to E$ be a continuous function. Then we say that f **splits locally with respect to** Γ if, for each $w \in \Gamma$, there exists a neighborhood U of w such that f splits over U with respect to Γ.

3.7.2 Corollary (to Theorem 1.5.4). *Let $U \subseteq \mathbb{C}$ be an open set with $U \cap \Gamma \neq \emptyset$, and let $f : \Gamma \cap U \to E$ be a continuous function. Suppose (f_+, f_-) and $(\widetilde{f}_+, \widetilde{f}_-)$ are two splittings of f with respect to Γ over U. Then there exists a holomorphic function $h : U \to E$ such that*

$$f_+ - \widetilde{f}_+ = h \quad \text{on } \overline{D}_+ \cap U \qquad \text{and} \qquad f_- - \widetilde{f}_- = h \quad \text{on } \overline{D}_- \cap U .$$

Proof. Since $f_+ - \widetilde{f}_+ = f_- - \widetilde{f}_-$ on $\Gamma \cap U$, then there is a well-defined continuous function $h : U \to E$ with

$$h = \begin{cases} f_+ - \widetilde{f}_+ & \text{on } \overline{D}_+ \cap U , \\ f_- - \widetilde{f}_- & \text{on } \overline{D}_- \cap U . \end{cases}$$

By Theorem 1.5.4, this function is holomorphic on U. □

The fact established by the following theorem will be called the **additive local principle**:

3.7.3 Theorem. *Let $f : \Gamma \to E$ be a continuous function, which locally splits with respect to Γ. Then f globally splits with respect to Γ.*

In particular, all E-valued functions, which are holomorphic in some neighborhood of Γ, globally split with respect to Γ.

Proof. Choose open sets $U_1, \dots, U_m \subseteq \mathbb{C}$ with $\Gamma \subseteq U_1 \cup \ldots \cup U_m$ such that, for each $1 \leq j \leq m$, over U_j there exists a local splitting (f_j^+, f_j^-) of f with respect to Γ. Moreover, set

$$U_0 = \mathbb{C} \setminus \Gamma \qquad \text{and} \qquad f_0^+ = f_0^- = 0 \quad \text{on } U_0 .$$

Then, for all $0 \leq j, k \leq m$ with $U_j \cap U_k \cap \Gamma \neq \emptyset$,

$$f_j^+ - f_k^+ = f_j^- - f_k^- \qquad \text{on } U_j \cap U_k \cap \Gamma. \tag{3.7.2}$$

Indeed, if $1 \leq j, k \leq m$, this is clear, since $f_j^+ + f_j^- = f = f_k^+ + f_k^-$ on $U_j \cap U_k \cap \Gamma$, and, for $j = 0$ or $k = 0$, this is trivial, as $U_0 \cap \Gamma = \emptyset$. Now, by (3.7.2), there is a well-defined family $g_{jk} \in \mathcal{C}^E(U_j \cap U_k)$, $0 \leq j, k \leq m$, such that, for all $0 \leq j, k \leq m$,

$$g_{jk} = \begin{cases} f_j^+ - f_k^+ & \text{on } U_j \cap U_k \cap \overline{D}_+ \text{ if } U_j \cap U_k \cap \overline{D}_+ \neq \emptyset\,, \\ f_j^- - f_k^- & \text{on } U_j \cap U_k \cap \overline{D}_- \text{ if } U_j \cap U_k \cap \overline{D}_- \neq \emptyset\,. \end{cases} \tag{3.7.3}$$

On $(U_j \cap U_k) \setminus \Gamma$, these functions are holomorphic, since the functions $f_j^\pm$ are holomorphic on $U_j \cap D_j^\pm$. Hence, by Theorem 1.5.4,

$$g_{jk} \in \mathcal{O}^E(U_j \cap U_k)\,, \qquad 0 \leq j, k \leq m.$$

Moreover it is clear from (3.7.3) that, for all $0 \leq j, k, l \leq m$ with $U_j \cap U_k \cap U_l \neq \emptyset$,

$$g_{jk} + g_{kl} = g_{jl} \qquad \text{on } U_j \cap U_k \cap U_l\,,$$

i.e.,, the family $\{g_{jk}\}_{0 \leq j,k \leq m}$ is a $(\{U_0, \ldots, U_m\}, \mathcal{O}^E)$-cocycle (Definition 2.4.1). Hence, from Theorem 2.4.2 we get a family $h_j \in \mathcal{O}^E(U_j)$, $0 \leq j \leq m$, with

$$g_{jk} = h_j - h_k \qquad \text{on } U_j \cap U_k\,,\ 0 \leq j, k \leq m.$$

Then it follows from (3.7.3) that, for all $0 \leq j, k \leq m$,

$$h_j - h_k = \begin{cases} f_j^+ - f_k^+ & \text{on } U_j \cap U_k \cap \overline{D}_+ \text{ if } U_j \cap U_k \cap \overline{D}_+ \neq \emptyset, \\ f_j^- - f_k^- & \text{on } U_j \cap U_k \cap \overline{D}_- \text{ if } U_j \cap U_k \cap \overline{D}_- \neq \emptyset, \end{cases}$$

and therefore

$$h_j - f_j^+ = h_k - f_k^+ \quad \text{on } U_j \cap U_k \cap \overline{D}_+ \quad \text{if} \quad U_j \cap U_k \cap \overline{D}_+ \neq \emptyset$$

and

$$h_j - f_j^- = h_k - f_k^- \quad \text{on } U_j \cap U_k \cap \overline{D}_- \quad \text{if} \quad U_j \cap U_k \cap \overline{D}_- \neq \emptyset\,.$$

Hence, there are well-defined continuous functions $\widetilde{f}_\pm : \overline{D}_\pm \to E$ such that

$$\widetilde{f}_+\big|_{U_j \cap \overline{D}_+} = h_j - f_j^+\,, \qquad 0 \leq j \leq m\,, \tag{3.7.4}$$

$$\widetilde{f}_-\big|_{U_j \cap \overline{D}_-} = h_j - f_j^+\,, \qquad 0 \leq j \leq m\,. \tag{3.7.5}$$

Since $U_0, \ldots, U_m$ is an open covering of $\mathbb{C}$, in view of the corresponding properties of the functions h_j and $f_j^\pm$, it follows from (3.7.4) and (3.7.5) that $\widetilde{f}_+$ is holomorphic in D_+, and $\widetilde{f}_-$ is holomorphic in D_-. Since $f = f_j^+ + f_j^-$ on $U_j \cap \Gamma$, $1 \leq j \leq m$, it follows from (3.7.4) and (3.7.5) that

$$f = \widetilde{f}_+ - \widetilde{f}_- \qquad \text{on } \Gamma\,. \tag{3.7.6}$$

Now we take a radius $0 < R < \infty$ such that $\{z \in \mathbb{C} \mid |z| > R\} \subseteq D_-$, and let

$$\widetilde{f}_-(z) = \sum_{n=-\infty}^{\infty} \widetilde{f}_n z^n$$

be the Laurent expansion of $\widetilde{f}_-$ with respect to $\{z \in \mathbb{C} \mid |z| > R\}$. Set

$$h(z) = \sum_{n=0}^{\infty} \widetilde{f}_n z^n \quad \text{for } z \in \mathbb{C} \quad \text{and} \quad f_-(z) = \sum_{n=-\infty}^{-1} \widetilde{f}_n z^n \quad \text{for } |z| > R.$$

Then

$$f_-(z) = \widetilde{f}_-(z) - h(z) \qquad \text{for } |z| > R.$$

As $\widetilde{f}_-$ is continuous on $\overline{D}_-$ and holomorphic in D_- and h is holomorphic in $\mathbb{C}$, this shows that f_- admits a continuous extension to $\overline{D}_-$, which is holomorphic in D_-. Moreover, by definition of f_-, it is clear that f_- extends holomorphically to ∞, where $f_-(\infty) = 0$. As $\widetilde{f}_+$ is continuous on $\overline{D}_+$ and holomoprhic in D_+, and h is holomorphic in $\mathbb{C}$, in the same way we see that

$$f_+ := \widetilde{f}_+ - h$$

is continuous on $\overline{D}_+$ and holomorphic in D_+. Since $\widetilde{f}_- = h + f_-$, from (3.7.6) it follows that

$$f = \widetilde{f}_+ - \widetilde{f}_- = \widetilde{f}_+ - h - f_- = f_+ - f_- \qquad \text{on } \Gamma .$$

□

3.8 Factorization of scalar functions with respect to a contour. First remarks

In this section, $D_+ \subseteq \mathbb{C}$ is a bounded open set with piecewise $\mathcal{C}^1$-boundary Γ oriented by D_+ (Section 1.4.1) such that $0 \in D_+$, and $D_- = \mathbb{C} \setminus \overline{D}_+$. We denote by $\Gamma_0, \ldots, \Gamma_m$ the connected components of Γ, endowed with the orientation of Γ, so that $-\Gamma_0$ is the boundary of the unbounded component of D_-.

3.8.1 Definition. Let $f : \Gamma \to \mathbb{C}^*$ be a continuous function. We say that f admits a **factorization** with respect to Γ if there exist continuous functions $f_- : \overline{D}_- \cup \{\infty\} \to \mathbb{C}^*$ and $f_+ : \overline{D}_+ \to \mathbb{C}^*$, which are holomorphic in $D_- \cup \{\infty\}$ (Section 3.1.1) and D_+, respectively, such that, for some integer κ,

$$f(z) = z^\kappa f_+(z) f_-(z) \qquad \text{for } z \in \Gamma . \tag{3.8.1}$$

To underline the difference with Definition 3.11.1 below, in this case we also say that f admits a **global** factorization with respect to Γ.

If f is a scalar rational function which is holomorphic and $\neq 0$ on $\Gamma_0 \cup \ldots \cup \Gamma_m$, then a factorization of f with respect to $\Gamma_0 \cup \ldots \cup \Gamma_m$ is easy to find. Indeed, then f is of the form

$$f = \frac{p_- p_+}{q_- q_+}$$

where p_-, q_-, p_+, q_+ are polynomials such that p_-, q_- have no zeros on $\overline{D}_-$ and p_+, q_+ have no zeros on $\overline{D}_+$. Let $\deg p_-$ and $\deg q_-$ be the degrees of p_- and q_-, respectively, and set

$$\kappa = \deg p_- - \deg q_- .$$

Then the functions p_-/q_- and p_+/q_+ are holomorphic on $\overline{D}_-$ and $\overline{D}_+$, respectively,

$$a := \lim_{z \to \infty} z^{-\kappa} \frac{p_-(z)}{q_-(z)}$$

exists and $a \neq 0$. Therefore we obtain a factorization of f with respect to $\Gamma_0 \cup \ldots \cup \Gamma_m$, by setting

$$f_-(z) = z^{-\kappa} \frac{p_-(z)}{q_-(z)} \quad \text{for } z \in \overline{D}_- \quad \text{and} \quad f_-(\infty) = a,$$

$$f_+(z) = \frac{p_+(z)}{q_+(z)} \quad \text{for } z \in \overline{D}_+ .$$

Not every continuous function $f : \Gamma_0 \cup \ldots \cup \Gamma_m \to \mathbb{C}^*$ admits a factorization with respect to $\Gamma_0 \cup \ldots \cup \Gamma_m$ (see Remark 3.11.4 below). To study the factorization problem in the general case, now, also for continuous functions, we define the index, which was introduced for holomorphic functions already in Definition 2.5.3.

For that let Γ' be the union of some of the connected components $\Gamma_0, \ldots, \Gamma_m$ of Γ, oriented in an arbitrary way. First consider two holomorphic functions $f, g : \Gamma' \to \mathbb{C}^*$ such that

$$\left| f(z) g(z)^{-1} - 1 \right| < 1$$

for all $z \in \Gamma'$. Then this estimate holds also in some neighborhood of Γ'. Therefore $\log\left(fg^{-1}\right)$ is defined in this neighborhood, where log is the main branch of the logarithm, and

$$\frac{\left(fg^{-1}\right)'}{fg^{-1}} = \frac{\left(e^{\log\left(fg^{-1}\right)}\right)'}{fg^{-1}} = \left(fg^{-1}\right)' .$$

Hence

$$\operatorname{ind}_{\Gamma'} f - \operatorname{ind}_{\Gamma'} g = \operatorname{ind}_{\Gamma'} \left(fg^{-1}\right) = \frac{1}{2\pi i} \int_{\Gamma'} \left(f(z) g^{-1}(z)\right)' dz = 0 .$$

Therefore the following definition is correct (and agrees with Definition 2.5.3 when f is holomorphic):

3.8.2 Definition. Let Γ' be the union of some of the connected components $\Gamma_0, \ldots$, Γ_m of Γ, oriented in an arbitrary way. Let $f : \Gamma' \to \mathbb{C}^*$ be a continuous function. Then we define

$$\operatorname{ind}_{\Gamma'} f = \operatorname{ind}_{\Gamma'} \widetilde{f} \tag{3.8.2}$$

where $\widetilde{f} : \Gamma' \to \mathbb{C}^*$ is a holomorphic function which is sufficiently close to f, uniformly on Γ'. (By Corollary 3.3.3 such a function $\widetilde{f}$ exists.) This number $\operatorname{ind}_{\Gamma'} f$ is called the **index** of f with respect to Γ'. By Proposition 2.5.4 (iv), it is an integer.

3.8.3 Proposition. (i) *Let Γ' be the union of some of the connected components $\Gamma_0, \ldots, \Gamma_m$ of Γ, oriented in an arbitrary way. Let $f, g : \Gamma' \to \mathbb{C}^*$ be two continuous functions. Then*

$$\operatorname{ind}_{\Gamma'}(fg) = \operatorname{ind}_{\Gamma'} f + \operatorname{ind}_{\Gamma'} g\,. \tag{3.8.3}$$

(ii) *Let $f : \Gamma \to \mathbb{C}^*$ be a continuous function which admits an extension to $\overline{D}_+$ which is continuous on Γ and meromorphic in D_+. Denote this extension also by f, let N be the number of zeros of f in D_+, counted according to their multiplicities, and let P be the number of poles of f in D, also counted according to their multiplicities. Then (recall that Γ is oriented by D_+)*

$$\operatorname{ind}_{\Gamma} f = N - P\,. \tag{3.8.4}$$

(iii) *Let $f : \Gamma \to \mathbb{C}^*$ be a continuous function which admits an extension to $\overline{D}_- \cup \{\infty\}$ which is continuous on Γ and meromorphic in $D_- \cup \{\infty\}$ (Section* 3.1.1). *Denote this extension also by f, let N be the number of zeros of f in $D_- \cup \{\infty\}$, counted according to their multiplicities, and let P be the number of poles of f in $D_- \cup \{\infty\}$, also counted according to their multiplicities. Then (recall that Γ is oriented by D_+) then*

$$-\operatorname{ind}_{\Gamma} f = N - P\,. \tag{3.8.5}$$

Proof. By our definition of the index in the case of a continuous function, part (i) follows immediately from part (i) of Proposition 2.5.4.

Consider part (ii). Using the Mergelyan approximation Theorem 2.2.1 and Cauchy's theorem it is easy to see that D_+ can be replaced by a slightly smaller set. Then the assertion follows from part (ii) of Proposition 2.5.4.

In the same way part (iii) follows from part (iii) of Proposition 2.5.4. □

3.8.4 Corollary. *Let $f : \Gamma \to \mathbb{C}$ be a continuous function, and let $f(z) = z^\kappa f_+(z) f_-(z)$ be a factorization of f with respect to Γ. Then the integer κ is uniquely determined, namely (recall that Γ is oriented by D_+)*

$$\kappa = \operatorname{ind}_{\Gamma} f.$$

Proof. Since $0 \in D_+$, it follows from Proposition 3.8.3 that

$$\operatorname{ind}_\Gamma f = \operatorname{ind}_\Gamma z^\kappa + \operatorname{ind}_\Gamma f_+ + \operatorname{ind}_\Gamma f_- = \operatorname{ind}_\Gamma z^\kappa = \kappa. \qquad \square$$

3.8.5 Proposition. *Let $0 \le j \le m$, and let $f : \Gamma_j \to \mathbb{C}^*$ be a continuous function with $\operatorname{ind}_{\Gamma_j} f = 0$. Then there is a continuous function $\log f : \Gamma_j \to \mathbb{C}$ with*

$$e^{\log f} = f \qquad \text{on } \Gamma_j.$$

Note that $\log f$ is uniquely determined up to an additive constant of the form $k2\pi i$, $k \in \mathbb{Z}$.

Proof. Choose a holomorphic function $\widetilde{f} : \Gamma_j \to \mathbb{C}^*$ which is so close to f that

$$\left| f(z)\widetilde{f}^{-1}(z) - 1 \right| < 1 \qquad \text{for } z \in \Gamma_j,$$

and

$$\operatorname{ind}_{\Gamma_j} \widetilde{f} = \operatorname{ind}_{\Gamma_j} f = 0, \qquad 0 \le j \le m.$$

By the first relation $\log(f\widetilde{f}^{-1})$ is well defined on Γ_j, and, by Theorem 2.5.5, from the second relation it follows that there exists a holomorphic function $h : \Gamma \to \mathbb{C}$ with $e^h = \widetilde{f}$. It remains to set $\log f = h + \log(f\widetilde{f}^{-1})$, where the latter log denotes the main branch of the complex logarithm. $\square$

3.8.6. Let $f : \Gamma \to \mathbb{C}$ be a continuous function, and let $f(z) = z^\kappa f_+(z) f_-(z)$ be a factorization of f with respect to Γ. By multiplying by a constant we can always achieve that $f_-(\infty) = 1$. With this additional property, the factorization of f with respect to Γ is uniquely determined. Indeed, let $f = z^\kappa \widetilde{f}_+ \widetilde{f}_-$ be a second factorization of f. Then

$$f_+ \widetilde{f}_+^{-1} = \widetilde{f}_- f_-^{-1} \qquad \text{on } \Gamma.$$

Therefore (cf. Theorem 1.5.4) the function defined by

$$h = \begin{cases} f_+ \widetilde{f}_+^{-1} & \text{on } \overline{D}_+, \\ \widetilde{f}_- f_-^{-1} & \text{on } \overline{D}_- \cup \{\infty\}, \end{cases}$$

is a well-defined holomorphic function on $\mathbb{C} \cup \{\infty\}$ which is equal to 1 at ∞. Hence this function is identically equal to 1, i.e., $f_\pm = \widetilde{f}_\pm$.

3.8.7 Theorem. *If the contour Γ is not connected, i.e., $m \ge 1$, then, for $1 \le j \le m$, we denote by U_j the bounded connected component of D_- with boundary $-\Gamma_j$ and we fix some point $p_j \in U_j$.*

Let $f : \Gamma \to \mathbb{C}$ be a continuous function. Set

$$\kappa_j := \operatorname{ind}_{\Gamma_j} f, \qquad 0 \le j \le m.$$

Further, let $g : \Gamma \to \mathbb{C}$ be one of the continuous functions with

$$e^{g(z)} = \begin{cases} z^{-\kappa_0} f(z)\,, & \text{for } z \in \Gamma_0, \\ (z-p_j)^{\kappa_j} f(z)\,, & \text{for } z \in \Gamma_j,\ 1 \le j \le m, \end{cases}$$

(which exists by Proposition 3.8.5*). Then the following are equivalent:*

(i) *The function f admits a factorization with respect to Γ.*

(ii) *The function g splits with respect to Γ.*

Proof. First assume that there is a factorization $f(z) = f_-(z) z^\kappa f_+(z)$ of f. Then

$$\begin{aligned} \operatorname{ind}_{\Gamma_j} f_- &= 0 && \text{for } 0 \le j \le m, \\ \operatorname{ind}_{\Gamma_0} f_+ &= \kappa_0 - \kappa, && \\ \operatorname{ind}_{\Gamma_j} f_+ &= \kappa_j && \text{for } 1 \le j \le m. \end{aligned}$$

Therefore it follows from the Mergelyan approximation Theorem 2.2.1 and Theorem 2.5.5 that there exists a continuous function $g_- : \overline{D}_- \cup \{\infty\} \to \mathbb{C}$, which is holomorphic in $D_- \cup \{\infty\}$, such that $f_- = e^{g_-}$ on $\overline{D}_- \cup \{\infty\}$. Moreover, set

$$\widetilde{f}_+(z) = \begin{cases} z^{\kappa-\kappa_0} f_+(z), & \text{for } z \in \Gamma_0, \\ z^\kappa (z-p_j)^{\kappa_j} f_+(z), & \text{for } z \in \Gamma_j,\ 1 \le j \le m. \end{cases}$$

Then $\operatorname{ind}_{\Gamma_j} \widetilde{f}_j = 0$ for $0 \le j \le m$, and from Theorem 2.5.5 and the Mergelyan Theorem 2.2.1 we get a contiunous function $g_+ : \overline{D}_+ \to \mathbb{C}$, which is holomorphic in D_+, such that $\widetilde{f}_+ = e^{g_+}$ on $\overline{D}_+$. It follows

$$e^{g(z)} = z^{-\kappa_0} f(z) = z^{-\kappa_0} f_-(z) z^\kappa f_+(z) = \widetilde{f}_-(z) \widetilde{f}_+(z) = e^{g_-(z)+g_+(z)}$$

for $z \in \Gamma_0$, and

$$e^{g(z)} = (z-p_j)^{\kappa_j} f(z) = (z-p_j)^{\kappa_j} f_-(z) z^\kappa f_+(z) = \widetilde{f}_-(z) \widetilde{f}_+(z) = e^{g_-(z)+g_+(z)}$$

for $z \in \Gamma_j$, $1 \le j \le m$. Hence, there are some integers μ_j with

$$g(z) = g_+(z) + g_-(z) + \mu_j \qquad \text{for } z \in \Gamma_j, \quad 0 \le j \le m.$$

Setting

$$\widetilde{g}_-(z) = g_-(z) + \mu_j \qquad \text{for } z \in \Gamma_j, \quad 0 \le j \le m,$$

we get a splitting $(g_+, \widetilde{g}_-)$ of g.

Now we assume that a splitting $g = g_+ + g_-$ of g is given. Then

$$\begin{aligned} z^{-\kappa_0} f(z) &= e^{g(z)} = e^{g_-(z)} e^{g_+(z)} && \text{for } z \in \Gamma_0, \\ (z-p_j)^{\kappa_j} f(z) &= e^{g(z)} = e^{g_-(z)} e^{g_+(z)} && \text{for } z \in \Gamma_j,\ 1 \le j \le m. \end{aligned}$$

Therefore it is sufficient to prove that the function φ defined by

$$\varphi(z) = \begin{cases} z^{\kappa_0} & \text{if } z \in \Gamma_0, \\ (z - p_j)^{-\kappa_j} & \text{if } z \in \Gamma_j,\ 1 \le j \le m, \end{cases}$$

admits a factorization. Let U_0 be the unbounded connected component of D_-, and let U_j, $1 \le j \le m$, be the bounded connected component of D_- with the boundary $-\Gamma_j$, $1 \le j \le m$. Set

$$\varphi_+(z) = (z - p_1)^{-\kappa_1} \cdot \ldots \cdot (z - p_m)^{-\kappa_m} \quad \text{for } \overline{D}_+,$$

and for $z \in \overline{D}_-$ we define

$$\varphi_-(z) = \begin{cases} z^{\kappa_0 - \kappa}(z - p_1)^{-\kappa_1} \cdot \ldots \cdot (z - p_m)^{-\kappa_m} & \text{if } z \in \overline{U}_0, \\ z^{-\kappa}(z - p_1)^{\kappa_1} \cdot \ldots \underset{j}{\widehat{}} \ldots \cdot (z - p_j)^{\kappa_m} & \text{if } z \in \overline{U}_j,\ 1 \le j \le m. \end{cases}$$

Since $p_j \in U - j$ and $0 \in D_+$, then ϕ_+ is holomorphic and $\neq 0$ on $\overline{D}_+$, φ_- is holomorphic and $\neq 0$ on $\overline{D}_- \cup \{\infty\}$ and

$$\varphi(z) = z^{\kappa}\varphi_-(z)\varphi_+(z) \qquad \text{for } z \in \Gamma. \qquad \square$$

3.9 Factorization of Hölder functions

Here we use the notations and definitions introduced in the preceding Section 3.8, and we prove:

3.9.1 Theorem. *Let $0 < \alpha < 1$, and let $f : \Gamma \to \mathbb{C}^*$ be Hölder continuous with exponent α (Def. 2.1.6).*

Then f admits a factorization with respect to Γ.

If $f = z^{\kappa} f_+ f_-$ is an arbitrary factorization of f, then f_+ is Hölder continuous with exponent α on $\overline{D}_+$, and f_- is Hölder continuous with exponent α on $\overline{D}_-$.

If, moreover, Γ is of class $\mathcal{C}^k$ and f is of class $\mathcal{C}^{k+\alpha}$ on Γ (Def. 3.4.1) for some $k \in \mathbb{N}^*$, and $f = z^{\kappa} f_+ f_-$ is an arbitrary factorization of f, then f_+ is of class $\mathcal{C}^{k+\alpha}$ on $\overline{D}_+$, and f_- is of class $\mathcal{C}^{k+\alpha}$ on $\overline{D}_-$.

Proof. Let $k \in \mathbb{N}$ and assume that if $k = 0$, then f is Hölder-α continuous, and if $k \ge 1$, then Γ is of class $\mathcal{C}^k$ and f is of class $\mathcal{C}^{k+\alpha}$. As observed in Section 3.8.6, up to a multiplicative constant, the solution of the factorization problem is uniquely determined. Therefore it is sufficient to prove the existence of at least one factorization $f = z^{\kappa} f_+ f_-$ such that the factors $f_\pm$ are of class $\mathcal{C}^{k+\alpha}$ on $\overline{D}_\pm$.

Set

$$\kappa_j = \operatorname{ind}_{\Gamma_j} f, \quad 0 \le j \le m, \qquad \text{and} \qquad \kappa = \kappa_0 + \kappa_1 + \ldots + \kappa_m.$$

Let $U_1, \dots, U_m$ be the bounded connected components of D_- so that $-\Gamma_j$ is the boundary of U_j. Choose points $p_j \in U_j$ and set

$$\widetilde{f}(z) = z^{-\kappa}(z-p_1)^{\kappa_1} \cdot \ldots \cdot (z-p_m)^{\kappa_m} f(z).$$

Since Γ_0 is the boundary of the simply connected open set $D_+ \cup \overline{U}_1 \cup \ldots \cup \overline{U}_m$ (Section 2.5.1) and $0, p_1, \dots, p_m \in D_+ \cup \overline{U}_1 \cup \ldots \cup \overline{U}_m$, it follows from Proposition 3.8.3 (ii) that

$$\operatorname{ind}_{\Gamma_0}\Big(z^{-\kappa}(z-p_1)^{\kappa_1} \cdot \ldots \cdot (z-p_m)^{\kappa_m}\Big) = -\kappa + \kappa_1 + \ldots + \kappa_m = -\kappa_0,$$

In the same way we get

$$\operatorname{ind}_{\Gamma_j}\Big(z^{-\kappa}(z-p_1)^{\kappa_1} \cdot \ldots \cdot (z-p_m)^{\kappa_m}\Big) = -\kappa_j \qquad \text{for } 1 \le j \le m.$$

Hence, by Proposition 3.8.3 (i),

$$\operatorname{ind}_{\Gamma_j} \widetilde{f} = 0 \qquad \text{for all } 0 \le j \le m.$$

Therefore, by Proposition 3.8.5, we can find a continuous function $g : \Gamma \to \mathbb{C}$ with

$$e^g = \widetilde{f}.$$

Since $\widetilde{f}$ is of class $\mathcal{C}^{k+\alpha}$ and, locally, the function g is of the form $g = \log f$, where log is a branch of the logarithm (which is a holomorphic function), it follows that also g is of class $\mathcal{C}^{k+\alpha}$. Therefore, by theorems 3.3.2 and (3.4.5), there exists a $\mathcal{C}^{k+\alpha}$-function $g_+ : \overline{D}_+ \to \mathbb{C}$, which is holomorphic in D_+, and a $\mathcal{C}^{k+\alpha}$-function $g_- : \overline{D}_- \cup \{\infty\} \to \mathbb{C}$, which is holomorphic in $D_- \cup \{\infty\}$, such that $g = g_+ + g_-$ on Γ. Then

$$\widetilde{f} = e^{g_+} e^{g_-} \qquad \text{on } \Gamma.$$

Hence, by definition of $\widetilde{f}$,

$$f = \frac{z^{\kappa} e^{g_+} e^{g_-}}{(z-p_1)^{\kappa_1} \ldots (z-p_m)^{\kappa_m}}$$

Setting

$$f_- = e^{g_-} \quad \text{and} \quad f_+ = \frac{e^{g_+}}{(z-p_1)^{\kappa_1} \ldots (z-p_m)^{\kappa_m}}$$

we get a factorization

$$f = z^{\kappa} f_+ f_- \tag{3.9.1}$$

of f with respect to Γ. Since the functions $g_\pm$ are of class $\mathcal{C}^{k+\alpha}$ on $\overline{D}_\pm$, also the functions $f_\pm$ are of class $\mathcal{C}^{k+\alpha}$ on $\overline{D}_\pm$. □

3.10 Factorization of Wiener functions

Here $\mathbb{T}$ is the unit circle, D_+ is the open unit disc, and $D_- = \mathbb{C} \setminus \overline{D}_+$. Denote by $W(\mathbb{C})$ the space of functions $f : \Gamma \longrightarrow \mathbb{C}$ of the form

$$f(z) = \sum_{n=-\infty}^{\infty} f_n z^n \qquad \text{with } \|f\|_W := \sum_{n=-\infty}^{\infty} |f_n| < \infty. \tag{3.10.1}$$

It follows from Cauchy's product theorem that if $f, g \in W(\mathbb{C})$, then $fg \in W(\mathbb{C})$ and

$$\|fg\|_W \leq \|f\|_W \|g\|_W.$$

Hence, $W(C)$ is a Banach algebra with the norm $\| \cdot \|_W$. Moreover:

3.10.1 Proposition. *If $f \in W(\mathbb{C})$ and $f(z) \neq 0$ for all $z \in \mathbb{T}$, then $f^{-1} \in W(\mathbb{T})$.*

Proof. Let $f \in W(\mathbb{C})$ be given, which is not an invertible element of $W(\mathbb{C})$. We have to find $\theta \in \mathbb{T}$ with $f(\theta) = 0$.

Since f is not invertible as an element of $W(A)$, by the theory of commutative Banach algebras, there exits a multiplicative functional Φ on $W(\mathbb{C})$ with

$$\Phi(f) = 0.$$

For all fixed complex numbers $\lambda \in \mathbb{C} \setminus \mathbb{T}$, the function $z - \lambda$ is an invertible element of $W(\mathbb{C})$ (as $(z - \lambda)^{-1}$ is holomorphic in a neighborhood of $\mathbb{T}$). Hence, for each fixed $\lambda \in \mathbb{C} \setminus \mathbb{T}$, $\Phi(z - \lambda) \neq 0$ and, therefore,

$$\Phi(z) - \lambda = \Phi(z - \lambda) \neq 0.$$

Hence $\theta := \Phi(z) \in \mathbb{T}$. Now let f_n be the coefficients with $f(z) = \sum_{n=-\infty}^{\infty} f_n z^n$. Since $\sum |f_n| < \infty$, and, by definition, $\|z^n\|_W = 1$, then it follows that

$$f(\theta) = \sum_{n=-\infty}^{\infty} f_n \theta^n = \sum_{n=-\infty}^{\infty} f_n \big(\Phi(z)\big)^n = \sum_{n=-\infty}^{\infty} f_n \Phi(z^n) = \Phi(f) = 0. \qquad \square$$

3.10.2 Theorem. *Let $f \in W(\mathbb{C})$ such that $f(z) \neq 0$ for all $z \in \Gamma$. Then f admits a factorization $f(z) = z^\kappa f_+(z) f_-(z)$ with respect to Γ such that $f_+, f_- \in W(\mathbb{C})$.*

Moreover, if $f(z) = z^\kappa f_+(z) f_-(z)$ is a factorization of f with respect to Γ, then the factors $f_\pm$ automatically belong to $W(\mathbb{C})$.

Proof. As observed in Section 3.8.6, up to a multiplicative constant, the solution of the factorization problem is uniquely determined. Therefore it is sufficient to prove the existence of at least one factorization $f = z^\kappa f_+ f_-$ such that $f_+, f_- \in W(\mathbb{C})$.

By Proposition 3.10.1, f^{-1} belongs to $W(\mathbb{C})$. Since the functions, which are holomorphic on Γ, are dense in $W(\mathbb{C})$, we can find a holomorphic function h on Γ with

$$\|h - f^{-1}\|_W < \frac{1}{\|f\|_W}.$$

As the set $f^{-1}(\mathbb{T})$ is a compact subset of $\mathbb{C}^*$ and $|f^{-1}(z) - h(z)| \le \|f^{-1} - h\|_W$ for all $z \in \mathbb{T}$, we can moreover achieve that $h(z) \neq 0$ for all $z \in \mathbb{T}$. Set

$$g = fh - 1.$$

Then

$$\|g\|_W = \|f(h - f^{-1})\|_W \le \|f\|_W \|h - f^{-1}\|_W < 1 \tag{3.10.2}$$

and

$$f = h^{-1}(1 + g).$$

From (3.10.2) it follows that the series

$$\sum_{n=1}^{\infty} \frac{(-1)^{n-1}}{n} g^n$$

converges absolutely with respect to the norm $\|\cdot\|_W$ to some element of $W(A)$, which we denote by $\log(1+g)$. Since convergence in $W(A)$ implies pointwise convergence, it follows that

$$\big(\log(1+g)\big)(z) = \log\big(1 + g(z)\big) \qquad \text{for all } z \in \mathbb{T},$$

where the log on the right-hand side denotes the main branch of the complex logarithm. Since $\log(1+g)$ is an element of $W(A)$, it can be written also in the form

$$\log\big(1 + g(z)\big) = \sum_{n=-\infty}^{\infty} z^n a_n \qquad \text{with } \sum_{n=-\infty}^{\infty} \|a_n\|_W < \infty.$$

Set

$$v_-(z) = \sum_{n=-\infty}^{-1} z^n a_n\,, \qquad v_+(z) = \sum_{n=0}^{\infty} z^n a_n,$$

$w_+ = e^{v_+}$ and $w_- = e^{v_-}$. So we get continuous functions $w_+ : \overline{D}_+ \to \mathbb{C}^*$, $w_- : \overline{D}_- \cup \{\infty\} \to \mathbb{C}^*$, which are holomorphic in D_+ and $D_- \cup \{\infty\}$, respectively, such that $w_+, w_- \in W(\mathbb{C})$, $1 + g = w_- w_+$ and therefore

$$f = h^{-1} w_- w_+ \qquad \text{on } \Gamma.$$

Since h is holomorphic on Γ, it follows from Theorem 3.9.1 that there exists a factorization

$$h^{-1}(z) = z^{\kappa} h_-(z) h_+(z) \tag{3.10.3}$$

of h with respect to Γ. From (3.10.3) and Theorem 1.5.4 then we get that h_+ and h_- are holomorphic on Γ. Hence h_+ and h_- belong to $W(\mathbb{C})$. Setting $f_\pm = w_\pm h_\pm$, we get a factorization $f(z) = z^{\kappa} f_-(z) f_+(z)$ of f with respect to Γ, where $f_+, f_- \in W(\mathbb{C})$. □

3.11 The multiplicative local principle

In this section we use the notations and definition introduced in Section 3.8.

In Section 3.7 we saw that there is a local principle for the splitting problem. Here we will show that there is a local principle also for the factorization problem (see Theorem 3.11.5 below).

3.11.1 Definition. Let $U \subseteq \mathbb{C}$ be an open set with $U \cap \Gamma \neq \emptyset$, and let $f : \Gamma \cap U \to \mathbb{C}^*$ be a continuous function. We say that f admits a **factorization over** U with respect to Γ if there exist $\mathbb{C}^*$-valued functions f_- and f_+, where f_+ is continuous on $U \cap \overline{D}_+$ and holomorphic in $U \cap D_+$, and f_- is continuous on $U \cap \overline{D}_-$ and holomorphic in $U \cap D_-$, such that

$$f = f_- f_+ \qquad \text{on } U \cap \Gamma\,. \tag{3.11.1}$$

Then the pair (f_-, f_+) or the representation $f = f_- f_+$ will be called a **factorization of f over U with respect to** Γ. Let $f : \Gamma \to \mathbb{C}^*$ be a continuous function. We say that f admits **local factorizations with respect to** Γ if, for each $w \in \Gamma$, there exists a neighborhood U of w such that f admits a factorization over U with respect to Γ.

3.11.2 Lemma. *Let $f : \Gamma \cap U \to E$ be a continuous function, and let $U \subseteq \mathbb{C}$ be an open set with $U \cap \Gamma \neq \emptyset$. If (f_-, f_+) and $(\widetilde{f}_-, \widetilde{f}_+)$ are two factorizations of f over U with respect to Γ, then there is a (uniquely determined) holomorphic function $h : U \to \mathbb{C}^*$ with*

$$\widetilde{f}_- = h f_- \quad \text{on } U \cap \overline{D}_- \qquad \text{and} \qquad \widetilde{f}_+ = h f_+ \quad \text{on } U \cap \overline{D}_+\,. \tag{3.11.2}$$

Proof. By hypothesis $f_- f_+ = f = \widetilde{f}_- \widetilde{f}_+$ on $U \cap \Gamma$. Then, setting

$$h := \begin{cases} \widetilde{f}_+ / f_+ & \text{on } U \cap \overline{D}_+, \\ \widetilde{f}_- / f_- & \text{on } U \cap \overline{D}_-\,, \end{cases}$$

we get a continuous function $h : U \to \mathbb{C}^*$, which satisfies (3.11.2) and which is holomorphic in $U \setminus \Gamma$. By Theorem 1.5.4, h is holomorphic on all of U. □

3.11.3 Lemma. *Let $g : \Gamma \to \mathbb{C}$ be a continuous function. Then the following are equivalent:*

(i) *The function e^g admits local factorizations with respect to Γ.*

(ii) *The function g locally splits (additively) with respect to Γ.*

Proof. (i)⇒(ii): Let $w \in \Gamma$ be given. As e^g admits local factorizations with respect to Γ, then there exist a neighborhood U of w and continuous functions $u_\pm : U \cap \overline{D}_\pm \to \mathbb{C}^*$, which are holomorphic in $U \cap D_\pm$, such that

$$e^g = u_+ u_- \qquad \text{on } \Gamma \cap U. \tag{3.11.3}$$

After shrinking U, we may assume that a certain branch of the logarithm is defined on the values of u_+, and a certain other branch of the logarithm is defined on the values of u_-, i.e., we have continuous functions $v_\pm : U \cap \overline{D}_\pm \to \mathbb{C}$, which are holomorphic in $U \cap D_\pm$, such that $u_\pm = e^{v_\pm}$. Together with (3.11.4) and (3.11.3) this implies

$$e^g = e^{v_+ + v_-} \qquad \text{on } \Gamma \cap U.$$

Hence, for some $k \in \mathbb{Z}$ (we may assume that $U \cap \Gamma$ is connected), $g = (k2\pi i + v_+) + v_-$ on $\Gamma \cap U$.

(ii)⇒(i): Let $w \in \Gamma$ be given. As g locally splits additively with respect to Γ, then there exist a neighborhood U of w and continuous functions $v_\pm : U \cap \overline{D}_\pm \to \mathbb{C}$, which are holomorphic in $U \cap D_\pm$, such that

$$g = v_+ + v_- \qquad \text{on } \Gamma \cap U.$$

Then $e^g = e^{v_+} e^{v_-}$. □

3.11.4 Remark. Together with the example from Section 3.6 this lemma shows that not any continuous function $f : \Gamma \to \mathbb{C}^*$ admits local factorizations with respect to Γ.

The fact stated by the following theorem will be called the **multiplicative local principle**:

3.11.5 Theorem. *Let $f : \Gamma \to \mathbb{C}^*$ be a continuous function which admits local factorizations with respect to Γ. Then f admits a global factorization with respect to Γ.*

Proof. Recall that $\Gamma = \partial D_+$ is oriented by D_+ (by Definition 2.1.6) and that $\Gamma_0, \ldots, \Gamma_m$ are oriented in the same way. Set

$$\kappa_j = \operatorname{ind}_{\Gamma_j} f, \quad 0 \le j \le m, \qquad \text{and} \qquad \kappa = \kappa_0 + \kappa_1 + \ldots + \kappa_m.$$

Let $U_1, \ldots, U_m$ be the bounded connected components of D_-. Choose points $p_j \in U_j$ and set

$$\widetilde{f}(z) = z^{-\kappa}(z - p_1)^{\kappa_1} \cdot \ldots \cdot (z - p_m)^{\kappa_m} f(z).$$

Since Γ_0 is the boundary of the simply connected open set $D_+ \cup \overline{U}_1 \cup \ldots \cup \overline{U}_m$ (Section 2.5.1), the orientation included, and since $0, p_1, \ldots, p_m \in D_+ \cup \overline{U}_1 \cup \ldots \cup \overline{U}_m$, it follows from Proposition 3.8.3 (ii) that

$$\operatorname{ind}_{\Gamma_0}\left(z^{-\kappa}(z - p_1)^{\kappa_1} \cdot \ldots \cdot (z - p_m)^{\kappa_m}\right) = -\kappa + \kappa_1 + \ldots + \kappa_m = -\kappa_0,$$

In the same way we get

$$\operatorname{ind}_{\Gamma_j}\left(z^{-\kappa}(z - p_1)^{\kappa_1} \cdot \ldots \cdot (z - p_m)^{\kappa_m}\right) = -\kappa_j \qquad \text{for } 1 \le j \le m.$$

Hence, by Proposition 3.8.3 (i),

$$\operatorname{ind}_{\Gamma_j} \widetilde{f} = 0 \qquad \text{for all } 0 \le j \le m.$$

Therefore, by Proposition 3.8.5, we can find a continuous function $g : \Gamma \to \mathbb{C}$ with

$$e^g = \widetilde{f}. \tag{3.11.4}$$

By hypothesis, f admits local factorizations with respect to Γ. Hence $\widetilde{f}$ admits local factorizations with respect to Γ. By Lemma 3.11.3, this implies that g locally splits additively with respect to Γ. Therefore, from Theorem 3.7.3 we get a continuous function $g_+ : \overline{D}_+ \to \mathbb{C}$, which is holomorphic in D_+, and a continuous function $g_- : \overline{D}_- \cup \{\infty\} \to \mathbb{C}$, which is holomorphic in $D_- \cup \{\infty\}$, such that $g = g_+ + g_-$ on Γ. Then

$$\widetilde{f} = e^{g_+} e^{g_-} \qquad \text{on } \Gamma.$$

Hence, by definition of $\widetilde{f}$,

$$f = \frac{z^\kappa e^{g_+} e^{g_-}}{(z - p_1)^{\kappa_1} \dots (z - p_m)^{\kappa_m}}.$$

Setting

$$f_- = e^{g_-} \quad \text{and} \quad f_+ = \frac{e^{g_+}}{(z - p_1)^{\kappa_1} \dots (z - p_m)^{\kappa_m}}$$

we get the required factorization $f = z^\kappa f_+ f_-$ of f with respect to Γ. □

3.12 Comments

In writing this chapter we used different sources. For sections 3.1–3.5, 3.8 and 3.9 see [Mu], [GKru], for Section 3.6 see [Bar]. The material of sections 3.7 and 3.11, in this form, probably appears here for the first time. For Section 3.10 see [K].

Chapter 4

The Rouché theorem for operator functions

In this chapter we generalize to finite meromorphic Fredholm operator functions the classical Rouché theorem from Complex analysis and the Smith factorization form. The proof is based on the local Smith form for matrix functions.

4.1 Finite meromorphic Fredholm functions

In this section E is a Banach space.

4.1.1 Definition. Let $w \in \mathbb{C}$, let U be a neighborhood of w, and let $A : U \setminus \{w\} \to L(E)$ be a holomorphic function which is meromorphic on U (Section 1.10.6).

Then we shall say that A is **finite meromorphic** at w if the Laurent expansion of A at w is of the form

$$A(z) = \sum_{n=m}^{\infty} (z-w)^n A_n \,,$$

where (if $m < 0$) the operators $A_m, \dots, A_{-1}$ are finite dimensional.

If, in addition, A_0 is a Fredholm operator, then A is called **finite meromorphic and Fredholm at** w. The index of A_0 then will be called the **index of** A **at** w.

4.1.2 Theorem. *Let $w \in \mathbb{C}$, and let W be a neighborhood of w. Let $A : W \setminus \{w\} \to L(E)$ be a holomorphic function which is finite meromorphic and Fredholm at w. Assume the index of A at w is zero. Then there exist a neighborhood $U \subseteq W$ of w and a finite dimensional projector*[1] *P in E such that, with $Q := I - P$, the following holds:*

[1] By a **projector** in E we always mean a bounded linear projector in E, i.e., an operator $P \in L(E)$ with $P^2 = P$. A projector P in E is called **finite dimensional** if $\dim \operatorname{Im} P < \infty$.

There exist holomorphic functions $S, T : U \to GL(E)$ and a holomorphic function $A_P : U \setminus \{w\} \to L(\operatorname{Im} P)$, which is meromorphic at w, such that

$$SAT = Q + PA_PP \qquad \text{on } U \setminus \{w\}. \tag{4.1.1}$$

Proof. Let

$$A(z) = \sum_{n=m}^{\infty} (z-w)^n A_n$$

be the Laurent expansion of A at w. Since A_0 is a Fredholm operator with index zero, by multiplication by an invertible operator (from the left or from the right), we may assume that A_0 is a projector with $\dim \operatorname{Ker} A_0 < \infty$. Using again that the operators $A_m, \dots, A_{-1}$ are finite dimensional, we can find finite dimensional projectors P and P_0 in E such that

$$P_0P = PP_0 = P_0\,, \tag{4.1.2}$$

and, with $Q := I - P$,

$$A_0 = Q + P_0\,, \tag{4.1.3}$$

$$A_jQ = QA_j = 0 \qquad \text{for } m \le j \le -1\,. \tag{4.1.4}$$

Choose a neighborhood $U \subseteq W$ of w so small that the Laurent expansion of A at w converges on $U \setminus \{w\}$. Setting

$$A_+(z) = Q + P_0 + \sum_{n=1}^{\infty}(z-w)^n A_n \quad \text{and} \quad V(z) = A_+(z)Q + P$$

for $z \in U$, we get holomorphic functions $A_+, V : U \to L(E)$. Then, by (4.1.3) and (4.1.4),

$$A_+Q = AQ \quad \text{on } U \setminus \{w\} \tag{4.1.5}$$

and hence

$$VQ = AQ \quad \text{on } U \setminus \{w\}. \tag{4.1.6}$$

Since $A_+(w) = Q + P_0$, it follows from (4.1.3) and (4.1.2) that $V(w) = I$. By shrinking U we can achieve that

$$V(z) \in GL(E) \qquad \text{for all } z \in U\,. \tag{4.1.7}$$

Then it follows from the definition of V that

$$V^{-1}P = P \quad \text{on } U, \tag{4.1.8}$$

and from (4.1.6) it follows that

$$Q = V^{-1}AQ \quad \text{on } U \setminus \{w\}. \tag{4.1.9}$$

Setting

$$S = I - QV^{-1}A_+P \qquad \text{on } U$$

and

$$A_P = PV^{-1}AP \qquad \text{on } U \setminus \{w\},$$

we obtain a holomorphic function $S : U \to L(E)$ and $A_P : U \setminus \{w\} \to L(\operatorname{Im} P)$. The values of S are invertible, namely:

$$S^{-1} = I + QV^{-1}A_+P \qquad \text{on } U.$$

Moreover,

$$V^{-1}AS = V^{-1}AQ + V^{-1}AP - V^{-1}AQV^{-1}A_+P \quad \text{on } U \setminus \{w\}.$$

By (4.1.9) this implies

$$\begin{aligned} V^{-1}AS &= Q + V^{-1}AP - QV^{-1}A_+P \\ &= Q + PV^{-1}AP + QV^{-1}AP - QV^{-1}A_+P \\ &= Q + PA_PP + QV^{-1}(A - A_+)P \quad \text{on } U \setminus \{w\}. \end{aligned} \tag{4.1.10}$$

It follows from (4.1.4) that

$$A(z) - A_+(z) = \sum_{n=m}^{-1} (z - w)A_n\,, \qquad z \in U \setminus \{w\}\,.$$

Since, by (4.1.4), $A_nP = PA_n$ for $m \le n \le -1$, this implies that

$$(A - A_+)P = P(A - A_+)P$$

and therefore

$$QV^{-1}(A - A_+)P = QV^{-1}P(A - A_+)P \qquad \text{on } U \setminus \{w\}.$$

Since, by (4.1.8), $V^{-1}P = P$, this further implies that

$$QV^{-1}(A - A_+)P = 0.$$

Together with (4.1.10) this gives

$$V^{-1}AS = Q + PA_PP \quad \text{on } U \setminus \{w\}. \tag{4.1.11}$$

Setting $T = V^{-1}$, we get the required relation (4.1.1). It remains to observe that from (4.1.1) it follows that A_P is meromorphic at w and holomorphic on $U \setminus \{w\}$, because A has these properties. □

The function A_P in (4.1.1) can be represented by a meromorphic matrix function. Since the inverse of such a function is again meromorphic (if it exists), we immediately obtain the following

4.1.3 Corollary (to Theorem 4.1.2). *Let $D \subseteq \mathbb{C}$ be an open set, and let $Z \subseteq D$ be a discrete and closed subset of D. Let A be a holomorphic $GL(E)$-valued function on $D \setminus Z$ which is finite meromorphic and Fredholm at each point of Z. Then A^{-1} is finite meromorphic and Fredholm at each point of Z.*

4.1.4 Proposition. *Let $D \subseteq \mathbb{C}$ be a connected open set, and let $Z \subseteq D$ be a discrete and closed subset of D. Let A be a holomorphic $L(E)$-valued function on $D \setminus Z$ which is finite meromorphic and Fredholm at each point of D. Suppose, there exists at least one point $z_0 \in D \setminus Z$ such that $A(z_0)$ is invertible. Then there exists a discrete and closed subset Z' of D with $Z' \supseteq Z$ such that $A(z)$ is invertible for each $z \in D \setminus Z'$, and the function $A^{-1} : D \setminus Z' \to GL(E)$ is finite meromorphic and Fredholm at each point of D.*

Proof. Denote by D' the set of all $w \in D \setminus Z$ such that there exists a neighborhood $U \subseteq D$ with $U \setminus \{w\} \subseteq D \setminus Z$ and $A(z) \in GL(E)$ for all $z \in U \setminus \{w\}$.

We claim that $D' = D$. Obviously, D' is open and, by hypothethis, $D' \supseteq \{z_0\} \neq \emptyset$. Since D is connected, it remains to prove that D' is relatively closed in D. Let $w \in D$ be a boundary point of D'. Since, by hypothesis, A is finite meromorphic and Fredholm at w, first we can find a neighborhood $W \subseteq D \setminus Z$ such that A is finite meromorphic and Fredholm at w. Then from Theorem 4.1.2 we get a neighborhood $U \subseteq W$ of w, a finite dimensional projector P in E, holomorphic functions $S, T : U \to GL(E)$ and a holomorphic function $A_P : U \setminus \{w\} \to L(\operatorname{Im} P)$, which is meromorphic at w, such that, with $Q := I - P$,

$$A = S(Q + PA_P P)T \qquad \text{on } U \setminus \{w\}. \tag{4.1.12}$$

Then the determinant $\det A_P$ is holomorphic on $U \setminus \{w\}$ and meromorphic at w. Hence, there is a neighborhood $V \subseteq U$ of w such that either

$$\det A_P \equiv 0 \qquad \text{on } V \setminus \{w\} \tag{4.1.13}$$

or

$$\det A_P(z) \neq 0 \qquad \text{for all } z \in V \setminus \{w\}. \tag{4.1.14}$$

By (4.1.12), from (4.1.13) it would follow that A is not invertible for all $z \in V \setminus \{w\}$. But this is impossible, as w is a boundary point of D'. Hence, we have (4.1.14). Again by (4.1.12) this means that $A(z)$ is invertible for all $z \in V \setminus \{w\}$, i.e., $w \in D'$.

Let Z' be the set of all $w \in D$ such that either $w \in Z$ or $w \in D \setminus Z$ and $A(w)$ is not invertible. Since $D = D'$, it follows, by definition of D', that Z' is discrete and closed in D. □

4.1.5 Proposition. *Let $D \subseteq \mathbb{C}$ be a bounded open set with piecewise $\mathcal{C}^1$-boundary. Let $Z \subseteq D$ be a finite set, and let $M : \overline{D} \setminus Z \to L(E)$ be a holomorphic function which is finite meromorphic at each point of Z, such that*

$$\|M(z)\| < 1 \qquad \textit{for all } z \in \partial D. \tag{4.1.15}$$

Then $I+M$ is finite meromorphic and Fredholm at each point of $\overline{D}$, and the index of $I+M$ is zero at each point of $\overline{D}$. Moreover, there exists a finite subset Z' of D with $Z' \supseteq Z$ such that $A(z)$ is invertible for each $z \in \overline{D} \setminus Z'$,

Proof. Since M is finite meromorphic at each point of Z, it is clear that $I + M$ is finite moromorphic at each point of Z. Let K the sum of the principal parts of M at the points of Z, and set $A = I + M - K$ on $\overline{D} \setminus Z$. Then A admits a holomorphic extension to D, which we also denote by A.

Let $\mathcal{F}(E)$ be the ideal of finite dimensional operators in E, and let $\mathcal{F}^\infty(E)$ be the closure of $\mathcal{F}(E)$ in $L(E)$ with respect to the operator norm. Consider the factor algebra $\widehat{L}(E) := L(E)/\mathcal{F}^\infty(E)$. For $T \in L(E)$, we denote by $\widehat{T}$ the class of T in $\widehat{L}(E)$. Then $\widehat{K} \equiv 0$, and it follows from (4.1.15) that

$$\|\widehat{A}(z) - \widehat{I}\| = \|\widehat{M}(z)\| \leq \|M(z)\| < 1 \quad \text{for all } z \in \partial D.$$

Hence, by the maximum principle,

$$\|\widehat{A}(z) - \widehat{I}\| < 1 \qquad \text{for all } z \in \overline{D}.$$

Therefore $\widehat{A}(z)$ is invertible for all $z \in \overline{D}$. Hence, $A(z)$ is a Fredholm operator with index zero for all $z \in \overline{D}$, which implies that $I + M = A + K$ is finite meromorphic and Fredholm at each point of $\overline{D}$, and the index of $I + M$ is zero at each point of $\overline{D}$. Since $I + M(z)$ is invertible for $z \in \partial D$, it follows from Proposition 4.1.4 that there exists a finite subset Z' of D with $Z' \supseteq Z$ such that $I + M(z)$ is invertible for each $z \in \overline{D} \setminus Z'$. □

4.2 Invertible finite meromorphic Fredholm functions

First recall the notion of the *trace* of a finite dimensional linear operator. For a complex matrix

$$A = \left(a_{jk}\right)_{1\leq j,k\leq n} \in L(n,\mathbb{C}),$$

the **trace** of A, which we denote by $\operatorname{tr} A$, is defined by

$$\operatorname{tr} A = \sum_{j=1}^{n} a_{jj}\,. \tag{4.2.1}$$

If

$$B = \left(B_{jk}\right)_{1\leq j,k\leq n} \in L(n,\mathbb{C})$$

is a second matrix, then

$$\operatorname{tr}(AB) = \sum_{j=1}^{n}\sum_{\nu=1}^{n} a_{j\nu}b_{\nu j} = \sum_{\nu=1}^{n}\sum_{j=1}^{n} b_{\nu j}a_{j\nu} = \operatorname{tr}(BA)\,. \tag{4.2.2}$$

In particular, if $A \in L(n,\mathbb{C})$ and $B \in GL(n,\mathbb{C})$, then

$$\operatorname{tr}(BAB^{-1}) = \operatorname{tr}(A). \tag{4.2.3}$$

Therefore, for each finite dimensional complex vector space F and each operator $A \in L(F)$, the trace of A is well defined: If $f_1, \dots, f_n$ is a Basis of F and $a = (a_{jk})_{j,k=1,\dots,n}$ is the matrix with

$$Af_k = \sum_{j=1}^{n} a_{jk} f_j ,$$

then

$$\operatorname{tr} A := \sum_{j=1}^{n} a_{jj} .$$

From (4.2.2) it follows that

$$\operatorname{tr}(AB) = \operatorname{tr}(BA) \qquad \text{for all } A, B \in L(F). \tag{4.2.4}$$

Now let F an H be two finite dimensional complex vector spaces such that F is a subspace of H.

If $A \in L(F)$, then we say that $\widetilde{A} \in L(H)$ is an **extension by zero** of A if $\widetilde{A}\big|_F = A$ and $\operatorname{rank} \widetilde{A} = \operatorname{rank} A$. If $A \in L(F)$ and $\widetilde{A} \in L(H)$ is an extension by zero of A, then

$$\operatorname{tr} A = \operatorname{tr} \widetilde{A}. \tag{4.2.5}$$

Indeed, since $\widetilde{A}\big|_F = A$ and $\operatorname{rank} \widetilde{A} = \operatorname{rank} A$, we can find a basis $f_1, \dots, f_n$ of H such that, for some $m \le n$, the vectors $f_1, \dots, f_m$ form a basis of F and $f_{m+1}, \dots, f_n \in \operatorname{Ker} \widetilde{A}$. Let $(a_{jk})_{j,k=1,\dots,n}$ be the matrix with

$$\widetilde{A} f_k = \sum_{j=1}^{n} a_{jk} f_j \qquad \text{for } 1 \le k \le n. \tag{4.2.6}$$

Since $f_{m+1}, \dots, f_n \in \operatorname{Ker} \widetilde{A}$ and therefore $a_{jk} = 0$ for $k \ge m+1$, this implies that

$$\operatorname{tr} \widetilde{A} = \sum_{j=1}^{m} a_{jj} . \tag{4.2.7}$$

Since $A = \widetilde{A}\big|_F$, it follows from (4.2.6) that

$$Af_k = \sum_{j=1}^{m} a_{jk} f_j \qquad \text{for } 1 \le k \le m. \tag{4.2.8}$$

Taking into account also (4.2.7), this implies (4.2.5).

Now let E be an arbitrary Banach space.

Let $\mathcal{F}(E)$ be the ideal of $L(E)$ which consists of the finite dimensional operators. For $A \in \mathcal{F}(E)$, we denote by $\mathcal{S}_A$ the set of all finite dimensional subspaces F of E such that $\operatorname{Im} A \subseteq F$ and $F \cap \operatorname{Ker} A = \{0\}$. Then, for each $A \in \mathcal{F}(E)$ and all $F \in \mathcal{S}_A$, we view $A\big|_F$ as an operator in F. We claim that

$$\operatorname{tr}\left(A\big|_F\right) = \operatorname{tr}\left(A\big|_G\right) \qquad \text{for all } A \in \mathcal{F}(E) \text{ and } F, G \in \mathcal{S}_A\,. \tag{4.2.9}$$

Indeed, let $A \in \mathcal{F}(E)$ and $F, G \in \mathcal{S}_A$ be given. Take $H \in \mathcal{S}_A$ such that $F \cup G \subseteq H$. Then $A\big|_H$ is an extension by zero both of $A\big|_F$ and $A\big|_G$. Therefore it follows from (4.2.5) that

$$\operatorname{tr}\left(A\big|_F\right) = \operatorname{tr}\left(A\big|_H\right) = \operatorname{tr}\left(A\big|_G\right).$$

In view of (4.2.9), the following definition is correct:

4.2.1 Definition. Let $A \in L(E)$ be finite dimensional. Then we take a finite dimensional subspace F of E with $\operatorname{Im} A \subseteq F$ and $F \cap \operatorname{Ker} A = \{0\}$, and define

$$\operatorname{tr} A = \operatorname{tr}\left(A\big|_F\right).$$

Note that, by (4.2.4), for all operators $A, B \in L(E)$ such that at least one of them is finite dimensional,

$$\operatorname{tr}(AB) = \operatorname{tr}(BA)\,. \tag{4.2.10}$$

4.2.2 Proposition. *Let $D \subseteq \mathbb{C}$ be a bounded open set with piecewise C^1-boundary, let $Z \subseteq D$ be a finite set, and let $A, B : \overline{D} \setminus Z \to L(E)$ be two holomorphic functions which are finite meromorphic at the points of Z. Then the integrals*

$$\int_{\partial D} A(z)B(z)dz \qquad \text{and} \qquad \int_{\partial D} B(z)A(z)dz \tag{4.2.11}$$

are finite dimensional, and

$$\operatorname{tr} \int_{\partial D} A(z)B(z)\,dz = \operatorname{tr} \int_{\partial D} B(z)A(z)\,dz\,. \tag{4.2.12}$$

Proof. Since the functions AB and BA are finite meromorphic at Z, by Cauchy's theorem, it is clear that the integrals (4.2.11) are finite dimensional. To prove (4.2.12), again by Cauchy's theorem, we may assume that Z consists of a single point w, and D is a disc centered at w. Since the functions A and B are finite meromorphic at w, their Laurent expansions are of the form

$$A(z) = \sum_{n=-m}^{\infty} (z-w)^n A_n\,, \quad B(z) = \sum_{n=-m}^{\infty} (z-w)^n B_n\,, \quad z \in \overline{D} \setminus \{w\},$$

where $m \in \mathbb{N}^*$ and the operators $A_{-m}, \ldots, A_{-1}$ and $B_{-m}, \ldots, B_{-1}$ are finite dimensional. Then

$$\int_{\partial D} A(z)B(z)\,dz = 2\pi i \sum_{-m \le j \le m-1} A_j B_{-j-1}$$

and

$$\int_{\partial D} B(z)A(z)\,dz = 2\pi i \sum_{-m \le j \le m-1} B_j A_{-j-1}\,.$$

Now it follows from (4.2.10) that

$$\begin{aligned} \operatorname{tr} \int_{\partial D} A(z)B(z)\,dz &= 2\pi i \sum_{-m \le j \le m-1} \operatorname{tr}(A_j B_{-j-1}) \\ &= 2\pi i \sum_{-m \le j \le m-1} \operatorname{tr}(B_j A_{-j-1}) = \operatorname{tr} \int_{\partial D} B(z)A(z)\,dz\,. \end{aligned}$$

□

4.2.3 Proposition. *Let $D \subseteq \mathbb{C}$ be a bounded open set with piecewise C^1-boundary, let $Z \subseteq D$ be a finite set, and let $A : \overline{D} \setminus Z \to GL(E)$ be a holomorphic function which is finite meromorphic and Fredholm at the points of Z. Then the operators*

$$\int_{\partial D} A'(z)A^{-1}(z)dz \qquad \textit{and} \qquad \int_{\partial D} A^{-1}(z)A'(z)dz \tag{4.2.13}$$

are finite dimensional and

$$\operatorname{tr} \int_{\partial D} A'(z)A^{-1}(z)dz = \operatorname{tr} \int_{\partial D} A^{-1}(z)A'(z)dz\,. \tag{4.2.14}$$

Proof. By hypothesis, A is finite meromorphic and Fredholm at the points of Z. This implies that A' is finite meromorphic at Z. By Corollary 4.1.3, also A^{-1} is finite meromorphic at Z. Hence $A'A^{-1}$ and $A^{-1}A'$ are finite meromorphic at Z. Hence the operators (4.2.13) are finite dimensional.

To prove that they have the same trace, again by Cauchy's theorem, we may assume that Z consists only of one point w. Since, on $D \setminus \{w\}$, the values of A are invertible and A is finite meromorphic and Fredholm at w, it is clear that the index of A at w is zero. Therefore, by Theorem 4.1.2, there exist a neighborhood $U \subseteq D$ of w, a finite dimensional projector P in E, holomorphic functions $S, T : U \to GL(E)$ and a holomorphic function $A_P : U \setminus \{w\} \to L(\operatorname{Im} P)$, which is meromorphic at w, such that, with $Q := I - P$,

$$A = S(Q + PA_P P)T \qquad \text{on } U \setminus \{w\}.$$

From this representation it follows that the values of A_P are invertible on $U \setminus \{w\}$ and

$$A'A^{-1} = PA'_P A_P^{-1} P\,, \qquad A^{-1}A' = PA_P^{-1} A'_P P\,.$$

Therefore, it follows from (4.2.10) that

$$\begin{aligned} \operatorname{tr} \int_{\partial D} A'(z)A^{-1}(z)\,dz &= \int_{\partial D} \operatorname{tr}\left(PA'_P(z)A_P^{-1}(z)P\right) dz \\ &= \int_{\partial D} \operatorname{tr}\left(PA_P^{-1}(z)A'_P(z)P\right) dz = \operatorname{tr} \int_{\partial D} A^{-1}(z)A'(z)\,dz\,. \end{aligned}$$

□

4.2.4 Definition. Let $D \subseteq \mathbb{C}$ be a bounded open set with piecewise $\mathcal{C}^1$-boundary, let $Z \subseteq D$ be a finite set, and let $A : \overline{D} \setminus Z \to GL(E)$ be a holomorphic function which is finite meromorphic and Fredholm at the points of Z. Then, by Proposition 4.2.3, the following definition is correct:

$$\operatorname{ind}_{\partial D} A := \frac{1}{2\pi i} \operatorname{tr} \int_{\partial D} A'(z)A^{-1}(z)dz = \frac{1}{2\pi i} \operatorname{tr} \int_{\partial D} A^{-1}(z)A'(z)dz\,.$$

The number $\operatorname{ind}_{\partial D} A$ will be called the **index of A with respect to the contour ∂D**

In Section 4.4, we shall prove that this index is always an integer, using the Smith factorization theorem from the following Section 4.3. The present section is concluded with the following proposition:

4.2.5 Proposition. *Let $D \subseteq \mathbb{C}$ be a bounded open set with piecewise $\mathcal{C}^1$-boundary, let $Z \subseteq D$ be a finite set, and let $A, B : \overline{D} \setminus Z \to GL(E)$ be two holomorphic functions which are finite meromorphic and Fredholm at the points of Z. Then*

$$\operatorname{ind}_{\partial D}(AB) = \operatorname{ind}_{\partial D} A + \operatorname{ind}_{\partial D} B\,. \tag{4.2.15}$$

Proof. Since $(AB)'(AB)^{-1} = A'A^{-1} + AB'B^{-1}A^{-1}$, we have

$$\operatorname{ind}_{\partial D}(AB) = \operatorname{ind}_{\partial D} A + \frac{1}{2\pi i} \operatorname{tr} \int_{\partial D} AB'B^{-1}A^{-1}\,dz\,.$$

Since the functions A, B and B' are finite meromorphic, and, by Corollary 4.1.3, also the functions A^{-1} and B^{-1} are finite meromorphic, it follows from Proposition 4.2.2 that

$$\frac{1}{2\pi i} \operatorname{tr} \int_{\partial D} AB'B^{-1}A^{-1}\,dz = \frac{1}{2\pi i} \operatorname{tr} \int_{\partial D} B'B^{-1}\,dz = \operatorname{ind}_{\partial D} B.$$

Together this proves (4.2.15). □

4.3 Smith factorization

Recall that, in this book, we denote by $L(n,\mathbb{C})$ the algebra of complex $n \times n$ matrices, and by $GL(n,\mathbb{C})$ we denote the group of invertible elements in $L(n,\mathbb{C})$.

4.3.1 Theorem (Smith factorization). *Let $w \in \mathbb{C}$, let W be a neighorhood of w, and let A be an $n \times m$-matrix of scalar meromorphic functions on W such that at least one of these functions does not identically vanish on $W \setminus \{w\}$. Then there exist uniquely determined integers $\kappa_1 \geq \ldots \geq \kappa_r$, $1 \leq r \leq \min(n,m)$, a neighborhood $U \subseteq W$ of w and holomorphic functions $E : U \to GL(n,\mathbb{C})$, $F : U \to GL(m,\mathbb{C})$ such that*

$$EAF = \begin{pmatrix} \Delta & 0 \\ 0 & 0 \end{pmatrix}$$

where Δ is the $r \times r$ diagonal matrix with the diagonal

$$(z-w)^{\kappa_1}, \ldots, (z-w)^{\kappa_r}\ .$$

4.3.2 Definition. The integers $\kappa_1 \geq \ldots \geq \kappa_r$ from the preceding definition will be called the **powers** of A at z_0.[2]

Proof. Uniqueness: The number r is the rank of the matrix EAF over $\mathbb{C} \setminus \{z_0\}$. Since the values of E and F are invertible, this implies that r is the rank of A over $U \setminus \{z_0\}$. Hence r is uniquely determined by A.

Now we assume that there are given two vectors of integers $\kappa_1 \geq \ldots \geq \kappa_r$ and $\widetilde{\kappa}_1 \geq \ldots \geq \widetilde{\kappa}_r$ such that, for some $1 \leq p \leq r$,

$$\kappa_p > \widetilde{\kappa}_p, \tag{4.3.1}$$

and such that, for some holomorphic functions $E, \widetilde{E} : U \to GL(n,\mathbb{C})$, $F, \widetilde{F} : U \to GL(m,\mathbb{C})$ in a neighborhood $U \subseteq D$ of z_0,

$$EAF = \begin{pmatrix} \Delta & 0 \\ 0 & 0 \end{pmatrix} \quad \text{and} \quad \widetilde{E}A\widetilde{F} = \begin{pmatrix} \widetilde{\Delta} & 0 \\ 0 & 0 \end{pmatrix} \tag{4.3.2}$$

where Δ and $\widetilde{\Delta}$ are the $r \times r$ diagonal matrices with the diagonals

$$(z-z_0)^{\kappa_1}, \ldots, (z-z_0)^{\kappa_r} \quad \text{and} \quad (z-z_0)^{\widetilde{\kappa}_1}, \ldots, (z-z_0)^{\widetilde{\kappa}_r}\ ,$$

respectively. Since $\kappa_1 \geq \ldots \geq \kappa_p$ and $\widetilde{\kappa}_1 \geq \ldots \geq \widetilde{\kappa}_p$, then it follows from (4.3.1) that

$$\kappa_\nu > \widetilde{\kappa}_\mu \qquad \text{for } 1 \leq \nu \leq p \leq \mu \leq r. \tag{4.3.3}$$

[2] This vector of powers is not identical with the numerical characteristic of A at w which we introduce below (Def. 11.3.6), but it is closely related: The vector of powers and the numerical characteristic uniquely determine each other (see the beginning of Section 11.5 for this relation.)

Moreover it follows from (4.3.2) that

$$E^{-1}\begin{pmatrix}\Delta & 0\\ 0 & 0\end{pmatrix}F^{-1} = A = \widetilde{E}^{-1}\begin{pmatrix}\widetilde{\Delta} & 0\\ 0 & 0\end{pmatrix}\widetilde{F}^{-1}$$

and therefore

$$\widetilde{E}E^{-1}\begin{pmatrix}\Delta & 0\\ 0 & 0\end{pmatrix} = \begin{pmatrix}\widetilde{\Delta} & 0\\ 0 & 0\end{pmatrix}\widetilde{F}^{-1}F \qquad \text{on } U\setminus\{z_0\}\,. \tag{4.3.4}$$

Let $\alpha_{\mu\nu}$ be the elements of $\widetilde{E}E^{-1}$, and $\beta_{\mu\nu}$ the elements of $\widetilde{F}^{-1}F$, where μ is the row index and ν is the column index. Then by (4.3.2)

$$(z-z_0)^{\kappa_\nu}\alpha_{\mu\nu} = (z-z_0)^{\widetilde{\kappa}_\mu}\beta_{\mu\nu} \quad \text{on } U\setminus\{z_0\} \quad \text{if } 1\le\mu,\nu\le r\,.$$

By (4.3.3) this implies that $\beta_{\mu\nu}(z_0) = 0$ for $\nu \le p \le \mu \le r$. This is impossible, because $\widetilde{F}^{-1}(z_0)F(z_0)$ is invertible.

Existence: Let $\mathcal{M}$ be the ring of germs of scalar meromorphic functions at z_0, and let $\mathcal{O}$ be the subring of holomorphic germs. Let $\mathcal{M}(k)$ be the algebra of $k\times k$-matrices with elements from $\mathcal{M}$, and let $\mathcal{O}(k)$ be the subalgebra matrices with elements from $\mathcal{O}$. Denote by $G\mathcal{O}(k)$ the group of invertible elements in $\mathcal{O}(k)$. Let $[z-z_0]$ be the element in $\mathcal{O}$ defined by the function $z-z_0$. Further, let $\mathcal{M}(n,m)$ be the space of $n\times m$-matrices with elements from $\mathcal{M}$.

Then A can be viewed as an element of $\mathcal{M}(n,m)$ and we have to find $E\in G\mathcal{O}(n)$ and $F\in G\mathcal{O}(m)$ such that, for some integers $\kappa_1\ge\ldots\ge\kappa_r$, EAF is of the block form such that EAF is of the block form

$$EAF = \begin{pmatrix}\Delta & 0\\ 0 & 0\end{pmatrix}$$

where Δ is the $r\times r$ diagonal matrix with the diagonal

$$[z-z_0]^{\kappa_1},\ldots,[z-z_0]^{\kappa_r}\,.$$

Consider the following three row operations, which we will apply to matrices from $\mathcal{M}(n,m)$:

(I) Multiplying a row by an element $g\in G\mathcal{O}$.

(II) Interchanging two rows.

(III) Taking two different rows a and b and an arbitrary element $f\in\mathcal{O}$ and replacing the row b by $b+fa$.

Operation (I) can be realized multiplying from the left by the diagonal matrix obtained from the unit matrix after replacing the corresponding element of the diagonal by g. Operation (II) can be realized multiplying from the left by the matrix obtained from the unit matrix after interchanging the corresponding columns.

Operation (III) can be realized multiplying from the left by the matrix obtained from the unit matrix after replacing one of the zero elements outside the diagonal by f. All these matrices belong to the group $G\mathcal{O}(n)$.

Multiplying the corresponding matrices of the group $G\mathcal{O}(m)$ from the right we obtain the corresponding column operations, which we denote by (I′), (II′) and (III′), respectively.

Therefore it is sufficient to prove that, by a finite number of the operations (I), (II), (III), (I′), (II′) and (III′), the matrix A can be transformed to a matrix of the required diagonal form.

If $a \in \mathcal{M}$ and $a \neq 0$, then there is a uniquely determined entire number N such that the Laurent series of a is of the form

$$a = \sum_{n=N}^{\infty} a_n [z - z_0]^n \qquad \text{with} \qquad a_N \neq 0.$$

We call this number the **order** of a and denote it by $\operatorname{ord} a$. For the zero element $0 \in \mathcal{M}$ we write $\operatorname{ord} 0 = \infty$. For $a \in \mathcal{M}$ with $a \neq 0$, we define

$$\widehat{a} = [z - z_0]^{-\operatorname{ord} a} a.$$

Then $\widehat{a} \in G\mathcal{O}$ for all $a \in \mathcal{M}$ with $a \neq 0$.

Denote by N_1 the minimal order of the elements of A. Since A is not the zero matrix, $N_1 < \infty$. Choose $1 \leq p \leq n$ and $1 \leq q \leq m$ such that at the place (p, q) of A we have an element of order N_1. Denote this element by a.

Now we proceed as follows: 1. We multiply the p-th row of A by $\widehat{a}^{-1}$. 2. We interchange the $q-th$ column of the obtained matrix with the first column. 3. We interchange the $p-th$ row of the now obtained matrix with the first row. After these operations, the orders of all elements of the matrix are still $\geq N_1$, and at the place $(1, 1)$ we have the element $[z - z_0]^{N_1}$.

Now consider the element b at the place $(1, 2)$ (first row and second column). Since $\operatorname{ord} b \geq N_1$, then $[z - z_0]^{\operatorname{ord} b - N_1} \widehat{b}$ belongs to $\mathcal{O}$ and we may multiply the first column by $[z - z_0]^{\operatorname{ord} b - N_1} \widehat{b}$ and then substract the result from the second column. So we get a matrix with a zero at the place $(1, 2)$. Doing the same with the elements at the places $(1, 3), \ldots, (1, m)$, we end up with a matrix with $[z - z_0]^{N_1}$ at the place $(1, 1)$ and with zeros at the places $(1, 2), \ldots, (1, m)$. Then the same procedure with row operations leads to a matrix of the form

$$\begin{pmatrix} [z - z_0]^{N_1} & 0 & \ldots & 0 \\ 0 & & & \\ \vdots & & B & \\ 0 & & & \end{pmatrix}$$

where B is an $(n - 1) \times (m - 1)$-matrix with elements from $\mathcal{M}$ and the orders of all elements of B are still $\geq N_1$. If $B = 0$, the proof is complete. If not we apply the same procedure to the matrix B. If N_2 is the minimal order of the elements of B, then this gives a matrix of the form

$$\begin{pmatrix} [z-z_0]^{N_1} & 0 & 0 & \dots & 0 \\ 0 & [z-z_0]^{N_2} & 0 & \dots & 0 \\ 0 & 0 & & & \\ \vdots & \vdots & & C & \\ 0 & 0 & & & \end{pmatrix}$$

where C is an $(n-2)\times(m-2)$-matrix. If $C=0$, we set $\kappa_1 = N_2$ and $\kappa_2 = N_1$, interchange the first two columns and then the first two rows, and the proof is complete. If not we proceed in this way and, for some $3 \le r \le \min(n,m)$, we end up with a block matrix of the form

$$\begin{pmatrix} \Delta & 0 \\ 0 & 0 \end{pmatrix}$$

where Δ is the $r \times r$ diagonal matrix with the diagonal

$$(z-z_0)^{N_1}, \dots, (z-z_0)^{N_r}.$$

This completes the proof of Lemma 4.3.1. □

4.3.3 Corollary. *Let $w \in \mathbb{C}$, and let $K : \mathbb{C} \setminus L(n,\mathbb{C})$ be a rational matrix function of the form*

$$K(z) = \sum_{n=-m}^{-1} (z-w)^n K_n, \tag{4.3.5}$$

where $K_{-m}, \dots, K_{-1}$, $1 \le m < \infty$, are constant complex $n\times n$-matrices. Then there exists a neighborhood U of w and a holomorphic matrix function $A : U \to L(n,\mathbb{C})$ such that $A(z)$ is invertible for all $z \in U \setminus \{w\}$ and K is the principal part of the Laurent expansion of A^{-1} at w.

Proof. For $K \equiv 0$ this is trivial. Let $K \not\equiv 0$. Then, by the Smith factorization Theorem 4.3.1, there are integers $\kappa_1 \ge \dots \ge \kappa_r$, $1 \le r \le n$, a neighborhood U of w and holomorphic functions $E, F : U \to GL(n,\mathbb{C})$, such that

$$K = E\Delta F$$

where Δ is the $n \times n$ diagonal matrix with the diagonal

$$(z-w)^{\kappa_1}, \dots, (z-w)^{\kappa_r}, \underbrace{0,\dots,0}_{(n-r)\text{ times}}.$$

Let Δ_+ be the $n\times n$-diagonal matrix with the diagonal $d_1, \dots, d_n$ where

$$d_j = \begin{cases} (z-w)^{-\kappa_j} & \text{if } 1 \le j \le r \text{ and } \kappa_j < 0, \\ 1 & \text{if } r+1 \le j \le n \text{ or } 1 \le j \le r \text{ and } \kappa_j \ge 0. \end{cases}$$

Then Δ_+ is holomorphic on U, invertible on $U \setminus \{w\}$, and $\Delta_+^{-1} - \Delta$ extends holomorphically to w. Set $A = F^{-1}\Delta_+ E^{-1}$. Then A is holomorphic on U, invertible on $U \setminus \{w\}$, and

$$\begin{aligned} A^{-1}(z) &= E(z)\Delta_+^{-1}(z)F(z) \\ &= E(z)\Delta(z)F(z) + E(z)\Big(\Delta_+^{-1}(z) - \Delta(z)\Big)F(z) \\ &= K(z) + E(z)\Big(\Delta_+^{-1}(z) - \Delta(z)\Big)F(z) \quad \text{for } z \in U \setminus \{w\}. \end{aligned}$$

Since $E(\Delta_+^{-1} - \Delta)F$ extends holomorphically to w, this shows that K is the principal part of the Laurent expansion of A^{-1} at w. □

It is impossible to replace in Corollary 4.3.3 the prescribed function (4.3.5) by an arbitrary function of the form

$$K(z) = \sum_{n=-m}^{0} (z-w)^n K_n \,,$$

with matrices $K_{-m}, \ldots, K_0$, $1 \le m < \infty$ where $K_0 \neq 0$. We give a counterexample:

4.3.4 Example. Let

$$K(z) := \begin{pmatrix} 1 & z^{-1} \\ 0 & 1 \end{pmatrix}. \tag{4.3.6}$$

Then it is impossible to find a neighborhood U of $0 \in \mathbb{C}$ and holomorphic functions $A, B : U \to L(2, \mathbb{C})$ such that $A(z)$ is invertible for $z \in U \setminus \{0\}$ and

$$A^{-1}(z) = \begin{pmatrix} 1 & z^{-1} \\ 0 & 1 \end{pmatrix} + zB(z)\,, \qquad z \in U \setminus \{0\}.$$

Indeed, assume this is possible. Then

$$\begin{pmatrix} 1 & 0 \\ 0 & 1 \end{pmatrix} = \begin{pmatrix} 1 & z^{-1} \\ 0 & 1 \end{pmatrix} A(z) + zB(z)A(z)\,, \qquad z \in U \setminus \{0\}. \tag{4.3.7}$$

Let

$$A(z) = \begin{pmatrix} a_{11}(z) & a_{12}(z) \\ a_{21}(z) & a_{22}(z) \end{pmatrix} \quad \text{and} \quad B(z) = \begin{pmatrix} b_{11}(z) & b_{12}(z) \\ b_{21}(z) & b_{22}(z) \end{pmatrix}.$$

Then it follows from (4.3.7) that

$$0 = a_{12}(z) + z^{-1}a_{22}(z) + zb_{11}(z)a_{12}(z) + zb_{12}(z)a_{22}(z)$$

and

$$1 = a_{22}(z) + zb_{21}(z)a_{12}(z) + zb_{22}(z)a_{22}(z)$$

for $z \in U \setminus \{0\}$. This is impossible, as from the first relation it follows that $a_{22}(0) = 0$ and from the second relation it follows that $a_{22}(0) = 1$.

4.4 The Rouché theorem

4.4.1 Theorem. *Let $D \subseteq \mathbb{C}$ be a bounded open set with piecewise C^1-boundary, let $Z \subseteq D$ be a finite set, and let $A : \overline{D} \setminus Z \to GL(E)$ be a holomorphic function which is finite meromorphic and Fredholm at the points of Z. Then* $\operatorname{ind}_{\partial D} A$ *(Def.* 4.2.4) *is an integer.*

Proof. By Cauchy's theorem, we may assume that Z consists only of a single point w. Since, on $D \setminus \{w\}$, the values of A are invertible and A is finite meromorphic and Fredholm at w, it is clear that the index of A at w is zero. Therefore, by Theorem 4.1.2, there exist a neighborhood $U \subseteq D$ of w, a finite dimensional projector P in E, holomorphic functions $S, T : U \to GL(E)$ and a holomorphic function $A_P : U \setminus \{w\} \to L(\operatorname{Im} P)$, which is meromorphic at w, such that, with $Q := I - P$,

$$A = S(Q + PA_P P)T \qquad \text{on } U \setminus \{w\}. \tag{4.4.1}$$

We may assume that U is a disc. Then, again by Cauchy's theorem,

$$\operatorname{ind}_{\partial D} A = \operatorname{ind}_{\partial U} A\,.$$

Since $\operatorname{ind}_{\partial U} T = \operatorname{ind}_{\partial U} S = 0$, this further implies, by (4.4.1) and Proposition 4.2.5, that

$$\operatorname{ind}_{\partial D} A = \operatorname{ind}_{\partial U}(Q + PA_P P).$$

By Cauchy's theorem this implies that

$$\operatorname{ind}_{\partial D} A = \operatorname{ind}_{\partial U} A_P\,. \tag{4.4.2}$$

By the Smith factorization Theorem 4.3.1, there exist uniquely determined integers $\kappa_1 \geq \ldots \geq \kappa_r$, a neighborhood $U \subseteq W$ of w and holomorphic functions $E, F : U \to GL(\operatorname{Im} P)$ such that

$$EA_P F = \begin{pmatrix} \Delta & 0 \\ 0 & 0 \end{pmatrix} \tag{4.4.3}$$

where, with respect to an appropriate basis of $\operatorname{Im} P$, Δ can be represented by the $r \times r$ diagonal matrix with the diagonal

$$(z-w)^{\kappa_1}, \ldots, (z-w)^{\kappa_r}\,.$$

Note that

$$\operatorname{ind}_{\partial U} \Delta = \frac{1}{2\pi i} \int\limits_{\partial U} \operatorname{tr}(\Delta' \Delta^{-1})\, dz = \frac{1}{2\pi i} \int\limits_{\partial U} \sum_{j=1}^{r} \frac{\kappa_j}{z-w}\, dz = \sum_{j=1}^{r} \kappa_j.$$

By Proposition 4.2.5 and (4.4.3), this implies that

$$\operatorname{ind}_{\partial U} A_P = \sum_{j=1}^{r} \kappa_j\,.$$

Together with (4.4.2) this proves the theorem □

4.4.2 Lemma. *Let $D \subseteq \mathbb{C}$ be a bounded open set with piecewise C^1-boundary. Let $Z \subseteq D$ be a finite set, and let $M : \overline{D} \setminus Z \to L(E)$ be a holomorphic function which is finite meromorphic at each point of Z, such that*

$$\|M(z)\| < 1 \qquad \text{for all } z \in \partial D. \tag{4.4.4}$$

Then $I + M$ is finite meromorphic and Fredholm at each point of $\overline{D}$, and the index of $I + M$ is zero at each point of $\overline{D}$. Moreover, there exists a finite subset Z' of D with $Z' \supseteq Z$ such that $I + M(z)$ is invertible for each $z \in \overline{D} \setminus Z'$. Moreover,

$$\operatorname{ind}_{\partial D}(I + M) = 0\,. \tag{4.4.5}$$

Proof. Except for relation (4.4.5), this is precisely the statement of propositon 4.1.5. Moreover, by the same Proposition 4.1.5, for each $0 \le t \le 1$, the function $I + tM$ has the same properties. Obviously, the function

$$[0, 1] \ni t \longrightarrow \operatorname{ind}_{\partial D}(I + tM)$$

is continuous. Since, by Theorem 4.4.1, the values of this function are integers and since $\operatorname{ind}_{\partial D} I = 0$, it follows that $\operatorname{ind}_{\partial D}(I + M) = 0$. □

4.4.3 Theorem. *Let $D \subseteq \mathbb{C}$ be a bounded open set with piecewise C^1-boundary, let $Z \subseteq D$ be a finite set, let $A : \overline{D} \setminus Z \to GL(E)$ be a holomorphic function which is finite meromorphic and Fredholm at the points of Z, and let $S : \overline{D} \setminus Z \to GL(E)$ be a holomorphic function which is finite meromorphic at each points of Z such that*

$$\|A^{-1}(z)S(z)\| < 1 \qquad \text{for } z \in \partial D\,. \tag{4.4.6}$$

Then the function $A + S$ is finite meromorphic and Fredholm at each point of Z, and

$$\operatorname{ind}_{\partial D}(A + S) = \operatorname{ind}_{\partial D} A\,. \tag{4.4.7}$$

Proof. By Lemma 4.4.2, the function $I + A^{-1}S$ is finite meromorphic and Fredholm at each point of $\overline{D}$, the index of $I + A^{-1}$ is zero at each point of $\overline{D}$, there exists a finite subset Z' of D with $Z' \supseteq Z$ such that $I + A^{-1}S(z)$ is invertible for each $z \in \overline{D} \setminus Z'$, and

$$\operatorname{ind}_{\partial D}(I + A^{-1}S) = 0\,.$$

Since $A + S = A(I + A^{-1}S)$, by Proposition 4.2.5, this implies

$$\operatorname{ind}_{\partial D}(A + S) = \operatorname{ind}_{\partial D} A + \operatorname{ind}_{\partial D}(I + A^{-1}S) = \operatorname{ind}_{\partial D} A\,.$$ □

4.5 Comments

The material of this chapter is mostly taken from [GS].

Chapter 5

Multiplicative cocycles ($\mathcal{O}^G$-cocycles)

Let A be a Banach algebra with unit 1, and let G be an open subgroup of the group of invertible elements of A. The Runge approximation Theorem 2.5.6 for invertible scalar functions admits the following generalization:

5.0.1 Theorem. *Let $D \subseteq \mathbb{C}$ be a bounded open set with piecewise $\mathcal{C}^1$-boundary (possibly not connected), and let $f : \overline{D} \to G$ be a continuous function which is holomorphic in D, such that all values of f belong to the same connected component of G.*

(i) *If $\mathbb{C} \setminus \overline{D}$ is connected, then f can be approximated uniformly on $\overline{D}$ by G-valued functions defined and holomorphic on $\mathbb{C}$.*

(ii) *Suppose $\mathbb{C} \setminus \overline{D}$ is not connected. Let $U_1, \ldots, U_N$ be the bounded connected components of $\mathbb{C} \setminus \overline{D}$, and assume that, for each $1 \le j \le N$, a point $p_j \in U_j$ is given. Then f can be approximated uniformly on $\overline{D}$ by G-valued functions defined and holomorphic on $\mathbb{C} \setminus \{p_1, \ldots, p_N\}$.*

Part (i) of this theorem will be proved at the end of Section 5.4, and part (ii) will be obtained in Section 5.10 as a special case of Theorem 5.10.5. (Theorem 5.10.5 is the formulation of Theorem 5.0.1 on the Riemann sphere.)

Theorem 5.0.1 is one of the main statements of the theory developed in the present chapter. The proof is much more difficult than the proof of Theorem 2.5.6, and can be obtained only in the framework of the theory of $\mathcal{O}^G$-cocycles, developed in this chapter.[1] The reason is that the group G, in general, is not commutative. After some preparation in Section 5.3, in Section 5.4 we only prove part (i) of Theorem 5.0.1. Then, in Section 5.5, we prove the Cartan lemma, which can be viewed as a special case of Theorem 0.0.1 stated in the introduction to this book.

[1] At least the authors do not see another way.

Theorem 0.0.1 in turn can be considered as a special result on the *triviality of* $\mathcal{O}^{GL(E)}$*-cocycles* (see Definition 5.6.1 below), because a $GL(E)$-valued function, defined and holomorphic in the intersection of two open sets, may be viewed as a special $\mathcal{O}^{GL(E)}$-cocycle. Using the language introduced in Definition 5.6.1, Theorem 0.0.1 then says that this cocycle is $\mathcal{O}^{GL(E)}$*-trivial.*

To prove Theorem 0.0.1, however we have to pass to general multiplicative cocycles (Def. 5.6.1). In Section 5.6 we introduce the language of multiplicative cocycles and state the main result on such cocycles. This is Theorem 5.6.3, which contains Theorem 0.0.1 as a special case. In Section 5.9 we prove it in the simply connected case, whence also Theorem 0.0.1 is proved in the case when $D_1 \cup D_2$ is simply connected. Using this, then in Section 5.10 we prove the Runge approximation Theorem 5.0.1 in its general form. At the end, using the Runge approximation Theorem 5.0.1 in its general form, in Section 5.11 we prove Theorem 5.6.3 in its general form, whence also Theorem 0.0.1 is proved in its general form.

In Section 5.13 we prove the generalized Weierstrass Theorem 0.0.2, stated in the introduction to this book, as well as the corresponding right-sided and two-sided versions. In Section 5.12 we prove Weierstrass theorems for functions of the form $I + K$, where the values of K are compact.

5.1 Topological properties of $GL(E)$

5.1.1. Until now it was not important whether the group of invertible elements of a Banach algebra is connected or not. This changes beginning with the present chapter. Therefore we devote this section to this question.

First recall the following example of a Banach algebra with a non-connected group of invertible elements: Let E be an infinite dimensional Banach space, and let $\mathcal{F}^\omega(E)$ be the ideal of compact operators in $L(E)$. Then the group of invertible elements of $L(E)/\mathcal{F}^\omega(E)$ is not connected, because this group is the image under the canonical map

$$L(E) \longrightarrow L(E)/\mathcal{F}^\omega(E)$$

of the set of Fredholm operators in $L(E)$ (see, e.g., [GGK2]), where Fredholm operators with different indicies define different connected components of the group $G\big(L(E)/\mathcal{F}^\omega(E)\big)$.

For many Banach spaces, the group of invertible operators is connected. This is the case, for example, for all Hilbert spaces, which we prove below (Theorem 5.1.5).

Note without proofs that for each of the following Banach spaces the group of invertible operators is not only connected but even contractible:

Hilbert spaces of infinite dimension,

c_0,

l_p, $1 \le p \le \infty$,

$L_p[0,1]$, $1 \le p \le \infty$,

$\mathcal{C}^0(K)$ (K a compact metric space of continuum cardinality),

$\mathcal{C}^k(M)$, $k \in \mathbb{N}$, (M a smooth compact manifold).

The proofs and more examples can be found in the original papers [A, Ku, N1, Mi, MiE, EMS1, EMS2]. The interested reader is recommended to consult also the review about these papers written by G. Neubauer [N2] .

But there are also quite natural Banach spaces with non-connected groups of invertible operators. At the end of this section, we give an example.

We begin with the following simple lemma.

5.1.2 Lemma. *Let E be a Banach space, let $A \in GL(E)$, and let $\sigma(A)$ be the spectrum of A. Assume there exists a continuous curve $\gamma : [0,1] \to \mathbb{C} \setminus \sigma(A)$ such that $\gamma(0) = 0$ and $|\gamma(1)| > \|A\|$. Then A belongs to the connected component of the unit operator in $GL(E)$.*

Proof. Since $\gamma(t) \notin \sigma(A)$ and therefore $A - \gamma(t) \in GL(E)$ for all $0 \le t \le 1$, the operators $A = A - \gamma(0)I$ and $A - \gamma(1)I$ belong to the same connected component of $GL(E)$. Since $|\gamma(1)| > \|A\|$, setting

$$\alpha(t) = -\gamma(1)\left(I - \frac{t}{\gamma(1)}\right), \qquad 0 \le t \le 1,$$

we obtain a continuous curve $\alpha : [0,1] \to GL(E)$ which connects the operator $A - \gamma(1)I = \alpha(1)$ with $-\gamma(1)I = \alpha(0)$. As $-\gamma(1)I$ and I belong to the same connected component of $GL(E)$, this completes the proof. □

5.1.3 Corollary. *For each $n \in \mathbb{N}^*$, the group $GL(n, \mathbb{C})$ of invertible complex $n \times n$-matrices is connected.*

Proof. This follows from Lemma 5.1.2, because the spectrum of a matrix is finite. □

5.1.4. Now let H be a Hilbert space of infinite dimension. As already mentioned above, then $GL(H)$ is even contractible [Ku] (in distinction to the groups $GL(n, \mathbb{C})$, $n \in \mathbb{N}^*$). In the present book we need only the simpler fact that $GL(H)$ is connected.[2] Again using Lemma 5.1.2, this can be easily deduced from the spectral representation of unitary operators:

5.1.5 Theorem. *For each Hilbert space H, the group $GL(H)$ of invertible operators is connected.*

---

[2] This is due to the fact that we deal with functions of one complex variable and that the topology of a domain in the complex plane is simple compared to the topology of domains in higher dimensional spaces. That $GL(H)$ is even contractible becomes important if we pass to several complex variables (see [Bu]).

Proof. Let $GL_I(H)$ be the connected component of the unit operator in $GL(H)$. We have to prove that $GL(H) = GL_I(H)$. Let $A \in GL(H)$ be given. Then $A = US$, where $S := (A^*A)^{1/2}$ is positive definite, and $U := AS^{-1}$ is unitary. Since the spectrum $\sigma(S)$ of S is contained in the real line and $0 \notin \sigma(S)$, it follows from Lemma 5.1.2 that $S \in GL_I(H)$.

It remains to prove that $U \in GL_I(H)$. It follows from the spectral representation of unitary operators (see, e.g., [GGK1]) that H can be written as an orthogonal sum $H = H_1 \oplus H_2$, where H_1 and H_2 are invariant subspaces of U such that the spectrum of $U\big|_{H_1}$ lies on the half circle $|z| = 1$, $\operatorname{Im} z \geq 0$, and the spectrum of $U\big|_{H_2}$ lies on the half circle $|z| = 1$, $\operatorname{Im} z < 0$. Therefore, by Lemma 5.1.2, $U\big|_{H_1}$ belongs to the connected component of the unit operator of H_1, and $U\big|_{H_2}$ belongs to the connected component of the unit operator of H_2. Hence $U \in GL_I(H)$. □

Next we give an example for a Banach space E such that $GL(E)$ is not connected:

5.1.6 Theorem. [3] *Let c_0 be the Banach space of sequences $\xi = \{\xi_n\}_{n\in\mathbb{N}}$ of complex numbers tending to zero with the norm*

$$\|\xi\|_\infty = \max_{n\in\mathbb{N}} |\xi_n|,$$

and let l_p, $1 \leq p < \infty$, be the Banach space of sequences $\{\xi_n\}_{n\in\mathbb{N}}$ of complex numbers with the norm

$$\|\xi\|_p = \left(\sum_{n=0}^{\infty} |\xi_n|^p\right)^{1/p} < \infty.$$

Then the group $GL(c_0 \oplus l_p)$ consists of an infinite number of connected components.

To prove this theorem, we need the following three lemmas:

5.1.7 Lemma. *Let c_0 and l_p, $1 \leq p < \infty$ be as in Theorem 5.1.6, and let P_n be the projector in c_0, defined by*

$$P_n\Big(\{\xi_n\}_{n\in\mathbb{N}}\Big) = (\xi_1, \ldots, \xi_n, 0, \ldots). \tag{5.1.1}$$

For $j \in \mathbb{N}$, we denote by $f_j : l_p \to \mathbb{C}$ the functional defined by

$$f_j\big(\{\xi_n\}_{n\in\mathbb{N}}\big) = \xi_j.$$

Let $A \in L(c_0, l_p)$ and $j \in \mathbb{N}$ be fixed. Then

$$\lim_{n\to\infty} \big\|f_j A(I - P_n)\big\| = 0 \tag{5.1.2}$$

[3] This example is due to A. Douady [Do1]. As observed in [Do1] instead of c_0 and l_p here one could take an arbitrary pair of Banach spaces F and G such that each operator from $L(F, G)$ is compact, and F and G are isomorphic to their hyperplanes.

where

$$\big\|f_j A(I-P_n)\big\| := \sup_{\xi\in c_0, \|\xi\|_\infty \le 1} \Big| f_j\Big(A(\xi - P_n\xi)\Big)\Big|.$$

Proof. Assume the contrary. Since

$$\big\|f_j A(I-P_n)\big\| \ge \big\|f_j A(I-P_{n+1})\big\| \qquad \text{for all } n\in\mathbb{N},$$

then there exists $\varepsilon > 0$ such that

$$\big\|f_j A(I-P_n)\big\| > \varepsilon \qquad \text{for all } n\in\mathbb{N}. \tag{5.1.3}$$

Choose a number $n_1 \in \mathbb{N}$ (arbitrarily). Then, by (5.1.3), we can find a vector $\widetilde{\eta}^{(1)} \in \operatorname{Im}(I-P_{n_1})$ with $\|\widetilde{\eta}^{(1)}\|_\infty \le 1$ and

$$\Big|(f_jA)(\widetilde{\eta}^{(1)})\Big| > \varepsilon. \tag{5.1.4}$$

Since $\widetilde{\eta}^{(1)}$ tends to zero, we have

$$\lim_{n\to\infty} \big\|(I-P_n)\widetilde{\eta}^{(1)}\big\|_\infty = 0.$$

Therefore it follows from (5.1.4) that there exists $n_2 \in \mathbb{N}$ so large that

$$\Big|(f_jA)(P_{n_2}\widetilde{\eta}^{(1)})\Big| > \varepsilon.$$

Set

$$\eta^{(1)} = \frac{\Big|(f_jA)(P_{n_2}\widetilde{\eta}^{(1)})\Big|}{(f_jA)(P_{n_2}\widetilde{\eta}^{(1)})} P_{n_2}\widetilde{\eta}^{(1)}.$$

Then

$$\eta^{(1)} \in \operatorname{Im}(P_{n_2} - P_{n_1}), \quad \|\eta^{(1)}\|_\infty \le 1 \quad \text{and} \quad (f_jA)(\eta^{(1)}) > \varepsilon.$$

If this way we get by induction an increasing sequence $n_\mu \in \mathbb{N}$, $\mu \in \mathbb{N}^*$, and a sequence of vectors $\eta^{(\mu)} \in c_0$, $\mu \in \mathbb{N}^*$, such that

$$\eta^{(\mu)} \in \operatorname{Im}(P_{n_{\mu+1}} - P_{n_\mu}), \quad \|\eta^{(\mu)}\|_\infty \le 1 \quad \text{and} \quad (f_jA)(\eta^{(\mu)}) > \varepsilon$$

for all μ. Now, for each $N \in \mathbb{N}^*$, we set

$$\phi^{(N)} = \eta^{(1)} + \ldots + \eta^{(N)}.$$

Then

$$\|\phi^{(N)}\|_\infty = \max_{1\le\mu\le N} \|\eta^{(\mu)}\|_\infty \le 1$$

and

$$(f_jA)(\phi^{(N)}) = (f_jA)(\eta^{(1)}) + \ldots + (f_jA)(\eta^{(N)}) > N\varepsilon.$$

This is a contradiction to the boundedness of f_jA. □

5.1.8 Lemma. *Let c_0 and l_p, $1 \le p < \infty$, be as in Theorem 5.1.6. Then each operator $A \in L(c_0, l_p)$ can be approximated in the operator norm by finite dimensional operators, namely if P_n is the projector in c_0 defined by (5.1.1), then*

$$\lim_{n\to\infty} \|A - AP_n\| = 0. \tag{5.1.5}$$

Proof. Assume that there exists a bounded linear operator $A : c_0 \to l_p$ such that (5.1.5) is violated. Since

$$\|A - AP_n\| \ge \|A - AP_{n+1}\| \qquad \text{for all } n \in \mathbb{N},$$

then there exists $\varepsilon > 0$ with

$$\|A - AP_n\| > \varepsilon \qquad \text{for all } n \in \mathbb{N}. \tag{5.1.6}$$

Let $\|A\|$ be the operator norm on A.

Choose $\widetilde{\alpha} \in c_0$ with

$$\|\widetilde{\alpha}\|_\infty \le 1 \qquad \text{and} \qquad \|A\widetilde{\alpha}\|_p > \|A\| - \frac{\varepsilon}{10}. \tag{5.1.7}$$

As $\widetilde{\alpha}$ tends to zero, then $\lim_{n\to\infty} \|\widetilde{\alpha} - P_n\widetilde{\alpha}\| = 0$. Therefore and by (5.1.7) we can find $m \in \mathbb{N}$ with

$$\|AP_m\widetilde{\alpha}\|_p > \|A\| - \frac{\varepsilon}{10}.$$

Set $\alpha = P_m\widetilde{\alpha}$. Then

$$\alpha \in \operatorname{Im} P_m, \qquad \|\alpha\|_\infty \le 1 \qquad \text{and} \qquad \|A\alpha\|_p > \|A\| - \frac{\varepsilon}{10}. \tag{5.1.8}$$

Let $R_n : l_p \to l_p$ be the projector defined by

$$R_n\big(\{\xi_n\}_{n\in\mathbb{N}}\big) = \xi_j.$$

Since $1 \le p < \infty$, then $\lim_{n\to\infty} \|(I - R_k)A\alpha\|_p = 0$. Hence, we can find $k \in \mathbb{N}^*$ such that

$$\|(I - R_k)A\alpha\|_p \le \frac{\varepsilon}{10}. \tag{5.1.9}$$

Together with (5.1.8) this implies that

$$\|R_kA\alpha\|_p \ge \|A\| - \frac{\varepsilon}{5}. \tag{5.1.10}$$

From Lemma 5.1.7 it follows that $\lim_{n\to\infty} \|R_kA(I - P_n)\| = 0$. Therefore, we can find an integer $l > m$ with

$$\|R_kA(I - P_l)\| \le \frac{\varepsilon}{10}. \tag{5.1.11}$$

By (5.1.6) there exists $\beta \in \operatorname{Im}(I - P_l)$ such that

$$\|\beta\|_\infty \leq 1 \qquad \text{and} \qquad \|A\beta\|_p \geq \varepsilon. \tag{5.1.12}$$

Together with (5.1.11) this implies that

$$\|R_k A\beta\|_p \leq \frac{\varepsilon}{10} \tag{5.1.13}$$

and

$$\|(I - R_k)A\beta\|_p \geq \varepsilon - \frac{\varepsilon}{10}. \tag{5.1.14}$$

Since $\alpha \in \operatorname{Im} P_m$, $\beta \in \operatorname{Im}(I - P_l)$ and $l > m$, now we have

$$\|\alpha + \beta\|_\infty = \max\left(\|\alpha\|_\infty, \|\beta\|_\infty\right) \leq 1,$$

and from (5.1.9), (5.1.10), (5.1.13) and (5.1.14) it follows that

$$\|A(\alpha+\beta)\|_p \geq \|R_k A\alpha\|_p - \|(I-R_k)A\alpha\|_p + \|(I-R_k)A\beta\|_p - \|R_k A\beta\|_p \geq \|A\| + \frac{\varepsilon}{2}.$$

This is a contradiction to the definition of $\|A\|$. □

5.1.9 Lemma. *Let c_0 and l_p, $1 \leq p < \infty$, be as in Theorem* 5.1.6, *and let P be the projector in $c_0 \oplus l_p$ with $\operatorname{Im} P = c_0$ and $\operatorname{Ker} P = l_p$. Set $Q = I - P$. Let $A \in GL(c_0 \oplus l_p)$. Then $PA\big|_{c_0}$ is a Fredholm operator in c_0 and $QA\big|_{l_p}$ is a Fredholm operator in l_p.*

Proof. We have

$$P = PAA^{-1}P = PAPA^{-1}P + PAQA^{-1}P$$

and

$$P = PA^{-1}AP = PA^{-1}PAP + PA^{-1}QAP.$$

Since, by Lemma 5.1.8, the operators $PAQA^{-1}P$ and $PA^{-1}QAP$ are compact, this implies that both $PAPA^{-1}\big|_{c_0}$ and $PA^{-1}PA\big|_{c_0}$ are Fredholm operators. Hence (see., e.g., [GGK2]), $PA\big|_{c_0}$ is a Fredholm operator. In the same way we see that $QA\big|_{l_p}$ is a Fredholm operator. □

Proof of Theorem 5.1.6. We consider the space $c_0 \oplus l_p$ as the space of families $\{\xi_n\}_{h=-\infty}^{\infty}$, $\xi_m \in \mathbb{C}$, such that

$$\lim_{n \to -\infty} \xi_n = 0 \qquad \text{and} \qquad \sum_{n=0}^{\infty} |\xi_n|^p < \infty.$$

Then, for $k \in \mathbb{Z}$, we have the shift operator A_k defined by

$$A_k\Big(\{\xi_n\}_{j=-\infty^\infty}\Big) = \{\xi_{n+k}\}_{j=-\infty^\infty}\}.$$

Clearly, then $A_k \in GL(c_0 \oplus l_p)$ for all $k \in \mathbb{Z}$. Moreover, if P is the projector from Lemma 5.1.9, $PA_k\big|_{c_0}$ is a Fredholm operator with index k in c_0.

Assume A_k and A_m belong to the same connected component of $GL(c_0 \oplus l_p)$. Then $A_{k-m} = A_k A_m^{-1}$ belongs to the connected component of $GL(c_0 \oplus l_p)$ which contains the unit operator. Hence, there is a continuous curve $A : [0,1] \to GL(c_0 \oplus l_p)$ with $A(0) = I$ and $A(1) = A_{k-m}$. Then, by Lemma 5.1.9, $\{PA(t)\big|_{c_0}\}_{0 \le t \le t}$, is a continuous family of Fredholm operators in c_0. Since $A(0) = I$ has the index zero, then also $A(1) = A_{k-m}\big|_{c_0^k}$ has the index zero. Hence $k = m$.

Hence we proved that, for all $k, m \in \mathbb{Z}$ with $k \neq m$, the operators A_k and A_m belong to different connected components of $GL(c_0 \oplus l_p)$. □

5.2 Two factorization lemmas

The following lemma will be used several times in this book:

5.2.1 Lemma. *Let A be a Banach algebra with unit 1 and the norm $\|\cdot\|$, and let A_1, A_2 be two algebraic subalgebras of A with $1 \in A_1 \cap A_2$, which are Banach algebras with respect to their own norms $\|\cdot\|_1$ and $\|\cdot\|_2$, respectively. Assume that:*

(i) $\|x\|_j \ge \|x\|$ *for each* $x \in A_j$, $j = 1, 2$.

(ii) *Each element $x \in A$ can be written in the form $x = x_1 + x_2$ with $x_j \in A_j$, $j = 1, 2$.*

Let $C < \infty$ be the smallest constant such that, for all $x \in A$, there exist $x_j \in A_j$ with $x = x_1 + x_2$ and

$$\|x_j\|_j \le C\|x\|, \qquad j = 1, 2. \tag{5.2.1}$$

(Such a constant then exists by the Banach open mapping theorem.)

Then, for each $a \in A$ with

$$\|a\| < \frac{1}{2C},$$

there exist elements $a_1 \in A_1$ and $a_2 \in A_2$ such that $1 - a_1$ is an invertible element of A_1, $1 - a_2$ is an invertible element of A_2, and

$$(1 - a) = (1 - a_1)^{-1}(1 - a_2). \tag{5.2.2}$$

Moreover

$$\|a_1\|_1, \|a_2\|_2 \le 2C\|a\| < 1. \tag{5.2.3}$$

Proof. By definition of C, we can find maps $P_j : A \to A_j$ (possibly nonlinear) such that, for all $x \in A$,

$$P_1 x + P_2 x = x \qquad \text{and} \qquad \|P_j x\|_j \le C\|x\|, \quad j = 1, 2.$$

Now let $a \in A$ with $\|a\| < 1/2C$ be given. Then $C\|a\| < 1/2$. Hence, the series

$$a_1 := -P_1 a - P_1(aP_1 a) - P_1\big(aP_1(aP_1 a)\big) - \dots$$

and

$$a_2 := P_2 a + P_2(aP_1 a) + P_2\big(aP_1(aP_1 a)\big) + \dots$$

converge absolutely in A_1 and A_2, respectively, and we have the estimates (5.2.3):

$$\|a_1\|_1, \|a_2\|_2 \leq \sum_{n=1}^{\infty} \big(C\|a\|\big)^n = \frac{C\|a\|}{1 - C\|a\|} < 2C\|a\| < 1.$$

Then $1 - a_1$ is an invertible element of A_1, and $1 - a_2$ is an invertible element of A_2. Moreover

$$\begin{aligned}(1 - a_1)(1 - a) &= 1 + P_1 a + P_1(aP_1 a) + P_1\big(aP_1(aP_1 a)\big) + \dots \\ &\qquad - a \quad - aP_1 a \qquad - aP_1(aP_1 a) \qquad - \dots \\ &= 1 - P_2 a - P_2(aP_1 a) - P_2\big(aP_1(aP_1 a)\big) - \dots \\ &= 1 - a_2,\end{aligned}$$

i.e., we have (5.2.2). □

5.2.2 Lemma. *Let A be a Banach algebra with unit 1, and let $f : [0, 1] \to A$ be continuous such that all $f(t)$, $0 \leq t \leq 1$, are right (left) invertible*[4] *and $f(0)$ is invertible. Then all $f(t)$, $0 \leq t \leq 1$, are invertible.*

Proof. It is sufficient to consider the case of right invertibility. The case of left invertibility follows from changing the order of multiplication in A.

Assume there exists $0 \leq t \leq 1$ such that $f(t)$ is not invertible. Since the group of invertible elements of A is open and $f(0)$ is invertible, then there exists $0 < t_0 \leq 1$ such that $f(t_0)$ is not invertible, but all $f(t)$ with $0 \leq t < t_0$ are invertible. Since $f(t_0)$ is right invertible, we have $b \in A$ with $f(t_0)b = 1$. Fix $0 < t < t_0$ such that

$$\|f(t) - f(t_0)\| < \frac{1}{\|b\|}.$$

Then

$$c := f(t)b = \Big(f(t_0) + f(t) - f(t_0)\Big)b = 1 + \Big(f(t) - f(t_0)\Big)b,$$

where

$$\Big\|\Big(f(t) - f(t_0)\Big)b\Big\| < 1.$$

Hence, c is invertible. Since also $f(t)$ is invertible, it follows that $b = \big(f(t)\big)^{-1}c$ is invertible. Moreover, as $f(t_0)b = 1$, we see that $f(t_0)$ is the inverse of b. Hence $f(t_0)$ is invertible, which is a contradiction. □

[4] An element $a \in A$ is called **right (left) invertible** if there exists $b \in A$ with $ab = 1$ ($ba = 1$).

5.2.3 Lemma. *Let A be a Banach algebra with unit 1 and the norm $\|\cdot\|$, and let A_1, A_2 be two closed subalgebras of A such that A is the direct sum of A_1 and A_2. Let P_1 be the linear projector from A to A_1, let $P_2 := I - P_1$, and let $a \in A$ such that*

$$\|a\| < \frac{1}{C} \qquad \text{where } C := \max\{\|P_1\|, \|P_2\|\}. \tag{5.2.4}$$

Then there exist elements $a_1 \in A_1$ and $a_2 \in A_2$ such that $1 - a_1$ and $1 - a_2$ are invertible elements of A, and

$$(1 - a) = (1 - a_1)^{-1}(1 - a_2). \tag{5.2.5}$$

Moreover, $1 - a_1$ and $1 - a_2$ belong to the connected component of the unit element in the group of invertible elements of A, and

$$\|a_1\|, \|a_2\| < \frac{C\|a\|}{1 - C\|a\|}. \tag{5.2.6}$$

Proof. By (5.2.4), for all $0 \le t \le 1$, the series

$$\begin{aligned} f_1(t) &:= -tP_1 a - t^2 P_1(aP_1 a) - t^3 P_1\big(aP_1(aP_1 a)\big) - \dots, \\ f_2(t) &:= tP_2 a + t^2 P_2(aP_1 a) + t^3 P_2\big(aP_1(aP_1 a)\big) + \dots \end{aligned} \tag{5.2.7}$$

converge absolutely in A_1 and A_2, respectively, where

$$\|f_1(t)\|, \|f_2(t)\| \le \sum_{n=1}^{\infty} t^n \big(C\|a\|\big)^n = \frac{tC\|a\|}{1 - tC\|a\|} < \infty. \tag{5.2.8}$$

Moreover, for all $0 \le t \le 1$,

$$\begin{aligned} \big(1 - f_1(t)\big)(1 - ta) &= 1 + tP_1 a + t^2 P_1(aP_1 a) + t^3 P_1\big(aP_1(aP_1 a)\big) + \dots \\ &\qquad - ta \quad - t^2 aP_1 a \qquad - t^3 aP_1(aP_1 a) \qquad - \dots \\ &= 1 - tP_2 a - t^2 P_2(aP_1 a) - t^3 P_2\big(aP_1(aP_1 a)\big) - \dots \\ &= 1 - f_2(t). \end{aligned} \tag{5.2.9}$$

As the convergence of the series (5.2.7) is uniformly in t, so we found continuous functions $f_1 : [0, 1] \to A_1$ and $f_2 : [0, 1] \to A_2$ such that

$$\big(1 - f_1(t)\big)(1 - ta) = 1 - f_2(t) \qquad \text{for all } 0 \le t \le 1, \tag{5.2.10}$$

and $a_1(0) = a_2(0) = 1$. Changing the order of multiplication in A and the role of A_1 and A_2, in the same way we get continuous functions $g_2 : [0, 1] \to A_2$ and $g_1 : [0, 1] \to A_1$ such that

$$(1 - ta)\big(1 - g_2(t)\big) = 1 - g_1(t) \qquad \text{for all } 0 \le t \le 1, \tag{5.2.11}$$

and $g_2(0) = g_1(0) = 1$. (5.2.11) can be written also in the form

$$(1 - ta)^{-1}\big(1 - g_1(t)\big) = 1 - g_2(t)\,, \qquad 0 \le t \le 1. \tag{5.2.12}$$

Multiplying (5.2.10) and (5.2.12) we obtain

$$\big(1 - f_1(t)\big)\big(1 - g_1(t)\big) = \big(1 - f_2(t)\big)\big(1 - g_2(t)\big)\,, \qquad 0 \le t \le 1,$$

or

$$f_1(t)g_1(t) - f_1(t) - g_1(t) = f_2(t)g_2(t) - f_2(t) - g_2(1)\,, \qquad 0 \le t \le 1.$$

As $A_1 \cap A_2 = \{0\}$, this implies that

$$\big(1 - g_1(t)\big)\big(1 - g_1(t)\big) = 1 \text{ and } \big(1 - f_2(t)\big)\big(1 - g_2(t)\big) = 1\,, \qquad 0 \le t \le 1.$$

Hence all $1 - f_1(t)$ and all $1 - f_2(t)$, $0 \le t \le 1$, are right invertible. As $f_1(0) = f_2(0) = 1$, this implies by Lemma 5.2.2 that all $1 - f_1(t)$ and all $1 - f_2(t)$, $0 \le t \le 1$, are invertible. In particular, setting $a_1 = f_1(1)$ and $a_2 = f_2(1)$, we get elements $a_1 \in A_1$ and $a_2 \in A_2$ such that $1 - a_1$ and $1 - a_2$ belong to the connected component of the unit element in the group of invertible elements of A. It follows from (5.2.10) that these elements solve (5.2.5), and estimate (5.2.6) follows from (5.2.8). □

5.3 $\overline{\mathcal{O}}^E$-cocycles

In this section E is a Banach space.

In this chapter we need a version of the theory of $\mathcal{O}^E$-cocycles (see Section 2.6) for holomorphic functions which admit continuous extensions to the boundary of their domain of definition. This is the content of the present section.

5.3.1 Definition. Let $X \subseteq \mathbb{C}$ be an arbitrary non-empty set. Then we denote by $\mathcal{C}^E(X)$ the space of continuous E-valued functions on X. By $\overline{\mathcal{O}}^E(X)$ we denote the subspace all functions from $\mathcal{C}^E(X)$ which are holomorphic in the inner (with respect to $\mathbb{C}$) points of X. Moreover we set $\mathcal{C}^E(\emptyset) := \overline{\mathcal{O}}^E(\emptyset) := \{0\}$ where 0 is the zero vector of E.

5.3.2 Definition. Let $D \subseteq \mathbb{C}$ be an open set, and let $\mathcal{U} = \{U_j\}_{j\in I}$ be a covering of $\overline{D}$ by relatively open subsets of $\overline{D}$. Then we denote by $C^1(\mathcal{U}, \overline{\mathcal{O}}^E)$ the space of families $f = \{f_{jk}\}_{j,k\in I}$ where $f_{jk} \in \overline{\mathcal{O}}^E(U_j \cap U_k)$, and by $Z^1(\mathcal{U}, \overline{\mathcal{O}}^E)$ we denote the subspace of all $f \in C^1(\mathcal{U}, \overline{\mathcal{O}}^E)$ with

$$f_{jk} + f_{kl} = f_{jl} \qquad \text{on } U_j \cap U_k \cap U_l, \tag{5.3.1}$$

for all $j, k, l \in I$ with $U_j \cap U_k \cap U_l \neq \emptyset$. The elements of $Z^1(\mathcal{U}, \overline{\mathcal{O}}^E)$ will be called **$(\mathcal{U}, \overline{\mathcal{O}}^E)$-cocycles**. If the covering $\mathcal{U}$ is not specified, then we speak also about **$\overline{\mathcal{O}}^E$-cocycles** on $\overline{D}$.

5.3.3 Theorem. *Let $D \subseteq \mathbb{C}$ a bounded open set, let $\mathcal{U} = \{U_j\}_{j\in I}$ be a covering of $\overline{D}$ by relatively open subsets of $\overline{D}$, and let $f \in Z^1(\mathcal{U};\overline{\mathcal{O}}^E)$. Then there exists a family $\{h_j\}_{j\in I}$ of functions $h_j \in \overline{\mathcal{O}}^E(U_j)$ with*

$$f_{jk} = h_j - h_k \qquad on U_j \cap U_k \tag{5.3.2}$$

for all $j,k \in I$ with $U_j \cap U_k \neq \emptyset$.

Proof. Take a family $\widetilde{\mathcal{U}} = \{\widetilde{U}_j\}_{j\in I}$ of open (in $\mathbb{C}$) sets such that $U_j = \overline{D} \cap \widetilde{U}_j$ for all $j \in I$. Choose a $\mathcal{C}^\infty$-partition of unity $\{\chi_j\}_{j\in I}$ subordinated to $\widetilde{\mathcal{U}}$. By setting

$$\varphi_j = -\sum_{\nu\in I} \chi_\nu f_{\nu j},$$

we obtain functions $\varphi_j \in \mathcal{C}^E(U_j)$ which are of class $\mathcal{C}^\infty$ in $D\cap U_j$. Since f satisfies (5.3.2) and $\sum_{\nu\in I}\chi_\nu = 1$ on $\overline{D}$, we have

$$\varphi_j - \varphi_k = -\sum_{\nu\in I} \chi_\nu (f_{\nu j} - f_{\nu k}) = \left(\sum_{\nu\in I}\chi_\nu\right) f_{jk} = f_{jk} \qquad \text{on } U_j \cap U_k \tag{5.3.3}$$

for all $j,k \in I$ with $U_j \cap U_k \neq \emptyset$. Since f_{jk} is holomorphic on $D \cap U_j \cap U_k$, it follows that

$$\overline{\partial}\varphi_j = \overline{\partial}\varphi_k \qquad \text{on } D \cap U_j \cap U_k.$$

Hence there is a well-defined $\mathcal{C}^\infty$-function $\psi : D \to E$ with $\psi = \overline{\partial}\varphi_j$ on $D \cap U_j$. By Proposition 2.1.2, we have

$$\psi = \overline{\partial}\varphi_j = -\sum_{\nu\in I} (\overline{\partial}\chi_\nu) f_{\nu j} \qquad \text{on } D \cap U_j$$

for all $j \in I$. Hence, ψ admits a continuous extension to $\overline{D}$, and, by theorems 2.1.10 and 2.1.9, we can find a continuous function $u : \overline{D} \to E$ with $\psi = \overline{\partial}u$ on D. Set $h_j = \varphi_j - u$ on U_j. Then h_j is continuous on U_j, and, on $D \cap U_j$, we have

$$\overline{\partial}h_j = \overline{\partial}\varphi_j - \overline{\partial}u = \psi - \overline{\partial}u = 0\,.$$

Hence $h_j \in \overline{\mathcal{O}}^E(U_j)$, and it follows from (5.3.3) that

$$h_j - h_k = (\varphi_j - u) - (\varphi_k - u) = \varphi_j - \varphi_k = f_{jk} \quad \text{on } U_j \cap U_k$$

for all $j,k \in I$ with $U_j \cap U_k \neq \emptyset$. □

5.3.4 Proposition. *Assume that under the hypotheses of Theorem 5.3.3 moreover the following holds: For all $j,k \in I$ with $U_j \cap U_k \neq \emptyset$, the function f_{jk} admits a continuous extension to $\overline{U_j \cap U_k}$.*

If then $\{h_j\}_{j\in I}$ is a family of functions $h_j \in \overline{\mathcal{O}}^E(U_j)$, which solves (5.3.2), then $h_j \in \overline{\mathcal{O}}^E(\overline{U}_j)$ for all $j \in I$.

Proof. Let some $j \in I$ be given and let $(z_n)_{n\in\mathbb{N}}$ be a sequence in U_j which converges to some point $z \notin U_j$. Since $\overline{D}$ is closed and covered by $\mathcal{U}$, then there exists at least one index $k \in I$ such that $z \in U_k$. Since U_k is relatively open in $\overline{D}$, it follows that, for some $n_0 \in \mathbb{N}$, $z_n \in U_k$ if $n \geq n_0$. Since h_k is continuous on U_k,

$$\lim_{n_0 \leq n \to \infty} h_k(z_n) = h_k(z).$$

Since $z_n \in U_j \cap U_k$ for $n \geq n_0$ and f_{jk} admits a continuous extension to $\overline{U_j \cap U_k}$, moreover

$$\lim_{n_0 \leq n \to \infty} f_{jk}(z_n)$$

exists. Together this implies by (5.3.2) that

$$\lim_{n\to\infty} h_j(z_n) = \lim_{n_0 \leq n \to \infty} f_{jk}(z_n) + h_k(z)$$

exists. □

For many purposes the following immediate corollary of Theorem 5.3.3 and Proposition 5.3.4 is sufficient:

5.3.5 Corollary. *Let $D \subseteq \mathbb{C}$ be a bounded open set, let $U_1, U_2 \subseteq \overline{D}$ be two relatively open subsets of $\overline{D}$ such that $\overline{D} = U_1 \cup U_2$ and $U_1 \cap U_2 \neq \emptyset$, and let $f \in \overline{\mathcal{O}}^E(U_1 \cap U_2)$. Then there exist $f_j \in \overline{\mathcal{O}}^E(U_j)$, $j = 1, 2$, with*

$$f = f_1 - f_2 \qquad \text{on } U_1 \cap U_2. \tag{5.3.4}$$

Moreover, if $f \in \overline{\mathcal{O}}^E(\overline{U_1 \cap U_2})$ and if two functions $f_j \in \overline{\mathcal{O}}^E(U_j)$, $j = 1, 2$, solve (5.3.4), then automatically $f_j \in \overline{\mathcal{O}}^E(\overline{U}_j)$, $j = 1, 2$.

5.4 Runge approximation of G-valued functions First steps

5.4.1 Proposition. *Let A be a Banach algebra with unit 1, and let G_1A be the connected component of the group of invertible elements of A which contains the unit element. Then, for any $\varepsilon > 0$, each $f \in G_1A$ can be written as a finite product of the form*

$$f = (1 + g_1) \cdot \ldots \cdot (1 + g_n) \tag{5.4.1}$$

where $g_j \in A$ with $\|g_j\| < \varepsilon$ for $1 \leq j \leq n$.

Proof. Let Θ be the set of all $f \in G_1A$ of the form (5.4.1). We have to prove that Θ is open and closed in G_1A.

Let $f \in \Theta$ be written in the form (5.4.1), and let U_f be the neighborhood of f which consists of all $g \in G_1A$ with $\|f^{-1}g - 1\| < \varepsilon$. Then, for each $g \in U_f$,

$$g = ff^{-1}g = (1 + g_1) \ldots (1 + g_n)\Big(1 + (f^{-1}g - 1)\Big)$$

is also of this form, i.e., $g \in \Theta$. Hence Θ is open.

Now let $(f_j)_{j\in\mathbb{N}}$ be a sequence in Θ which converges to some $f \in G_1A$. Then we can find $j_0 \in \mathbb{N}$ such that

$$\left\| f_{j_0}^{-1} f - 1 \right\| < \varepsilon.$$

Since f_{j_0} is of the form

$$f_{j_0} = (1+g_1)\dots(1+g_n) \qquad \text{with } \|g_j\| < \varepsilon,$$

then also

$$f = f_{j_0} f_{j_0}^{-1} f = (1+g_1)\dots(1+g_n)\Big(1 + (f_{j_0}^{-1} f - 1)\Big)$$

is of this form. Hence Θ is also closed. □

5.4.2 Corollary. *Let A be a Banach algebra with unit 1, let GA be the group of invertible elements of A, and let G be an open subgroup of GA. For $a \in GA$, we denote by G_a the connected component of a in GA. Then*

$$G_a \subseteq G \qquad \textit{for all } a \in G. \tag{5.4.2}$$

Proof. Since G is an open subset of GA, there exists $\varepsilon > 0$ such that

$$U_\varepsilon(1) := \Big\{ x \in A \,\Big|\, \|1 - x\| < \varepsilon \Big\} \subseteq G.$$

By Proposition 5.4.1, each element in G_1A is a finite product of elements from $U_\varepsilon(1)$. Since G is a subgroup of GA, it follows that $G_1A \subseteq G$, and hence $aG_1A \subseteq G$ for all $a \in G$. Since $G_a = aG_1$ for all $a \in GA$, this implies (5.4.2). □

5.4.3. If A is a Banach algebra with unit 1, and GA is the group of invertible elements of A, then we define

$$\exp a = e^a = \sum_{n=0}^{\infty} \frac{a^n}{n!} \qquad \text{for all } a \in A \tag{5.4.3}$$

and

$$\log a = -\sum_{n=1}^{\infty} \frac{(1-a)^n}{n} \qquad \text{for all } a \in A \text{ with } \|1 - a\| < 1. \tag{5.4.4}$$

5.4.4 Lemma. *Let A be a Banach algebra with unit 1, let GA be the group of invertible elements of A, and let G_1A be the connected component of GA which contains the unit element. Then:*

(i) *$e^a \in G_1A$ for all $a \in A$, where $\big(e^a\big)^{-1} = e^{-a}$.*

(ii) *If $a \in A$ with $\|a - 1\| < 1$, then $e^{\log a} = a$.*

Proof. If A is commutative, this is well known. In the general case, we can pass to the smallest closed subalgebra of A which contains a and the unit element. This subalgebra is commutative. □

5.4.5 Definition. Let $D \subseteq \mathbb{C}$ be a bounded open set, let A be a Banach algebra with unit 1, and let GA be the group of invertible elements of A.

Then we denote by $\overline{\mathcal{O}}^A(\overline{D})$ the algebra of continuous A-valued functions on $\overline{D}$ which are holomorphic in D. By setting

$$\|f\|_{\overline{\mathcal{O}}^A(\overline{D})} := \max_{z\in\overline{D}} \|f(z)\|_A\,, \qquad f \in \overline{\mathcal{O}}^A(\overline{D}),$$

we introduce a norm in $\overline{\mathcal{O}}^A(\overline{D})$, where $\|\cdot\|_A$ is the norm of A. In this way, also $\overline{\mathcal{O}}^A(\overline{D})$ becomes a Banach algebra with unit. If it is clear what we mean we simply write $\|\cdot\|$ instead of $\|\cdot\|_{\overline{\mathcal{O}}^A(\overline{D})}$.

If G is an open subgroup of GA, then we denote by $\overline{\mathcal{O}}^G(\overline{D})$ the subset of $\overline{\mathcal{O}}^A(\overline{D})$ which consists of the functions with values in G. Note that then $\overline{\mathcal{O}}^{GA}(\overline{D})$ is the group of invertible elements of $\overline{\mathcal{O}}^A(\overline{D})$, and, if G is an open subgroup of GA, then $\overline{\mathcal{O}}^G(\overline{D})$ is an open subgroup of $\overline{\mathcal{O}}^{GA}(\overline{D})$.

5.4.6 Proposition. *Let $D \subseteq \mathbb{C}$ be a bounded open set with piecewise $\mathcal{C}^1$-boundary, and let $P \subseteq \mathbb{C} \setminus \overline{D}$ be a set which contains at least one point of each bounded connected component of $\mathbb{C} \setminus \overline{D}$ (if there is any[5]). Let A be a Banach algebra with unit 1, and let G be an open subgroup of the group of invertible elements of A. Then, for each $f \in \overline{\mathcal{O}}^G(\overline{D})$, the following two conditions are equivalent:*

(i) *f can be approximated uniformly on $\overline{D}$ by functions from $\mathcal{O}^G(\mathbb{C} \setminus P)$.*

(ii) *There exists $\widetilde{f} \in \mathcal{O}^G(\mathbb{C} \setminus P)$ such that f and $\widetilde{f}$ belong to the same connected component of $\overline{\mathcal{O}}^G(\overline{D})$.*

Proof. Since the connected components of $\overline{\mathcal{O}}^G(\overline{D})$ are open, it is clear that (i) ⇒ (ii).

Suppose (ii) is satisfied. Since f and $\widetilde{f}$ belong to the same connected component of $\overline{\mathcal{O}}^G(\overline{D})$, the function $f\widetilde{f}^{-1}$ belongs to the connected component of the unit element in $\overline{\mathcal{O}}^G(\overline{D})$. Therefore, by Proposition 5.4.1, it can be written in the form

$$f\widetilde{f}^{-1} = (1 + g_1) \dots (1 + g_n) \tag{5.4.5}$$

where $g_j \in \overline{\mathcal{O}}^A(\overline{D})$ with $\|g_j\| < 1$. Set

$$h_j = \log(1 + g_j) = \sum_{\nu=1}^{\infty} \frac{(-1)^{\nu+1}}{\nu} g_j^\nu\,, \qquad 1 \le j \le n.$$

[5] If $\mathbb{C} \setminus \overline{D}$ is connected, then $P = \emptyset$ is possible.

Since, for any j, the series on the right-hand side converges uniformly on $\overline{D}$, so we obtain functions $h_j \in \overline{\mathcal{O}}^A(\overline{D})$. By Lemma 5.4.4 (ii), $1+g_j = e^{h_j}$ and therefore, by (5.4.5),

$$f = e^{h_1} \dots e^{h_n} \widetilde{f}.$$

Now, by the Runge approximation Theorem 2.2.2, for each j, we can find a sequence $(h_{j,\nu})_{\nu \in \mathbb{N}}$ which converges to h_j uniformly on $\overline{D}$. Then

$$g_\nu := e^{h_{1,\nu}} \dots e^{h_{n,\nu}} \widetilde{f}\,, \qquad \nu \in \mathbb{N},$$

is a sequence of functions from $\mathcal{O}^A(\mathbb{C} \setminus P)$ which converges to f uniformly on $\overline{D}$. By Lemma 5.4.4 (i) the values of each $e^{h_{j,\nu}}$ belong to $G_1 A$, the connected component of GA which contains the unit element. Hence the values of

$$g \widetilde{f}^{-1} = e^{h_1} \dots e^{h_n}$$

belong to $G_1 A$. Since the values of $\widetilde{f}$ belong to G, this means that also the values of g belong to G. □

Using the Riemann mapping theorem, now we prove:

5.4.7 Lemma. *Let $D \subseteq \mathbb{C}$ be a bounded, simply connected open set (Section 1.4.3) with piecewise $\mathcal{C}^1$-boundary. Let A be a Banach algebra with unit 1, and let $G_1 A$ be the connected component of the unit element in the group of invertible elements of A. Then the group $\overline{\mathcal{O}}^{G_1 A}(\overline{D})$ is connected (with respect to uniform convergence on $\overline{D}$).*

Proof. Let some $f \in \overline{\mathcal{O}}^{G_1 A}(\overline{D})$ be given. We have to find a continuous curve in $\overline{\mathcal{O}}^{G_1 A}(\overline{D})$ which connects f and the constant function with value 1. Since $G_1 A$ is an open subset of A, we have $\varepsilon > 0$ such that the ball

$$B_\varepsilon(f) := \left\{ g \in \overline{\mathcal{O}}^A(\overline{D}) \;\middle|\; \|f - g\|_{\overline{\mathcal{O}}^A(\overline{D})} < \varepsilon \right\}$$

is contained in $\overline{\mathcal{O}}^{G_1 A}(\overline{D})$. Moreover, by the Mergelyan approximation Theorem 2.2.1, there exists a holomorphic function $\widetilde{f} : U \to A$ defined in a neighborhood U of $\overline{D}$ such that $\widetilde{f}|_{\overline{D}} \in B_\varepsilon(f)$. Therefore it is sufficient to find a continuous curve in $\overline{\mathcal{O}}^{G_1 A}(\overline{D})$ which connects $\widetilde{f}|_{\overline{D}}$ with the constant function with value 1.

Since D is bounded, simply connected and with piecewise $\mathcal{C}^1$-boundary, after shrinking U, we may assume that also U is bounded and simply connected. Therefore, by the Riemann mapping theorem, we can find a biholomorphic map Φ from U onto the unit disc. Set

$$\big(\lambda(t)\big)(z) := \widetilde{f}\Big(\Phi^{-1}\big((1-t)\Phi(z)\big)\Big)$$

for $t \in [0,1]$, $z \in \overline{D}$. This defines a continuous curve $\lambda : [0,1] \to \overline{\mathcal{O}}^{G_1 A}(\overline{D})$ with $\lambda(0) = \widetilde{f}|_{\overline{D}}$. The function $\lambda(1)$ has the constant value $f\big(\Phi^{-1}(0)\big) \in G_1 A$.

Since G_1A is connected, we can find a continuous curve $\gamma : [1,2] \to G_1A$ with $\gamma(1) = f\big(\Phi^{-1}(0)\big)$ and $\gamma(2) = 1$. Setting for $z \in \overline{D}$,

$$\big(\gamma(t)\big)(z) = \begin{cases} \big(\lambda(t)\big)(z) & \text{if } t \in [0,1]\,, \\ \big(\gamma(t)\big)(z) & \text{if } t \in [1,2]\,, \end{cases}$$

we get a continuous curve with the required properties. □

Now we can prove part (i) of Theorem 5.0.1.

Proof of Theorem 5.0.1 (i). Let G_1A be the connected component of the group of invertible elements of A which contains the unit element. Since all values of f belong to the same connected component of G, after multiplying by a constant element of G we may assume that f belongs to the group $\overline{\mathcal{O}}^{G_1A}(\overline{D})$. By the preceding Lemma 5.4.7, this group is connected. In particular, condition (ii) in Proposition 5.4.6 is satisfied (with $P = \emptyset$). Therefore it follows from this proposition that f can be approximated uniformly on $\overline{D}$ by functions from $\mathcal{O}^{G_1A}(\mathbb{C})$. □

5.5 The Cartan lemma

5.5.1 Definition. A pair (D_1, D_2) of bonded open sets $D_1, D_2 \subseteq \mathbb{C}$ with piecewise $\mathcal{C}^1$-boundaries will be called a **Cartan pair** if:

- The intersection $D_1 \cap D_2$ is not empty and has piecewise $\mathcal{C}^1$-boundary, and $\mathbb{C} \setminus \overline{D_1 \cap D_2}$ is connected.
- The union $D_1 \cup D_2$ has piecewise $\mathcal{C}^1$-boundary.
- $\big(\overline{D}_1 \setminus D_2\big) \cap \big(\overline{D}_2 \setminus D_1\big) = \emptyset$.

For example, if $D_1, D_2 \subseteq \mathbb{C}$ are two open rectangles such that also $D_1 \cup D_2$ and $D_1 \cap D_2$ are non-empty rectangles, then (D_1, D_2) is a Cartan pair in the sense of this definition.

If (D_1, D_2) are a Cartan pair, then the closures of the connected components of $D_1 \cap D_2$ are pairwise disjoint. This follows from the fact that $D_1 \cap D_2$ has piecewise $\mathcal{C}^1$-boundary. Moreover, since $\big(\overline{D}_1 \setminus D_2\big) \cap \big(\overline{D}_2 \setminus D_1\big) = \emptyset$, then $\overline{D_1 \cap D_2} = \overline{D}_1 \cap \overline{D}_2$.

In this section we prove the following

5.5.2 Lemma (Cartan lemma). *Let (D_1, D_2) be a Cartan pair, let A be a Banach algebra with unit* 1, *and let G be an open subgroup of the group of invertible elements of A. Further, let $f \in \overline{\mathcal{O}}^G(\overline{D}_1 \cap \overline{D}_2)$ (Def.* 5.4.5) *such that all values of f belong to the same connected component of G. Then there exist $f_j \in \overline{\mathcal{O}}^G(\overline{D}_j)$, $j = 1, 2$, such that*

$$f = f_1 f_2 \qquad \text{on} \quad \overline{D}_1 \cap \overline{D}_2. \tag{5.5.1}$$

Moreover, then, for each $\varepsilon > 0$, the following two assertions are true:

(i) *There exist* $f_j \in \overline{\mathcal{O}}^G(\overline{D}_j)$, $j = 1, 2$, *with* (5.5.1) *and*

$$\max_{z \in \overline{D}_1} \|f_1(z) - 1\| < \varepsilon. \tag{5.5.2}$$

(ii) *There exist* $f_j \in \overline{\mathcal{O}}^G(\overline{D}_j)$, $j = 1, 2$, *with* (5.5.1) *and*

$$\max_{z \in \overline{D}_2} \|f_2(z) - 1\| < \varepsilon. \tag{5.5.3}$$

We prove this lemma in two steps.

5.5.3 Lemma. *Let* (D_1, D_2) *be a Cartan pair, and let* A *be a Banach algebra with unit* 1. *Then, for each* $\varepsilon > 0$, *there exists* $\delta > 0$ *such that, for each* $g \in \overline{\mathcal{O}}(\overline{D}_1 \cap \overline{D}_2, A)$ *(for the notation cf.* 5.4.5*) with*

$$\max_{z \in \overline{D}_1 \cap \overline{D}_2} \|g(z)\| < \delta,$$

there exist $g_j \in \overline{\mathcal{O}}(\overline{D}_j, A)$ *with*

$$\max_{z \in \overline{D}_j} \|g_j(z)\| < \varepsilon\,, \qquad j = 1, 2,$$

such that

$$1 + g = (1 + g_1)(1 + g_2) \qquad \text{on } \overline{D}_1 \cap \overline{D}_2.$$

Proof. We consider $\overline{\mathcal{O}}^A(\overline{D}_1 \cap \overline{D}_2)$, $\overline{\mathcal{O}}^A(\overline{D}_1)$ and $\overline{\mathcal{O}}^A(\overline{D}_2)$ as Banach algebras endowed with the maximum norm. Then, by Corollary 5.3.5, each $f \in \overline{\mathcal{O}}^A(\overline{D}_1 \cap \overline{D}_2)$ can be written in the form $f = f_1 + f_2$ with $f_j \in \overline{\mathcal{O}}^A(\overline{D}_j)$. Therefore the assertion follows from Lemma 5.2.1. □

Proof of Lemma 5.5.2. Let G_1A be the connected component of the unit element in the group of invertible elements A. Since all values of f belong to the same connected component of G, by multiplication by a constant element, we may achieve that $f(z) \in G_1A$ for all $z \in W$. Since $D_1 \cap D_2$ has piecewise $\mathcal{C}^1$ boundary and $\mathbb{C} \setminus \overline{D_1 \cap D_2}$ is connected, we can apply part (i) of the Runge approximation Theorem 5.0.1 (which was proved at the end of Section 2.2.2). So, for each $\delta > 0$, we can find $f_\delta \in \mathcal{O}^{G_1A}(\mathbb{C})$ with

$$\max_{z \in \overline{D}_1 \cap \overline{D}_2} \|f(z) f_\delta^{-1}(z) - 1\| < \delta \quad \text{and} \quad \max_{z \in \overline{D}_1 \cap \overline{D}_2} \|f_\delta^{-1}(z) f(z) - 1\| < \delta.$$

Therefore, by Lemma 5.5.3, for sufficiently small δ, we can find $g_j^{(1)}, g_j^{(2)} \in \overline{\mathcal{O}}^A(\overline{D}_j)$ such that

$$\|g_j^{(k)}\|_{\overline{\mathcal{O}}^A(\overline{D}_j)} < \varepsilon\,, \qquad j = 1, 2,\ k = 1, 2,$$

and

$$f f_\delta^{-1} = \left(1 + g_1^{(1)}\right)\left(1 + g_2^{(1)}\right) \quad \text{and} \quad f_\delta^{-1} f = \left(1 + g_1^{(2)}\right)\left(1 + g_2^{(2)}\right)$$

on $\overline{D_1 \cap D_2}$. To prove assertion (i) we set $f_1 = (1 + g_1^{(1)})$ and $f_2 = (1 + g_2^{(1)}) f_\delta$. To prove assertion (ii) we set $f_1 = f_\delta(1 + g_1^{(2)})$ and $f_2 = 1 + g_2^{(2)}$. □

5.6 $\mathcal{O}^G$-cocycles. Definitions and statement of the main result

Here we introduce the notion of multiplicative cocycles and state the main result on them, which will be proved in the subsequent sections.

Throughout this section, A is a Banach algebra with unit 1, and G is an open subgroup of the group of invertible elements of A. If $U \subseteq \mathbb{C}$ is a non-empty open set, then we denote by $\mathcal{O}^G(U)$ the group of holomorphic G-valued functions defined on U (as everywhere in this book). For practical reasons, we set $\mathcal{O}^G(\emptyset) = \{1\}$ where 1 is the unit element of A.

5.6.1 Definition. Let $D \subseteq \mathbb{C}$ an open set, and let $\mathcal{U} = \{U_j\}_{j\in I}$ be an open covering of D.

(i) We denote by $Z^1(\mathcal{U}, \mathcal{C}^G)$ the set of all families $f = \{f_{jk}\}_{j,k\in I}$ of functions $f_{jk} \in \mathcal{C}^G(U_j \cap U_k)$ with

$$f_{jk} f_{kl} = f_{jl} \qquad \text{on } U_j \cap U_k \cap U_l, \tag{5.6.1}$$

for all $j, k, l \in I$ with $U_j \cap U_k \cap U_l \neq \emptyset$. The elements of $Z^1(\mathcal{U}, \mathcal{C}^G)$ will be called **$(\mathcal{U}, \mathcal{C}^G)$-cocycles**. Condition (5.6.1) is called the (multiplicative) **cocycle condition**. Note that it in particular implies that

$$f_{jj} = 1 \qquad \text{and} \qquad f_{jk} = f_{kj}^{-1}.$$

(ii) We denote by $Z^1(\mathcal{U}, \mathcal{O}^G)$ the set of all cocycles $f \in Z^1(\mathcal{U}, \mathcal{C}^G)$ such that $f_{jk} \in \mathcal{O}^G(U_j \cap U_k)$ for all $j, k \in I$. The elements of $Z^1(\mathcal{U}, \mathcal{O}^G)$ will be called **$(\mathcal{U}, O^G)$-cocycles.**

(iii) Two cocycles $f, g \in Z^1(\mathcal{U}, \mathcal{C}^G)$ will be called **$\mathcal{C}^G$-equivalent** or **continuously equivalent** if there exists a family $\{f_j\}_{j\in I}$ of functions $f_j \in \mathcal{C}^G(U_j)$ such that

$$g_{jk} = h_j f_{jk} h_k^{-1} \quad \text{on } U_j \cap U_k$$

for all $j, k \in I$ with $U_j \cap U_k \neq \emptyset$.

(iv) Two cocycles $f, g \in Z^1(\mathcal{U}, \mathcal{O}^G)$ will be called **$\mathcal{O}^G$-equivalent** or **holomorphically equivalent** if there exists a family $\{f_j\}_{j\in I}$ of functions $f_j \in \mathcal{O}^G(U_j)$ such that

$$g_{jk} = h_j f_{jk} h_k^{-1} \quad \text{on } U_j \cap U_k$$

for all $j, k \in I$ with $U_j \cap U_k \neq \emptyset$.

(v) A cocycle $f \in Z^1(\mathcal{U}, \mathcal{C}^G)$ will be called **$\mathcal{C}^G$-trivial** or **continuously trivial** if there exists a family $\{f_j\}_{j\in I}$ of functions $f_j \in \mathcal{C}^G(U_j)$ such that

$$f_{jk} = f_j f_k^{-1} \quad \text{on } U_j \cap U_k$$

for all $j, k \in I$ with $U_j \cap U_k \neq \emptyset$. In other words, f is called $\mathcal{C}^G$-trivial if it is $\mathcal{C}^G$-equivalent to the cocycle $e \in Z^1(\mathcal{U}, \mathcal{C}^G)$ defined by $e_{jk} \equiv 1$, $j, k \in I$. In this case we say also that f **splits** continuously.

(vi) A cocycle $f \in Z^1(\mathcal{U}, \mathcal{O}^G)$ will be called **$\mathcal{O}^G$-trivial** or **holomorphically trivial** if there exists a family $\{f_j\}_{j\in I}$ of functions $f_j \in \mathcal{O}^G(U_j)$ such that

$$f_{jk} = f_j f_k^{-1} \quad \text{on } U_j \cap U_k$$

for all $j, k \in I$ with $U_j \cap U_k \neq \emptyset$. In this case we say also that f **splits** holomorphically.

Due to P. Cousin the elements of $Z^1(\mathcal{U}, \mathcal{O}^G)$ are also called **multiplicative Cousin problems**. By Theorem 2.3.1, the additive Cousin problems have always a solution. This is no more true for multiplicative Cousin problems. We give an example:

5.6.2 Example. Let $0 < r < R < \infty$. Denote by D the annulus

$$D := \left\{ z \in \mathbb{C} \;\middle|\; r < |z| < R \right\},$$

which we cover by the open sets

$$U_1 := \left\{ z \in D \;\middle|\; \operatorname{Im} z < \frac{r}{2} \right\} \qquad \text{and} \qquad U_2 := \left\{ z \in D \;\middle|\; \operatorname{Im} z > -\frac{r}{2} \right\}.$$

Then the intersection $U_1 \cap U_2$ consists of two connected components

$$V_1 := \left\{ z \in U_1 \cap U_2 \;\middle|\; \operatorname{Re} z < 0 \right\}$$

and

$$V_2 := \left\{ z \in U_1 \cap U_2 \;\middle|\; \operatorname{Re} z > 0 \right\}.$$

Take a Banach space X such that $GL(X)$ is not connected. (By Theorem 5.1.6 such Banach spaces exist). Let $GL_I(E)$ be the connected component of $GL(E)$ which contains the unit operator I. Take any operator $A \in GL(X) \setminus GL_I(X)$ and define

$$F(z) = \begin{cases} I & \text{if } z \in V_1, \\ A & \text{if } z \in V_2. \end{cases}$$

We interpret F as a $(\{U_1, U_2\}, \mathcal{O}^{GL(X)})$-cocycle, setting $F_{12} = F$ and $F_{21} = F^{-1}$ on $U_1 \cap U_2$, $F_{11} = I$ on U_1 and $F_{22} = I$ on U_2. Then F is not $\mathcal{C}^{GL(X)}$-trivial.

Indeed, assume it is $\mathcal{C}^{GL(X)}$-trivial, i.e., $F = C_1 C_2^{-1}$ on $U_1 \cap U_2$ for certain continuous functions $C_1 : U_1 \to GL(X)$ and $C_2 : U_2 \to GL(X)$. Take $r < \rho < R$ and set

$$\gamma(t) = C_1\left(\rho e^{-it}\right) C_2^{-1}\left(\rho e^{it}\right) \qquad \text{for } 0 \leq t \leq \pi.$$

Then γ is a continuous curve in $GL(X)$ connecting I and A, which is a contradiction.

On the positive side, there is the following theorem, which is the main result of the present chapter:

5.6.3 Theorem. *Let $D \subseteq \mathbb{C}$ be an arbitrary open set, let $\mathcal{U}$ be an arbitrary open covering of D, and let f be a $(\mathcal{U}, \mathcal{O}^G)$-cocycle. Assume that at least one of the following conditions is satisfied:*

(i) *D is simply connected.*

(ii) *G is connected.*

(iii) *f is $\mathcal{C}^G$-trivial.*

Then f is $\mathcal{O}^G$-trivial.

We point out again the special case of a covering by two open sets, which is sufficient for many applications:

5.6.4 Corollary. *Let $D_1, D_2 \subseteq \mathbb{C}$ be two open sets, and let $f : D_1 \cap D_2 \to G$ be holomorphic. Assume that at least one of the following conditions is satisfied:*

(i) *D is simply connected.*

(ii) *G is connected.*

(iii) *There exist continuous functions $c_j : D_j \to G$ with $f = c_1 c_2^{-1}$ on $D_1 \cap D_2$.*

Then there exist holomorphic functions $f_j : D_j \to G$ with $f = f_1 f_2^{-1}$ on $D_1 \cap D_2$.

Note that by well-known topological results, each of the conditions (i) and (ii) in Theorem 5.6.3 implies condition (iii), but we do not use this topological fact. We prove directly that each of the conditions (i), (ii) or (iii) yields $\mathcal{O}^G$-triviality of f. In the case of condition (i) this will be done in Section 5.9, and in the case of conditions (ii) and (iii) we prove this in Section 5.11.

Recall that, by Theorem 5.1.5, condition (ii) in Theorem 5.6.3 is satisfied if $G = GL(H)$ where H is an Hilbert space. Therefore we have

5.6.5 Corollary (to Theorem 5.6.3). *For each Hilbert space H and each open set $D \subseteq \mathbb{C}$, any $\mathcal{O}^{GL(H)}$-cocycle over D is $\mathcal{O}^{GL(H)}$-trivial.*

In particular, for each $n \in \mathbb{N}^*$ and each open set $D \subseteq \mathbb{C}$, any $\mathcal{O}^{GL(n,\mathbb{C})}$-cocycle over D is $\mathcal{O}^{GL(n,\mathbb{C})}$-trivial.

5.7 Refinement of the covering

In this section we develop a technique which allows us to compare cocycles with different coverings. Throughout this section, A is a Banach algebra with unit 1, G a subgroup of the group of invertible elements of A, and $D \subseteq \mathbb{C}$ is an arbitrary open set.

5.7.1 Definition. Let $\mathcal{U} = \{U_j\}_{j\in I}$ and $\mathcal{V} = \{V_\mu\}_{\mu\in J}$ be two open coverings of D such that $\mathcal{V}$ is a refinement of $\mathcal{U}$. Then (by definition of a refinement) there is a map $\tau : J \to I$ with $V_\mu \subseteq U_{\tau(\mu)}$ for all $\mu \in J$. For any such map τ and each cocycle $f \in \mathcal{Z}^1(\mathcal{U}, \mathcal{C}^G)$ (Def. 5.6.1), we define a cocycle $\tau^* f \in \mathcal{Z}^1(\mathcal{V}, \mathcal{C}^G)$ setting

$$(\tau^* f)_{\mu\nu} = f_{\tau(\mu)\tau(\nu)}\Big|_{V_\mu \cap V_\nu},$$

for all $\mu, \nu \in J$ with $V_\mu \cap V_\nu \neq \emptyset$.

We shall say that a cocycle $g \in Z^1(\mathcal{V}, \mathcal{C}^G)$ is **induced** by a cocycle $f \in \mathcal{Z}^1(\mathcal{U}, \mathcal{C}^G)$ if there exists a map $\tau : J \to I$ with $V_\mu \subseteq U_{\tau(\mu)}$, $\mu \in J$, such that $g = \tau^* f$.

Note that in this case $g \in Z^1(\mathcal{V}, \mathcal{O}^G)$ if $f \in \mathcal{Z}^1(\mathcal{U}, \mathcal{O}^G)$.

Note also that, in general, for a cocycle $f \in \mathcal{Z}^1(\mathcal{U}, \mathcal{O}^G)$, there exist different cocycles in $Z^1(\mathcal{V}, \mathcal{C}^G)$, which are induced by f. However, there is the following

5.7.2 Proposition. *Let $D \subseteq \mathbb{C}$ be an open set, let $\mathcal{U} = \{U_j\}_{j\in I}$ and $\mathcal{V} = \{V_\nu\}_{\nu\in J}$ be two open coverings of D such that $\mathcal{V}$ is a refinement of $\mathcal{U}$, and let $\mathcal{F} = \mathcal{C}^G$ or $\mathcal{F} = \mathcal{O}^G$. Further, let $f, g \in Z^1(\mathcal{U}, \mathcal{F})$ and $\widetilde{f}, \widetilde{g} \in Z^1(\mathcal{V}, \mathcal{F})$ such that $\widetilde{f}$ is induced by f and $\widetilde{g}$ is induced by g. Then the following are equivalent:*

(i) *f and g are $\mathcal{F}$-equivalent.*

(ii) *$\widetilde{f}$ and $\widetilde{g}$ are $\mathcal{F}$-equivalent.*

In particular, the following are equivalent:

(i′) *f is $\mathcal{F}$-trivial.*

(ii′) *$\widetilde{f}$ is $\mathcal{F}$-trivial.*

Proof. By hypothesis, we have some maps $\tau, \varphi : J \to I$ with $\widetilde{f} = \tau^* f$ and $\widetilde{g} = \varphi^* g$.

We first assume that f and g are $\mathcal{F}$-equivalent. Then there is a family of functions $h_j \in \mathcal{F}(U_j)$, $j \in I$, with

$$h_j^{-1} f_{jk} h_k = g_{jk} \qquad \text{on } U_j \cap U_k, \tag{5.7.1}$$

for all $j, k \in I$ with $U_j \cap U_k \neq \emptyset$. Then we define a family of functions $\widetilde{h}_\mu \in \mathcal{F}(V_\mu)$, $\mu \in J$, setting

$$\widetilde{h}_\mu = h_{\tau(\mu)} g_{\tau(\mu),\varphi(\mu)} \qquad \text{on } V_\mu, \quad \mu \in J.$$

Then

$$\widetilde{h}_\nu^{-1} \widetilde{f}_{\nu\mu} \widetilde{h}_\mu = g_{\tau(\nu),\varphi(\nu)}^{-1} h_{\tau(\nu)}^{-1} f_{\tau(\nu),\tau(\mu)} h_{\tau(\mu)} g_{\tau(\mu),\varphi(\mu)} \qquad \text{on } V_\mu \cap V_\nu\,,$$

for all $\mu, \nu \in J$ with $V_\mu \cap V_\nu \neq \emptyset$. By (5.7.1) this implies

$$\widetilde{h}_\nu^{-1} \widetilde{f}_{\nu\mu} \widetilde{h}_\mu = g_{\tau(\nu),\varphi(\nu)}^{-1} g_{\tau(\nu),\tau(\mu)} g_{\tau(\mu),\varphi(\mu)} \qquad \text{on } V_\nu \cap V_\mu\,,$$

for all $\mu, \nu \in J$ with $V_\mu \cap V_\nu \neq \emptyset$. Since g satisfies the cocycle condition this further implies

$$\widetilde{h}_\nu^{-1} \widetilde{f}_{\nu\mu} \widetilde{h}_\mu = g_{\varphi(\nu),\varphi(\mu)} = \widetilde{g}_{\nu\mu} \qquad \text{on } V_\nu \cap V_\mu ,$$

for all $\mu, \nu \in J$ with $V_\mu \cap V_\nu \neq \emptyset$, i.e., $\widetilde{f}$ and $\widetilde{g}$ are $\mathcal{F}$-equivalent.

Now we assume that $\widetilde{f}$ and $\widetilde{g}$ are $\mathcal{F}$-equivalent. Then there exists a family of functions $h_\mu \in \mathcal{F}(V_\mu)$, $\mu \in J$, with

$$\widetilde{h}_\nu^{-1} \widetilde{f}_{\nu\mu} \widetilde{h}_\mu = \widetilde{g}_{\nu\mu} \qquad \text{on } V_\mu \cap V_\nu ,$$

for all $\mu, \nu \in J$, with $V_\mu \cap V_\nu \neq \emptyset$. Since $\widetilde{f} = \tau^* f$ and $\widetilde{g} = \varphi^* g$, then

$$\widetilde{h}_\nu^{-1} f_{\tau(\nu),\tau(\mu)} \widetilde{h}_\mu = g_{\varphi(\nu),\varphi(\mu)} \qquad \text{on } V_\nu \cap V_\mu ,$$

for all $\mu, \nu \in J$, with $V_\mu \cap V_\nu \neq \emptyset$. Since f and g satisfy the cocycle condition, this implies

$$\widetilde{h}_\nu^{-1} f_{\tau(\nu),j} f_{j,\tau(\mu)} \widetilde{h}_\mu = g_{\varphi(\nu),j} g_{j,\varphi(\mu)} \qquad \text{on } V_\nu \cap V_\mu \cap U_j ,$$

and further,

$$f_{j,\tau(\mu)} \widetilde{h}_\mu g_{\varphi(\mu),j} = f_{j,\tau(\nu)} \widetilde{h}_\nu g_{\varphi(\nu),j} \qquad \text{on } V_\nu \cap V_\mu \cap U_j ,$$

for all $j \in I$ and $\nu, \mu \in J$ with $V_\nu \cap V_\mu \cap U_j \neq \emptyset$. Therefore we can define a family of functions $h_j \in \mathcal{F}(U_j)$, $j \in I$, by setting

$$h_j = f_{j,\tau(\mu)} \widetilde{h}_\mu g_{\varphi(\mu),j} \qquad \text{on } V_\mu \cap U_j ,$$

for all $j \in I$ and $\mu \in J$ with $V_\mu \cap U_j \neq \emptyset$. Using again that f and g satisfy the cocycle condition, we obtain

$$h_j^{-1} f_{jk} h_k = g_{j,\varphi(\mu)} \widetilde{h}_\mu^{-1} f_{\tau(\mu),j} f_{jk} f_{k,\tau(\mu)} \widetilde{h}_\mu g_{\varphi(\mu),k} = g_{jk} \qquad \text{on } V_\mu \cap U_j \cap U_k ,$$

for all $j, k \in I$ and $\mu \in J$ with $V_\mu \cap U_j \cap U_k \neq \emptyset$. Hence f and g are $\mathcal{F}$-equivalent. □

5.7.3 Definition. Let $D \subseteq \mathbb{C}$ be an open set. By a **$\mathcal{C}^G$-cocycle over** D we mean a $(\mathcal{U}, \mathcal{C}^G)$-cocycle such that $\mathcal{U}$ is an open covering of D. Correspondingly, by an **$\mathcal{O}^G$-cocycle over** D we mean a $(\mathcal{U}, \mathcal{O}^G)$-cocycle such that $\mathcal{U}$ is an open covering of D. The open covering $\mathcal{U}$ then will be called the **covering of this cocycle**.

In Definition 5.6.1 we introduced the notion of *equivalence* for cocycles with the same covering. In view of Proposition 5.7.2, the following definition is correct.

5.7.4 Definition. Let $D \subseteq \mathbb{C}$ be an open set.

(i) Two $\mathcal{C}^G$-cocycles f and g over D will be called **$\mathcal{C}^G$-equivalent** or **continuously equivalent** over D if there exists an open covering $\mathcal{W}$ of D which is a refinement both of the covering of f and of the covering of g such that at least one $(\mathcal{W}, \mathcal{C}^G)$-cocycle induced by f is $\mathcal{C}^G$-equivalent to at least one $(\mathcal{W}, \mathcal{C}^G)$-cocycle induced by g (or, what is the same (by Proposition 5.7.2), such that any $(\mathcal{W}, \mathcal{C}^G)$-cocycle induced by f is $\mathcal{C}^G$-equivalent to any $(\mathcal{W}, \mathcal{C}^G)$-cocycle induced by g).

(ii) Two $\mathcal{O}^G$-cocycles f and g over D will be called **$\mathcal{O}^G$-equivalent** or **holomorphically equivalent** over D if there exists an open covering $\mathcal{W}$ of D which is a refinement both of the covering of f and of the covering of g such that at least one $(\mathcal{W}, \mathcal{O}^G)$-cocycle induced by f is $\mathcal{O}^G$-equivalent to at least one $(\mathcal{W}, \mathcal{O}^G)$-cocycle induced by g.

5.7.5 Definition. Let $D \subseteq \mathbb{C}$ an open set, let $\mathcal{U} = \{U_j\}_{j\in I}$ be an open covering of D, and let Y be an open subset of D. Set

$$\mathcal{U} \cap Y = \left\{ U_j \cap Y \,\middle|\, j \in I \right\},$$

and let $\mathcal{F} = \mathcal{C}^G$ or $\mathcal{F} = \mathcal{O}^G$. Then we define:

(i) Let f be an $(\mathcal{U}, \mathcal{F})$-cocycle over D.

Then we denote by $f\big|_Y$ the $(\mathcal{U} \cap Y, \mathcal{F})$-cocycle defined by

$$(f|_Y)_{jk} = f_{jk}\big|_{U_j \cap U_k \cap Y}$$

for all $j, k \in I$ with $U_j \cap U_k \cap Y \neq \emptyset$. This cocycle $f\big|_Y$ will be called the **restriction** of f to Y. We shall say that f **is $\mathcal{F}$-trivial over** Y if $f\big|_Y$ is $\mathcal{F}$-trivial.

(ii) Let f, g be two $\mathcal{F}$)-cocycles over D. Then we shall say that f **and** g **are $\mathcal{F}$-equivalent over** Y if the restricted cocycles $f\big|_Y$ and $g\big|_Y$ are $\mathcal{F}$-equivalent.

5.7.6 Proposition. *Let $D \subseteq \mathbb{C}$ be an open set, let $U = \{U_j\}_{j\in I}$ be an open covering of D, let $\mathcal{F} = \mathcal{C}^G$ or $\mathcal{F} = \mathcal{O}^G$, and let f be an $\mathcal{F}$-cocycle over D, which is $\mathcal{F}$-trivial over each U_j.*

Then f is $\mathcal{F}$-equivalent to some $(\mathcal{U}, \mathcal{F})$-cocycle.

Proof. Let $\mathcal{V}$ be the covering of f. For each $j \in I$, we take an open covering $\mathcal{W}_j = \{W_{j\nu}\}_{\nu\in\mathbb{N}}$ of U_j so fine that $\mathcal{W} := \{W_{j\nu)}\}_{(j,\nu)\in I\times\mathbb{N}}$ is a refinement of $\mathcal{V}$. Let $f' = \{f_{j\mu,k\nu}\}_{(j,\mu),(k,\nu)\in I\times\mathbb{N}}$ be a $(\mathcal{W}, \mathcal{F})$-cocycle induced by f. Then f' is $\mathcal{F}$-equivalent to f (by Definition 5.7.4), and therefore it is sufficient to prove that f' is $\mathcal{F}$-equivalent to some $(\mathcal{U}, \mathcal{F})$-cocycle.

Since f is $\mathcal{F}$-equivalent to f', each $f|_{U_j}$ is $\mathcal{F}$-equivalent to $f'|_{U_j}$. Since each $f|_{U_j}$ is $\mathcal{F}$-trivial, it follows that also each $f'|_{U_j}$ is $\mathcal{F}$-trivial. Hence, for all $j \in I$, we have a family $h_{j\nu} \in \mathcal{F}(W_{j\nu})$, $\nu \in \mathbb{N}$, such that, for each $j \in I$,

$$h_{j\nu}^{-1} f'_{j\nu,j\mu} h_{j\mu} = 1 \qquad \text{on } W_{j\nu} \cap W_{j\mu} \tag{5.7.2}$$

for all $\nu, \mu \in \mathbb{N}$ with $W_{j\nu} \cap W_{j\mu} \neq \emptyset$. Setting

$$f''_{j\nu,k\mu} = h_{j\nu}^{-1} f'_{j\nu,k\mu} h_{k\mu} \qquad \text{on } W_{j\nu} \cap W_{k\mu}\,,$$

for all $(j, \nu), (k, \mu) \in I \times \mathbb{N}$ with $W_{j\nu} \cap W_{k\mu} \neq \emptyset$, we define a $(\mathcal{W}, \mathcal{F})$-cocycle f''. By its definition, f'' is $\mathcal{F}$-equivalent to f'. Hence f'' is $\mathcal{F}$-equivalent to f. Since f'

satisfies the cocycle condition, we obtain that

$$f''_{j\nu,k\mu} = h_{j\nu}^{-1} f'_{j\nu,j\mu'} f'_{j\mu',k\mu'} f'_{k\mu',k\mu} h_{k\mu}$$
$$= \left(h_{j\nu}^{-1} f'_{j\nu,j\mu'} h_{j\mu'}\right) h_{j\mu'}^{-1} f'_{j\mu',k\mu'} h_{k\mu'} \left(h_{k\mu'}^{-1} f'_{k\mu',k\mu} h_{k\mu}\right)$$
$$\text{on } W_{j\nu} \cap W_{k\mu} \cap W_{j\nu'} \cap W_{k\mu'}$$

for all $j, k \in I$ and $\nu, \mu, \nu', \mu' \in \mathbb{N}$ with $W_{j\nu} \cap W_{k\mu} \cap W_{j\nu'} \cap W_{k\mu'} \neq \emptyset$. By (5.7.2) this implies

$$f''_{j\nu,k\mu} = h_{j\mu'}^{-1} f'_{j\mu',k\mu'} h_{k\mu'} = f''_{j\nu',k\mu'} \qquad \text{on } W_{j\nu} \cap W_{k\mu} \cap W_{j\nu'} \cap W_{k\mu'}$$

for all $j, k \in I$ and $\nu, \mu, \nu', \mu' \in \mathbb{N}$ with $W_{j\nu} \cap W_{k\mu} \cap W_{j\nu'} \cap W_{k\mu'} \neq \emptyset$. Hence there is a well-defined family of functions $f'''_{jk} \in \mathcal{F}(U_j \cap U_k)$, $j, k \in I$, defined by

$$f'''_{jk} = f''_{j\nu,k\mu} \qquad \text{on } W_{j\nu} \cap W_{k\mu}$$

for all $j, k \in I$ and $\mu, \nu \in \mathbb{N}$ with $W_{j\nu} \cap W_{k\mu} \neq \emptyset$. Since f'' satisfies the cocycle condition, this implies

$$f'''_{jk} f'''_{kl} = f''_{j\nu,k\mu} f''_{k\mu,l\lambda} = f''_{j\nu,l\lambda} \qquad \text{on } W_{j\nu} \cap W_{k\mu} \cap W_{l\lambda}$$

for all $j, k, l \in I$ and $\nu, \mu, \lambda \in \mathbb{N}$ with $W_{j\nu} \cap W_{k\mu} \cap W_{l\lambda} \neq \emptyset$. Hence

$$f'''_{jk} f'''_{kl} = f'''_{jl} \qquad \text{on } U_j \cap U_k \cap U_l$$

for all $j, k, l \in I$ with $U_j \cap U_k \cap W_l \neq \emptyset$, i.e.,

$$f''' \in Z^1(\mathcal{U}, \mathcal{F}).$$

Let $\tau : I \times \mathbb{N} \to I$ be the map defined by $\tau(j, \mu) = j$ for $(j, \mu) \in I \times \mathbb{N}$. Since

$$W_{j\nu} \subseteq U_j = U_{\tau(j,\nu)} \qquad \text{for all } (j, \nu) \in I \times \mathbb{N}$$

and since, by definition of f''',

$$f''_{j\nu,k\mu} = f'''_{jk}\Big|_{W_{j\nu} \cap W_{k\mu}} = f'''_{\tau(j\nu)\tau(k\mu)}\Big|_{W_{j\nu} \cap W_{k\mu}}, \qquad (j, \nu), (k, \mu) \in I \times \mathbb{N},$$

then f'' is induced by f'''. Since f'' is $\mathcal{F}$-equivalent to f, this implies that f is $\mathcal{F}$-equivalent to f'''. □

5.8 Exhausting by compact sets

Here we prove the following technical lemma which we shall use several times in this book:

5.8.1 Lemma. *Let A be a Banach algebra with unit 1, let G be an open subgroup of the group of invertible elements of A. Let $D \subseteq \mathbb{C}$ be an open set (possibly unbounded), let f be an $\mathcal{O}^G$-cocycle over D, and let $(D_n)_{n\in\mathbb{N}}$ be a sequence of bounded open sets $D_n \subseteq D$ such that:*

(1) *$\overline{D}_n \subseteq D_{n+1}$ for all $n \in \mathbb{N}$.*

(2) *$\bigcup_{n\in\mathbb{N}} D_n = D$.*

(3) *For each $n \in \mathbb{N}$, any function from $\mathcal{O}^G(D_{n+1})$ can be approximated uniformly on $\overline{D}_n$ by functions from $\mathcal{O}^G(D)$.*

(4) *The cocycle f is $\mathcal{O}^G$-trivial over each D_n, $n \in \mathbb{N}$.*

Then f is $\mathcal{O}^G$-trivial over D.

Proof. Let $\|\cdot\|$ be the norm of A and set

$$\operatorname{dist}(a, A \setminus G) = \inf_{b\in A\setminus G} \|a - b\| \qquad \text{for } a \in G.$$

Let $\mathcal{U} = \{U_j\}_{j\in I}$ be the covering associated to f. By Proposition 5.7.2, after passing to a refinement, we may assume that each U_j is a relatively compact open disc in D and $f_{jk} \in \overline{\mathcal{O}}^G(\overline{U}_j \cap \overline{U}_k)$ for all $j, k \in I$. Note that then, for each $j \in I$, there exists $n_j \in \mathbb{N}$ with

$$\overline{U}_j \subseteq D_n \qquad \text{if } n \geq n_j. \tag{5.8.1}$$

Moreover we may assume that

$$\begin{cases} \text{for each compact set } K \subseteq D \text{ there exists only a} \\ \text{finite number of indices } j \in I \text{ with } U_j \cap K \neq \emptyset. \end{cases} \tag{5.8.2}$$

To prove the lemma it is sufficient to find a sequence $\left(f_n\right)_{n\in\mathbb{N}}$ of families $f_n = \left\{f_{nj}\right\}_{j\in I}$ of functions $f_{nj} \in \overline{\mathcal{O}}^G(D_{n+1} \cap \overline{U}_j)$ as well as a sequence $(\varepsilon_n)_{n\in\mathbb{N}}$ of positive numbers, such that, for all $n \in \mathbb{N}$,

$$f_{jk} = f_{nj}^{-1} f_{nk} \qquad \text{on } D_{n+1} \cap \overline{U}_j \cap \overline{U}_k\,, \quad j, k \in I, \tag{5.8.3}$$

$$\varepsilon_n < \frac{1}{4} \min_{z\in\overline{D}_n\cap\overline{U}_j} \operatorname{dist}\Big(f_{nj}(z), A \setminus G\Big)\,, \quad j \in I, \tag{5.8.4}$$

$$\max_{z\in\overline{D}_n\cap\overline{U}_j} \|f_{nj}(z) - f_{n-1,j}(z)\| < \varepsilon_{n-1} \text{ if } n \geq 1\,, \quad j \in I, \text{ and} \tag{5.8.5}$$

$$\varepsilon_n < \frac{\varepsilon_{n-1}}{2} \text{ if } n \geq 1\,. \tag{5.8.6}$$

Indeed, then it follows from (5.8.1), (5.8.5) and (5.8.6) that, for all $j \in I$ and $n, m \in \mathbb{N}$ with $n_j \leq n < m$,

$$\max_{z\in\overline{U}_j} \|f_{mj}(z) - f_{nj}(z)\| < \varepsilon_n + \frac{\varepsilon_n}{2} + \ldots + \frac{\varepsilon_n}{2^{m-n-1}} < 2\varepsilon_n,$$

which implies that, for each $j \in I$, the sequence $\left(f_{nj}\right)_{n \geq n_j}$ converges uniformly on $\overline{U}_j$ to some function $f_j \in \overline{\mathcal{O}}^A(\overline{U}_j)$ where

$$\max_{z \in \overline{U}_j} \|f_j(z) - f_{n,j}(z)\| \leq 2\varepsilon_n \qquad \text{for } n \geq n_j.$$

By (5.8.4), the latter inequality implies that

$$\max_{z \in \overline{U}_j} \|f_j(z) - f_{nj}(z)\| < \frac{1}{2} \inf_{z \in \overline{U}_j} \operatorname{dist}\Big(f_{nj}(z), G \setminus A\Big) \qquad \text{for } n \geq n_j.$$

Hence $f_j \in \overline{\mathcal{O}}^G(\overline{U}_j)$, $j \in I$. It remains to observe that now we can pass to the limit for $n \to \infty$ in (5.8.3), which gives $f_{jk} = f_j^{-1} f_k$ on $\overline{U}_j \cap \overline{U}_k$ for all $j, k \in I$. Hence f is $\mathcal{O}^G$-trivial.

To prove the existence of such sequences, we first recall that, by hypothesis of the lemma, each $f|_{D_{n+2}}$ is $\mathcal{O}^G$-trivial. Therefore we can find a sequence $\big(\widetilde{f}_n\big)_{n \in \mathbb{N}}$ of families $\widetilde{f}_n = \{\widetilde{f}_{nj}\}_{j \in I}$ of functions $\widetilde{f}_{nj} \in \mathcal{O}^G(D_{n+2} \cap U_j)$ such that

$$f_{jk} = \widetilde{f}_{nj}^{-1} \widetilde{f}_{nk} \qquad \text{on } D_{n+2} \cap U_j \cap U_k \tag{5.8.7}$$

for all $n \in \mathbb{N}$ and $j, k \in I$. We claim that

$$\widetilde{f}_{nj} \in \overline{\mathcal{O}}^G(D_{n+2} \cap \overline{U}_j) \qquad \text{for all } j \in I. \tag{5.8.8}$$

Indeed, let $(z_\nu)_{\nu \in \mathbb{N}}$ be a sequence in $D_{n+2} \cap U_j$ which converges to some point $z \in D_{n+2} \cap \overline{U}_j$. Since $\mathcal{U}$ covers D, we can find $k \in I$ with $z \in U_k$. Since U_k is open, then $z_\nu \in D_{n+2} \cap \overline{U}_j \cap U_k$ for sufficiently large ν, where, by (5.8.7),

$$\widetilde{f}_{nj}(z_\nu) = \widetilde{f}_{nk}(z_\nu)\, f_{jk}^{-1}(z_\nu).$$

Since both $\widetilde{f}_{nk}$ and f_{jk} are continuous on $D_{n+2} \cap \overline{U}_j \cap U_k$ and since $z \in D_{n+2} \cap \overline{U}_j \cap U_k$, this implies that $\lim_{\nu \to \infty} \widetilde{f}_{nj}(z_\nu)$ exists.

Now we proceed by induction.

Beginning of the induction: Since $\widetilde{f}_{0j} \in \overline{\mathcal{O}}^G(D_2 \cap \overline{U}_j)$ we can define $f_{0j} = \widetilde{f}_{0j}\big|_{D_1 \cap \overline{U}_j}$. It follows from condition (5.8.2) that $\bigcup_{j \in I} f_{0j}(\overline{D_0 \cap U_j})$ is a compact subset of G. Hence we can find $\varepsilon_0 > 0$ such that

$$\varepsilon_0 < \frac{1}{4} \min_{z \in \overline{D_0 \cap U_j}} \operatorname{dist}\Big(f_{0,j}(z), G \setminus A\Big) \qquad \text{for all } j \in I.$$

With this choice of the family $\{f_{0j}\}_{j \in I}$ and the number ε_0 conditions (5.8.3)–(5.8.6) are satisfied for $n = 0$.

Hypothesis of induction: Assume, for some $m \in \mathbb{N}$, we already have families $f_0 = \{f_{0j}\}_{j \in I}, \ldots, f_m = \{f_{mj}\}_{j \in I}$ of functions

$$f_{0j} \in \overline{\mathcal{O}}^G(D_0 \cap \overline{U}_j), \ldots, f_{mj} \in \overline{\mathcal{O}}^G(D_m \cap \overline{U}_j)$$

as well as positive numbers $\varepsilon_0, \dots, \varepsilon_m$ such that (5.8.5) -(5.8.4) hold for $n = 0, \dots, m$.

Step of induction: Since the compact set $\overline{D}_m \cap \overline{U}_j$ is contained in $D_{m+1} \cap \overline{U}_j$ and f_{mj} is continuous on $D_{m+1} \cap \overline{U}_j$, the function f_{mj} is bounded on $\overline{D}_m \cap \overline{U}_j$. By condition (5.8.2), this implies that

$$\max_{j \in I} \max_{z \in \overline{D}_m \cap \overline{U}_j} \|f_{mj}(z)\| < \infty. \tag{5.8.9}$$

By (5.8.7),

$$f_{jk} = \widetilde{f}_{mj}^{-1} \widetilde{f}_{mk} \qquad \text{on } D_{m+2} \cap U_j \cap U_k.$$

Moreover, by hypothesis of induction,

$$f_{jk} = f_{mj}^{-1} f_{mk} \qquad \text{on } D_{m+1} \cap U_j \cap U_k.$$

Since $D_{m+1} \subseteq D_{m+2}$, this yields

$$\widetilde{f}_{mk} f_{mk}^{-1} = \widetilde{f}_{mj} f_{m,j}^{-1} \qquad \text{on } D_{m+1} \cap U_j \cap U_k.$$

Hence, there is a well-defined function $\Phi \in \mathcal{O}^G(D_{m+1})$ with

$$\Phi = \widetilde{f}_{mj} f_{mj}^{-1} \tag{5.8.10}$$

on $D_{m+1} \cap U_j$ for all $j \in I$. Note that, since f_{mj} is continuous on $D_{m+1} \cap \overline{U}_j$ and, by (5.8.8), $\widetilde{f}_{mj}$ is continuous on $D_{m+2} \cap \overline{U}_j$, (5.8.10) even holds on $D_{m+1} \cap \overline{U}_j$, $j \in I$. By hypothesis of the lemma, Φ can be approximated uniformly on $\overline{D}_m$ by functions from $\mathcal{O}^G(D)$. Therefore and by (5.8.9), we can find $\Psi \in \mathcal{O}^G(D)$ such that

$$\max_{\overline{D}_m} \|\Psi\Phi - 1\| < \frac{\varepsilon_m}{\max\limits_{\overline{D}_m \cap \overline{U}_j} \|f_{mj}\|} \qquad \text{for all } j \in I.$$

Since (5.8.10) holds over $D_{m+1} \cap \overline{U}_j$ and $\overline{D}_m \cap \overline{U}_j \subseteq D_{m+1} \cap \overline{U}_j$, this implies that

$$\max_{\overline{D}_m \cap \overline{U}_j} \|\Psi \widetilde{f}_{mj} f_{mj}^{-1} - 1\| < \frac{\varepsilon_m}{\max\limits_{\overline{D}_m \cap \overline{U}_j} \|f_{mj}\|} \qquad \text{for all } j \in I. \tag{5.8.11}$$

Setting

$$f_{m+1,j} = \Psi \widetilde{f}_{mj} \qquad \text{on } D_{m+2} \cap \overline{U}_j,$$

now we obtain a family $f_{m+1} = \{f_{m+1,j}\}_{j \in I}$ of functions $f_{m+1,j} \in \overline{\mathcal{O}}^G(D_{m+2} \cap \overline{U}_j)$. Further, it follows from condition (5.8.2) that $\bigcup_{j \in I} f_{m+1,j}(\overline{D}_{m+1} \cap \overline{U}_j)$ is a compact subset of G. Hence we can find $\varepsilon_{m+1} > 0$ so small that condition (5.8.4) is satisfied for $n = m + 1$. As $\varepsilon_m > 0$, we may moreover assume that (5.8.6) holds for $n = p + 1$. From (5.8.7) we get

$$f_{m+1,j}^{-1} f_{m+1,k} = \widetilde{f}_{mj}^{-1} \Psi^{-1} \Psi \widetilde{f}_{mk} = \widetilde{f}_{mj}^{-1} \widetilde{f}_{mk} = f_{jk}$$

on $D_{m+2} \cap \overline{U}_j \cap \overline{U}_k$, i.e., (5.8.3) holds for $n = m+1$. From (5.8.11) it follows that

$$\max_{\overline{D}_m \cap \overline{U}_j} \|f_{m+1,j} - f_{mj}\| = \max_{\overline{D}_m \cap \overline{U}_j} \left\| \left(\Psi \widetilde{f}_{mj} f_{mj}^{-1} - 1 \right) f_{mj} \right\| < \varepsilon_m .$$

Hence also (5.8.5) holds for $n = m+1$. □

5.9 Proof of Theorem 5.6.3 for simply connected open sets

In this section A is a Banach algebra with unit 1, and G is an open subgroup of the group of invertible elements of A.

The first step in the proof of Theorem 5.6.3 is the following

5.9.1 Lemma (Cartan lemma. Second version). *Let (D_1, D_2) be a Cartan pair such that $\mathbb{C} \setminus \overline{D_1 \cup D_2}$ is connected, and let $f \in \mathcal{O}^G(D_1 \cap D_2)$ such that all values of f belong to the same connected component of G. Then there exist $f_j \in \mathcal{O}^G(D_j)$, $j = 1, 2$, such that*

$$f = f_1^{-1} f_2 \qquad \text{on} \quad D_1 \cap D_2. \tag{5.9.1}$$

Proof. Take a sequence of Cartan pairs $\big((D_{n,1}, D_{n,2})\big)_{n \in \mathbb{N}}$ such that

- $D_{n,j} \subseteq D_{n+1,j} \subseteq D_j$ for $j = 1, 2$ and all $n \in \mathbb{N}$,
- $\bigcup_{n \in \mathbb{N}} D_{n,j} = D_j$ for $j = 1, 2$ and
- For each $n \in \mathbb{N}$, $\mathbb{C} \setminus (\overline{D}_{1,n} \cup \overline{D}_{2,n})$ is connected.

Set $D = D_1 \cup D_2$ and $D_n = D_{n,1} \cup D_{n,2}$. Then, it follows from part (i) of the Runge approximation Theorem 5.0.1 (which we already proved at the end of Section 5.4) that the functions from $\mathcal{O}^G(D_{n+1})$ can be approximated uniformly on $\overline{D}_n$ by functions from $\mathcal{O}^G(D)$, i.e., we have the situation considered in Lemma 5.8.1.

Now we consider the $\big(\{D_1, D_2\}, \mathcal{O}^G\big)$-cocycle F defined by $F_{12} = f$. We have to prove that F is $\mathcal{O}^G$-trivial. By the Cartan Lemma 5.5.2, F is $\mathcal{O}^G$-trivial over each D_n. Therefore, by Lemma 5.8.1, F is $\mathcal{O}^G$-trivial. □

Using this lemma and propositions 5.7.2 and 5.7.6, now we can prove the following special case of Theorem 5.6.3:

5.9.2 Lemma. *Let*

$$K := \Big\{ z \in \mathbb{C} \;\Big|\; 0 < \operatorname{Re} z < 1 \text{ and } 0 < \operatorname{Im} z < 1 \Big\},$$

and let Ω an open neighborhood of $\overline{K}$. Then each $\mathcal{O}^G$-cocycle over Ω is $\mathcal{O}^G$-trivial over K.

Proof. Let an $\mathcal{O}^G$-cocycle f over Ω be given, let $\mathcal{V} = \{V_\nu\}_{\nu \in I}$ be the covering of f, and let $\mathcal{V} \cap K := \{V_\nu \cap K\}_{\nu \in I}$. We choose $n \in \mathbb{N}$ sufficiently large and denote by U_{jk}, $j, k = 1, \dots, n$, the open rectangle of all $z \in K$ with

$$\left(k - 1 - \frac{1}{3}\right)\frac{1}{n} < \operatorname{Re} z < \left(k + \frac{1}{3}\right)\frac{1}{n}$$

and

$$\left(j - 1 - \frac{1}{3}\right)\frac{1}{n} < \operatorname{Im} z < \left(j + \frac{1}{3}\right)\frac{1}{n}.$$

Then $\mathcal{U} := \{U_{jk}\}_{1 \le j,k \le n}$ is a refinement of $\mathcal{V} \cap K$. Let f' be a $(\mathcal{U}, \mathcal{O})$-cocycle induced by f. By Proposition 5.7.2, it is sufficient to prove that f' is $\mathcal{O}^G$-trivial.

To do this we give the family $\mathcal{U}$ an order saying that $U_{jk} < U_{j'k'}$, if and only if, either $j < j'$ or $j = j'$ and $k < k'$. Let $U_1, \dots, U_{n^2}$ be the family $\mathcal{U}$ numbered in this way.

Now we prove by induction that, for all $1 \le j \le N^2$, the cocycle f' is $\mathcal{O}^G$-trivial over $U_1 \cup \ldots \cup U_j$. (For $j = n^2$ this is the assertion which we have to prove.)

Since U_1 belongs to the covering associated to f', it is clear that f' is $\mathcal{O}^G$-trivial over U_1.

Assume, for some $1 \le j \le N^2 - 1$, we already know that f' is $\mathcal{O}^G$-trivial over $U_1 \cup \ldots \cup U_j$. As f' is also $\mathcal{O}^G$-trivial over U_{j+1}, then it follows from Proposition 5.7.6 that f' is $\mathcal{O}^G$-equivalent to some $\left(\{U_1 \cup \ldots \cup U_j \,,\, U_{j+1}\}, \mathcal{O}^G\right)$-cocycle f''. Since, clearly, $(U_1 \cup \ldots \cup U_j \,,\, U_{j+1})$ is a Cartan pair (Definition 5.5.1) such that $\mathbb{C} \setminus (U_1 \cup \ldots \cup U_j) \cap U_{j+1}$ is connected, it follows from Lemma 5.9.1 that f'' is $\mathcal{O}^G$-trivial. Hence f' is $\mathcal{O}^G$-trivial. □

Proof of Theorem 5.6.3 *if D is simply connected.* Since D is simply connected, by the Riemann mapping theorem we may assume that either $D = \mathbb{C}$ or D is the open unit disc. In both cases, there exists a sequence $(D_n)_{n \in \mathbb{N}}$ of open discs $D_n \subseteq \mathbb{C}$ such that:

- $\overline{D}_n \subseteq D$,
- $\overline{D}_n \subseteq D_{n+1}$ for all $n \in \mathbb{N}$,
- $\bigcup_{n \in \mathbb{N}} D_n = D$,
- by part (i) of the Runge approximation Theorem 5.0.1 (which we already proved at the end of Section 5.4), for all $n \in \mathbb{N}$, any $f \in \mathcal{O}^G(D_{n+1})$ can be approximated uniformly on $\overline{D}_n$ by functions from $\mathcal{O}^G(D)$.

Hence, by Lemma 5.8.1, it is sufficient to prove that any $\mathcal{O}^G$-cocycle over D becomes $\mathcal{O}^G$-trivial after restriction to each D_n, which is indeed the case by Lemma 5.9.2 (again using the Riemann mapping theorem). □

Finally we point out the following special case of Theorem 5.6.3, which is now proved:

5.9.3 Corollary. *Let $D_1, D_2 \subseteq \mathbb{C}$ be two open sets such that $D_1 \cup D_2$ is simply connected, and let $f : D_1 \cap D_2 \to G$ be holomorphic. Then there exist holomophic functions $f_j : D_j \to G$, $j = 1, 2$, such that $f = f_1^{-1} f_2$ on $D_1 \cap D_2$.*

In particular, this proves the assertion of Theorem 0.0.1 if $D_1 \cup D_2$ is simply connected.

5.10 Runge approximation of G-valued functions General case

The aim of this section is to prove part (ii) of the Runge approximation Theorem 5.0.1. Recall that part (i) of Theorem 5.0.1 was proved already at the end of Section 5.4. Then, using part (i) of Theorem 5.0.1, at the end of the preceding section we obtained Corollary 5.9.3, which now, in the present section, will be combined with part (i) of Theorem 5.0.1 to prove part (ii) of Theorem 5.0.1.

Doing this, it is convenient to pass to the Riemann sphere and to prove the slightly more general Runge approximation Theorem 5.10.5 below.[6] First let us shortly recall the notion of the Riemann sphere. The Riemann sphere is given by the set $\mathbb{P}^1 := \mathbb{C} \cup \{\infty\}$ endowed with a topology and a notion of holomorphic functions what we now are going to explain:

Definition of the topology: A subset U of $\mathbb{P}^1$ is called **open** if either $\infty \notin U$ and U is open as a subset of $\mathbb{C}$ (with respect to the usual Euclidean topology of $\mathbb{C}$) or $\infty \in U$ and $\mathbb{C} \setminus U$ is compact (again with respect to the Euclidean topology of $\mathbb{C}$). It is easy to see that in this way $\mathbb{P}^1$ becomes a topological space, which is, by stereographic projection[7], homeomorphic to the 2-dimensional sphere.

Definition of holomorphic functions: Set

$$\frac{1}{0} = \infty\,, \qquad \frac{1}{\infty} = 0 \qquad \text{and} \qquad z \pm \infty = \infty \pm z = \infty \text{ for all } z \in \mathbb{C}.$$

Then, for each $a \in \mathbb{P}^1$, a map $T_a : \mathbb{P}^1 \to \mathbb{P}^1$ is well-defined, by setting

$$T_a(z) = \begin{cases} 1/(z-a) & \text{if } a \neq \infty, \\ z & \text{if } a = \infty. \end{cases}$$

[6]Theorem 5.0.1 is the special case $\infty \notin \overline{D}$ of Theorem 5.10.5 where part (i) corresponds to the case of connected $\mathbb{P}^1 \setminus \overline{D}$ and part (ii) corresponds to the case of non-connected $\mathbb{P}^1 \setminus \overline{D}$.

[7]Let us identify the complex plane $\mathbb{C}$ with the plane in $\mathbb{R}^3$ which consists of the points of the form $(x, y, 0)$, $x, y \in \mathbb{R}$, and let S be the sphere of radius 1 centered at $(0,0,1) \in \mathbb{R}^3$. Let $\mathcal{N} := (0,0,2) \in S$ be the "north pole" of S. For each $p \in S \setminus \{\mathcal{N}\}$, we denote by $L(p)$ the line through p and $\mathcal{N}$. Then $L(p)$ intersects $\mathbb{C}$ in precisely one point. Let us denote this point by $\mathbb{C} \cap L(p)$. The bijective map $\pi : S \to \mathbb{P}^1$ defined by

$$\pi(p) = \begin{cases} \mathbb{C} \cap L(p) & \text{if } p \in S \setminus \{\mathcal{N}\}, \\ \infty & \text{if } p = \infty, \end{cases}$$

is called the **stereographic projector**.

It is easy to see that, for each $a \in \mathbb{P}^1$, T_a is a homeomorphism of $\mathbb{P}^1$ with $T_a(a) = \infty$ and $T_a(\mathbb{P}^1 \setminus \{a\}) = \mathbb{C}$. If there is no danger of confusion, we use also the following notation:

- The letter z denotes the identical map T_∞.
- If $a \in \mathbb{P}^1$ and $a \neq \infty$, then the expression $\frac{1}{z-a}$ denotes the map T_a.

Now let $U \subseteq \mathbb{P}^1$ be open, $U \neq \emptyset$, and let $f : U \to \mathbb{C}$ be a function. If $\infty \notin U$, i.e., $U \subseteq \mathbb{C}$, then we already know what it means that f is holomorphic on U. If $\infty \in U$, then we define: f is **holomorphic** on U if the following two conditions are fulfilled:

- f is holomorphic on $U \setminus \{\infty\}$.
- $f \circ T_0 = f\left(\frac{1}{z}\right)$ is holomorphic on $T_0^{-1}(U)$.

Note that, by Riemann's removability theorem, this is the case, if and only if, f is continuous on U, and the restriction of f to $U \setminus \{\infty\}$ is holomorphic. The ring of all holomorphic functions defined on U will be again denoted by $\mathcal{O}(U)$.

We summarize:

The **Riemann sphere** $\mathbb{P}^1$ is given by the triplet $\left(\mathbb{C} \cup \{\infty\}, \mathcal{T}, \mathcal{O}\right)$ where $\mathcal{T}$ is the family of open subsets of $\mathbb{C} \cup \{\infty\}$ defined above, and $\mathcal{O}$ is the map $\mathcal{T} \setminus \{\emptyset\} \ni U \to \mathcal{O}(U)$.

Now let E be a Banach space, and let $U \subseteq \mathbb{P}^1$ be an open set, $U \neq \emptyset$. If $\infty \notin U$, i.e., $U \subseteq \mathbb{C}$, then we already defined what a holomorphic E-valued function is. If $\infty \in U$, then, as in the case of scalar functions, we define: f is **holomorphic on** U, if and only if,

- f is holomorphic on $U \setminus \{\infty\}$.
- $f\left(\frac{1}{z}\right)$ is holomorphic on $T_0^{-1}(U)$.

The space of all holomorphic E-valued functions on U will be again denoted by $\mathcal{O}^E(U)$.

Now let $D \subset \mathbb{P}^1$ be a non-empty open set, and let f be a map from D to $\mathbb{P}^1$. If $\infty \notin f(D)$, i.e., $f(D) \subseteq \mathbb{C}$, then we just defined what it means that f is holomorphic. If $\infty \in f(D)$, then we define: f is **holomorphic** on D if the restriction of f to $D \setminus f^{-1}(\infty)$ is holomorphic and, moreover, the function

$$f \circ T_{f^{-1}(\infty)} = f\left(\frac{1}{z - f^{-1}(\infty)}\right)$$

is holomorphic on $T^{-1}_{f^{-1}(\infty)}$. If D and D' are two open sets in $\overline{\mathbb{C}}$, then we say that f is **biholomorphic from** D **onto** D', if and only if, f is bijective from D onto D' such that f is holomorphic on D and f^{-1} is holomorphic on D'.

Note that, for each $a \in \mathbb{P}^1$, T_a is biholomorphic from $\mathbb{P}^1$ onto $\mathbb{P}^1$ where $\mathbb{P}^1 \setminus \{a\}$ is mapped onto $\mathbb{C}$.

An open set $D \subseteq \mathbb{P}^1$ will be called an open set with **piecewise $\mathcal{C}^1$-boundary** if $\overline{D} \neq \mathbb{P}^1$ and, for $a \in \mathbb{P}^1 \setminus \overline{D}$, the open set $T_a(D)$ (which is contained in $\mathbb{C}$) is an open set with piecewise $\mathcal{C}^1$-boundary as defined in Section 1.4.1.

Since, for each $a \in \mathbb{P}^1$, T_a is biholomorphic between $\mathbb{P}^1 \setminus \{a\}$ and the complex plane, from the theory of holomorphic functions in the complex plane we get a corresponding theory on each set of the form $\mathbb{P}^1 \setminus \{a\}$ where a is an arbitrary point in $\mathbb{P}^1$. For example, from part (i) of the Runge approximation Theorem 5.0.1 (which is already proved) we immediately get:

5.10.1 Proposition. *Let D be an open set with piecewise $\mathcal{C}^1$ boundary in $\mathbb{P}^1$ such that $\mathbb{P}^1 \setminus \overline{D}$ is connected, let G be an open subgroup of the group of invertible elements of a Banach algebra with unit, and let $f : \overline{D} \to G$ be a continuous function which is holomorphic in D such that all values of f belong to the same connected component of G. Then, for each $a \in \mathbb{P}^1 \setminus \overline{D}$, f can be approximated uniformly on $\overline{D}$ by functions from $\mathcal{O}^G(\mathbb{P}^1 \setminus \{a\})$.*

Moreover, from Corollary 5.9.3 we immediately get:

5.10.2 Proposition. *Let D_1 and D_2 be two open sets in $\mathbb{P}^1$ such that $\mathbb{P}^1 \setminus (D_1 \cup D_2)$ is connected and not empty. Then, for each holomorphic function $f : D_1 \cap D_2 \to G$ there exist holomorphic functions $f_j : D_j \to G$, $j = 1, 2$, such that $f = f_1^{-1} f_2$ on $D_1 \cap D_2$.*

5.10.3 Definition. If X is a subset of $\mathbb{P}^1$ and G an open subgroup of the group of invertible elements of G, then we denote by $\overline{\mathcal{O}}^G(X)$ the group of continuous G-valued functions on X which are holomorphic in the interior of X.

5.10.4 Lemma. *Let $D \subseteq \mathbb{P}^1$ be an open set with piecewise $\mathcal{C}^1$-boundary, and let $U_1, \dots, U_n$ be the connected components of $\mathbb{P}^1 \setminus \overline{D}$. Let $n \geq 2$ and let some points $a_j \in U_j$, $1 \leq j \leq n$, be chosen. Further, let G be an open subgroup of the group of invertible elements of a Banach algebra with unit, and let $f \in \overline{\mathcal{O}}^G(\overline{D})$. Then there exist functions $f_j \in \overline{\mathcal{O}}^G(\mathbb{P}^1 \setminus U_j)$, $1 \leq j \leq n$, and a function $h \in \mathcal{O}^G\big(\mathbb{P}^1 \setminus \{a_1, \dots, a_n\}\big)$ such that $f = h f_n \dots f_1$ on $\overline{D}$.*

Proof. For $1 \leq k \leq n$, we consider the following statement:

$A(k)$: There exist functions $f_j \in \overline{\mathcal{O}}^G\big(\mathbb{P}^1 \setminus U_j\big)$, $1 \leq j \leq k$, and a function $h_k \in \overline{\mathcal{O}}^G\Big(\overline{D} \cup \big(U_1 \setminus \{a_1\}\big) \cup \dots \cup \big(U_k \setminus \{a_k\}\big)\Big)$ such that $f = h_k f_k \dots f_1$ on $\overline{D}$.

Since

$$\overline{D} \cup \big(U_1 \setminus \{a_1\}\big) \cup \dots \cup \big(U_n \setminus \{a_n\}\big) = \mathbb{P}^1 \setminus \{a_1, \dots, a_n\},$$

then $A(n)$ is the assertion of the lemma. Therefore it is sufficient to prove $A(1)$ and the conclusions $A(k) \Rightarrow A(k+1)$, $1 \leq k \leq n-1$.

Proof of $A(1)$: Since

$$\big(\mathbb{P}^1 \setminus \overline{U}_1\big) \cup \big(D \cup (\overline{U}_1 \setminus \{a_1\})\big) = \mathbb{P}^1 \setminus \{a_1\}$$

and

$$\left(\mathbb{P}^1 \setminus \overline{U}_1\right) \cap \left(D \cup \left(\overline{U}_1 \setminus \{a_1\}\right)\right) = D,$$

from Proposition 5.10.2 we get functions $f_1 \in \mathcal{O}^G\left(\mathbb{P}^1 \setminus \overline{U}_1\right)$ and $h_1 \in \mathcal{O}^G\left(D \cup \left(\overline{U}_1 \setminus \{a_1\}\right)\right)$ such that

$$f = h_1 f_1 \tag{5.10.1}$$

on D. Since f is continuous and with values in G on $\overline{D}$, since h_1 is continuous and with values in G on $D \cup \partial U_1$, since f_1 is continuous and with values in G on $\overline{D} \setminus \partial U_1$ and since (5.10.1) holds in D, it follows that $f_1 \in \overline{\mathcal{O}}^G(\mathbb{P}^1 \setminus U_1)$, $h_1 \in \overline{\mathcal{O}}^G\left(\overline{D} \cup (U_1 \setminus \{a_1\})\right)$ and (5.10.1) holds on $\overline{D}$, i.e., assertion $A(1)$ is valid.

Proof of $A(k) \Rightarrow A(k+1)$: Let $1 \le k \le n-1$ be given, assume that statement $A(k)$ is valid, and let $f_1, \ldots, f_k$ and h_k be as in this statement. Since

$$\left(\mathbb{P}^1 \setminus \overline{U}_{k+1}\right) \cup \left(D \cup \left(\overline{U}_1 \setminus \{a_1\}\right) \cup \ldots \cup \left(\overline{U}_{k+1} \setminus \{a_{k+1}\}\right)\right) = \mathbb{P}^1 \setminus \{a_{k+1}\}$$

and

$$\begin{aligned}\left(\mathbb{P}^1 \setminus \overline{U}_{k+1}\right) \cap \Big(D \cup \left(\overline{U}_1 \setminus \{a_1\}\right) \cup \ldots \cup \left(\overline{U}_{k+1} \setminus \{a_{k+1}\}\right)\Big)& \\ = D \cup \left(\overline{U}_1 \setminus \{a_1\}\right) \cup \ldots \cup \left(\overline{U}_k \setminus \{a_k\}\right)&,\end{aligned}$$

from Proposition 5.10.2 we get functions

$$f_{k+1} \in \mathcal{O}^G_{Z,m}\left(\mathbb{P}^1 \setminus \overline{U}_{k+1}\right)$$

and

$$h_{k+1} \in \mathcal{O}^G_{Z,m}\Big(D \cup \left(\overline{U}_1 \setminus \{a_1\}\right) \cup \ldots \cup \left(\overline{U}_{k+1} \setminus \{a_{k+1}\}\right)\Big)$$

such that

$$h_k = h_{k+1} f_{k+1} \tag{5.10.2}$$

on $D \cup \left(\overline{U}_1 \setminus \{a_1\}\right) \cup \ldots \cup \left(\overline{U}_k \setminus \{a_k\}\right)$. Since h_k is continuous and with values in G on $\overline{D}$, since h_{k+1} is continuous and with values in G on $D \cup \partial U_{k+1}$, since f_{k+1} is continuous and with values in G on $\overline{D} \setminus \partial U_{k+1}$ and since (5.10.2) holds in D, it follows that

$$f_{k+1} \in \overline{\mathcal{O}}^G_{Z,m}\left(\mathbb{P}^1 \setminus U_{k+1}\right),$$

$$h_{k+1} \in \mathcal{O}^G_{Z,m}\Big(\overline{D} \cup \left(U_1 \setminus \{a_1\}\right) \cup \ldots \cup \left(U_{k+1} \setminus \{a_{k+1}\}\right)\Big)$$

and (5.10.2) holds on $\overline{D}$. Since $f = h_k f_k \cdots f_1$ on $\overline{D}$, this implies that

$$f = h_{k+1} f_{k+1} f_k \cdots f_1$$

on $\overline{D}$, i.e., assertion $A(k+1)$ is valid. □

5.10.5 Theorem (Runge approximation). *Let G be an open subgroup of the group of invertible elements of a Banach algebra with unit. Let $D \subseteq \mathbb{P}^1$ be an open set with piecewise $\mathcal{C}^1$-boundary, and let $f : \overline{D} \to G$ be a continuous function which is holomorphic in D such that all values of f belong to the same connected component of G. Further, let $U_1, \dots, U_n$ be the connected components of $\mathbb{P}^1 \setminus \overline{D}$, and let some points $a_j \in U_j$, $1 \le j \le n$, be given. Then f can be approximated uniformly on $\overline{D}$ by functions from $\mathcal{O}^G\big(\mathbb{P}^1 \setminus \{a_1, \dots, a_n\}\big)$.*

Proof. If $n = 1$, the assertion of the theorem is that of Proposition 5.10.1. If $n \ge 2$, then, by Lemma 5.10.4, f can be written in the form

$$f = hf_n \dots f_1 \qquad \text{on } \overline{D}, \tag{5.10.3}$$

where $f_j \in \overline{\mathcal{O}}^G(\mathbb{P}^1 \setminus U_j)$, $1 \le j \le n$, and $h \in \mathcal{O}^G\big(\mathbb{P}^1 \setminus \{a_1, \dots, a_n\}\big)$. Let V be the interior of $\mathbb{P}^1 \setminus U_j$. Since the boundary of U_j is piecewise $\mathcal{C}^1$ (as a part of the boundary of D), also the boundary of V is piecewise $\mathcal{C}^1$ and $\overline{V} = \mathbb{P}^1 \setminus U_j$. Since U_j is connected (as a connected component of some set), $\mathbb{P}^1 \setminus \overline{V} = U_j$ is connected. Therefore, Proposition 5.10.1 can be applied to each V_j. Hence, each f_j can be approximated uniformly on $\overline{V} = \mathbb{P}^1 \setminus U_j$ by functions from $\mathcal{O}^G\big(\mathbb{P}^1 \setminus \{a_j\}\big)$. Since $\mathcal{O}\big(\mathbb{P}^1 \setminus \{a_j\}, G\big) \subseteq \mathcal{O}^G\big(\mathbb{P}^1 \setminus \{p_1, \dots, p_n\}\big)$ and $\overline{D} \subseteq \mathbb{P}^1 \setminus U_j$), this means in particular that each f_j can be approximated uniformly on $\overline{D}$ by functions from $\mathcal{O}^G\big(\mathbb{P}^1 \setminus \{p_1, \dots, p_n\}\big)$. Since h belongs to $\mathcal{O}^G\big(\mathbb{P}^1 \setminus \{p_1, \dots, p_n\}\big)$ and by (5.10.3), this implies the assertion of the theorem. □

5.11 Proof of Theorem 5.6.3 in the general case

Here we prove Theorem 5.6.3. Recall that the sufficiency of condition (i) in Theorem 5.6.3 (that D is simply connected) was already proved at the end of Section 5.9. So it remains to prove that condition (ii) and condition (iii) in Theorem 5.6.3 are sufficient.

Throughout this section, A is a Banach algebra with unit 1, GA is the group of invertible elements of A, G_1A is the connected component of the unit element in GA, and G is an open subgroup of GA.

5.11.1 Lemma. *Let (D_1, D_2) be a Cartan pair (Definition 5.5.1) such that D_1 and D_2 are connected, and let $F \in \overline{\mathcal{O}}^G\big(\overline{D_1 \cap D_2}\big)$ (Def. 5.4.5). Then the following are equivalent:*

(i) *All values of F belong to the same connected component of G.*

(ii) *There exist functions $F_j \in \overline{\mathcal{O}}^G(\overline{D}_j)$, $j = 1, 2$, such that $F = F_1^{-1}F_2$ on $\overline{D}_1 \cap \overline{D}_2$.*

(iii) *There exist functions $C_j \in \mathcal{C}^G(D_j)$, $j = 1, 2$, such that $F = C_1^{-1}C_2$ on $\overline{D}_1 \cap \overline{D}_2$.*

Proof. The conclusion (i)⇒(ii) is the assertion of the Cartan Lemma 5.5.2. The conclusion (ii)⇒(iii) is trivial.

To prove (iii)⇒(i), we consider two arbitrary points $z, w \in \overline{D}_1 \cap \overline{D}_2$. Since $\overline{D}_1$ and $\overline{D}_2$ are connected, we can find continuous curves $\gamma_j : [0,1] \to \overline{D}_j$, $j = 1, 2$, such that $\gamma_j(0) = z$ and $\gamma_j(1) = w$, $j = 1, 2$. If now condition (iii) is satisfied, then, by setting

$$\gamma(t) = C_1^{-1}\big(\gamma_1(t)\big)C_2\big(\gamma_(t)\big)\,, \qquad 0 \le t \le 1,$$

we can define a continuous curve in G, where

$$F(z) = C_1^{-1}(z)C_2(z) = C_1^{-1}\big(\gamma_1(0)\big)C_2\big(\gamma_2(0)\big) = \gamma(0)$$

and

$$F(w) = C_1^{-1}(w)C_2(w) = C_1^{-1}\big(\gamma_1(1)\big)C_2\big(\gamma_2(1)\big) = \gamma(1),$$

i.e., γ connects $F(z)$ and $F(w)$. □

5.11.2 Lemma. *Let $D \subseteq \mathbb{C}$ be a bounded, connected open set with piecewise $\mathcal{C}^1$-boundary such that $\mathbb{C} \setminus \overline{D}$ consists of n connected components, $n \ge 2$. Then there exists a Cartan pair (D_1, D_2) with $D = D_1 \cup D_2$ and satisfying the following conditions:*

(1) *D_1 is simply connected;*

(2) *D_2 is connected;*

(3) *$\mathbb{C} \setminus \overline{D}_2$ consists of $n-1$ connected components.*

Proof. Let $U_1, \dots, U_n$ be the connected components of $\mathbb{C} \setminus \overline{D}$ where U_1 denotes the unbounded component. Let ∂U_j be the boundary of U_j. Choose points $a_1 \in \partial U_1$ and $a_2 \in \partial U_2$ which are smooth points of the boundary ∂D of D. Since D is connected and hence (as an open set) arcwise connected and since ∂D is piecewise $\mathcal{C}^1$, $\overline{D}$ is arcwise connected. Therefore we can find a continuous curve $\varphi : [0,1] \to \overline{D}$ with $\varphi(0) = a_1$ and $\varphi(1) = a_2$. Moreover, since a_1 and a_2 are smooth points of ∂D, we can achieve that φ is a $\mathcal{C}^\infty$-diffeomorphism between $[0,1]$ and $\varphi([0,1])$ and $\varphi([0,1])$ meets ∂D transversally (in both points, a_1 and a_2). Choose a neighborhood V in $\mathbb{C}$ of the interval $[0,1]$ so small that there exists an extension of φ to some $\mathcal{C}^\infty$-diffeomorphism Φ from V onto some neighborhood W of $\gamma([0,1])$. If $\delta > 0$ and $\varepsilon < \varepsilon'$, then we set

$$K(\delta; \varepsilon, \varepsilon') = \Big\{ z = x + iy \in \mathbb{C} \,\Big|\, -\delta < x < 1 + \delta \text{ and } \varepsilon < y < \varepsilon' \Big\}.$$

Choose $\delta_0 > 0$ and ε_0 so small that $\overline{K}(\delta_0; -\varepsilon_0, \varepsilon_0) \subseteq V$. Since $\gamma([0,1])$ meets ∂D transversally in the smooth points a_1 and a_2 and since D is connected, further then we can choose $0 < \varepsilon_1 < \varepsilon_0$ so small that, for all $-\varepsilon_1 < \varepsilon < \varepsilon' < \varepsilon_1$,

$$D_1(\varepsilon, \varepsilon') := D \cap \Phi\big(K(\delta_0; \varepsilon, \varepsilon')\big) \quad \text{and} \quad D_2(\varepsilon, \varepsilon') := D \setminus \overline{D_1(\varepsilon, \varepsilon')}$$

are connected bounded open sets with piecewise $\mathcal{C}^1$-boundary. Since $U_1 \cup U_2 \cup \gamma([0,1])$ is connected, then U_1 and U_2 are contained in the same connected component of $\mathbb{C} \setminus \overline{D_2(\varepsilon, \varepsilon')}$ and the number of connected components of $\mathbb{C} \setminus \overline{D_2(\varepsilon, \varepsilon')}$ is $n-1$. Moreover, choosing this ε_1 sufficiently small, we can achieve that, for all $-\varepsilon_1 < \varepsilon < \varepsilon' < \varepsilon_1$, the open set

$$\Phi^{-1}\big(D_1(\varepsilon, \varepsilon')\big)$$

is star shaped with respect to point $1/2$ and therefore simply connected. Hence, $D_1(\delta_0; \varepsilon, \varepsilon')$ is simply connected for all $-\varepsilon_1 < \varepsilon < \varepsilon' < \varepsilon_1$. We summarize:

For all $-\varepsilon_1 < \varepsilon < \varepsilon' < \varepsilon_1$, $D_1(\varepsilon, \varepsilon')$ and $D_2(\varepsilon, \varepsilon')$ are connected bounded open sets with piecewise $\mathcal{C}^1$- boundary where the open set $D_1(\varepsilon, \varepsilon')$ is simply connected and $\mathbb{C} \setminus \overline{D_2(\varepsilon, \varepsilon)}$ consists of $n-1$ connected components.

Now we fix some $0 < \varepsilon < \varepsilon_1/2$ and set

$$D_1 := D_1(-2\varepsilon, 2\varepsilon) \quad \text{and} \quad D_2 := D_2(-\varepsilon, \varepsilon).$$

Then the intersection $D_1 \cap D_2$ consists of the two connected components $D_1(-2\varepsilon, -\varepsilon)$ and $D_1(\varepsilon, 2\varepsilon)$ and (as mentioned in the summary above) each of these components is simply connected and has piecewise $\mathcal{C}^1$-boundary. Moreover, it is clear that $D_1 \cup D_2 = D$ and $(\overline{D}_1 \setminus D_2) \cap (\overline{D}_2 \setminus D_1) = \emptyset$. Hence (D_1, D_2) is a Cartan pair with $D_1 \cup D_2$. That this Cartan pair has the properties (1), (2) and (3) was also already mentioned in the above summary. □

5.11.3 Lemma. *Let $D \subseteq \mathbb{C}$ be an open set with piecewise $\mathcal{C}^1$-boundary, let U be a neighborhood of $\overline{D}$ and f a $\mathcal{O}^G$-cocycle over U. Suppose at least one of the following two conditions is fulfilled:*

(a) *G is connected.*

(b) *f is $\mathcal{C}^G$-trivial over U.*

Then f is $\mathcal{O}^G$-trivial over D.

Proof. We may assume that D is connected. Further we proceed by induction over the number of connected components of $\mathbb{C} \setminus \overline{D}$.

Beginning of induction: Suppose this number is 1, i.e., $\mathbb{C} \setminus \overline{D}$ is connected. As the boundary of D is piecewise $\mathcal{C}^1$, then also $\mathbb{C} \setminus D$ is connected, which means that D is simply connected. Therefore the assertion of the lemma follows from the fact that the claim of Theorem 5.6.3 was already proved (at the end of Section 5.9) if condition (i) in Theorem 5.6.3 is satisfied.

Hypothesis of induction: Assume, for some $n \in \mathbb{N}$ with $n \geq 2$, the assertion of the lemma is already proved if the number of connected components of $\mathbb{C} \setminus \overline{D}$ is $n-1$.

Step of induction: Assume that the number of connected components of $\mathbb{C} \setminus \overline{D}$ is equal to n. Then, by Lemma 5.11.2, we can find a Cartan pair (D_1, D_2) with

$D = D_1 \cup D_2$ satisfying conditions (1), (2), (3) (of Lemma 5.11.2). Since the boundaries of D_1, D_2, $D_1 \cap D_2$ and D are piecewise $\mathcal{C}^1$, we can find a Cartan pair (D_1', D_2') satisfying the same conditions (1), (2), (3) such that $\overline{D}_j \subseteq D_j'$ and $\overline{D}_1' \cup \overline{D}_2' \subseteq U$.

Then f is $\mathcal{O}^G$-trivial over D_1', again by the fact that the claim of Theorem 5.6.3 was already proved (at the end of Section 5.9) if condition (i) in Theorem 5.6.3 is satisfied. Moreover, since the number of connected components of $\mathbb{C} \setminus \overline{D}_2'$ is equal to $n-1$, it follows from the hypothesis of induction that f is also $\mathcal{O}^G$-trivial over D_2'. By Proposition 5.7.6, this implies that $f\big|_{D_1' \cup D_2'}$ is $\mathcal{O}^G$-equivalent to a certain $(\{D_1', D_2'\}, \mathcal{O}^G)$-cocycle f'. Since $\overline{D_1 \cap D_2} \subseteq D_1' \cap D_2'$, setting

$$F := f_{12}'\big|_{\overline{D_1 \cap D_2}}$$

we obtain a function $F \in \overline{\mathcal{O}}^G(\overline{D_1 \cap D_2})$. We claim that all values of F belong to the same connected component of G.

If G is connected, this is trivial. If not, then condition (b) in the lemma under proof is satisfied, i.e., f is $\mathcal{C}^G$-trivial over U. As $D_1' \cup D_2' \subseteq U$, then f is also $\mathcal{C}^G$-trivial over $D_1' \cup D_1'$. Since $f\big|_{D_1' \cup D_2'}$ is $\mathcal{C}^G$-equivalent to f' (it is even $\mathcal{O}^G$-equivalent to f'), this implies that also f' is $\mathcal{C}^G$-trivial, i.e., we can find $C_j \in \mathcal{C}^G(D_j')$, $j = 1, 2$, with

$$f_{12}' = C_1^{-1} C_2 \qquad \text{on } D_1' \cap D_2'.$$

Hence condition (iii) in Lemma 5.11.1 is satisfied, and it follows from Lemma 5.11.1 that all values of F belong to the same connected component of G.

Since all values of F belong to the same connected component of G, it follows from the Cartan Lemma 5.5.2 (or from Lemma 5.11.1) that there exist functions $F_j \in \overline{\mathcal{O}}^G(\overline{D}_j)$, $j = 1, 2$, with

$$F = F_1^{-1} F_2 \qquad \text{on } \overline{D}_j \cap \overline{D}_j.$$

Since $F|_{D_1 \cap D_2} = f_{12}'|_{D_1 \cap D_2}$, this means in particular that $f'\big|_D$ is $\mathcal{O}^G$-trivial. Finally, as $f\big|_{D_1' \cup D_2'}$ and f' are $\mathcal{O}^G$-equivalent and therefore $f|_D$ and $f'|_D$ are $\mathcal{O}^G$-equivalent, it follows that f is $\mathcal{O}^G$-trivial over D. □

Proof of Theorem 5.6.3. Since the sufficiency of condition (i) in Theorem 5.6.3 (that D is simply connected) was already proved at the end of Section 5.9, we may assume that at least one of the following two conditions is fulfilled:

(a) G is connected.

(b) f is $\mathcal{C}^G$-trivial over U.

Take a sequence $(\Omega_n)_{n \in \mathbb{N}}$ of bounded open sets with $\mathcal{C}^1$-boundaries such that $\overline{\Omega}_n \subseteq \Omega_{n+1}$ for all $n \in \mathbb{N}$ and $\bigcup_{n \in \mathbb{N}} \Omega_n = D$. Let U_n be the union of all bounded connected components of $\mathbb{C}\setminus\overline{\Omega}_n$ which are subsets of D (if there are any – otherwise $U_n := \emptyset$), and set

$$D_n = \Omega_n \cup \overline{U}_n.$$

Then also $(D_n)_{n\in\mathbb{N}}$ is a sequence of bounded open sets with $\mathcal{C}^1$-boundaries such that $\overline{D}_n \subseteq D_{n+1}$ for all $n \in \mathbb{N}$ and $\bigcup_{n\in\mathbb{N}} D_n = D$. Moreover this sequence has the important property that each bounded connected component of $\mathbb{C} \setminus \overline{D}_n$ (if there are any) contains at least one point of $\mathbb{C} \setminus D$. By the Runge approximation Theorem 5.0.1, this implies that, for each n, the functions from $\overline{\mathcal{O}}^G(\overline{D}_n)$ can be approximated uniformly on $\overline{D}_n$ by functions from $\mathcal{O}^G(D)$. In particular we see that the sequence $(D_n)_{n\in\mathbb{N}}$ has the properties (1), (2), (3) of Lemma 5.8.1. Therefore, by this lemma, for the $\mathcal{O}^G$-triviality of f it is sufficient that each $f|_{D_n}$ is $\mathcal{O}^G$-trivial, which is indeed the case, by Lemma 5.11.3. □

5.12 $\mathcal{O}^{\mathcal{G}^\infty(E)}$-cocycles

In this section E is a Banach space.

5.12.1 Definition. (i) We denote by $\mathcal{F}^\omega(E)$ the ideal of all compact operators from $L(E)$, and by $\mathcal{F}^\infty(E)$ we denote the ideal of all operators from $\mathcal{F}^\omega(E)$, which can be approximated by finite dimensional operators.

(ii) Let $\mathcal{G}^\omega(E)$ be the group of invertible operators of the form $I + K$ with $K \in \mathcal{F}^\omega(E)$, and let $\mathcal{G}^\infty(E)$ be the group of invertible operators of the form $I + K$ with $K \in \mathcal{F}^\infty(E)$.

(iii) We denote by $\mathcal{F}^\omega_I(E)$ the subalgebra of $L(E)$ of operators of the form $\lambda I + K$ with $K \in \mathcal{F}^\omega(E)$ and $\lambda \in \mathbb{C}$, and by $\mathcal{F}^\infty_I(E)$ we denote the subalgebra of $L(E)$ of operators of the form $\lambda I + K$ with $K \in \mathcal{F}^\infty(E)$ and $\lambda \in \mathbb{C}$.

(iv) We set

$$G\mathcal{F}^\omega_I(E) = GL(E) \cap \mathcal{F}^\omega_I(E) \quad \text{and} \quad G\mathcal{F}^\infty_I(E) = GL(E) \cap \mathcal{F}^\infty_I(E).$$

In the following proposition we collect some well-known facts on these algebras and groups, which will be used without further reference throughout this book.

5.12.2 Proposition. *The operators from $\mathcal{F}^\infty(E)$ are compact. Hence, zero is the only possible accumulation point of the spectrum of such an operator, and all other points of the spectrum are eigenvalues of finite multiplicity. Therefore it is easy to see that:*

(i) *For $\dim E < \infty$, the factor algebras $\mathcal{F}^\infty_I(E)/\mathcal{F}^\infty(E)$ and $\mathcal{F}^\omega_I(E)/\mathcal{F}^\omega(E)$ are isomorphic to $\mathbb{C}$.*

(ii) *If $\dim E = \infty$ and $\lambda I + K = \lambda' I + K'$ where $K, K' \in \mathcal{F}^\omega(E)$ and $\lambda, \lambda' \in \mathbb{C}$, then $\lambda = \lambda'$ and $K = K'$.*

(iii) *The algebras* $\mathcal{F}_I^\infty(E)$ *and* $\mathcal{F}_I^\omega(E)$ *are closed in* $L(E)$ *with respect to the operator norm.*

(iv) *Let* $\dim E = \infty$, *let* $D \subseteq \mathbb{C}$ *be an open set, and let* $f : D \to \mathcal{F}_I^\aleph(E)$ *be a holomorphic (continuous) function, where* $\aleph$ *stands for one of the symbols* ∞ *or* ω. *Then* f *is of the form*

$$f(z) = \lambda(z) + K(z)\,, \qquad z \in D,$$

where $\lambda : D \to \mathbb{C}$ *and* $K : D \to \mathcal{F}^\aleph(E)$ *are uniquely determined holomorphic (continuous) functions.*

(v) *The sets* $G\mathcal{F}_I^\infty(E)$, $\mathcal{G}^\infty(E)$, $G\mathcal{F}_I^\omega(E)$ *and* $\mathcal{G}^\omega(E)$ *are connected and closed subgroups of* $GL(E)$. *The group* $\mathcal{G}\mathcal{F}_I^\infty(E)$ *is the group of invertible elements of the Banach algebra* $\mathcal{F}_I^\infty(E)$, *and the group* $\mathcal{G}\mathcal{F}_I^\omega(E)$ *is the group of invertible elements of the Banach algebra* $\mathcal{F}_I^\omega(E)$.

5.12.3 Remark. It is easy to see that there is no Banach algebra which contains $\mathcal{G}^\infty(E)$ or $\mathcal{G}^\omega(E)$ as an open subgroup of its group of invertible elements. Therefore we cannot immediately apply to $\mathcal{G}^\infty(E)$ and $\mathcal{G}^\omega(E)$ the theory of cocycles as developed above. We circumvent this problem passing to the groups $G\mathcal{F}_I^\infty(E)$ and $G\mathcal{F}_I^\omega(E)$, which are groups of invertible elements of a Banach algebra. Note however that $\mathcal{G}^\infty(E)$ and $\mathcal{G}^\omega(E)$ are *complex Banach Lie groups* and that the theory of Grauert and Bungart mentioned in the introduction to this book is developed for such groups. So, the theory of $\mathcal{O}^G$-cocycles, presented in this chapter, is valid for arbitrary Banach Lie groups. But, since $\mathcal{G}^\infty(E)$ and $\mathcal{G}^\omega(E)$ are the only examples of *true* Banach Lie groups, which we meet in this book, for simplicity, we avoid the notion of a general Banach Lie group.

5.12.4 Definition. Let $D \subseteq \mathbb{C}$ be an arbitrary open set, let $\mathcal{U} = \{U_j\}_{j\in I}$ be an open covering of D, let $A = \{A_{jk}\}_{j,k\in I}$ be a $\big(\mathcal{U}, \mathcal{O}^{GL(E)}\big)$-cocycle (Def. 5.6.1), and let $\aleph$ stand for one of the symbols ∞ or ω.

(i) The cocycle A will be called a $\big(\mathcal{U}, \mathcal{O}^{\mathcal{G}^\aleph(E)}\big)$**-cocycle** or simply an $\mathcal{O}^{\mathcal{G}^\aleph(E)}$**-cocycle**, if for all $j,k \in I$ with $U_j \cap U_k \neq \emptyset$,

$$A_{jk}(z) \in \mathcal{G}^\aleph(E) \qquad \text{for all } z \in U_j \cap U_k\,.$$

(ii) The cocycle A will be called $\mathcal{O}^{\mathcal{G}^\aleph(E)}$**-trivial**, if there exists a family of holomorphic functions $A_j : U_j \to \mathcal{G}^\aleph(E)$, $j \in I$, such that

$$A_{jk} = A_j A_k^{-1} \qquad \text{on } U_j \cap U_k$$

for all $j,k \in I$ with $U_j \cap U_k \neq \emptyset$.

5.12.5 Theorem. *Let* $D \subseteq \mathbb{C}$ *be an open set, let* $\mathcal{U}$ *be an open covering of* D, *and let* A *be a* $\big(\mathcal{U}, \mathcal{O}^{\mathcal{G}^\aleph(E)}\big)$*-cocycle, where* $\aleph$ *stands for one of the symbols* ∞ *or* ω. *Then* A *is* $\mathcal{O}^{\mathcal{G}^\aleph(E)}$*-trivial.*

Proof. Since $\mathcal{G}^{\aleph}(E)$ is contained in $G\mathcal{F}_I^{\aleph}(E)$, the cocycle A can be viewed as a $\big(\mathcal{U}, \mathcal{O}^{G\mathcal{F}_I^{\aleph}(E)}\big)$-cocycle. Since $G\mathcal{F}_I^{\aleph}(E)$ is connected, it follows from Theorem 5.6.3 that A is $\mathcal{O}^{G\mathcal{F}_I^{\aleph}(E)}$-trivial. Hence we have a family of holomorphic functions $\widetilde{A}_j : U_j \to G\mathcal{F}_I^{\aleph}(E)$, $j \in I$, such that

$$A_{jk} = \widetilde{A}_j \widetilde{A}_k^{-1} \qquad \text{on } U_j \cap U_k \tag{5.12.1}$$

for all $j, k \in I$ with $U_j \cap U_k \neq \emptyset$. For $\dim E < \infty$ this completes the proof. Let $\dim E = \infty$, and let $\lambda_j : U_j \to \mathbb{C}^*$ and $K_j : U_j \to \mathcal{F}^{\aleph}(E)$ be the holomorphic functions with

$$\widetilde{A}_j = \lambda_j I + K_j\,, \qquad j \in I.$$

Passing to the factor algebra $L(E)/\mathcal{F}^{\aleph}(E)$, then it follows from (5.12.1) that

$$\lambda_j = \lambda_k \qquad \text{on } U_j \cap U_k$$

for all $j, k \in I$ with $U_j \cap U_k \neq \emptyset$. Therefore it remains to set $A_j = \lambda_j^{-1} \widetilde{A}_j$. □

We point out again the special case of Theorem 5.12.5 for coverings by two open sets, which is sufficient for many applications:

5.12.6 Corollary. *Let $D_1, D_2 \subseteq \mathbb{C}$ be two open sets, and let $A : D_1 \cap D_2 \to \mathcal{G}^{\aleph}(E)$ be holomorphic, where $\aleph$ stands for one of the symbols ∞ or ω. Then there exist holomorphic functions $A_j : D_j \to \mathcal{G}^{\aleph}(E)$ with $A = A_1 A_2^{-1}$ on $D_1 \cap D_2$.*

5.13 Weierstrass theorems

Here we prove Theorem 0.0.2 stated in the introduction to this book. We do this in a more general setting of Banach algebras. Throughout this section, A is a Banach algebra with unit 1, and G is an open subgroup of the group of invertible elements of A. We prove:

5.13.1 Theorem. *Let $D \subseteq \mathbb{C}$ be an open set, and let Z be a discrete and closed subset of D. Suppose, for each $w \in Z$, there are given a neighborhood U_w of w with $U_w \cap Z = \{w\}$ and a holomorphic function $f_w : U_w \setminus \{w\} \to G$. Moreover, we assume that at least one of the following conditions is fulfilled:*

(i) *D is simply connected.*

(ii) *G is connected.*

Then there exist a holomorphic function $h : D \setminus Z \to G$ and holomorphic functions $h_w : U_w \to G$, $w \in Z$, such that

$$h_w f_w = h \qquad \text{on } U_w \setminus \{w\}\,. \tag{5.13.1}$$

The topological conditions (i) and (ii) in Theorem 5.13.1 can be replaced by the more general condition that *the problem can be solved continuously*, i.e., there is the following Oka-Grauert principle:

5.13.2 Theorem. *Let $D \subseteq \mathbb{C}$ be an open set, and let Z be a discrete and closed subset of D. Suppose, for each $w \in Z$, there are given a neighborhood U_w of w with $U_w \cap Z = \{w\}$ and a holomorphic function $f_w : U_w \setminus \{w\} \to G$. Assume that:*

(iii) *There exist a continuous function $c : D \setminus Z \to G$ and continuous functions $c_w : U_w \to G$, $w \in Z$, such that $c_w f_w = c$ on $U_w \setminus \{w\}$, $w \in Z$.*

Then there exist a holomorphic function $h : D \setminus Z \to G$ and holomorphic functions $h_w : U_w \to G$, $w \in Z$, such that

$$h_w f_w = h \qquad \text{on } U_w \setminus \{w\}\,. \tag{5.13.2}$$

Proof of Theorems 5.13.1 *and* 5.13.2. We choose neighborhoods $V_w \subseteq U_w$, $w \in Z$, so small that $V_w \cap V_v = \emptyset$ for all $v, w \in Z$ with $w \neq v$.

It is sufficient to find $h \in \mathcal{O}^G(D \setminus Z)$ and $h_w \in \mathcal{O}^G(V_w)$, $w \in Z$, such that

$$h_w f_w = h \tag{5.13.3}$$

on $V_w \setminus \{w\}$. Indeed, since $V_w \cap Z = U_w \cap Z = \{w\}$, then, by (5.13.3), each h_w admits an extension to a function from $\mathcal{O}^G(U_w)$, which we also denote by h_w, such that (5.13.2) holds.

Set $D_1 = \bigcup_{w \in Z} V_w$ and $D_2 = D \setminus Z$. Since the sets V_w are pairwise disjoint and $V_w \cap Z = \{w\}$, the family of functions f_w can be interpreted as a single holomorphic function $f \in \mathcal{O}^G(D_1 \setminus Z) = \mathcal{O}^G(D_1 \cap D_2)$. Now, by Corollary 5.6.4, there exist $h_j \in \mathcal{O}^G(D_j)$, $j = 1, 2$, with $f = h_1^{-1} h_2$ on $D_1 \cap D_2$. Setting $h_w = h_1|_{V_w}$ and $h = h_2$, we complete the proof. □

Since, in theorems 5.13.1 and 5.13.2, the multiplication by the functions h_w is carried out from the left, we call these theorems *left-sided* Weierstrass theorems. There are also *right-* and *two-sided* versions.

If the multiplication in A is denoted by "$\cdot$", then we can pass to the Banach Algebra $\widetilde{A}$ which consists of the same additive group A but with the multiplication "$\tilde{\cdot}$" defined by $a \,\tilde{\cdot}\, b = b \cdot a$. In this way, from theorems 5.13.1 and 5.13.2 we get the following *right-sided* Weierstrass theorems:

5.13.3 Theorem. *Let $D \subseteq \mathbb{C}$ be an open set, and let Z be a discrete and closed subset of D. Suppose, for each $w \in Z$, there are given a neighborhood U_w of w with $U_w \cap Z = \{w\}$ and a holomorphic function $f_w : U_w \setminus \{w\} \to G$. Moreover, we assume that at least one of the following conditions is fulfilled:*

(i) *D is simply connected.*

(ii) *G is connected.*

Then there exist a holomorphic function $h : D \setminus Z \to G$ and holomorphic functions $h_w : U_w \to G$, $w \in Z$, such that

$$f_w h_w = h \qquad \text{on } U_w \setminus \{w\}\,. \tag{5.13.4}$$

5.13.4 Theorem. *Let $D \subseteq \mathbb{C}$ be an open set, and let Z be a discrete and closed subset of D. Suppose, for each $w \in Z$, there are given a neighborhood U_w of w with $U_w \cap Z = \{w\}$ and a holomorphic function $f_w : U_w \setminus \{w\} \to G$. Assume that:*

(iii) *There exist a continuous function $c : D \setminus Z \to G$ and continuous functions $c_w : U_w \to G$, $w \in Z$, such that $f_w c_w = c$ on $U_w \setminus \{w\}$, $w \in Z$.*

Then there exist a holomorphic function $h : D \setminus Z \to G$ and holomorphic functions $h_w : U_w \to G$, $w \in Z$, such that

$$f_w h_w = h \qquad \text{on } U_w \setminus \{w\}. \tag{5.13.5}$$

Finally, we present a *two-sided* Weierstrass theorem:

5.13.5 Theorem. *Let $D \subseteq \mathbb{C}$ be an open set, and let Z be a discrete and closed subset of D. Suppose, for each $w \in Z$, there are given a neighborhood U_w of w with $U_w \cap Z = \{w\}$ and two holomorphic functions $f_w, g_w : U_w \setminus \{w\} \to G$. Moreover, we assume that at least one of the following conditions is fulfilled:*

(i) *D is simply connected.*

(ii) *G is connected.*

Then there exist a holomorphic function $h : D \setminus Z \to G$ and holomorphic functions $h_w : U_w \to G$, $w \in Z$, such that

$$f_w h_w g_w = h \qquad \text{on } U_w \setminus \{w\}. \tag{5.13.6}$$

Proof. Let a family of positive integers m_w, $w \in Z$, be given. Then from the left-sided Weierstrass Theorem 10.1.1 we get a holomorphic function $h^l : D \setminus Z \to G$ and holomorphic functions $h_w^l : U_w \to G$, $w \in Z$, such that

$$h_w^l g_w = h^l \qquad \text{on } U_w \setminus \{w\}. \tag{5.13.7}$$

From the right-sided Weierstrass Theorem 10.2.1 we get a holomorphic function $h^r : D \setminus Z \to G$ and holomorphic functions $h_w^r : U_w \to G$, $w \in Z$, such that

$$f_w h_w^r = h^r \qquad \text{on } U_w \setminus \{w\}. \tag{5.13.8}$$

Set $h = h^l h^r$ and $h_w = h_w^r h_w^l$, $w \in Z$. Then $h \in \mathcal{O}^G(D \setminus Z)$, $h_w \in \mathcal{O}^G(U_w)$ and

$$f_w h_w g_w = f_w h_w^r h_w^l g_w = h^r h^l = h \qquad \text{on } U_w \setminus \{w\}.$$

□

5.13.6 Remark. Instead of conditions (i) or (ii) in Theorem 5.13.5 also the following condition would be sufficient (Oka-Grauert principle):

(iii) There exist a continuous function $c : D \setminus Z \to G$ and continuous functions $c_w : U_w \to G$, $w \in Z$, such that $f_w c_w g_w = c$ on $U_w \setminus \{w\}$, $w \in Z$.

But to prove this we would need a generalization of the theory of multiplicative cocycles where the group G in Definition 9.1.2 is replaced by a *fiber bundle* of groups with characteristic fiber G. This generalization is well known in Complex analysis of several variables, as a part of the theory of Grauert [Gr1, Gr2, Gr3] and Bungart [Bu] mentioned in the introduction to the present book. To keep this book simpler and shorter we omit this.

5.14 Weierstrass theorems for $\mathcal{G}^\infty(E)$ and $\mathcal{G}^\omega(E)$-valued functions

In this section, we use the notations introduced in Definition 5.12.1 and the simple well-known facts listed in the subsequent Proposition 5.12.2, and $\aleph$ will stand for one of the symbols ∞ or ω.

We first prove the following left-sided version of the Weierstrass theorem for the group $\mathcal{G}^\aleph(E)$:

5.14.1 Theorem. *Let $D \subseteq \mathbb{C}$ be an open set, and let Z be a discrete and closed subset of D. Suppose, for each $w \in Z$, there are given a neighborhood U_w of w with $U_w \cap Z = \{w\}$ and a holomorphic function $A_w : U_w \setminus \{w\} \to \mathcal{G}^\aleph(E)$. Then there exist a holomorphic function $H : D \setminus Z \to \mathcal{G}^\aleph(E)$ and holomorphic functions $H_w : U_w \to \mathcal{G}^\aleph(E)$, $w \in Z$, such that*

$$H_w A_w = H \qquad \text{on } U_w \setminus \{w\}. \tag{5.14.1}$$

Proof. Let a family of positive integers m_w, $w \in Z$, be given. Since $\mathcal{G}^\aleph(E) \subseteq G\mathcal{F}_I^\aleph(E)$, the functions A_w can be interpreted as functions with values in $G\mathcal{F}_I^\aleph(E)$. Since the latter group is the group of invertible elements of a Banach algebra and since this group is connected, we can apply Theorem 5.13.1 to it and obtain a holomorphic function $H : D \setminus Z \to G\mathcal{F}_I^\aleph(E)$ and holomorphic functions $H_w : U_w \to G\mathcal{F}_I^\aleph(E)$, $w \in Z$, such that

$$H_w A_w = H \qquad \text{on } U_w \setminus \{w\}. \tag{5.14.2}$$

If $\dim E < \infty$ and therefore $\mathcal{G}^\aleph(E) = GL(E) = G\mathcal{F}_I^\aleph(E)$, this completes the proof.

Let $\dim E = \infty$, and let $\lambda_w : U_w \to \mathbb{C}$, $\lambda : D \setminus Z \to \mathbb{C}$, $K_w : U_w \to \mathcal{F}^\aleph$ and $K : D \setminus Z \to \mathcal{F}^\aleph$ be the holomorphic functions with $H_w = \lambda_w I + K_w$ and $H = \lambda I + K$. Then, passing to the factor algebra $\mathcal{F}_I(E)/\mathcal{F}^\aleph(E) \cong \mathbb{C}$, we see: Since H and H_w are invertible, the functions λ and λ_w have no zeros, and from (5.14.2) it follows that $\lambda_w = \lambda$ on U_w. Hence the H_w/λ and H/λ are $\mathcal{G}^\aleph$-valued functions with the required properties. □

Precisely in the same way, solely replacing the left-sided Theorem 5.13.1 by the right-sided Theorem 5.13.3, we get the corresponding right-sided result:

5.14.2 Theorem. *Let $D \subseteq \mathbb{C}$ be an open set, and let Z be a discrete and closed subset of D. Suppose, for each $w \in Z$, there are given a neighborhood U_w of w with $U_w \cap Z = \{w\}$ and a holomorphic function $A_w : U_w \setminus \{w\} \to \mathcal{G}^{\aleph}(E)$. Then there exist a holomorphic function $H : D \setminus Z \to \mathcal{G}^{\aleph}(E)$ and holomorphic functions $H_w : U_w \to \mathcal{G}^{\aleph}(E)$, $w \in Z$, such that*

$$A_w H_w = H \qquad \text{on } U_w \setminus \{w\}. \tag{5.14.3}$$

Both theorems together again give a two-sided version:

5.14.3 Theorem. *Let $D \subseteq \mathbb{C}$ be an open set, and let Z be a discrete and closed subset of D. Suppose, for each $w \in Z$, there are given a neighborhood U_w of w with $U_w \cap Z = \{w\}$ and two holomorphic functions $F_w, G_w : U_w \setminus \{w\} \to \mathcal{G}^{\aleph}(E)$. Then there exist a holomorphic function $H : D \setminus Z \to \mathcal{G}^{\aleph}(E)$ and holomorphic functions $H_w : U_w \to \mathcal{G}^{\aleph}(E)$, $w \in Z$, such that*

$$F_w H_w G_w = H \qquad \text{on } U_w \setminus \{w\}. \tag{5.14.4}$$

Proof. Let a family of positive integers m_w, $w \in Z$, be given. Then from the left-sided Theorem 5.14.1 we get a holomorphic function $H^l : D \setminus Z \to \mathcal{G}^{\aleph}(E)$ and holomorphic functions $H_w^l : U_w \to \mathcal{G}^{\aleph}$, $w \in Z$, such that

$$H_w^l G_w = H^l \qquad \text{on } U_w \setminus \{w\}.$$

From the right-sided Theorem 5.14.2 we get a holomorphic function $H^r : D \setminus Z \to \mathcal{G}^{\aleph}(E)$ and holomorphic functions $H_w^r : U_w \to \mathcal{G}^{\aleph}(E)$, $w \in Z$, such that

$$F_w H_w^r = H^r \qquad \text{on } U_w \setminus \{w\}.$$

Setting $H = H^l H^r$ and $H_w = H_w^r H_w^l$, $w \in Z$, we get holomorphic functions $H : D \setminus Z \to \mathcal{G}^{\aleph}(E))$ and $H_w : U_w \to \mathcal{G}^{\aleph}(E)$, $w \in Z$, such that

$$F_w H_w G_w = F_w H_w^r H_w^l G_w = H^r H^l = H \qquad \text{on } U_w \setminus \{w\}. \qquad \square$$

5.15 Comments

This chapter is based on the theory of of multiplicative cocycles (fiber bundles) on Stein manifolds (any domain in $\mathbb{C}$ is a Stein manifold), which was developed in the 1950s by H. Grauert [Gr1, Gr2, Gr3] for cocycles with values in a (finite dimensional) complex Lie group and generalized by L. Bungart [Bu] to cocycles with values in a *Banach* Lie group. The Oka-Grauert principle was first discovered by K. Oka [Ok] for cocycles of several complex variables with values in $\mathbb{C}^*$. That the Oka-Grauert principle holds also for non-commutative groups is a very deep result due to H. Grauert [Gr1, Gr2, Gr3] (not easy even for one variable).

In the papers [GL1, GL2, GL3], the authors presented some of these results, which are relevant for operator theory. Direct proofs for the case of one variable and also some new results were given there. This chapter is an extension and completion of the material contained in these papers. Here style and presentation are essentially improved. The first versions of the Weierstrass theorems from section 3.8 were published in the papers of I. Gohberg and L. Rodman in [GR2] (for matrices) and [GR2] (for infinite dimensional operators).

Chapter 6

Families of subspaces

Let E be a Banach space. We denote by $G(E)$ the set of closed subspaces of E. In this chapter we study functions with values in $G(E)$.

Consider first the case $E = \mathbb{C}^n$, $n \in \mathbb{N}^*$. Let $G(k,n)$ be the set of k-dimensional subspaces of $\mathbb{C}^n$, $0 \le k \le n$. Then $G(\mathbb{C}^n)$ is the disjoint union of the sets $G(k,n)$, and it is well known (see, e.g., [HaGr]) that each of them is a complex manifold called the (complex) *Grassmann manifold of k-dimensional subspaces of* $\mathbb{C}^n$. From this general point of view it is therefore clear what a holomorphic function with values in $G(\mathbb{C}^n)$ is. For a more direct definition (not using Grassmann manifolds) of holomorphic $G(\mathbb{C}^n)$-valued functions, we refer to section 18.1 of the book [GLR]. Note that all values of such a holomorphic function have the same dimension, except for the case that the domain of definition is not connected.

If $\dim E = \infty$, then there are different reasonable definitions of holomorphic $G(E)$-valued functions. We will discuss them in this chapter. First we introduce the notion of *continuous* $G(E)$-valued functions. For that we use the gap metric on $G(E)$.

6.1 The gap metric

Let E be a Banach space. Recall that for two non-empty subsets $X, Y \subset E$, the number

$$\operatorname{dist}(X,Y) = \inf_{x \in X, y \in Y} \|x - y\| \tag{6.1.1}$$

is called the **distance** between X and Y (here $\|\cdot\|$ is the norm of E). Note that the distance between two subspaces of E is always zero, because any subspace contains the zero-vector. For subspaces, there is another "distance" which is called the **gap** (in order to avoid confusion with the distance):

6.1.1 Definition. Let E be a Banach space. If X is a subspace of E (not necessarily closed), then we denote by

$$S(X) := \left\{ x \in X \;\middle|\; \|x\| = 1 \right\}$$

the unit sphere of X. If X, Y are two subspaces, then we define the **gap** $\Theta(X, Y)$ between X and Y as follows: If $X \neq \{0\}$ and $Y \neq \{0\}$, then

$$\theta(X,Y) := \max \left\{ \sup_{v \in S(X)} \operatorname{dist}\big(v, S(Y)\big)\,,\; \sup_{v \in S(Y)} \operatorname{dist}\big(v, S(X)\big) \right\}. \tag{6.1.2}$$

If $X = \{0\}$ and $Y \neq \{0\}$ or if $X \neq \{0\}$ and $Y = \{0\}$, then we set $\theta(X,Y) = 1$, and if $X = Y = \{0\}$, then we set $\theta(X, Y) = 0$. From this definition it is immediately clear that

$$\theta(X,Y) \leq 2 \qquad \text{for all } X, Y \in G(E). \tag{6.1.3}$$

If X is a subspace of a Banach space E and $\overline{X}$ is the closure of X, then it is clear that $\Theta(X, \overline{X}) = 0$. Therefore, on the set of all subspaces of E, the gap is not a metric. However we have:

6.1.2 Theorem. *Let E be a Banach space, and $G(E)$ the set of all closed subspaces of E. Then the gap Θ defined by* (6.1.1) *is a complete metric on $G(E)$.*

Proof. First we check the three axioms of a metric.

I. Suppose $X, Y \in G(E)$ and $\theta(X, Y) = 0$. If at least one of the spaces X and Y is the zero space, then this clearly implies (by the definition of $\theta(X, Y)$ in this case) that also the other one is the zero space and hence $X = Y$. Now let $X \neq \{0\}$ and $Y \neq \{0\}$. Then, in particular,

$$\sup_{x \in S(X)} \operatorname{dist}\big(x, S(Y)\big) = 0,$$

i.e., $\operatorname{dist}\big(x, S(Y)\big) = 0$ for all $x \in S(X)$. Since $S(Y)$ is closed, this is possible, if and only if, $S(X) \subseteq S(Y)$. In the same way we see that $S(Y) \subseteq S(X)$. Hence $S(X) = S(Y)$, which means that $X = Y$.

II. If we interchange the letters X and Y in the expression of the right-hand side of definition (6.1.2), then the value of this expression does not change. Therefore it is clear that $\theta(X, Y) = \theta(Y, X)$ for all $X, Y \in G(E)$ with $X \neq \{0\}$ and $Y \neq \{0\}$.

III. We prove the triangle inequality. Let $X, Y, Z \in G(E)$.

First let $Z = \{0\}$. If $X \neq \{0\}$ and $Y \neq \{0\}$, then $\theta(X, Z) = 1$ and $\theta(Z, Y) = 1$ (by definition). Since $\theta(X, Y) \leq 2$ (see (6.1.3)), this implies that

$$\theta(X,Y) \leq 2 = \theta(X,Z) + \theta(Z,Y).$$

If $X = \{0\}$ and $Y \neq \{0\}$, then $\theta(X, Y) = 1$, $\theta(X, Z) = 0$ and $\theta(Z, Y) = 1$ which implies that

$$\theta(X,Y) = 1 = 0 + 1 = \theta(X,Z) + \theta(Z,Y).$$

Correspondingly we proceed if $X \neq \{0\}$ and $Y = \{0\}$.

Now let $Z \neq \{0\}$ and at least one of the spaces X, Y is the zero space. If $X = \{0\}$ and $Y \neq \{0\}$, then $\theta(X,Y) = 1$ and $\theta(X,Z) = 1$ and therefore

$$\theta(X,Y) = \theta(X,Z) \leq \theta(X,Z) + \theta(Z,Y).$$

Correspondingly we proceed if $X \neq \{0\}$ and $Y = \{0\}$. If $X = Y = \{0\}$, then $\theta(X,Y) = 0$ and therefore

$$\theta(X,Y) = 0 \leq \theta(X,Z) + \theta(Z,Y).$$

Finally we consider the case when none of the spaces X, Y, Z is the zero space. Then at least one of the following relations is true:

$$\theta(X,Y) = \sup_{v \in S(X)} \operatorname{dist}\big(v, S(Y)\big) \qquad \text{or} \qquad \theta(X,Y) = \sup_{v \in S(Y)} \operatorname{dist}\big(v, S(X)\big).$$

We may assume that this is the first one (otherwise we have to change the roles of X and Y). Now let $\varepsilon > 0$. Then we can choose $x \in S(X)$ such that

$$\theta(X,Y) \leq \operatorname{dist}\big(x, S(Y)\big) + \varepsilon. \tag{6.1.4}$$

Moreover, then we can take $z \in S(Z)$ with

$$\operatorname{dist}\big(x, S(Z)\big) + \varepsilon \geq \|x - z\|. \tag{6.1.5}$$

Then, by the triangle inequality in E,

$$\begin{aligned}\operatorname{dist}\big(x, S(Y)\big) &= \inf_{v \in S(Y)} \|x - v\| \leq \inf_{v \in S(Y)} \big(\|x - z\| + \|z - v\|\big) \\ &\leq \|x - z\| + \inf_{v \in S(Y)} \|z - v\| = \|x - z\| + \operatorname{dist}\big(z, S(Y)\big).\end{aligned}$$

By (6.1.4), this implies

$$\theta(X,Y) \leq \|x - z\| + \operatorname{dist}\big(z, S(Y)\big) + \varepsilon,$$

and, by (6.1.5), we further get

$$\begin{aligned}&\theta(X,Y) \leq \operatorname{dist}\big(x, S(Z)\big) + \operatorname{dist}\big(z, S(Y)\big) + 2\varepsilon \\ &\leq \sup_{v \in S(X)} \operatorname{dist}\big(v, S(Z)\big) + \sup_{v \in S(Z)} \operatorname{dist}\big(v, S(Y)\big) + 2\varepsilon \leq \theta(X,Z) + \theta(Z,Y) + 2\varepsilon.\end{aligned}$$

Since $\varepsilon > 0$ can be chosen arbitrarily small, this completes the proof of the triangle inequality for θ. Hence it is proved that θ is a metric.

We prove the completeness. Let $(X_n)_{n \in \mathbb{N}}$ be a Cauchy sequence in $(G(E), \theta)$. Passing to a subsequence we may assume that

$$\theta(X_n, X_{n+m}) < \frac{1}{2^n} \qquad \text{for all } n, m \in \mathbb{N}. \tag{6.1.6}$$

This implies that either $X_n = \{0\}$ for all n or $X_n \neq \{0\}$ for all n. The first case is trivial. Consider the second one.

Then, in particular,

$$\operatorname{dist}\big(v, S(X_n)\big) < \frac{1}{2^n} \qquad \text{for all } v \in S(X_{n+m}) \text{ and } n, m \in \mathbb{N}. \tag{6.1.7}$$

Denote by X the set of all vectors $x \in E$ such that there exists a sequence $(x_n)_{n\in\mathbb{N}}$ with

$$x_n \in X_n \quad \text{for all } n \in \mathbb{N} \qquad \text{and} \qquad \lim_{n\to\infty} x_n = x.$$

Obviously X is a linear subspace of E. Let $\overline{X}$ be the topological closure of X in E. We will now prove that $\overline{X}$ is the required limit of the sequence $(X_n)_{n\in\mathbb{N}}$.

Let $v \in S(X)$ and $n \in \mathbb{N}$. Then, by definition of X, there is a sequence $(x_m)_{m\in\mathbb{N}}$ with $x_m \in X_{n+m}$ and $\lim x_m = v$. Since $\|v\| \neq 1$, then also

$$\lim_{m\to\infty} \frac{x_m}{\|x_m\|} = v.$$

Since, by (6.1.7),

$$\operatorname{dist}\left(\frac{x_m}{\|x_m\|}, S(X_n)\right) < \frac{1}{2^n},$$

this implies that

$$\operatorname{dist}(v, S(X_n)) \leq \frac{1}{2^n}.$$

Since $S(\overline{X})$ is the closure of $S(X)$, this implies that

$$\sup_{v\in S(\overline{X})} \operatorname{dist}(v, S(X_n)) \leq \frac{1}{2^n} \qquad \text{for all } n \in \mathbb{N}. \tag{6.1.8}$$

Now let $v \in S(X_n)$, $n \in \mathbb{N}$. Then, by (6.1.6), we can find a sequence $x_m \in X_{n+m}$, $m \in \mathbb{N}$, such that $x_0 = v$ and, for each $m \in \mathbb{N}$, we have the inequality

$$\|x_m - x_{m+1}\| < \frac{1}{2^m}.$$

Then $x := \lim x_m$ exists (E is complete) and

$$\|x - v\| = \|x - x_0\| \leq \frac{1}{2^{n-1}}. \tag{6.1.9}$$

Since $\|v\| = 1$, this further yields

$$\big|\|x\| - 1\big| = \big|\|x\| - \|v\|\big| \leq \|x - v\| \leq \frac{1}{2^{n-1}}$$

and therefore

$$\left\| x - \frac{x}{\|x\|} \right\| = \left\| (\|x\| - 1)\frac{x}{\|x\|} \right\| = \big| \|x\| - 1 \big| \le \frac{1}{2^{n-1}}.$$

Since, by definition of X, $x \in X$, and therefore $x/\|x\| \in S(X) \subseteq S(\overline{X})$, this implies that

$$\operatorname{dist}\big(x, S(\overline{X})\big) \le \frac{1}{2^{n-1}}$$

and further, by (6.1.9),

$$\operatorname{dist}\big(v, S(\overline{X})\big) \le \|v - x\| + \operatorname{dist}\big(x, S(\overline{X})\big) \le \frac{1}{2^{n-2}}.$$

Hence, it is proved that

$$\sup_{v \in S(X_n)} \operatorname{dist}\big(v, S(\overline{X})\big) \le \frac{1}{2^{n-2}} \qquad \text{for all } n \in \mathbb{N}.$$

Together with (6.1.8) this implies that

$$\theta(X_n, \overline{X}) \le \frac{1}{2^{n-2}} \qquad \text{for all } n \in \mathbb{N}.$$

Hence $\lim \theta(X_n, \overline{X}) = 0$. □

In the following propositions we collect some further useful estimates and relations for the gap.

6.1.3 Proposition. *Let E be a Banach space, and let X, Y be two subspaces of E such that X is closed, $X \subseteq Y$ and $X \ne Y$. Then*

$$\Theta(X, Y) \ge 1. \tag{6.1.10}$$

Proof. Since X is closed and a proper subspace of Y, for each $\varepsilon > 0$ we can find $y \in S(Y)$ with

$$\operatorname{dist}(y, X) > 1 - \varepsilon.$$

Hence

$$\sup_{y \in S(Y)} \operatorname{dist}\big(y, S(X)\big) \ge \sup_{y \in S(Y)} \operatorname{dist}(y, X) \ge 1,$$

which implies (6.1.10) by definition of $\Theta(X, Y)$. □

6.1.4 Proposition. *Let E be a Banach space, and let X, Y be two closed subspaces of E, $X \ne \{0\}$, $Y \ne \{0\}$. Then*

$$\sup_{v \in S(X)} \operatorname{dist}(v, Y) \le \sup_{v \in S(X)} \operatorname{dist}\big(v, S(Y)\big) \le 2 \sup_{v \in S(X)} \operatorname{dist}(v, Y) \tag{6.1.11}$$

and

$$\max\left\{\sup_{v\in S(X)} \operatorname{dist}(v,Y)\,,\ \sup_{v\in S(Y)} \operatorname{dist}(v,X)\right\}$$
$$\le \theta(X,Y) \le 2\max\left\{\sup_{v\in S(X)} \operatorname{dist}(v,Y)\,,\ \sup_{v\in S(Y)} \operatorname{dist}(v,X)\right\}. \tag{6.1.12}$$

Further, if there exist bounded linear projectors P_X and P_Y from E to X and Y, respectively, then

$$\max\left\{\sup_{v\in S(X)} \operatorname{dist}(v,Y)\,,\ \sup_{v\in S(Y)} \operatorname{dist}(v,X)\right\} \le \|P_X - P_Y\| \tag{6.1.13}$$

and therefore, by the right inequality in (6.1.12),

$$\theta(X,Y) \le 2\|P_X - P_Y\|. \tag{6.1.14}$$

Finally, if E is a Hilbert space and P_X, P_Y are the orthogonal projectors from E to X and Y, respectively, then

$$\|P_X - P_Y\| = \max\left\{\sup_{v\in S(X)} \operatorname{dist}(v,Y)\,,\ \sup_{v\in S(Y)} \operatorname{dist}(v,X)\right\}. \tag{6.1.15}$$

and hence, by (6.1.12),

$$\|P_X - P_Y\| \le \theta(X,Y) \le 2\|P_X - P_Y\|. \tag{6.1.16}$$

Proof. Let $\varepsilon > 0$ be given. Then we can find $x \in S(X)$ such that

$$\sup_{v\in S(X)} \operatorname{dist}\big(v, S(Y)\big) < \operatorname{dist}\big(x, S(Y)\big) + \varepsilon. \tag{6.1.17}$$

Further, then we can find $y \in Y$ such that

$$\|x - y\| < \operatorname{dist}(x,Y) + \varepsilon. \tag{6.1.18}$$

Since $\|x\| = 1$, then also

$$\big|1 - \|y\|\big| < \operatorname{dist}(x,Y) + \varepsilon$$

and therefore

$$\left\|\frac{y}{\|y\|} - y\right\| = \left\|\frac{y}{\|y\|}\big(1 - \|y\|\big)\right\| \le \operatorname{dist}(x,Y) + \varepsilon. \tag{6.1.19}$$

Since $y/\|y\| \in S(Y)$ and therefore

$$\operatorname{dist}\big(x, S(Y)\big) \le \left\|x - \frac{y}{\|y\|}\right\|,$$

now it follows from (6.1.17) that

$$\sup_{v\in S(X)} \operatorname{dist}\big(v, S(Y)\big) < \left\| x - \frac{y}{\|y\|} \right\| + \varepsilon \le \|x-y\| + \left\| y - \frac{y}{\|y\|} \right\| + \varepsilon.$$

Together with (6.1.18) and (6.1.19) this yields

$$\sup_{v\in S(X)} \operatorname{dist}\big(v, S(Y)\big) < 2\operatorname{dist}(x,Y) + 3\varepsilon \le 2 \sup_{v\in S(X)} \operatorname{dist}(v,Y) + 3\varepsilon.$$

Since $\varepsilon > 0$ can be chosen arbitrarily small, this proves the right inequality in (6.1.11). The left inequality in (6.1.11) is trivial.

(6.1.12) follows from (6.1.11) and the inequality obtained from (6.1.11) interchanging X and Y.

Now we assume that there exist bounded linear projectors P_X and P_Y from E to X and Y, respectively. Let $\varepsilon > 0$. Then we can choose $x \in S(X)$ such that

$$\sup_{v\in S(X)} \le \operatorname{dist}(x,Y) + \varepsilon.$$

Since $\operatorname{dist}(x,Y) \le \|x - P_Y x\|$, $P_X x = x$ and $\|x\| = 1$, this yields

$$\sup_{v\in S(X)} \operatorname{dist}(v,Y) \le \|x - P_Y x\| = \|(P_X - P_Y)x\| + \varepsilon \le \|P_X - P_Y\| + \varepsilon.$$

Since $\varepsilon > 0$ can be chosen arbitrarily small, this means that

$$\sup_{v\in S(X)} \operatorname{dist}(v,Y) \le \|P_X - P_Y\|.$$

In the same way we get

$$\sup_{v\in S(Y)} \operatorname{dist}(v,X) \le \|P_X - P_Y\|.$$

Together this implies (6.1.13).

Finally we assume that E is a Hilbert space and P_X, P_Y are the orthogonal projectors from E to X and Y, respectively. In view of the general inequality (6.1.13), we only have to prove that

$$\|P_X - P_Y\| \le \max\left\{ \sup_{v\in S(X)} \operatorname{dist}(v,Y)\, ,\ \sup_{v\in S(Y)} \operatorname{dist}(v,X) \right\}. \tag{6.1.20}$$

First note that, for all $v \in E$ with $P_Y v \neq 0$,

$$\|(I - P_X)P_Y v\| = \operatorname{dist}(P_Y v, X) = \|P_Y v\| \operatorname{dist}\left(\frac{P_Y v}{\|P_Y v\|}, X \right).$$

Hence

$$\|(I-P_X)P_Yv\| \le \|P_Yv\| \sup_{w\in S(Y)} \operatorname{dist}(w,X) \qquad \text{for all } v\in E. \tag{6.1.21}$$

In the same way we get

$$\|(I-P_Y)P_Xv\| \le \|P_Xv\| \sup_{w\in S(X)} \operatorname{dist}(w,Y) \qquad \text{for all } v\in E. \tag{6.1.22}$$

Let $\langle\cdot\,,\,\cdot\rangle$ be the scalar product in E. Since the projectors P_X and P_Y are orthogonal, then

$$\begin{aligned}\|P_Y(I-P_X)v\|^2 &= \langle P_Y(I-P_X)v, P_Y(I-P_X)v\rangle = \langle (I-P_X)v, P_Y(I-P_X)v\rangle \\ &= \langle (I-P_X)v, (I-P_X)P_Y(I-P_X)v\rangle \le \|(I-P_X)v\|\|(I-P_X)P_Y(I-P_X)v\|\end{aligned}$$

for all $v\in E$. Together with (6.1.21) this yields

$$\|P_Y(I-P_X)v\|^2 \le \|(I-P_X)v\|\|P_Y(I-P_X)v\| \sup_{w\in S(Y)} \operatorname{dist}(w,X)$$

and therefore

$$\|P_Y(I-P_X)v\| \le \|(I-P_X)v\| \sup_{w\in S(Y)} \operatorname{dist}(w,X) \qquad \text{for all } v\in E. \tag{6.1.23}$$

Since $P_Y-P_X = P_Y(I-P_X)-(I-P_Y)P_X$ and P_Y is orthogonal, we get

$$\|(P_Y-P_X)v\|^2 = \|P_Y(I-P_X)v\|^2 + \|(I-P_Y)P_Xv\|^2 \qquad \text{for all } v\in E.$$

In view of (6.1.23) and (6.1.22) this implies

$$\begin{aligned}&\|(P_Y-P_X)v\|^2 \\ &\quad\le \|(I-P_X)v\|^2\left(\sup_{w\in S(Y)} \operatorname{dist}(w,X)\right)^2 + \|P_Xv\|^2\left(\sup_{w\in S(X)} \operatorname{dist}(w,Y)\right)^2 \\ &\qquad\qquad \le \|v\|^2\left(\max\left\{\sup_{w\in S(Y)} \operatorname{dist}(w,X),\ \sup_{w\in S(X)} \operatorname{dist}(w,Y)\right\}\right)^2\end{aligned}$$

for all $v\in E$. Taking the square root we get (6.1.20) □

6.1.5 Proposition. *The set of complemented subspaces*[1] *of a Banach space is open with respect to the gap metric. More precisely: Let E be a Banach space, and let*

[1] A subspace X of a Banach space E is called **complemented** if there exists a closed subspace Y of E such that E is the direct sum of X and Y.

X, Y be two closed subspaces of E such that E is the direct sum of X and Y. Let P be the linear projector from E to X parallel to Y, and let $Q = I - P$. Set

$$\varepsilon = \frac{1}{8\|P\|}.$$

Then, for all closed subspaces X', Y' of E with

$$\theta(X, X') < \varepsilon \qquad \text{and} \qquad \theta(Y, Y') < \varepsilon, \tag{6.1.24}$$

E is the direct sum of X' and Y'. Moreover, if P' is the projector from E to X' parallel to Y', then

$$\|P'\| \le 4\|P\|. \tag{6.1.25}$$

Proof. We may assume that $X \neq \{0\}$ and $Y \neq \{0\}$, because if, for example $X = \{0\}$, then the inequality $\theta(X', X) < 1$ means, by definition, that $X' = \{0\}$.

Now let $X', Y' \in G(E)$ with (6.1.24) be given. First prove that $X' \cap Y' = \{0\}$. Assume the contrary. Then we can find $v \in X' \cap Y'$ with $\|v\| = 1$ and, by definition of θ,

$$\operatorname{dist}\big(v, S(X)\big) \le \sup_{w \in S(X')} \operatorname{dist}\big(w, S(X)\big) \le \theta(X', X) < \varepsilon.$$

Therefore we can find $x \in S(X)$ with

$$\|v - x\| < \varepsilon. \tag{6.1.26}$$

In the same way, we find $y \in S(Y)$ with

$$\|v - y\| < \varepsilon.$$

Together with (6.1.26) this gives

$$\|y - x\| < 2\varepsilon. \tag{6.1.27}$$

Since $\|x\| = 1$, this implies

$$\|x + y\| = \|2x + (y - x)\| \ge \|2x\| - \|y - x\| > 2 - 2\varepsilon$$

and further, as $\varepsilon = 1/8\|P\| \le 1/8$,

$$\|x + y\| > 2 - \frac{1}{4}. \tag{6.1.28}$$

On the other hand

$$\|x + y\| = \|Px + Qy\| = \|P(x - y) + y\| \le \|P\|\|x - y\| + 1,$$

which implies, by (6.1.27),

$$\|x + y\| < \|P\| 2\varepsilon + 1.$$

Since $\varepsilon = 1/8\|P\|$, this means that

$$\|x+y\| < \frac{1}{4} + 1,$$

which contradicts (6.1.28).

Since $X' \cap Y' = \{0\}$, we have the linear projector from $X' + Y'$ to X' parallel to Y'. We next prove that this projector is bounded where

$$\|P'\| \leq 4\|P\|. \tag{6.1.29}$$

Assume the contrary. Then we can find $v \in X' + Y'$ such that

$$\|v\| = 1 \qquad \text{and} \qquad \|P'v\| > 4\|P\|. \tag{6.1.30}$$

Set

$$x' = \frac{P'v}{\|P'v\|} \qquad \text{and} \qquad y' = \frac{v - P'v}{\|P'v\|}.$$

Take $x \in S(X)$ with

$$\|x' - x\| < \text{dist}\,\big(x', S(X)\big) + \varepsilon.$$

Then it follows from the definition of θ that

$$\|x' - x\| \leq \sup_{w \in S(X')} \text{dist}\,\big(w, S(X)\big) + \varepsilon \leq \theta(X', X) + \varepsilon. \tag{6.1.31}$$

Since, by hypothesis $\theta(X', X') < \varepsilon$, this yields

$$\|x' - x\| < 2\varepsilon. \tag{6.1.32}$$

From (6.1.29) we get

$$\|y'\| = \frac{\|v - P'v\|}{\|P'v\|} \leq \frac{\|v\| + \|P'v\|}{\|p'v\|} = 1 + \frac{1}{\|P'v\|} < 1 + \frac{1}{2\|P\|}. \tag{6.1.33}$$

By the left inequality in (6.1.4) and since $\theta(Y', Y) < \varepsilon$, we obtain

$$\text{dist}(y', Y) \leq \|y'\| \sup_{w \in S(Y')} \text{dist}(w, Y)\|y'\|\theta(Y', Y) < \|y'\|\varepsilon.$$

Together with (6.1.33) this yields

$$\text{dist}(y', Y) < \varepsilon + \frac{\varepsilon}{2\|P\|}. \tag{6.1.34}$$

Therefore we can find $y \in Y$ with

$$\|y' - y\| < 2\varepsilon + \frac{\varepsilon}{2\|P\|}. \tag{6.1.35}$$

From the definition of x', y' and (6.1.30) it follows that

$$\|x' + y'\| = \frac{\|v\|}{\|P'v\|} < \frac{1}{4\|P\|}.$$

Together with (6.1.35) and (6.1.31) this further gives

$$\|y - x\| \leq \|y - y'\| + \|x' + y'\| + \|x - x'\| < 2\varepsilon + \frac{\varepsilon}{2\|P\|} + \frac{1}{4\|P\|} + \theta(X', X) + \varepsilon,$$

which implies, as $\theta(X', X) < \varepsilon$,

$$\|y - x\| \leq \frac{1}{4\|P\|} + \frac{\varepsilon}{2\|P\|} + 4\varepsilon.$$

Since $\varepsilon = 1/8\|P\|$ and $\|P\| \geq 1$, this implies that

$$\|y - x\| \leq \frac{1}{4\|P\|} + \frac{1}{16\|P\|^2} + \frac{1}{2\|P\|} < \left(\frac{1}{4} + \frac{1}{16} + \frac{1}{2}\right)\frac{1}{\|P\|},$$

which gives the contradiction

$$1 = \|x\| = \|P(x - y)\| \leq \|P(y - x)\| \leq \|P\|\|y - x\| < \frac{1}{4} + \frac{1}{16} + \frac{1}{2}.$$

Since the projector P' is bounded, $X' + Y'$ is topologically closed in E. Indeed, let $(v_n)_{n\in\mathbb{N}}$ be a sequence in $X' + Y'$, which converges to some $v \in E$. Since P' is bounded, then also the sequences $P'v_n \in X'$ and $v_n - P'v_n \in Y'$ converge in E where, as the spaces X' and Y' are topologically closed in E,

$$\lim_{n\to\infty} P'v_n \in X' \qquad \text{and} \qquad \lim_{n\to\infty} (v_n - P'v_n) \in Y'.$$

Hence

$$\lim_{n\to\infty} v_n = \lim_{n\to\infty} P'v_n + \lim_{n\to\infty} (v_n - P'v_n) \in X' + Y'.$$

To complete the proof of the proposition, now it remains to show that $X + Y' = E$. ((6.1.25) then follows from (6.1.29).) Assume it is not. Since $X' + Y'$ is closed, then we can find $x \in E$ with $\|x\| = 1$ and

$$\operatorname{dist}(x, X' + Y') > \frac{1}{2}. \tag{6.1.36}$$

Since $Px \in X$ and $x - Px \in Y$, it follows from the left inequality in (6.1.12) that

$$\operatorname{dist}(Px, X') \leq \|Px\| \sup_{v\in S(X)} \operatorname{dist}(v, X') \leq \|Px\|\theta(X', X)$$

and

$$\operatorname{dist}(x - Px, Y') \leq \|x - Px\| \sup_{v\in S(Y)} \operatorname{dist}(v, Y') \leq \|x - Px\|\theta(Y', Y).$$

Since $\theta(X', X) < \varepsilon$, $\theta(Y', Y) < \varepsilon$ and $\|x\| = 1$, this further yields

$$\operatorname{dist}(Px, X') < \|P\|\varepsilon$$

and

$$\operatorname{dist}(x - Px, Y') < (1 + \|P\|)\varepsilon.$$

Therefore we can find $x' \in X'$ and $y' \in Y'$ such that

$$\|Px - x'\| < \|P\|\varepsilon \quad \text{and} \quad \|x - Px - y'\| < (1 + \|P\|)\varepsilon.$$

Then

$$\|x - (x' + y')\| = \|Px - x' + (x - Px) - y'\| \le (1 + 2\|P\|)\varepsilon.$$

Taking into account that $\varepsilon = 1/8\|P\|$ and $\|P\| \ge 1$, from this we get

$$\|x - (x' + y')\| \le \frac{1}{8} + \frac{1}{4} < \frac{1}{2}.$$

Since $x' + y' \in X' + Y'$, this contradicts (6.1.36). □

6.1.6 Proposition. *Let E be a Banach space. Then the following sets are open with respect to the gap metric:*

(i) *the set of infinite dimensional closed subspaces of E,*

(ii) *for each $k \in \mathbb{N}$, the set of k-dimensional subspaces of E,*

(iii) *for each $k \in \mathbb{N}$, the set of closed subspaces of E which are of codimension k in E.*

Proof. (i) Let Y be an infinite dimensional closed subspace of E. It is sufficient to prove that, for any finite dimensional subspace X of E,

$$\Theta(X, Y) > \frac{1}{2}. \tag{6.1.37}$$

So let a finite dimensional subspace X of E be given, and let $\varepsilon > 0$. Then $S(X)$ is compact. Therefore we can find a finite number of vectors $x_1, \ldots, x_N \in S(X)$ with

$$\min_{1 \le j \le N} \left\|x - x_j\right\| < \varepsilon \qquad \text{for all } x \in S(X). \tag{6.1.38}$$

Further, by definition of $\Theta(X, Y)$, we can find $y_1, \ldots, y_N \in S(Y)$ with

$$\left\|y_j - x_j\right\| < \Theta(X, Y) + \varepsilon \qquad \text{for } 1 \le j \le N. \tag{6.1.39}$$

Let F be the span of $y_1, \ldots, y_N$. Since Y is of infinite dimension, this is a proper subspace of Y. Therefore we can find $y_0 \in S(Y)$ with

$$\operatorname{dist}\left(y_0, F\right) > 1 - \varepsilon.$$

In particular,

$$\|y_0 - y_j\| > 1 - \varepsilon \qquad \text{for } 1 \le j \le N. \tag{6.1.40}$$

Since, again by definition of $\Theta(X,Y)$, $\operatorname{dist}\big(y_0, S(X)\big) \le \Theta(X,Y)$, we can find $x_0 \in S(X)$ with

$$\|y_0 - x_0\| < \operatorname{dist}\big(y_0, s(X)\big) + \varepsilon \le \Theta(X,Y) + \varepsilon. \tag{6.1.41}$$

Further, by (6.1.38), we can find an index $1 \le j_0 \le N$ such that

$$\|x_0 - x_{j_0}\| < \varepsilon. \tag{6.1.42}$$

From (6.1.39)–(6.1.42) now it follows that

$$\begin{aligned}\Theta(X,Y) + \varepsilon \ge \|y_0 - x_0\| &\ge \|y_0 - y_{j_0}\| - \|y_{j_0} - x_{j_0}\| - \|x_{j_0} - x_0\| \\ &> 1 - \varepsilon - \Theta(X,Y) - \varepsilon - \varepsilon = 1 - 3\varepsilon - \Theta(X,Y),\end{aligned}$$

i.e.,

$$\Theta(X,Y) > \frac{1}{2} - 2\varepsilon.$$

Since $\varepsilon > 0$ is arbitrary, this proves (6.1.37).

(ii) Let $k \in \mathbb{N}$ and let X be a k-dimensional subspace of E. Since X is of finite dimension, there is a continuous linear projector P from E onto X. Then k is the codimension of $\operatorname{Ker} P$ in E. Now let X' be an arbitrary closed subspace of E with

$$\Theta(X,X') < \frac{1}{8\|P\|}.$$

Then it follows from Proposition 6.1.5 (with $Y = Y' = \operatorname{Ker} P$) that E is the direct sum of X' and $\operatorname{Ker} P$. Therefore $\dim X'$ is the codimension of $\operatorname{Ker} P$ in E, i.e., $\dim X' = k$.

(iii) Let $k \in \mathbb{N}$ and let X be a closed subspace of E which is of codimension k in E. Since X is closed and of finite codimension in E, there is a continuous linear projector P from E onto X. Then $\dim \operatorname{Ker} P = k$. Now let X' be an arbitrary closed subspace of E with

$$\Theta(X,X') < \frac{1}{8\|P\|}.$$

Then it follows from Proposition 6.1.5 (with $Y = Y' = \operatorname{Ker} P$) that E is the direct sum of X' and $\operatorname{Ker} P$. Therefore, the codimension of X' in E is equal to $\dim \operatorname{Ker} P = k$. □

6.1.7 Proposition. *Let E be a Banach space, let X, Y be closed subspaces of E, and let $v \in E$. Then*

$$\Big|\operatorname{dist}(v,Y) - \operatorname{dist}(v,X)\Big| \le 2\Theta(X,Y)\|v\|. \tag{6.1.43}$$

Proof. It is sufficient to prove that

$$\operatorname{dist}(v,Y) \le \operatorname{dist}(v,X) + 2\Theta(X,Y)\|v\| \tag{6.1.44}$$

and then to change the roles of X and Y. Let $\varepsilon > 0$ and $x \in X$ with

$$\|v - x\| < \operatorname{dist}(v,X) + \varepsilon. \tag{6.1.45}$$

Since $\operatorname{dist}(v,X) \le \|v\|$, then

$$\|x\| \le 2\|v\| + \varepsilon. \tag{6.1.46}$$

Further, by definition of $\Theta(X,Y)$, there exists $y \in Y$ such that

$$\|x - y\| < \Theta(X,Y)\|x\| + \varepsilon,$$

which yields, by (6.1.46),

$$\|x - y\| < \Theta(X,Y)(2\|v\| + \varepsilon) + \varepsilon.$$

Together with (6.1.45) this implies that

$$\operatorname{dist}(v,Y) \le \|v - y\| \le \operatorname{dist}(v,X) + \varepsilon + \Theta(X,Y)(2\|v\| + \varepsilon) + \varepsilon.$$

As ε can be chosen arbitrarily small, this proves (6.1.44). □

6.1.8 Proposition. *Let E be a Banach space, let $(X_n)_{n\in\mathbb{N}}$ be a sequence of subspaces of E (possibly not closed), and let $(x_n)_{n\in\mathbb{N}}$ be a sequence of vectors $x_n \in X_n$ which converges to some vector $y \in E$. If there exists a* closed *subspace Y of E such that*

$$\lim_{n\to\infty} \Theta(X_n, Y) = 0,$$

then $y \in Y$.

Proof. We may assume that $x_n \in S(X_n)$ for all $n \in \mathbb{N}$. Then, by definition of Θ, we can find a sequence $y_n \in S(Y)$, $n \in \mathbb{N}$, such that

$$\|x_n - y_n\| \le 2\Theta(X_n, Y), \qquad n \in \mathbb{N}.$$

Since $\lim\limits_{n\to\infty} \Theta(X_n,Y) = 0$ and $\lim \|x_n - y\| = 0$, then it follows that also the sequence (y_n) converges to Y. As all y_n belong to Y and Y is closed, this implies that $y \in Y$. □

6.2 Kernel and image of operator functions

6.2.1 Definition. Let E be a Banach space, let $D \subseteq \mathbb{C}$ (possibly not open), and let $M = \{M(z)\}_{z\in D}$ be a family of subspaces of E. M will be called **continuous** if, for each $z \in D$, the space $M(z)$ is closed in E, and if the map $D \ni z \to M(z)$ is continuous with respect to the gap matric (cf. Theorem 6.1.2). By a **section** of M over D we mean a vector function $f : D \to E$ such that $f(z) \in M(z)$ for all $z \in D$. Such a section will be called **continuous** if it is continuous as a vector function with values in E.

Now let E, F be two Banach spaces, let $D \subseteq \mathbb{C}$, and let $A : D \to L(E, F)$ be an operator function. Then we set

$$\operatorname{Im} A = \big\{ \operatorname{Im} A(z) \big\}_{z \in D} \qquad \text{and} \qquad \operatorname{Ker} A = \big\{ \operatorname{Ker} A(z) \big\}_{z \in D}.$$

The family $\operatorname{Im} A$ will be called the **image** of A, and $\operatorname{Ker} A$ will be called the **kernel** of A.

6.2.2. Let E, F be two Banach spaces, let $D \subseteq \mathbb{C}$, and let $A : D \to L(E, F)$ be an operator function. Suppose A is continuous and, moreover, $\operatorname{Im} A(z)$ is closed for all $z \in D$. Then nevertheless it is possible that the families $\operatorname{Im} A$ and $\operatorname{Ker} A$ are not continuous.

For example, assume that $D \subseteq \mathbb{C}$ is connected and $A : D \to L(\mathbb{C}^n, \mathbb{C}^m)$ is a continuous matrix function, $n, m \in \mathbb{N}$. If the rank of $A(z)$ is not the same for all $z \in D$, i.e., if the functions

$$D \ni z \longrightarrow \dim \operatorname{Im} A(z) \qquad \text{and} \qquad D \ni z \longrightarrow \dim \operatorname{Ker} A(z)$$

are not constant, then neither $\operatorname{Im} A$ nor $\operatorname{Ker} A$ is continuous. This follows from Proposition 6.1.6. Hence the constancy of the rank of A is a necessary condition for the continuity of $\operatorname{Im} A$ and $\operatorname{Ker} A$. This condition is also sufficient. We prove this in the more general setting of Theorem 6.2.8 below.

6.2.3 Definition. Let E, F be two Banach spaces, and let $A \in L(E, F)$. Then we define $k_A = 0$ if $A = 0$ and

$$k_A = \inf_{v \in E, \operatorname{dist}(v, \operatorname{Ker} A) = 1} \|Av\| \qquad \text{if } A \neq 0. \tag{6.2.1}$$

Recall the following fact which follows easily from the Banach open mapping theorem: $k_A > 0$ if and only if $A \neq 0$ and $\operatorname{Im} A$ is closed. If this is the case, then the operator $A_0 : E/\operatorname{Ker} A \to \operatorname{Im} A$ induced by A is an invertible operator between the factor space $E/\operatorname{Ker} A$ and $\operatorname{Im} A$ and

$$k_A = \frac{1}{\|A_0^{-1}\|}. \tag{6.2.2}$$

If E, F are two Banach spaces, $D \subseteq \mathbb{C}$ and $A : D \to L(E, F)$ is an operator function, then we denote by k_A the function defined by $k_A(z) = k_{A(z)}$, $z \in D$.

6.2.4 Lemma. *Let E, F be Banach spaces, let $A, B \in L(E,F)$, and let $k_B > 0$. Then*

$$\sup_{v \in S(\operatorname{Ker} A)} \operatorname{dist}\big(v, S(\operatorname{Ker} B)\big) \le 2\frac{\|A-B\|}{k_B}\,, \tag{6.2.3}$$

$$\sup_{v \in S(\operatorname{Im} B)} \operatorname{dist}\big(v, S(\operatorname{Im} A)\big) \le 2\frac{\|A-B\|}{k_B}\,, \tag{6.2.4}$$

$$k_A \ge \left(1 - 2\left(\frac{1}{k_B}+1\right)\|A-B\| - 4\Theta\big(\operatorname{Im} A, \operatorname{Im} B\big)\right) k_B\,, \tag{6.2.5}$$

$$k_A \ge \Big(1 - 2\Theta(\operatorname{Ker} A, \operatorname{Ker} B)\Big) k_B - \|A-B\|\,. \tag{6.2.6}$$

Proof of (6.2.3). Assume that

$$\sup_{v \in S(\operatorname{Ker} A)} \operatorname{dist}\big(v, S(\operatorname{Ker} B)\big) > 2\frac{\|A-B\|}{k_B}.$$

Then it follows from (6.1.11) in proposition (6.1.4) that

$$\sup_{v \in S(\operatorname{Ker} A)} \operatorname{dist}\big(v, \operatorname{Ker} B\big) > \frac{\|A-B\|}{k_B}.$$

Hence, we can find $v \in S(\operatorname{Ker} A)$ with

$$\operatorname{dist}(v, \operatorname{Ker} B) > \frac{\|A-B\|}{k_B}. \tag{6.2.7}$$

Since $\|v\| = 1$ and $Av = 0$, then

$$\|Bv\| = \|Bv - Av\| \le \|B - A\|.$$

On the other hand, (6.2.7) implies, by definition of k_B, that $\|Bv\| > \|A-B\|$. □

Proof of (6.2.4). Assume that

$$\sup_{v \in S(\operatorname{Im} B)} \operatorname{dist}\big(v, S(\operatorname{Im} A)\big) > 2\frac{\|A-B\|}{k_B}.$$

Then, it follows from (6.1.11) in proposition (6.1.4) that

$$\sup_{v \in S(\operatorname{Im} B)} \operatorname{dist}\big(v, \operatorname{Im} A\big) > \frac{\|A-B\|}{k_B}.$$

Hence, we can find $v \in S(\operatorname{Im} B)$ with

$$\operatorname{dist}(v, \operatorname{Im} A) > \frac{\|A-B\|}{k_B}. \tag{6.2.8}$$

Let $\varepsilon > 0$. Since $\|v\| = 1$ and $v \in \operatorname{Im} B$, then, by definition of k_B, we can find $w \in E$ with

$$Bw = v \qquad \text{and} \qquad \|w\| \le \frac{1}{k_B} + \varepsilon.$$

Then

$$\|v - Aw\| = \|(B-A)w\| \le \|B-A\|\|w\| \le \frac{\|B-A\|}{k_B} + \varepsilon\|B-A\|.$$

As ε can be chosen arbitrarily small, this implies that

$$\|v - Aw\| \le \frac{\|B-A\|}{k_B}.$$

Hence

$$\operatorname{dist}(v, \operatorname{Im} A) \le \frac{\|B-A\|}{k_B},$$

which contradicts (6.2.8). □

Proof of (6.2.5). Set

$$q = 2\left(\frac{1}{k_B} + 1\right)\|A-B\| + 4\Theta(\operatorname{Im} A, \operatorname{Im} B).$$

We may assume that $q < 1$, because otherwise (6.2.5) is trivial. We have to prove that then $k_A > 0$ and

$$\frac{1}{k_A} \le \frac{1}{1-q} \cdot \frac{1}{k_B}.$$

By definition (6.2.1) of k_A, for that it is sufficient to find, for all $\varepsilon > 0$ and $y \in \operatorname{Im} A$, a vector $x \in E$ such that

$$Ax = y \qquad \text{and} \qquad \|x\| \le \left(\frac{1}{k_B} + \varepsilon\right) \cdot \frac{1}{1-q}\|y\|.$$

For the latter it is sufficient to construct, for all $y \in \operatorname{Im} A$ and $\varepsilon > 0$, a sequence $(x_n)_{n\in\mathbb{N}}$ in E such that, for all $n \in \mathbb{N}$,

$$\begin{cases} x_0 = 0, \\ \|x_n\| \le (1/k_B + \varepsilon)\, q^{n-1}\|y\| \qquad \text{if } n \ge 1 \end{cases} \tag{6.2.9}$$

and

$$\left\| y - \sum_{j=0}^{n} Ax_j \right\| \le q^n \|y\|. \tag{6.2.10}$$

Indeed, as $0 \le q < 1$, then

$$x := \sum_{n=0}^{\infty} x_n \in E$$

exists and we have

$$\|x\| \le \sum_{n=0}^{\infty} \|x_n\| \le \left(\frac{1}{k_B} + \varepsilon\right)(1+q)\|y\| \sum_{n=1}^{\infty} q^{n-1} = \left(\frac{1}{k_B} + \varepsilon\right)\frac{1}{1-q}\|y\|$$

and

$$\|y - Ax\| = \lim_{n\to\infty} \left\| y - \sum_{j=0}^{n} Ax_j \right\| \le \lim_{n\to\infty} q^n \|y\| = 0,$$

i.e., $Ax = y$.

So let $y \in \operatorname{Im} A$ and $\varepsilon > 0$ be given. Proceeding by induction, we set $x_0 = 0$. It is clear that then (6.2.9) and (6.2.10) are valid for $n = 0$. Now we assume that, for some $k \in \mathbb{N}$, we already have $x_0, \dots, x_k$ such that (6.2.9) and (6.2.10) are valid for $0 \le n \le k$.

Then $y - \sum_{j=0}^{k} Ax_j \in \operatorname{Im} A$. Therefore, by the definitions of $\Theta(\overline{\operatorname{Im} A}, \operatorname{Im} B)$ and k_B, we can find $x_{k+1} \in E$ such that

$$\|x_{k+1}\| \le \left(\frac{1}{k_B} + \varepsilon\right) \left\| y - \sum_{j=0}^{k} Ax_j \right\|$$

and

$$\left\| y - \sum_{j=0}^{k} Ax_j - Bx_{k+1} \right\| \le 2\Theta(\operatorname{Im} A, \operatorname{Im} B) \left\| y - \sum_{j=0}^{k} Ax_j \right\|.$$

Since, by induction hypothesis, (6.2.10) holds for $n = k$, the first inequality proves (6.2.9) for $n = k+1$, and the second inequality implies that

$$\left\| y - \sum_{j=0}^{k} Ax_j - Bx_{k+1} \right\| \le 2\Theta(\operatorname{Im} A, \operatorname{Im} B) q^k \|y\|.$$

As, by definition of q, $2\Theta(\operatorname{Im} A, \operatorname{Im} B) \le q/2$, this further implies that

$$\left\| y - \sum_{j=0}^{k} Ax_j - Bx_{k+1} \right\| \le \frac{1}{2} q^{k+1} \|y\|. \tag{6.2.11}$$

Since (6.2.9) is already proved for $n = k+1$, we get

$$\|Ax_{k+1} - Bx_{k+1}\| \le \|A - B\| \|x_{k+1}\| \le \|A - B\| \left(\frac{1}{k_B} + \varepsilon\right) q^k \|y\|.$$

As, again by definition of q,

$$\|A - B\| \left(\frac{1}{k_B} + \varepsilon\right) \le \frac{q}{2}$$

(we may assume that $\varepsilon \le 1$), this implies that

$$\big\|Ax_{k+1} - Bx_{k+1}\big\| \le \frac{1}{2} q^{k+1} \|y\|.$$

Together with (6.2.11) this proves (6.2.10) for $n = k + 1$. □

Proof of (6.2.6). Note that $\operatorname{Ker} B \neq E$, because of $k_B > 0$. We may assume that also $\operatorname{Ker} A \neq E$, because otherwise $\Theta(\operatorname{Ker} A, \operatorname{Ker} B) = \Theta(E, \operatorname{Ker} B) \ge 1$ (cf. Proposition 6.1.3) and (6.2.6) is trivial. Hence k_A is defined by

$$k_A = \inf_{v \in E\,,\,\operatorname{dist}(v, \operatorname{Ker} A) = 1} \big\|Av\big\|.$$

Therefore, we have to prove that, for each $v \in E$ with

$$\operatorname{dist}(v, \operatorname{Ker} A) = 1, \tag{6.2.12}$$

$$\|Av\| \ge \Big(1 - 2\Theta(\operatorname{Ker} A, \operatorname{Ker} B)\Big) k_B - \|A - B\|. \tag{6.2.13}$$

So let $v \in E$ with (6.2.12) be given. Moreover let $\varepsilon > 0$. Then, by (6.2.12), we can find $w \in \operatorname{Ker} A$ with

$$\|v - w\| < 1 + \varepsilon. \tag{6.2.14}$$

As $w \in \operatorname{Ker} A$, it follows from (6.2.12) that also

$$\operatorname{dist}(v - w, \operatorname{Ker} A) = 1. \tag{6.2.15}$$

By Proposition 6.1.7,

$$\Big|\operatorname{dist}(v - w, \operatorname{Ker} B) - \operatorname{dist}(v - w, \operatorname{Ker} A)\Big| \le 2\Theta(\operatorname{Ker} A, \operatorname{Ker} B)\|v - w\|.$$

In view of (6.2.15), this implies that

$$\operatorname{dist}(v - w, \operatorname{Ker} B) \ge 1 - 2\Theta\big(\operatorname{Ker} A, \operatorname{Ker} B\big)\big\|v - w\big\|$$

and further, by definition of k_B,

$$\big\|B(v - w)\big\| \ge \Big(1 - 2\Theta(\operatorname{Ker} A, \operatorname{Ker} B)\big\|v - w\big\|\Big) k_B.$$

Hence

$$\begin{aligned} \|Av\| = \|A(v - w)\| &\ge \|B(v - w)\| - \|A - B\|\|v - w\| \\ &\ge \Big(1 - 2\Theta(\operatorname{Ker} A, \operatorname{Ker} B)\big\|v - w\big\|\Big) k_B - \|A - B\|\|v - w\|. \end{aligned}$$

By (6.2.14) this implies that

$$\|Av\| \ge \Big(1 - 2\Theta(\operatorname{Ker} A, \operatorname{Ker} B)(1 + \varepsilon)\Big) k_B - \|A - B\|(1 + \varepsilon).$$

As ε can be chosen arbitrarily small, this proves (6.2.13). □

6.2.5 Corollary. *Let E, E', F be Banach spaces, and let $A_n \in L(E,F)$, $A'_n \in L(E,F)$, $n \in \mathbb{N}$, be two sequences such that*

$$\lim_{n\to\infty} \|A_n - A_0\| = \lim_{n\to\infty} \|A'_n - A'_0\| = 0\,, \tag{6.2.16}$$

$$\operatorname{Ker} A_n = \{0\}\,, \qquad n \in \mathbb{N}\,, \tag{6.2.17}$$

$$\operatorname{Im} A_n \supseteq \operatorname{Im} A'_n\,, \qquad n \in \mathbb{N}\,, \tag{6.2.18}$$

and the spaces $\operatorname{Im} A_n$, $n \in \mathbb{N}$, *are closed. Let* $A_n^{(-1)} \in L(\operatorname{Im} A_n, E)$, $n \in \mathbb{N}$, *be the operators with* $A_n^{(-1)} A_n = I$. *Then*

$$\lim_{n\to\infty} A_n^{(-1)} A'_n = A_0^{(-1)} A'_0\,. \tag{6.2.19}$$

Proof. Assume the contrary. Then, passing to subsequences we may assume that, for some $\varepsilon > 0$,

$$\left\| A_n^{(-1)} A'_n - A_0^{(-1)} A'_0 \right\| > \varepsilon\,.$$

Take a sequence of vectors $x_n \in E'$ with $\|x_n\| = 1$ and

$$\left\| (A_n^{(-1)} A'_n - A_0^{(-1)} A'_0) x_n \right\| > \varepsilon \quad \text{and} \quad \|x_n\| = 1\,, \qquad n \in \mathbb{N}^*\,. \tag{6.2.20}$$

Then

$$\begin{aligned} \|(A'_n - A'_0)\, x_n\| &= \left\| \Big(A_n (A_n^{(-1)} A'_n - A_0^{(-1)} A'_0) + (A_n - A_0) A_0^{(-1)} A'_0 \Big)\, x_n \right\| \\ &\geq \left\| A_n (A_n^{(-1)} A'_n - A_0^{(-1)} A'_0) x_n \right\| - \|A_n - A_0\| \left\| A_0^{(-1)} A'_0 \right\|. \end{aligned}$$

Since $\operatorname{Ker} A_n = \{0\}$, this implies that

$$\|(A'_n - A'_0)\, x_n\| \geq k_{A_n} \left\| (A_n^{(-1)} A'_n - A_0^{(-1)} A'_0) x_n \right\| - \|A_n - A_0\| \left\| A_0^{(-1)} A'_0 \right\|,$$

and further, by (6.2.20),

$$\|(A'_n - A'_0)\, x_n\| \geq k_{A_n} \varepsilon - \|A_n - A_0\| \left\| A_0^{(-1)} A'_0 \right\|, \qquad n \in \mathbb{N}^*. \tag{6.2.21}$$

Since $\operatorname{Ker} A_n = \{0\}$, $n \in \mathbb{N}$, it follows from estimate (6.1.45) in Lemma 6.2.4 that

$$k_{A_n} \geq k_{A_0} - \|A_n - A_0\|.$$

Together with (6.2.21) this implies that

$$\|(A'_n - A'_0)\, x_n\| \geq k_{A_0} \varepsilon - \|A_n - A_0\| \left(\varepsilon + \left\| A_0^{(-1)} A'_0 \right\| \right), \qquad n \in \mathbb{N}^*,$$

which is a contradiction to (6.2.16). □

6.2.6 Theorem. *Let E be a Banach space and let $\{M(z\}_{z\in D}$, $D \subseteq \mathbb{C}$, be a family of subspaces of E such that, for each point $z_0 \in D$, there exist a neighborhood $U \subseteq D$ of z_0, Banach spaces X, Y and continuous operator functions $T : U \to L(X, E)$, $S : U \to L(E, Y)$ such that, for all $z \in U$, $\operatorname{Im} S(z)$ is closed and*

$$M(z) = \operatorname{Im} T(z) = \operatorname{Ker} S(z).$$

Then $\{M(z\}_{z\in D}$ is continuous.

Proof. Let z_0 be given, and let U, X, Y, T, S be as in the hypothesis of the theorem. As $\operatorname{Im} S(z_0)$ is closed, and $M(z) = \operatorname{Ker} S(z)$ for all $z \in U$, then it follows from (6.2.3) in Lemma 6.2.4 that, for all $z \in U$,

$$\sup_{v\in S(M(z))} \operatorname{dist}\Big(v, S\big(M(z_0)\big)\Big) \leq \frac{\|S(z) - S(z_0)\|}{k_S(z_0)}.$$

Since also $\operatorname{Im} T(z_0)$ is closed (as $\operatorname{Im} T(z_0) = \operatorname{Ker} S(z_0)$) and $M(z) = \operatorname{Im} T(z)$ for all $z \in U$, from (6.2.4) in Lemma 6.2.4 we get

$$\sup_{v\in S(M(z_0))} \operatorname{dist}\Big(v, S\big(M(z)\big)\Big) \leq \frac{\|T(z) - T(z_0)\|}{k_T(z_0)}.$$

Together this implies that

$$\Theta\big(M(z), M(z_0)\big) \leq \max\left\{\frac{\|S(z) - S(z_0)\|}{k_S(z_0)}, \frac{\|T(z) - T(z_0)\|}{k_T(z_0)}\right\}.$$

As both T and S are continuous at z_0, this implies that $\{M(z\}_{z\in D}$ is continuous at z_0. □

6.2.7 Theorem. *Let E, F be two Banach spaces, let $D \subseteq \mathbb{C}$, and let $A : D \to L(E, F)$ be a continuous operator function such that, for all $z \in D$, $\operatorname{Im} A(z)$ is closed. Then the following three conditions are equivalent:*

(i) *The function k_A is continuous.*

(ii) *The family* $\operatorname{Im} A$ *is continuous.*

(iii) *The family* $\operatorname{Ker} A$ *is continuous.*

(iv) *For each compact set $K \subseteq D$,*

$$\inf_{z\in K} k_A(z) > 0.$$

Proof. (i)⇒(iv): Since, by hypothesis of the theorem, $\operatorname{Im} A(z)$ is closed for all $z \in D$,

$$k_A(z) > 0 \qquad \text{for all } z \in D.$$

Therefore, if k_A is continuous, then, for each compact set $K \subseteq D$, $\min_{z\in K} k_A(z)$ exists and is > 0.

(iv)⇒(ii) and (iv)⇒(iii): It follows from (6.2.3) in Lemma 6.2.4 that, for all $z, w \in D$,

$$\Theta\big(\operatorname{Ker} A(z), \operatorname{Ker} A(w)\big) \le 2\|A - B\| \max\left\{\frac{1}{k_A(z)}, \frac{1}{k_A(w)}\right\},$$

and from (6.2.4) in the same lemma it follows that, for all $z, w \in D$,

$$\Theta\big(\operatorname{Im} A(z), \operatorname{Im} A(w)\big) \le 2\|A - B\| \max\left\{\frac{1}{k_A(z)}, \frac{1}{k_A(w)}\right\}.$$

If condition (iv) is satisfied, these two inequalities imply that the families $\operatorname{Ker} A$ and $\operatorname{Im} A$ are continuous.

(ii)⇒(i): Set

$$q(z,w) = 2\left(\frac{1}{k_A(w)} + 1\right)\|A(z) - A(w)\| + 4\Theta\big(\operatorname{Im} A(z), \operatorname{Im} A(w)\big)$$

for $z, w \in D$. Fix $z_0 \in D$. Then, by (6.2.5) in Lemma 6.2.4,

$$k_A(z) \ge \Big(1 - q(z, z_0)\Big) k_A(z_0), \qquad z \in D, \tag{6.2.22}$$

and

$$k_A(z_0) \ge \Big(1 - q(z_0, z)\Big) k_A(z), \qquad z \in D. \tag{6.2.23}$$

Since A is continuous and condition (ii) is satisfied, we have

$$\lim_{z\to z_0} q(z, z_0) = 0. \tag{6.2.24}$$

By (6.2.22) this implies that, for some $\varepsilon > 0$,

$$k_A(z) \ge \frac{k_A(z_0)}{2} \qquad \text{if } z \in D \text{ and } |z - z_0| < \varepsilon.$$

Hence

$$q(z_0, z) \le 2q(z, z_0) \qquad \text{if } z \in D \text{ and } |z - z_0| < \varepsilon,$$

which implies, by (6.2.24), that also

$$\lim_{z\to z_0} q(z_0, z) = 0. \tag{6.2.25}$$

(6.2.22)–(6.2.25) together imply that $\lim_{z\to z_0} k_A(z) = k_A(z_0)$.

(iii)⇒(i): Let $z_0 \in D$. Then, by (6.2.6) in Lemma 6.2.4, for all $z \in D$,

$$k_A(z) \geq k_A(z_0) - 2\Theta\big(\operatorname{Ker} A(z), \operatorname{Ker} A(z_0)\big) k_A(z_0) - \|A(z) - A(z_0)\| \qquad (6.2.26)$$

and

$$k_A(z_0) \geq k_A(z) - 2\Theta\big(\operatorname{Ker} A(z), \operatorname{Ker} A(z_0)\big) k_A(z) - \|A(z) - A(z_0)\|. \qquad (6.2.27)$$

Since

$$\lim_{z \to z_0} \|A(z) - A(z_0)\| = 0 \quad \text{and} \quad \lim_{z \to z_0} \Theta\big(\operatorname{Ker} A(z), \operatorname{Ker} A(z_0)\big) = 0, \qquad (6.2.28)$$

it follows from (6.2.26) that, for some $\varepsilon > 0$,

$$k_A(z) \geq \frac{k_A(z_0)}{2} \quad \text{if } z \in D \text{ and } |z - z_0| < \varepsilon.$$

Together with (6.2.27) this implies that

$$k_A(z_0) \geq k_A(z) - 4\Theta\big(\operatorname{Ker} A(z), \operatorname{Ker} A(z_0)\big) k_A(z_0) - \|A(z) - A(z_0)\| \qquad (6.2.29)$$

if $z \in D$ and $|z - z_0| < \varepsilon$. Now, from (6.2.26) and (6.2.28) it follows that

$$\liminf_{z \to z_0} k_A(z) \geq k_A(z_0)$$

and from (6.2.29) and (6.2.28) it follows that

$$\limsup_{z \to z_0} k_A(z) \leq k_A(z_0).$$

Hence

$$\lim_{z \to z_0} k_A(z) = k_A(z_0).$$

□

6.2.8 Theorem. *Let E, F be two Banach spaces, let $D \subseteq \mathbb{C}$, and let $A : D \to L(E,F)$ be a continuous operator function such that, for some $k \in \mathbb{N}$, at least one of the following conditions is fulfilled:*

(i) *For all $z \in D$, $\dim \operatorname{Ker} A(z) = k$ and $\operatorname{Im} A(z)$ is closed.*

(ii) *For all $z \in D$, the codimension of $\operatorname{Im} A(z)$ in F is equal to k.*

Then both the family $\operatorname{Im} A$ and the family $\operatorname{Ker} A$ are continuous on D.

Proof. First assume that condition (i) is satisfied. Let $z_0 \in D$. Since $\operatorname{Ker} A(z_0)$ is of finite dimension, there is a closed subspace X of E such that E is the direct sum of $\operatorname{Ker} A(z_0)$ and X. Since also $\operatorname{Im} A(z_0)$ is closed, then the restriction $A(z_0)\big|_X$ is a bounded linear isomorphism between X and $\operatorname{Im} A(z_0)$. Hence

$$c := \inf_{v \in S(X)} \big\|A(z_0)v\big\| > 0. \qquad (6.2.30)$$

Since A is continuous, we can find an open disc U centered at z_0 such that

$$\|A(z) - A(z_0)\| < \frac{c}{2} \qquad \text{if } z \in U \cap D.$$

Then, by (6.2.30),

$$\inf_{v \in S(X)} \|A(z)v\| > \frac{c}{2} \qquad \text{if } z \in U \cap D. \tag{6.2.31}$$

In particular,

$$\operatorname{Ker} A(z) \cap X = \{0\} \qquad \text{if } z \in U \cap D.$$

Since $\dim \operatorname{Ker} A(z) = \dim \operatorname{Ker} A(z_0) = \operatorname{codim} X$, this implies that E is the direct sum of $\operatorname{Ker} A(z)$ and X if $z \in U \cap D$. Hence, for $z \in U \cap D$, $A(z)\big|_X$ is a bounded linear isomorphism from X onto $\operatorname{Im} A$. Denote by $B(z)$ the inverse of this isomorphism. Then, by (6.2.31),

$$\|B(z)\| < \frac{2}{c} \qquad \text{for all } z \in U \cap D.$$

This implies that, for each $z \in U \cap D$ and all $v \in S\big(\operatorname{Im} A(z)\big)$,

$$\big\|v - A(z_0)B(z)v\big\| = \big\|A(z)B(z)v - A(z_0)B(z)v\big\| \le \frac{2}{c}\big\|A(z) - A(z_0)\big\|.$$

Hence

$$\sup_{v \in S(\operatorname{Im} A(z))} \operatorname{dist}\big(v, \operatorname{Im} A(z_0)\big) \le \frac{2}{c}\|A(z) - A(z_0)\| \qquad \text{if } z \in U \cap D.$$

Since, by (6.1.4) in Proposition 6.1.4,

$$\sup_{v \in S(\operatorname{Im} A(z))} \operatorname{dist}\big(v, S(\operatorname{Im} A(z_0))\big) \le 2 \sup_{v \in S(\operatorname{Im} A(z))} \operatorname{dist}\big(v, \operatorname{Im} A(z_0)\big),$$

this implies that

$$\sup_{v \in S(\operatorname{Im} A(z))} \operatorname{dist}\big(v, S(\operatorname{Im} A(z_0))\big) \le \frac{4}{c}\|A(z) - A(z_0)\| \qquad \text{if } z \in U \cap D.$$

On the other hand, by (6.2.4) in Lemma 6.2.4,

$$\sup_{v \in S(\operatorname{Im} A(z_0))} \operatorname{dist}\big(v, S(\operatorname{Im} A(z))\big) \le \frac{2}{k_A(z_0)}\|A(z) - A(z_0)\| \qquad \text{if } z \in U \cap D.$$

Together this yields

$$\Theta\big(\operatorname{Im} A(z), \operatorname{Im} A(z_0)\big) \le \max\left\{\frac{2}{k_A(z_0)}, \frac{4}{c}\right\}\big\|A(z) - A(z_0)\big\| \quad \text{for all } z \in U \cap D.$$

Since A is continuous, this implies that $\operatorname{Im} A$ is continuous at z_0. By Theorem 6.2.7 this means that also $\operatorname{Ker} A$ is continuous at z_0.

Now we assume that condition (ii) is satisfied. Then, by the Banach open mapping theorem, $\operatorname{Im} A(z)$ is closed for all $z \in D$.

Now let $z_0 \in D$ be given. As $\operatorname{Im} A(z_0)$ is of the finite codimension k and closed in F, we can find a continuous linear projector P from F onto $\operatorname{Im} A(z_0)$. Consider the continuous operator function $PA : D \to L(E, \operatorname{Im} P)$. Then $\operatorname{Im} PA(z_0) = \operatorname{Im} P$. Hence, as PA is continuous, we can find an open disc U centered at z_0 such that

$$\operatorname{Im} PA(z) = \operatorname{Im} P \qquad \text{for all } z \in U \cap D. \tag{6.2.32}$$

By Theorem 6.2.7 this implies that the family $\{\operatorname{Ker} PA(z)\}_{z \in U \cap D}$ is continuous. To complete the proof, therefore it is sufficient to prove that

$$\operatorname{Ker} A(z) = \operatorname{Ker} PA(z) \qquad \text{for all } z \in U \cap D.$$

To do this, we assume that, for some $z \in U \cap D$, $\operatorname{Ker} A(z) \neq \operatorname{Ker} PA(z)$. Since, clearly, $\operatorname{Ker} A(z) \subseteq \operatorname{Ker} PA(z)$, then

$$M := \operatorname{Ker} P \cap \operatorname{Im} A(z) \neq \{0\}.$$

Let $m = \dim M$. As $\dim \operatorname{Ker} P = k$ and $M \neq \{0\}$, then $1 \leq m \leq k < \infty$. Therefore ($\operatorname{Im} A(z)$ is closed) we can find a closed subspace X of $\operatorname{Im} A(z)$ such that we have the direct sum

$$\operatorname{Im} A(z) = X \dot{+} M. \tag{6.2.33}$$

Since X has codimension m in $\operatorname{Im} A(z)$ and $\operatorname{Im} A(z)$ has codimension k in F, the codimension of X in F is $k + m$. By (6.2.33) and (6.2.32), the sum $X + \operatorname{Ker} P$ is a direct sum $X \dot{+} \operatorname{Ker} P$, where, as $\dim \operatorname{Ker} P = k$, the codimension of $X \dot{+} \operatorname{Ker} P$ in F is m. Hence, there is an m-dimensional subspace V of F such that

$$F = X \dot{+} \operatorname{Ker} P \dot{+} V. \tag{6.2.34}$$

Since $M \subseteq \operatorname{Ker} P$, from (6.2.33) it follows that

$$PX = \operatorname{Im} PA(z)$$

and further, by (6.2.32),

$$PX = \operatorname{Im} P. \tag{6.2.35}$$

Since $V \cap \operatorname{Ker} P = \{0\}$ (because of (6.2.34)) and $V \neq \{0\}$ (because of $\dim V = m \geq 1$), now we can find $v \in V$ with $Pv \neq 0$. Moreover, by (6.2.35), there exists $x \in X$ with

$$Px = Pv. \tag{6.2.36}$$

Since $v \in V$, $x \in X$ and $v \neq 0$, it follows from (6.2.34) that $v - x \notin \operatorname{Ker} P$, which contradicts (6.2.36). □

6.3 Holomorphic sections of continuous families of subspaces

6.3.1 Definition. Let E be a Banach space, let $D \subseteq \mathbb{C}$ be an open set, and let $\{M(z)\}_{z\in D}$ be a continuous family of subspaces of E. A section $f : D \to M$ will be called **holomorphic** if it is holomorphic as an E-valued map. If $U \subset D$ is open and $\neq \emptyset$, then the space of all holomorphic sections of M over U will be denoted by $\mathcal{O}^M(U)$. Note that this is a Fréchet space with respect to uniform convergence on the compact subsets of U, as each of the spaces $M(z)$, $z \in D$, is closed (by Definition 6.2.1 of a *continuous* family of subspaces). Sometimes a Banach space will be more convenient. Therefore we also introduce the Banach space $\mathcal{O}^M_\infty(U)$ of all *bounded* sections from $\mathcal{O}^M(U)$ endowed with the norm

$$\|f\|_\infty := \sup_{z\in U} \|f(z)\|_E\,, \qquad f \in \mathcal{O}^M_\infty(U).$$

For practical reasons, we define also $\mathcal{O}^M(\emptyset) = 0$ where 0 is the zero vector of E.

6.3.2 Definition. Let E be a Banach space, let $D \subseteq \mathbb{C}$ be an open set, and let $M = \{M(z)\}_{z\in D}$ be a continuous family of subspaces of E. Then we define

$$\Phi(z)(f) = f(z) \qquad \text{for all } z \in D \text{ and } f \in \mathcal{O}^M_\infty(D)$$

and

$$\Big(\Psi(z)f\Big)(\zeta) = (\zeta - z)f(\zeta) \qquad \text{for all } z, \zeta \in D \text{ and } f \in \mathcal{O}^M_\infty(D).$$

6.3.3 Proposition. *Let E, D, M, Φ and Ψ be as in Definition* 6.3.2, *and assume additionally that D is bounded. Then:*

(i) *For each $z \in D$, $\Phi(z)$ is a bounded linear operator from $\mathcal{O}^M_\infty(D)$ to E, and the operator function*

$$\Phi : D \longrightarrow L\big(\mathcal{O}^M_\infty(D), E\big)$$

defined in this way, is holomorphic.

(ii) *For each $z \in D$, $\Psi(z)$ is a bounded linear operator in $\mathcal{O}^M_\infty(D)$, and the operator function*

$$\Psi : D \longrightarrow L\big(\mathcal{O}^M_\infty(D), \mathcal{O}^M_\infty(D)\big)$$

defined in this way, is linear (and hence holomorphic).

(iii) $\operatorname{Im}\Psi(z) = \operatorname{Ker}\Phi(z)$ *for all $z \in D$.*

(iv) $\operatorname{Ker}\Psi(z) = \{0\}$ *for all $z \in D$.*

Proof. (i) It is clear that $\Phi(z)$ is linear for each fixed $z \in D$, and it follows from the maximum principle that $\Phi(z)$ is bounded for each fixed $z \in D$. As, for each fixed f in $\mathcal{O}^M_\infty(D)$, the function $D \ni z \to \Phi(z)f = f(z)$ is holomorphic, it follows that Φ is holomorphic (Theorem 1.7.1).

(ii) Since D is bounded and each of the spaces $M(\zeta)$, $\zeta \in D$, is linear, it is clear that, for each $z \in D$, $\Psi(z)$ is a bounded linear operator in $\mathcal{O}^M_\infty(D)$. For the same reason, setting

$$(Af)(\zeta) = \zeta f(\zeta) \qquad \text{for } \zeta \in D,$$

we get a bounded linear operator A in $\mathcal{O}^M_\infty(D)$. As $\Psi(z) = A - zI$, this shows that Ψ is linear.

(iii) For each $f \in \mathcal{O}^M_\infty(D)$ we have

$$\Big(\Phi(z)\big(\Psi(z)f\big)\Big)(\zeta) = (\zeta - z)f(\zeta)\Big|_{\zeta=z} = 0 \qquad \text{for all } \zeta, z \in D.$$

Hence

$$\operatorname{Im}\Psi(z) \subseteq \operatorname{Ker}\Phi(z) \qquad \text{for all } z \in D.$$

It remains to prove that

$$\operatorname{Im}\Psi(z) \supseteq \operatorname{Ker}\Phi(z) \qquad \text{for all } z \in D.$$

For that, let $z \in D$ and $f \in \operatorname{Ker}\Phi(z)$ be given. Then $f(z) = 0$ and therefore

$$u(\zeta) := \frac{f(\zeta)}{\zeta - z}, \qquad \zeta \in D,$$

is a holomorphic function on D. As f and D are bounded, also u is bounded. Moreover, since $f(\zeta) \in M(\zeta)$ and each of the spaces $M(\zeta)$ is a linear subspace of E, it is clear that

$$u(\zeta) \in M(\zeta) \qquad \text{if } \zeta \in D \setminus \{z\}.$$

Since M is a *continuous* family of subspaces, this implies (cf. Proposition 6.1.8) that also $u(z) \in M(z)$. Hence $u \in \mathcal{O}^M_\infty(D)$. Clearly, $\Psi(z)u = f$. This proves that $f \in \operatorname{Im}\Psi(z)$.

(iv) Let $z \in D$ and $f \in \mathcal{O}^M_\infty(D)$ be given. Now we assume that $\Psi(z)f = 0$.

$$(\zeta - z)f(\zeta) = 0 \qquad \text{for all } \zeta \in D.$$

Then, for $\zeta \neq 0$, we can divide by $\zeta - z$ and obtain that

$$f(\zeta) = 0 \qquad \text{for all } \zeta \in D \setminus \{z\}.$$

As z is an inner point of D and f is continuous, this implies that $f \equiv 0$ on D. □

6.4 Holomorphic families of subspaces

6.4.1 Definition. Let E be a Banach space, let $D \subseteq \mathbb{C}$ be an open set, and let $M = \{M(z\}_{z\in D}$ be a family of subspaces of E. The family M will be called **holomorphic** if

(i) M is continuous (Def. 6.2.1);

(ii) for each $z_0 \in D$, there exist a neighborhood $U \subseteq D$ of z_0, a Banach space X and a holomorphic operator function $A : U \to L(X, E)$ such that $M(z) = \operatorname{Im} A(z)$ for all $z \in U$.

The function A then will be called a **resolution** of M over U. If $U \neq D$, then we also speak about a **local** resolution of M. By a **global** resolution of M we mean a resolution of M over D.

Note that, by this definition, Theorem 6.2.8 immediately implies the following corollary:

6.4.2 Corollary. *Let E, F be two Banach spaces, let $D \subseteq \mathbb{C}$, and let $A : D \to L(E, F)$ be a holomorphic operator function such that, for some $k \in \mathbb{N}$, at least one of the following conditions is fulfilled:*

(i) *For all $z \in D$, $\dim \operatorname{Ker} A(z) = k$ and $\operatorname{Im} A(z)$ is closed.*

(ii) *For all $z \in D$, the codimension of $\operatorname{Im} A(z)$ in F is equal to k.*

Then both the family $\operatorname{Im} A$ and and the family $\operatorname{Ker} A$ are holomorphic on D.

In this section we again study additive Cousin problems (Def. 2.4.1), but with the additional property that the functions which form the Cousin problem are sections of a holomorphic family of subspaces. Here is the definition:

6.4.3 Definition. Let E be a Banach space, let $M = \{M(z\}_{z\in D}$ be a holomorphic family of subspaces of E, and let $\mathcal{U} = \{U_j\}_{j\in I}$ be an open covering of D.

Denote by $C^1(\mathcal{U}, \mathcal{O}^M)$ the space of families $f = \{f_{jk}\}_{j,k\in I}$ with $f_{jk} \in \mathcal{O}^M(U_j \cap U_k)$ (Def. 6.3.1). A family $f = \{f_{jk}\}_{j,k\in I} \in C^1(\mathcal{U}, \mathcal{O}^M)$ will be called an $(\mathcal{U}, \mathcal{O}^M)$**-cocycle** if, for all $j, k, l \in I$ with $U_j \cap U_k \cap U_l \neq \emptyset$,

$$f_{jk} + f_{kl} = f_{jl} \qquad \text{on } U_j \cap U_k \cap U_l. \tag{6.4.1}$$

Note that then, in particular,

$$f_{jk} = -f_{kj} \qquad \text{on } U_j \cap U_k \qquad \text{and} \qquad f_{jj} = 0 \qquad \text{on } U_j. \tag{6.4.2}$$

The space of all $(\mathcal{U}, \mathcal{O}^M)$-cocycles will be denoted by $Z^1(\mathcal{U}, \mathcal{O}^M)$. If the covering $\mathcal{U}$ is not specified, then we speak also about $\mathcal{O}^M$**-cocycles** over D.

Due to P. Cousin the elements of $Z^1(\mathcal{U}, \mathcal{O}^M)$ are also called **additive Cousin problems**.

In this section we prove the following two theorems:

6.4.4 Theorem. *Let E, F be Banach spaces, and let $D \subseteq \mathbb{C}$ be an open set. Suppose $A : D \to L(E, F)$ is holomorphic such that $\operatorname{Im} A = \{\operatorname{Im} A(z)\}_{z\in D}$ is a continuous family of subspaces of F. (This means, by our Definition 6.4.1, we suppose that $A : D \to L(E, F)$ is holomorphic such that $\operatorname{Im} A = \{\operatorname{Im} A(z)\}_{z\in D}$ is a holomorphic family of subspaces of F.) Then:*

(i) $\operatorname{Ker} A$ *is a holomorphic family of subspaces of* E.

(ii) *For each holomorphic section* $f : D \to \operatorname{Im} A$, *there exists a holomorphic function* $u : D \to E$ *such that*

$$Au = f \qquad \text{on } D. \tag{6.4.3}$$

6.4.5 Theorem. *Let* F *be a Banach space, let* $D \subseteq \mathbb{C}$ *be an open set, and let* $M = \{M(z)\}_{z\in D}$ *be a holomorphic family of subspaces of* F. *Then:*

(i) *Let* $K \subseteq D$ *be a compact set such that each connected component of* $D \setminus K$ *contains at least one point of* $\mathbb{C} \setminus D$. *Then any section of* M *defined and holomorphic in a neighborhood of* K *can be approximated uniformly on* K *by sections from* $\mathcal{O}^M(D)$.

(ii) *For each open covering* $\mathcal{U} = \{U_j\}_{j\in I}$ *of* D *and each* $(\mathcal{U}, \mathcal{O}^M)$*-cocycle* $\{f_{jk}\}_{j,k\in I}$, *there exists a family* $\{f_j\}_{j\in I}$ *of sections* $f_j \in \mathcal{O}^M(U_j)$ *such that*

$$f_{jk} = f_j - f_k \qquad \text{on } U_j \cap U_k, \quad j,k \in I. \tag{6.4.4}$$

The rest of this section is devoted to the proof of theorems 6.4.4 and 6.4.5. We begin with the following local version of Theorem 6.4.4 (ii):

6.4.6 Lemma. *Let* E, F *be Banach spaces, and let* $D \subseteq \mathbb{C}$ *be an open set. Suppose* $A : D \to L(E,F)$ *is holomorphic such that* $\operatorname{Im} A = \{\operatorname{Im} A(z)\}_{z\in D}$ *is a continuous family of subspaces of* F, *and let* $z_0 \in D$. *(This means, by our Definition* 6.4.1, *we suppose that* $A : D \to L(E,F)$ *is holomorphic such that* $\operatorname{Im} A = \{\operatorname{Im} A(z)\}_{z\in D}$ *is a holomorphic family of subspaces of* F.*) Then there exists a neighborhood* $U \subseteq D$ *of* z_0 *such that, for each holomorphic section* $f : D \to \operatorname{Im} A$, *there exists a holomorphic function* $u : U \to E$ *with*

$$Au = f \qquad \text{on } U. \text{ [2]}$$

Proof. Take $0 < r < 1$ so small that $\overline{K}_r(z_0) \subseteq D$. Multiplying by a constant, we may assume that

$$\max_{|z-z_0|\le r} \|A(z)\| < 1. \tag{6.4.5}$$

Let

$$A(z) = \sum_{n=0}^{\infty} A_n (z-z_0)^n$$

be the Taylor expansion of A at z_0. Then, by Cauchy's inequality[3] and (6.4.5),

$$\|A_n\| \le \frac{1}{r^n}, \qquad n \in \mathbb{N}. \tag{6.4.6}$$

[2] We point out that U depends on A, but *not* on f.

[3] which follows, by means of the Hahn-Banach theorem, immediately from Cauchy's inequality in the scalar case.

Note that $\operatorname{Im} A(z_0) = M(z_0)$ is closed and therefore $k_{A(z_0)} > 0$ (Def. 6.2.3). Set

$$C = \frac{8}{\min\{k_{A(z_0)}, 1\}}$$

and let U be the open disc with radius r/C centered at z_0. Now let a holomorphic section $f : D \to \operatorname{Im} A$ be given. Let

$$f(z) = \sum_{n=0}^{\infty} f_n (z - z_0)^n$$

be the Taylor expansion of f at z_0, and

$$K := \max_{|z-z_0| \leq r} \|f(z)\|.$$

Then, by Cauchy's inequality,

$$\|f_n\| \leq \frac{K}{r^n}, \qquad n \in \mathbb{N}. \tag{6.4.7}$$

Now it is sufficient to construct a sequence $(u_n)_{n\in\mathbb{N}}$ of vectors $u_n \in E$ such that, for all $n \in \mathbb{N}$,

$$\sum_{k=0}^{n} A_{n-k} u_k = f_n \tag{6.4.8}$$

and

$$\|u_n\| \leq K \frac{C^{n+1}}{r^n}. \tag{6.4.9}$$

Indeed, then, by (6.4.9),

$$u(z) = \sum_{k=0}^{\infty} u_k (z - z_0)^k, \qquad z \in U,$$

is a well-defined holomorphic vector function $u : U \to E$, and if

$$A(z)u(z) = \sum_{n=0}^{\infty} b_n (z - z_0)^n$$

is the Taylor expansion of Au at z_0, then from (6.4.8) it follows that

$$\begin{aligned} b_n = \frac{(Au)^{(n)}(z_0)}{n!} &= \frac{1}{n!} \sum_{k=0}^{n} \binom{n}{k} A^{(n-k)}(z_0) u^{(k)}(z_0) \\ &= \sum_{k=0}^{n} \frac{A^{(n-k)}(z_0)}{(n-k)!} \frac{u^{(k)}(z_0)}{k!} = \sum_{k=0}^{n} A_{n-k} u_k = f_n \qquad \text{for all } n \in \mathbb{N}. \end{aligned}$$

Hence $Au = f$ on U.

To construct the sequence $(u_n)_{n\in\mathbb{N}}$ we proceed by induction.

Beginning of induction: Since, by hypothesis, $k_{A(z_0)} > 0$ and $f_0 = f(z_0) \in \operatorname{Im} A(z_0)$, we can find $u_0 \in E$ such that

$$A_0 u_0 = A(z_0)u_0 = f_0 \qquad \text{and} \qquad \|u_0\| < \frac{2}{k_{A(z_0)}} < C.$$

Clearly, then (6.4.8) and (6.4.6) are valid for $n = 0$.

Hypothesis of induction: Assume that, for some $m \in \mathbb{N}$, vectors $u_0, \ldots, u_m \in E$ are already constructed such that (6.4.8) and (6.4.6) are valid for $0 \le n \le m$.

Step of induction: Set

$$g(z) = f(z) - A(z)\sum_{k=0}^{m} u_k(z - z_0)^k \qquad \text{for } z \in D.$$

Then g is holomorphic on D and $g(z) \in \operatorname{Im} A(z)$ for all $z \in D$. Let

$$g(z) = \sum_{n=0}^{\infty} g_n(z - z_0)^n$$

be the Taylor expansion of g at z_0. Then

$$g_n = f_n - \sum_{k=0}^{\min\{n,m\}} A_{n-k}u_k \qquad \text{for all } n \in \mathbb{N}. \tag{6.4.10}$$

By hypothesis of induction this implies that $g_n = 0$ if $0 \le n \le m$. Hence there is a holomorphic vector function $h : D \to F$ with

$$h(z) = \frac{g(z)}{(z - z_0)^{m+1}} \qquad \text{for } z \in D \setminus \{z_0\}.$$

As $g(z) \in \operatorname{Im} A(z)$ for all $z \in D$, then $h(z) \in \operatorname{Im} A(z)$ for all $z \in D \setminus \{z_0\}$. Since, by hypothesis, $\operatorname{Im} A$ is continuous, this implies that

$$g_{m+1} = h(z_0) \in \operatorname{Im} A(z_0) = \operatorname{Im} A_0$$

(cf. Proposition 6.1.8). Hence, by definition of $k_{A(z_0)}$, we can find a vector $u_{m+1} \in E$ such that

$$A_0 u_{m+1} = A(z_0)u_{m+1} = g_{m+1} \tag{6.4.11}$$

and

$$\|u_{m+1}\| \le \frac{2}{k_{A(z_0)}}\|g_{m+1}\| \le \frac{C}{4}\|g_{m+1}\|. \tag{6.4.12}$$

By (6.4.10), we have

$$g_{m+1} = f_{m+1} - \sum_{k=0}^{m} A_{m-k} u_k,$$

which implies, by (6.4.7) and (6.4.6), that

$$\|g_{m+1}\| \le \frac{F}{r^{m+1}} + \sum_{k=0}^{m} \frac{1}{r^{m-k}} \|u_k\|.$$

As, by induction hypothesis, (6.4.9) is valid for $0 \le n \le m$, from this we further get

$$\|g_{m+1}\| \le \frac{F}{r^{m+1}} + \sum_{k=0}^{m} \frac{1}{r^{m-k}} F \frac{C^{k+1}}{r^k} = \frac{F}{r^{m+1}} + \frac{F}{r^m} \sum_{k=0}^{m} C^{k+1}.$$

Since $r < 1$, this implies that

$$\|g_{m+1}\| \le \frac{F}{r^{m+1}} \left(1 + \sum_{k=0}^{m} C^{m+1} \right) = \frac{F}{r^{m+1}} \frac{C^{m+2} - 1}{C - 1} \le \frac{F}{C-1} \frac{C^{m+2}}{r^{m+1}}.$$

Together with (6.4.12) this implies that

$$\|u_{m+1}\| \le F \frac{C}{4(C-1)} \frac{C^{m+2}}{r^{m+1}}$$

and further, as $C \ge 8$,

$$\|u_{m+1}\| \le F \frac{C^{m+2}}{r^{m+1}},$$

i.e., (6.4.9) is valid also for $n = m + 1$. □

6.4.7. Proof of statement (i) in Theorem 6.4.4. Since, by hypothesis, $\operatorname{Im} A$ is a continuous family of subspaces of F, it follows from Theorem 6.2.7 that also $\operatorname{Ker} A$ is continuous. Therefore it remains to prove that, for each $z_0 \in D$, there exist a neighborhood U of z_0, a Banach space X and a holomorphic operator function $\Phi : U \to L(X, E)$ such that

$$\operatorname{Im} \Phi(z) = \operatorname{Ker} A(z) \qquad \text{for all } z \in U. \tag{6.4.13}$$

Let $z_0 \in D$ be given. Then, by Lemma 6.4.6, we can find a neighborhood U' of z_0 such that:

$$\begin{cases} \text{For each holomorphic section } f : D \to \operatorname{Im} A, \text{ there exists} \\ \text{a holomorphic function } u : U' \to X \text{ with } Au = f|_{U'}. \end{cases} \tag{6.4.14}$$

Let $U \subseteq U'$ be a second neighborhood of z_0 which is relatively compact in U'. Since $\operatorname{Ker} A$ is continuous, we have the Banach space

$$B := \mathcal{O}_\infty^{\operatorname{Ker} A}(U)$$

of bounded holomorphic sections of $\operatorname{Ker} A$ over U (cf. Definition 6.3.1). Further, let

$$\Phi : U \to L(B, X)$$

be the holomorphic operator function defined by

$$\Phi(z)f = f(z)\,, \qquad z \in U, \quad f \in B,$$

(cf. Proposition 6.3.3). It remains to prove that, with this choice of B and Φ, relation (6.4.13) is valid. Since the relation $\operatorname{Im} \Phi(z) \subseteq \operatorname{Ker} A(z)$ is trivial, we only have to prove that $\operatorname{Im} \Phi(z) \supseteq \operatorname{Ker} A(z)$ for all $z \in U$.

Let $z \in U$ and $v \in \operatorname{Ker} A(z)$ be given. We have to find $\varphi \in B$ with $\Phi(z)\varphi = v$. Setting

$$f(\zeta) = A(\zeta)v \qquad \text{for all } \zeta \in D,$$

we get a holomorphic vector function $f : D \to F$ with

$$f(z) = A(z)v = 0.$$

Hence, there is a well-defined holomorphic vector function $g : D \to F$ with

$$g(\zeta) = \frac{f(\zeta)}{\zeta - z} \qquad \text{for } \zeta \in D \setminus \{z\}.$$

Since, for each $\zeta \in D$, $f(\zeta) \in \operatorname{Im} A(\zeta)$ and $\operatorname{Im} A(\zeta)$ is a linear subspace of F, it is clear that

$$g(\zeta) \in \operatorname{Im} A(\zeta) \qquad \text{if } \zeta \in D \setminus \{z\}.$$

Since $\operatorname{Im} A$ is continuous, this implies (cf. Proposition 6.1.8) that also

$$g(z) \in \operatorname{Im} A(z).$$

Hence g is a holomorphic section of $\operatorname{Im} A$ over D. Now from (6.4.14) we get a holomorphic function $u : U' \to X$ with

$$A(\zeta)u(\zeta) = g(\zeta) \qquad \text{for all } \zeta \in U'. \tag{6.4.15}$$

Setting

$$\varphi(\zeta) = v - (\zeta - z)u(\zeta) \qquad \text{for } \zeta \in U,$$

we get a holomorphic function $\varphi : U \to X$ with $\varphi(z) = v$. As U is relatively compact in U', φ is bounded. Moreover, by definition of f and g,

$$\begin{aligned} A(\zeta)\varphi(\zeta) &= A(\zeta)v - (\zeta - z)A(\zeta)u(\zeta) \\ &= f(\zeta) - (\zeta - z)g(\zeta) = f(\zeta) - f(\zeta) = 0 \qquad \text{for all } \zeta \in U. \end{aligned}$$

Hence φ is a bounded holomorphic section of $\operatorname{Ker} A$ over U, i.e., $\varphi \in B$. Clearly $\Phi(z)\varphi = \varphi(z) = v$. □

6.4.8 Definition. Let F be a Banach space, let $D \subseteq \mathbb{C}$ an open set, and let $M = \{M(z)\}_{z\in D}$ be a holomorphic family of subspaces of F.

(i) A resolution $A : D \to L(E, F)$ of M will be called **injective** if $\operatorname{Ker} A(z) = \{0\}$ for all $z \in D$.

(ii) A resolution $A : D \to L(E, F)$ of M will be called **globally short** if $\operatorname{Ker} A$ [4] admits a global injective resolution.

Of course, any injective resolution is globally short.

6.4.9 Lemma. *Let E, F be Banach spaces, let $D \subseteq \mathbb{C}$ be an open set, and let $A : D \to L(E, F)$ be a function such that, for all $z \in D$, $\operatorname{Im} A(z)$ is closed and*

$$\operatorname{Ker} A(z) = \{0\}. \tag{6.4.16}$$

Further, let $f : D \to \operatorname{Im} A$ be a section of $\operatorname{Im} A$, and let $u : D \to E$ be the function with

$$A(z)u(z) = f(z) \qquad \text{for all} \qquad z \in D. \text{ [5]}$$

(i) Let $k \in \mathbb{N} \cup \{\infty\}$. Suppose A and f are of class $\mathcal{C}^k$ on D. Then u is of class $\mathcal{C}^k$ on D.

(ii) If A and f are holomorphic on D, then u is holomorphic on D.

Proof. (i) It is enough to prove the assertion for $k \in \mathbb{N}$. We do this by induction. For $k = 0$ the assertion follows from Corollary 6.2.5. Assume we have some $m \in \mathbb{N}$ such that the assertion is already proved for $k = m$, and let A and f be of class $\mathcal{C}^{m+1}$. Let x, y be the canonical real coordinates on $\mathbb{C}$, and let $\mu, \nu \in \mathbb{N}$ with $\mu + \nu = m + 1$. We have to prove that

$$\frac{\partial^{m+1} u}{\partial x^\mu \partial y^\nu}$$

exists and is continuous on D. At least one of the numbers μ and ν is ≥ 1. We may assume that $\mu \geq 1$. By hypothesis of induction, u is of class $\mathcal{C}^m$. Since $f = Au$ and f, A are of class $\mathcal{C}^{m+1}$, this implies that

$$\frac{\partial^m f}{\partial x^{\mu-1}\partial y^\nu} = \sum_{\kappa=0}^{\mu-1}\sum_{\lambda=0}^{\nu}\binom{\mu-1}{\kappa}\binom{\nu}{\lambda}\frac{\partial^{\kappa+\lambda} A}{\partial x^\kappa \partial y^\lambda}\,\frac{\partial^{\mu-1-\kappa+\nu-\lambda} u}{\partial x^{\mu-1-\kappa}\partial y^{\nu-\lambda}}$$

and therefore

$$A\frac{\partial^m u}{\partial x^{\mu-1}\partial y^\nu} = \frac{\partial^m f}{\partial x^{\mu-1}\partial y^\nu} - \sum_{\substack{0\le\kappa\le\mu-1,\,0\le\lambda\le\nu\\(\kappa,\lambda)\neq(0,0)}}\binom{\mu-1}{\kappa}\binom{\nu}{\lambda}\frac{\partial^{\kappa+\lambda} A}{\partial x^\kappa \partial y^\lambda}\,\frac{\partial^{\mu-1-\kappa-\nu-\lambda} u}{\partial x^{\mu-1-\kappa}\partial y^{\nu-\lambda}},$$

[4] By the just proved Theorem 6.4.4 (i), $\operatorname{Ker} A$ is holomorphic.

[5] By (6.4.16) this u exists and is uniquely determined by f.

where the right-hand side is of class $\mathcal{C}^1$. Hence the vector function

$$w := A\frac{\partial^m u}{\partial x^{\mu-1}\partial y^\nu}$$

is of class $\mathcal{C}^1$. In particular,

$$\frac{\partial w}{\partial x} \tag{6.4.17}$$

exists and is continuous on D. Set

$$v = \frac{\partial^m u}{\partial x^{\mu-1}\partial y^\nu}$$

and note that v is continuous on D (as u is of class $\mathcal{C}^m$). We have to prove that $\partial v/\partial x$ exists and is continuous on D.

Let z_0 be an arbitrary point in D, and let $(\Delta_n)_{n\in\mathbb{N}^*}$ be an arbitrary sequence of real numbers with $\Delta_n \neq 0$ and $\lim \Delta_n = 0$. We have to prove that

$$\lim_{n\to\infty}\frac{v(z_0+\Delta_n)-v(z_0)}{\Delta_n} \tag{6.4.18}$$

exists. Note that $w = Av$, by definition of v and w. Therefore

$$\begin{aligned}\frac{w(z_0+\Delta_n)-w(z_0)}{\Delta_n} &= \frac{A(z_0+\Delta_n)v(z_0+\Delta_n)-A(z_0)v(z_0)}{\Delta_n}\\ &= A(z_0+\Delta_n)\frac{v(z_0+\Delta_n)-v(z_0)}{\Delta_n} + \frac{A(z_0+\Delta_n)-A(z_0)}{\Delta_n}v(z_0).\end{aligned}$$

As $\partial w/\partial x$ and $\partial A/\partial x$ exist, this implies that

$$\lim_{n\to\infty} A(z_0+\Delta_n)\frac{v(z_0+\Delta_n)-v(z_0)}{\Delta_n} = \frac{\partial w}{\partial x}(z_0) - \frac{\partial A}{\partial x}(z_0)v(z_0). \tag{6.4.19}$$

By Corollary 6.2.5, this implies that (6.4.18) exists. Since both the point $z_0 \in D$ and the sequence $(\Delta_n)_{n\in\mathbb{N}}$ were chosen arbitrarily, this means that $\partial v/\partial x$ exists everywhere on D. Moreover, from (6.4.19) we get

$$A\frac{\partial v}{\partial x} = \frac{\partial w}{\partial x} - \frac{\partial A}{\partial x}v \qquad \text{on } D.$$

As the right-hand side of this relation is continuous, again from Corollary 6.2.5 it follows that $\partial v/\partial x$ is continuous.

(ii) If A and f are holomorphic, then from part (i) of the lemma we already know that u is of class $\mathcal{C}^\infty$. Since $f = Au$ and therefore, by the product rule,

$$\frac{\partial f}{\partial \bar z} = \frac{\partial A}{\partial \bar z}u + A\frac{\partial u}{\partial \bar z},$$

and since $\partial f/\partial\overline{z} = 0$ and $\partial A/\partial\overline{z} = 0$, this implies that

$$A\frac{\partial u}{\partial\overline{z}} = 0,$$

which means (as $\operatorname{Ker} A(z) = 0$ for all $z \in D$) that $\partial u/\partial\overline{z} = 0$. Hence u is holomorphic. □

6.4.10 Lemma. *Let F be a Banach space, let $D \subseteq \mathbb{C}$ be an open set, and let $M = \{M(z)\}_{z\in D}$ be a holomorphic family of subspaces of F, which admits a global injective resolution. Then the splitting statement* (ii) *of Theorem* 6.4.5 *is valid.*

Proof. Let an open covering $\mathcal{U} = \{U_j\}_{j\in I}$ of D and a $(\mathcal{U}, \mathcal{O}^M)$-cocycle $\{f_{jk}\}_{j,k\in I}$ be given. By hypothesis we can find a global injective resolution $A : D \to L(E, F)$ of M. As A is injective, then there is a uniquely determined family $\{u_{jk}\}_{j,k\in I}$ of vector functions $u_{jk} : U_j \cap U_k \to E$ such that

$$Au_{jk} = f_{jk} \qquad \text{on } U_j \cap U_k\,, \quad j, k \in I. \tag{6.4.20}$$

By Lemma 6.4.9, each u_{jk} is holomorphic. Moreover, since $\{f_{jk}\}_{j,k\in I}$ satisfies the cocycle condition, it follows from (6.4.20) and the injectivity of A that also $\{u_{jk}\}_{j,k\in I}$ satisfies the cocycle condition. Hence $\{u_{jk}\}_{j,k\in I}$ is a $(\mathcal{U}, \mathcal{O}^E)$-cocycle. Therefore, by Theorem 2.4.2, we can find a family $\{u_j\}_{j\in I}$ of holomorphic functions $u_j : U_j \to E$ with

$$u_{jk} = u_j - u_k \qquad \text{on } U_j \cap U_k\,, j, k \in I.$$

It remains to set $f_j = Au_j$. □

6.4.11 Lemma. *Let E, F be Banach spaces, let $D \subseteq \mathbb{C}$ an open set, let $M = \{M(z)\}_{z\in D}$ be a holomorphic family of subspaces of F, and let $A : D \to L(E, F)$ be a globally short resolution of M. Then the lifting statement* (ii) *of Theorem* 6.4.4 *is valid.*

Proof. Let $f : D \to M$ be a holomorphic section of M. As A is a resolution of M, it follows from Lemma 6.4.6 that, for each $z \in D$, there exist a neighborhood $U_z \subseteq D$ of z and a holomorphic function $u_z : U_z \to E$ with $Au_z = f$ on U_z. Set $f_{z,w} = u_z - u_w$ on $U_z \cap U_w$, $z, w \in D$. Recall that, by the already proved Theorem 6.4.4 (i) (see subSection 6.4.7), $\operatorname{Ker} A$ is a holomorphic family of subspaces of E. Since

$$Af_{z,w} = Au_z - Au_w = f - f = 0 \qquad \text{on } U_z \cap U_w,$$

$\{f_{z,w}\}_{z,w\in D}$ is a $\big(\{U_z\}_{z\in D}, \mathcal{O}^{\operatorname{Ker} A}\big)$-cocycle. Since $\operatorname{Ker} A$ admits a global injective resolution (as A is a globally short resolution), from Lemma 6.4.11 (ii) we get a family $\{f_z\}_{z\in D}$ of holomorphic sections $f_z : U_z \to \operatorname{Ker} A$ with

$$u_z - u_w = f_{z,w} = f_z - f_w \qquad \text{on } U_z \cap U_w\,, \quad z, w \in D.$$

It remains to set $u = f_z - u_z$ on U_z. □

6.4.12 Lemma. *Let F be a Banach space, let $D \subseteq \mathbb{C}$ be an open set, and let $M = \{M(z)\}_{z\in D}$ be a holomorphic family of subspaces of F, which admits a globally short resolution over D. Then the spitting statement* (ii) *of Theorem* 6.4.5 *is valid.*

Proof. [6] By hypothesis, we can find Banach spaces B, E and holomorphic operator functions $\Phi : D \to L(E,F)$, $\Psi : D \to L(B,E)$ such that Φ is a globally short resolution of M, and Ψ is a injective resolution of $\operatorname{Ker}\Phi$.

Let an open covering $\mathcal{U} = \{U_j\}_{j\in I}$ and a $(\mathcal{U}, \mathcal{O}^M)$-cocycle $f = \{f_{jk}\}_{j,k\in I}$ be given. It is sufficient to find a $(\mathcal{U}, \mathcal{O}^E)$-cocycle $\{g_{jk}\}_{jk\in I}$ such that

$$\Phi g_{jk} = f_{jk} \qquad \text{on } U_j \cap U_k\,, \quad j,k \in I. \tag{6.4.21}$$

Indeed, from Theorem 2.3.1 we then get a family $\{g_j\}_{j\in I}$ of holomorphic functions $g_j : U_j \to E$ with

$$g_{jk} = g_j - g_k \qquad \text{on } U_j \cap U_k\,, \quad j,k \in I,$$

and, setting $f_j = \Phi g_j$, we obtain a family $\{f_j\}_{j\in I}$ of holomorphic sections $f_j : U_j \to \operatorname{Im}\Phi = M$ with

$$f_j - f_k = \Phi g_j - \Phi g_k = \Phi(g_j - g_k) = \Phi g_{jk} = f_{jk} \qquad \text{on } U_j \cap U_k\,, \quad j,k \in I.$$

Now we are going to construct this cocycle. By the just proved Lemma 6.4.11, we can find a family $\{u_{jk}\}_{j,k\in I}$ of holomorphic vector functions $u_{jk} : U_j \cap U_k \to E$ such that

$$\Phi u_{jk} = f_{jk}. \tag{6.4.22}$$

Since $f_{jk} = -f_{kj}$, we may choose the u_{jk} in such a way that also

$$u_{jk} = -u_{kj} \qquad \text{for all } j,k \in I. \tag{6.4.23}$$

However, in general, $\{u_{jk}\}_{j,k\in I}$ is not yet a cocycle (except for the case $\operatorname{Ker}\Phi = 0)$), i.e.,, possibly, the family $\{u_{jkl}\}_{j,k,l\in I}$ of holomorphic vector functions $u_{jkl} : U_j \cap U_j \cap U_k \cap U_l \to E$ defined by

$$u_{jkl} = u_{jk} + u_{kl} + u_{lj}$$

contains non-zero elements. Take a $\mathcal{C}^\infty$-partition of unity $\{\chi_j\}_{j\in I}$ subordinated to $\mathcal{U}$, and set

$$c_{jk} = \sum_{\nu\in I} \chi_\nu u_{\nu jk} \qquad \text{on } U_{jk}, \quad j,k \in I.$$

[6] Note that, by the abstract Oka-Weil theorem in the theory of sheaves, this lemma follows immediately from Theorem 2.4.2. The proof given here is that what remains from the proof of the abstract Oka-Weil theorem in our special case.

(Observe that each $\chi_\nu u_{\nu jk}$ is well defined and of class $\mathcal{C}^\infty$ on U_{jk}, because $\chi_\nu \equiv 0$ outside a compact subset of U_ν.) In this way we get $\mathcal{C}^\infty$-functions $c_{jk} : U_{jk} \to E$. Since

$$\begin{aligned} c_{jk} + c_{kl} + c_{lj} &= \sum_{\nu\in I} \chi_\nu \Big(u_{\nu jk} + u_{\nu kl} + u_{\nu lj} \Big) \\ &= \sum_{\nu\in I} \chi_\nu \Big(u_{\nu j} + u_{jk} + u_{k\nu} + u_{\nu k} + u_{kl} + u_{l\nu} + u_{\nu l} + u_{lj} + u_{j\nu} \Big), \end{aligned}$$

and in view of (6.4.23), we see that

$$c_{jk} + c_{kl} + c_{lj} = \sum_{\nu\in I} \chi_\nu \Big(u_{jk} + u_{kl} + u_{lj} \Big) = \sum_{\nu\in I} \chi_\nu u_{jkl} = u_{jkl}. \tag{6.4.24}$$

Since f is a cocycle and therefore

$$\Phi u_{jkl} = \Phi u_{jk} + \Phi u_{kl} + \Phi u_{lj} = f_{jk} + f_{kl} + f_{lj} = 0,$$

we see that each u_{jkl} is a holomorphic section of $\operatorname{Ker}\Phi$. Since Ψ is injective, we have uniquely determined vector functions $u'_{jkl} : U_j \cap U_k \cap U_l \to B$ with $\Psi u'_{jkl} = u_{jkl}$. By part (ii) of Lemma 6.4.9, these functions are holomorphic.

Since each u_{jkl} is a section of $\operatorname{Im}\Psi$, it follows (by definition of the c_{jk}) that also each c_{jk} is is a section of $\operatorname{Ker}\Psi$. Since Ψ is injective, we have uniquely determined vector functions $c'_{jk} : U_j \cap U_k \to B$ with $\Psi c'_{jk} = c_{jk}$. By part (i) of Lemma 6.4.9, these functions are of class $\mathcal{C}^\infty$, and it follows from (6.4.24) that

$$c'_{jk} + c'_{kl} + c'_{lj} = u'_{jkl} \qquad \text{on } U_j \cap U_k \cap U_l, \quad j,k,l \in I. \tag{6.4.25}$$

Since the u'_{jkl} are holomorphic, this implies that the family $\{\overline{\partial} c'_{jk}\}_{j,k\in I}$ satisfies the cocycle condition:

$$\overline{\partial} c'_{jk} + \overline{\partial} c'_{kl} + \overline{\partial} c'_{lj} = 0 \qquad \text{on } U_j \cap U_k \cap U_l, \quad j,k,l \in I. \tag{6.4.26}$$

Set

$$v'_j = -\sum_{\nu\in I} \chi_\nu \overline{\partial} c'_{\nu j}.$$

Then

$$v'_j - v'_k = \sum_{\nu\in I} \chi_\nu \Big(-\overline{\partial} c'_{\nu j} + \overline{\partial} c'_{\nu k} \Big) \qquad \text{on } U_j \cap U_k \qquad j,k \in I.$$

In view of the cocycle condition (6.4.26), this implies that

$$v'_j - v'_k = \sum_{\nu\in I} \chi_\nu \overline{\partial} c'_{jk} = \overline{\partial} c'_{jk} \qquad \text{on } U_j \cap U_k \qquad j,k \in I. \tag{6.4.27}$$

By Theorem 2.3.1, now we can find $\mathcal{C}^\infty$-functions $w'_j : U_j \to B$ such that $\overline{\partial} w'_j = v'_j$ on U_j, $j \in I$. Set

$$h'_{jk} = c'_{jk} + w'_k - w'_j \qquad \text{on } U_{jk}\,, \quad j,k \in I.$$

Then, by (6.4.27),

$$\overline{\partial} h'_{jk} = \overline{\partial} c'_{jk} + \overline{\partial} w'_k - \overline{\partial} w'_j = \overline{\partial} c'_{jk} + v'_k - v'_j = 0 \qquad \text{on } U_{jk}\,, \quad j,k \in I.$$

Hence the functions h'_{jk} are holomorphic. Moreover,

$$\begin{aligned} &h'_{jk} + h'_{kl} + h'_{lj} \\ &\quad = (c'_{jk} + w'_k - w'_j) + (c'_{kl} + w'_l - w'_k) + (c'_{lj} + w'_j - w'_l) = c'_{jk} + c'_{kl} + c'_{lj} \end{aligned}$$

and further, by (6.4.24),

$$h'_{jk} + h'_{kl} + h'_{lj} = u'_{jkl} \qquad \text{on } U_j \cap U_k \cap U_l\,, \quad j,k,l \in I. \tag{6.4.28}$$

Now we set

$$g_{jk} = u_{jk} - \Psi h'_{jk} \qquad \text{on } U_j \cap U_k, \quad j,k \in I.$$

Then it follows from (6.4.28) and the definition of the functions u_{jkl} that

$$g_{jk} + g_{kl} + g_{lj} = u_{jkl} - \Psi u'_{jkl} = 0 \qquad \text{on } U_j \cap U_k, \quad j,k \in I,$$

i.e., $\{g_{jk}\}_{j,k\in I}$ is a $(\mathcal{U}, \mathcal{O}^E)$-cocycle. Moreover, since $\Phi \circ \Psi = 0$,

$$\Phi g_{jk} = \Phi u_{jk} - \Phi\Psi h'_{jk} = \Phi u_{jk} = f_{jk} \quad \text{on } U_{jk}\,, \quad j,k \in I,$$

i.e., (6.4.21) is satisfied. □

6.4.13 Proposition. *Let F be a Banach space, let $D \subseteq \mathbb{C}$ be an open set, and let $M = \{M(z)\}_{z\in D}$ be a holomorphic family of subspaces of F, which admits a global resolution over D. Moreover, let $\Omega \subseteq D$ be an open set which is relatively compact in D.*

Then, over Ω, M admits a globally short resolution.

Namely, if $B := \mathcal{O}^M_\infty(\Omega)$ and if

$$\Phi : \Omega \longrightarrow L(B,F) \qquad \text{and} \qquad \Psi : \Omega \longrightarrow L(B,B)$$

are the holomorphic operator functions (cf. Proposition 6.3.3) defined by

$$\Phi(z)(f) = f(z)\,, \qquad z \in \Omega\,, \quad f \in B,$$

and

$$\big(\Psi(z)(f)\big)(\zeta) = (\zeta - z) f(\zeta)\,, \qquad z \in \Omega\,, \quad f \in B,$$

then Φ is a globally short resolution of M, and Ψ is a global injective resolution of $\operatorname{Ker}\Phi$.

Proof. Let $z \in \Omega$ and $v \in M(z)$ be given. By hypothesis we have a Banach space E and holomorphic operator function $A : D \to L(E, F)$ which is a resolution of M over D, i.e., $\operatorname{Im} A(\zeta) = M(\zeta)$ for all $\zeta \in D$. In particular, there exists $w \in E$ with $A(z)w = v$. Consider the function

$$f(\zeta) := A(\zeta)w\,, \qquad \zeta \in \Omega.$$

Then it is clear that $f \in B = \mathcal{O}_\infty^M(\Omega)$ and

$$\Phi(z)f = f(z) = A(z)w = v.$$

Hence Φ is a resolution of M over Ω. From Proposition 6.3.3 (iii) and (iv) it follows that this resolution is even globally short over Ω where Ψ is a global injective resolution of $\operatorname{Ker}\Phi$. $\square$

6.4.14. Proof of the approximation statement (i) in Theorem 6.4.5. Let a holomorphic section f of M in a neighborhood of K be given. Take a neighborhood Ω_0 with $\mathcal{C}^1$-boundary of K which is so small that $\overline{\Omega}_0 \subseteq D$ and f is still defined and holomorphic in a neighborhood of $\overline{\Omega}_0$. Since each connected component of $\mathbb{C} \setminus K$ contains at least one point of $\mathbb{C} \setminus D$, we may assume that each connected component of $\mathbb{C} \setminus \overline{\Omega}_0$ contains at least one point of $\mathbb{C} \setminus D$. Further, choose a sequence $(\Omega_n)_{n \in \mathbb{N}^*}$ of open sets with $\mathcal{C}^1$-boundaries such that:

- $\overline{\Omega}_n \subseteq \Omega_{n+1} \subseteq D$, for all $n \in \mathbb{N}$.
- Each connected component of $\mathbb{C} \setminus \overline{\Omega}_n$ contains at least one point of $\mathbb{C} \setminus D$, for all $n \in \mathbb{N}$.
- $\bigcup_{n=0}^{\infty} \Omega_n = D$.

Now, clearly, it is sufficient to prove that, for each $n \in \mathbb{N}$, the following statement is valid:

- Each holomorphic section of M defined and holomorphic in a neighborhood of $\overline{\Omega}_n$ can be approximated uniformly on $\overline{\Omega}_n$ by holomorphic sections of M defined and holomorphic over a neighborhood of $\overline{\Omega}_{n+1}$.

So, let $n \in \mathbb{N}$ and a holomorphic section g of M in a neighborhood U of $\overline{\Omega}_n$ be given.

Since each connected component of $\mathbb{C} \setminus \overline{\Omega}_{n+2}$ contains at least one point of $\mathbb{C} \setminus D$ and M admits a resolution over D, it follows from Proposition 6.4.13 that, over Ω_{n+2}, there exists a globally short resolution $\Phi : \Omega_{n+2} \to L(E, F)$ of M. By Lemma 6.4.11 there exists a holomorphic vector function

$$h : U \cap \Omega_{n+2} \longrightarrow E$$

such that

$$\Phi h = g \qquad \text{on } U \cap \Omega_{n+2}.$$

Since each connected component of $\mathbb{C}\setminus\overline{\Omega}_n$ contains at least one point of $\mathbb{C}\setminus D$, from the Runge approximation Theorem 2.2.2 we get a sequence $(h_\nu)_{\nu\in\mathbb{N}}$ of holomorphic functions $h_\nu : D \to E$ such that

$$\lim_{\nu\to\infty} \max_{z\in\overline{\Omega}_n} \left\|h_\nu(z) - h(z)\right\| = 0.$$

Then $(\Phi h_\nu)_{\nu\in\mathbb{N}}$ is a sequence of holomorphic sections of M over Ω_{n+2} (which is a neighborhood of $\overline{\Omega}_{n+1}$) such that

$$\begin{aligned}\lim_{\nu\to\infty} \max_{z\in\overline{\Omega}_n} \left\|\Phi(z)h_\nu(z) - g(z)\right\| &= \lim_{\nu\to\infty} \max_{z\in\overline{\Omega}_n} \left\|\Phi(z)\big(h_\nu(z) - h(z)\big)\right\| \\ &\le \lim_{\nu\to\infty} \max_{z\in\overline{\Omega}_n} \left\|\Phi(z)\right\|\left\|h_\nu(z) - h(z)\right\| = 0.\end{aligned}$$

□

6.4.15. Proof of the splitting statement (ii) in Theorem 6.4.5. Choose a sequence $(\Omega_n)_{n\in\mathbb{N}}$ of open sets such that

- $\overline{\Omega}_n \subseteq \Omega_{n+1} \subseteq D$ for all $n \in \mathbb{N}$,
- Each connected component of $\mathbb{C} \setminus \overline{\Omega}_n$ contains at least one point of $\mathbb{C} \setminus D$, for all $n \in \mathbb{N}$.
- $\bigcup_{n=0}^{\infty} \Omega_n = D$.

Now let an open covering $=\{U_j\}_{j\in I}$ of D and a $(\mathcal{U}, \mathcal{O}^M)$-cocycle $f = \{f_{jk}\}_{j,k\in I}$ be given. For each $n \in \mathbb{N}$, we consider the open covering

$$\mathcal{U}^{(n)} := \left\{U_j \cap \Omega_n\right\}_{j\in I}$$

of Ω_n and the $\big(\mathcal{U}^{(n)}, \mathcal{O}^M\big)$-cocycle $f^{(n)} = \left\{f_{jk}^{(n)}\right\}_{j,k\in I}$ defined by

$$f_{jk}^{(n)} = f_{jk}\big|_{U_j\cap U_k\cap\Omega_n}, \qquad j,k \in I.$$

Since each Ω_n is relatively compact in D and M admits a resolution over D, it follows from Proposition 6.4.13 that, over each Ω_n, M admits a globally short resolution. Hence, by Lemma 6.4.12, for each $n \in \mathbb{N}$, we can find a family $\{f_j^{(n)}\}_{j\in I}$ of holomorphic sections $f_j^{(n)} : U_j^{(n)} \to M$ with

$$f_{jk} = f_{jk}^{(n)} = f_j^{(n)} - f_k^{(n)} \qquad \text{on } U_j \cap U_k \cap \Omega_n\,, \quad j,k \in I. \tag{6.4.29}$$

Hence

$$f_j^{(n+1)} - f_j^{(n)} = f_k^{(n+1)} - f_k^{(n)} \qquad \text{on } U_j \cap U_k \cap \Omega_n\,, \quad j,k \in I.$$

Therefore, setting

$$h^{(n)} = f_j^{(n+1)} - f_j^{(n)} \qquad \text{on } U_j \cap \Omega_n\,, \quad j \in I,$$

we get holomorphic sections $h_n : \Omega_n \to M$. By the approximation statement (i) of Theorem 6.4.4, we can find holomorphic sections $\widetilde{h}_n : D \to M$ such that

$$\|h^{(n)}(z) - \widetilde{h}^{(n)}(z)\| < \frac{1}{2^n} \qquad \text{for all } z \in \overline{\Omega}_{n-1}\,, \quad n \in \mathbb{N}^*. \tag{6.4.30}$$

Set

$$\widetilde{f}_j^{(n)} = f_j^{(n)} - \sum_{k=0}^{n-1} \widetilde{h}^{(k)} \qquad \text{on } U_j \cap \Omega_n.$$

Then

$$\begin{aligned}\widetilde{f}_j^{(n+1)} - \widetilde{f}_j^{(n)} &= \left(f_j^{(n+1)} - \sum_{k=0}^{n} \widetilde{h}^{(k)}\right) - \left(f_j^{(n)} - \sum_{k=0}^{n-1} \widetilde{h}^{(k)}\right) \\ &= f_j^{(n+1)} - f_j^{(n)} - \widetilde{h}^{(n)} = h^{(n)} - \widetilde{h}^{(n)}.\end{aligned}$$

By (6.4.30) this implies that

$$\left\|\widetilde{f}_j^{(n+1)}(z) - \widetilde{f}_j^{(n)}(z)\right\| < \frac{1}{2^n} \qquad \text{for } z \in \overline{\Omega}_{n-1}\,, \qquad j \in I, n \in \mathbb{N}^*.$$

Since, for each $j \in I$, each compact subset of U_j is contained in all $\overline{\Omega}_{n-1}$ with n sufficiently large, this implies that, for each $j \in I$, the sequence $\left(f_j^{(n)}\right)_{n\in\mathbb{N}}$ converges, uniformly on the compact subsets of U_j, to some holomorphic section $f_j : U_j \to M$. For these sections we have, by (6.4.29),

$$f_j - f_k = \lim_{n\to} \left(f_j^{(n)} - f_k^{(n)}\right) = \lim_{n\to} f_{j,k}^{(n)} = f_{jk} \qquad \text{on } U_j \cap U_k \quad j,k \in I,$$

which completes the proof of the splitting statement in Theorem 6.4.5. □

6.4.16. Proof of the lifting statement (ii) in Theorem 6.4.4. By Lemma 6.4.6 we can find an open covering $\mathcal{U} = \{U_j\}_{j\in I}$ of D and a family $\{u_j\}_{j\in I}$ of holomorphic vector functions $u_j : U_j \to E$ such that

$$Au_j = f \qquad \text{on } U_j\,, \qquad j \in I.$$

Setting

$$u_{jk} = u_j - u_k \qquad \text{on } U_j \cap U_k\,, \quad j,k \in I,$$

then we get a $\left(\mathcal{U}, \mathcal{O}^{\operatorname{Ker} A}\right)$-cocycle. Since, by Theorem 6.4.4 (i), $\operatorname{Ker} A$ is a holomorphic family of subspaces, from Theorem 6.4.5 (ii) we get a family $\{v_j\}_{j\in I}$ of holomorphic sections $v_j : U_j \to \operatorname{Ker} A$ such that

$$u_j - u_k = u_{jk} = v_j - v_k \qquad \text{on } U_j \cap U_k\,, \quad j,k \in I.$$

Therefore, setting

$$u = u_j - v_j \qquad \text{on } U_j\,, \quad j \in I,$$

we obtain a holomorphic vector function $u : D \to E$ such that

$$Au = A(u_j - v_j) = Au_j - Av_j = Au_j = f \qquad \text{on each } U_j$$

and, hence, $Au = f$ everywhere on D. □

6.5 Example: A holomorphic family of subspaces with jumping isomorphism type

In this section we construct a holomorphic family $\{M(z)\}_{z\in D}$ of subspaces of a Banach space E such that, for certain $z_0 \in D$, the space $M(z_0)$ is not isomorphic to the spaces $M(z)$ with $z \in D \setminus \{z_0\}$.

Let l_1 be the Banach space of summable complex sequences $\xi = \{\xi_n\}_{n\in\mathbb{N}}$ with the norm

$$\|\xi\|_1 := \sum_{n=0}^{\infty} |\xi_n|,$$

and let l_2 be the Hilbert space of square summable complex sequences $\xi = \{\xi_n\}_{n\in\mathbb{N}}$ with the norm

$$\|\xi\|_2 := \left(\sum_{n=0}^{\infty} |\xi_n|^2\right)^{1/2}.$$

6.5.1 Lemma. *Let E be a separable Banach space. Then there exists $A \in L(l_1, E)$ with* $\operatorname{Im} A = E$.

Proof. Let S be the unit sphere in E. Since E is separable, we can find a sequence $s = \{s_n\}_{n\in\mathbb{N}}$ in S which is dense in S. Then, setting

$$A\xi = \sum_{n=0}^{\infty} \xi_n s_n \qquad \text{for } \xi = \{\xi_n\}_{n\in\mathbb{N}} \in l_1,$$

an operator $A \in L(l_1, E)$ with $\|A\| = 1$ is well defined. It remains to prove that $\operatorname{Im} A = E$.

Let E_s be the set of all vectors $x \in E$ of the form $x = ts_n$, where $n \in \mathbb{N}$ and $t \in \mathbb{C}$. Then E_s is dense in E and has the following property:

$$\text{For each } x \in E_s, \text{ there exists } \xi \in l_1 \text{ with } \|\xi\| = \|x\| \text{ and } A\xi = x. \tag{6.5.1}$$

Indeed, by definition of E_s, for each $x \in E_s$, there exists $t \in \mathbb{C}$ and $m \in \mathbb{N}$ with $x = ts_m$. Let $\xi = \{\xi_j\}_{j\in\mathbb{N}} \in l_1$ be the sequence with

$$\xi_j = \begin{cases} t & \text{if } j = m, \\ 0 & \text{if } j \neq m. \end{cases}$$

Then $\|\xi\| = |t| = |t|\|s_m\| = \|x\|$ and $A\xi = ts_m = x$.

To prove that $\operatorname{Im} A = E$, now we consider an arbitrary vector $v \in E$. Since E_s is dense in E, then we can find a sequence $v_j \in E_s$, $j \in \mathbb{N}$, with

$$\sum_{j=0}^{\infty} \|v_j\| < \infty \qquad \text{and} \qquad v = \sum_{j=0}^{\infty} v_j.$$

By property (6.5.1), for each v_j we can find $\xi^{(j)} \in l_1$ with $\|\xi^{(j)}\| = \|v_j\|$ and $A\xi^{(j)} = v_j$. Then

$$\sum_{j=0}^{\infty} \|\xi^{(j)}\| = \sum_{j=0}^{\infty} \|v_j\| < \infty$$

and

$$A\Big(\sum_{j=0}^{\infty} \xi^{(j)}\Big) = \sum_{j=0}^{\infty} v_j = v. \qquad \square$$

6.5.2 Lemma. *There exists a closed subspace F of l_1 such that l_1 is not isomorphic to $F \oplus l_1/F$, where $F \oplus l_1/F$ is the direct sum of F and the factor space l_1/F.*

Proof. By Lemma 6.5.1 there exists $A \in L(l_1, l_2)$ with $\operatorname{Im} A = l_2$. We claim that $F := \operatorname{Ker} A$ has the required property.

Indeed, assume $F \oplus l_1/F$ and l_1 are isomorphic. Then l_1 has a closed subspace E which is isomorphic to l_1/F. As $\operatorname{Im} A = l_2$ and therefore $l_1/F = l_1/\operatorname{Ker} A$ is isomorphic to l_2 (as a Banach space), this implies that E is isomorphic to l_2.

But this is impossible, as l_1 (and hence E) has the property that each weakly convergent sequence is strongly convergent (see page 137 in [Ban]), whereas l_2 does not have this property. $\square$

The example. Let E be a Banach space, let F be a closed subspace of E such that E is not isomorphic to $F \oplus E/F$ (by lemma (6.5.2) this is possible), and let $\pi : E \to E/F$ be the canonical map. We define a holomorphic operator function $A : \mathbb{C} \to L\big(E \oplus E/F, E/F\big)$, by setting

$$\big(A(z)\big)(x,y) = \pi(x) + zy \qquad \text{for } (x,y) \in E \oplus E/F.$$

Then $\operatorname{Im} A(z) = E/F$ for all $z \in \mathbb{C}$, and it follows from Theorem 6.4.4 (i) that $\operatorname{Ker} A = \{\operatorname{Ker} A(z)\}_{z\in\mathbb{C}}$ is a holomorphic family of subspaces of $E \oplus E/F$. We have

$$\operatorname{Ker} A(0) = F \oplus E/F,$$

which is not isomorphic to E. However, for $z \in \mathbb{C}\setminus\{0\}$, then $\operatorname{Ker} A(z)$ is isomorphic to E. Indeed, if $z \neq 0$, then for each $(x,y) \in E \oplus E/F$,

$$(x,y) \in \operatorname{Ker} A(z) \Longleftrightarrow y = -\frac{\pi(x)}{z},$$

which yields that the map

$$E \ni x \longrightarrow \left(x, -\frac{\pi(s)}{z}\right) \in E \oplus E/F$$

is an isomorphism from E onto $\operatorname{Ker} A(z)$.

6.6 Injective families

In this section E is a Banach space.

6.6.1 Definition. Let $D \subseteq \mathbb{C}$ be an open set, and let $M = \{M(z)\}_{z\in D}$ be a holomorphic family of subspaces of E. We shall say that M is **injective** if, for each $z_0 \in D$, there exist a neighborhhood $U \subseteq D$ of z_0 and a holomorphic function $A : U \to L\big(M(z_0), E\big)$ such that $\operatorname{Ker} A(z) = \{0\}$ and $\operatorname{Im} A(z) = M(z)$ for all $z \in U$.

In other words, M is injective if and only if it admits an injective resolution (Def. 6.4.8) over some neighborhood of each point of D.

6.6.2 Remark. Let $M = \{M(z)\}_{z\in D}$ be a holomorphic family of subspaces of E. If D is connected, then the spaces of this family are pairwise isomorphic. This can be seen as follows:

Fix $z_0 \in D$ and denote by X the set of all $z \in D$ such that $M(z)$ is isomorphic to $M(z_0)$. Then $z_0 \in X$ and therefore $X \neq \emptyset$. Since M is injective, for each point $w \in D$, there is a neighborhood $U \subseteq D$ of w such that the spaces $M(z)$, $z \in U$, are isomorphic to $M(w)$. Therefore it is clear that X is open and relatively closed in D.

6.6.3 Theorem. *Let $D \subseteq \mathbb{C}$ be a connected open set, let $M = \{M(z)\}_{z\in D}$ be a holomorphic family of subspaces of E, which is injective, and let $z_0 \in D$. Suppose at least one of the following two conditions is satisfied:*

(i) *D is simply connected.*

(ii) *The group $GL\big(M(z_0)\big)$ is connected.*

Then there exists a holomorphic function $A\colon D \to L\big(M(z_0), E\big)$ such that $\operatorname{Ker} A(z) = \{0\}$ *and* $\operatorname{Im} A(z) = M(z)$ *for all $z \in D$.*

Proof. Since D is connected and M is injective, we can find an open covering $\mathcal{U} = \{U_j\}_{j\in I}$ of D and holomorphic functions $A_j : U_j \to L\big(M(z_0), E\big)$ with $\operatorname{Ker} A_j(z) = \{0\}$ and $\operatorname{Im} A_j(z) = M(z)$ for all $z \in U_j$, $j \in I$. Then $A_j(z)$ is an invertible operator from $M(z_0)$ onto $M(z)$; let by $A_j^{-1}(z)$ be the inverse of this operator, $j \in I$, $z \in U_j$.

Consider $j, k \in I$ with $U_j \cap U_k \neq \emptyset$. Then, for each $z \in U_j \cap U_k$,

$$G_{jk}(z) := A_j^{-1}(z) A_k(z)$$

is a well-defined operator from $GL\big(M(z_0)\big)$. Moreover, the so-defined function

$$G_{jk} : U_j \cap U_k \to GL\big(M(z_0)\big)$$

is holomorphic. Indeed, let v be an arbitrary vector from $M(z_0)$. Then the function $A_k(z)v$ is holomorphic, and it follows from Lemma 6.4.9 that $G_{jk}(z)v =$

$A_j^{-1}(z)A_k(z)v$ is holomorphic. Hence G_{jk} is holomorphic (Theorem 1.7.1). So we obtained a cocycle $\{G_{jk}\}_{j,k\in I} \in Z^1\left(\mathcal{U}, \mathcal{O}^{GL(M(z_0))}\right)$.

Since at least one of the conditions (i) or (ii) is satisfied, we can apply Theorem 5.6.3 and get holomorphic functions $B_j : U_j \to GL\bigl(M(z_0)\bigr)$, $j \in I$, with

$$A_j^{-1}(z)A_k(z) = G_{jk} = B_j B_k^{-1} \qquad \text{on } U_j \cap U_k$$

for all $j,k \in I$ with $U_j \cap U_k \neq \emptyset$. It remains to set $A = A_j B_j$ on U_j, $j \in I$. □

Moreover, there is the following Oka-Grauert principle:

6.6.4 Theorem. *Let $D \subseteq \mathbb{C}$ be a connected open set, let $M = \{M(z)\}_{z\in D}$ be a holomorphic family of subspaces of E, which is injective, and let $z_0 \in D$. Suppose the following condition is satisfied:*

(iii) *There exists a continuous function $C : D \to L\bigl(M(z_0), E\bigr)$ such that* $\operatorname{Ker} C(z) = \{0\}$ *and* $\operatorname{Im} C(z) = M(z)$ *for all $z \in D$.*

Then there exists a holomorphic function $A\colon D \to L\bigl(M(z_0), E\bigr)$ such that $\operatorname{Ker} A(z) = \{0\}$ *and* $\operatorname{Im} A(z) = M(z)$ *for all $z \in D$.*

Proof. This is a repetition of the proof of Theorem 6.6.3 until the moment where we use that one of conditions (i) or (ii) in Theorem 6.6.3 is satisfied. Instead here we use the function C from condition (iii) and consider the functions $C_j := A_j^{-1}C$, which are continuous on U_j, by Corollary 6.2.5. Then

$$C_j C_k^{-1} = A_j^{-1} C C^{-1} A_k = A_j^{-1} A_k = G_{jk} \quad \text{on } U_j \cap U_k$$

for all $j,k \in I$ with $U_j \cap U_k \neq \emptyset$. Hence the cocycle $\{G_{jk}\}$ is $\mathcal{C}^{GL(M(z_0))}$-trivial. Now, again from Theorem 5.6.3 we get holomorphic functions $B_j : U_j \to GL\bigl(M(z_0)\bigr)$, $j \in I$, with

$$A_j^{-1}(z)A_k(z) = G_{jk} = B_j B_k^{-1} \qquad \text{on } U_j \cap U_k$$

for all $j,k \in I$ with $U_j \cap U_k \neq \emptyset$. It remains to set $A = A_j B_j$ on U_j, $j \in I$. □

6.7 Shubin families

In this section E is a Banach space.

6.7.1 Definition. Let $D \subseteq \mathbb{C}$ be an open set, and let $M = \{M(z)\}_{z\in D}$ be a holomorphic family of subspaces of E. We shall say that M is a **Shubin family** if, for each $z_0 \in D$, there exist a neighborhhood $U \subseteq D$ of z_0 and a holomorphic function $A : U \to GL(E)$ such that $A(z)M(z_0) = M(z)$ for all $z \in U$.

6.7.2 Remark. Let $M = \{M(z)\}_{z\in D}$ be a holomorphic family of subspaces of E, which is a Shubin family. Suppose D is connected.

Then M is injective (Def. 6.6.1) and the spaces of the family M are pairwise isomorphic. Moreover, then, for any pair of points $z, w \in D$, there exists an operator $T \in GL(E)$ with $TM(z) = M(w)$.

This can be seen as follows: Fix $z_0 \in D$ and denote by X the set of all $z \in D$ such that $M(z) = TM(z_0)$ for some $T \in GL(E)$. Then $z_0 \in X$ and therefore $X \neq \emptyset$. Moreover, from the definition of a Shubin family it follows that, for each $w \in D$, there exists a neighborhood $W \subseteq D$ of w with the following property: If $z \in U$, then $M(z) = TM(w)$ for some $T \in GL(E)$. From this it is clear that X is open and relatively closed in D.

6.7.3 Theorem. *Let $D \subseteq \mathbb{C}$ be a connected open set, let D be a Banach space, let $M = \{M(z)\}_{z \in D}$ be a holomorphic family of subspaces of E, which is a Shubin family, and let $z_0 \in D$. Suppose at least one of the following two conditions is satisfied:*

(i) *D is simply connected.*

(ii) *The group $GL_{M(z_0)}(E) := \{T \in GL(E) \,|\, TM(z_0) = M(z_0)\}$ is connected.*

Then there exists a holomorphic function $A : D \to GL(E)$ such that $A(z)M(z_0) = M(z)$ for all $z \in D$.

Proof. First note that $GL_{M(z_0)}(E)$ is the group of invertible elements of a Banach algebra, namely of the closed subalgebra of $L(E)$ which consists of the operators T with $TM(z_0) \subseteq M(z_0)$.

Since D is connected and M is a Shubin family, we can find an open covering $\mathcal{U} = \{U_j\}_{j \in I}$ of D and holomorphic functions $A_j : U_j \to GL(E)$ with $A_j(z)M(z_0) = M(z)$ for all $z \in U_j$, $j \in I$. Setting

$$G_{jk} = A_j^{-1} A_k \qquad \text{on } U_j \cap U_k$$

for all $j, k \in I$ with $U_j \cap U_k \neq \emptyset$, we obtain a cocycle $\{G_{jk}\}_{j,k \in I} \in Z^1\big(\mathcal{U}, \mathcal{O}^{GL_{M(z_0)}(E)}\big)$. Since $GL_{M(z_0)}(E)$ is the group of invertible elements of a Banach algebra and at least one of the conditions (i) or (ii) is satisfied, we can apply Theorem 5.6.3. So we get holomorphic functions $B_j : U_j \to GL_{M(z_0)}(E)$, $j \in I$, with

$$A_j^{-1}(z)A_k(z) = G_{jk} = B_j B_k^{-1} \qquad \text{on } U_j \cap U_k$$

for all $j, k \in I$ with $U_j \cap U_k \neq \emptyset$. It remains to set $A = A_j B_j$ on U_j, $j \in I$. □

There is also the following Oka-Grauert principle:

6.7.4 Theorem. *Let $D \subseteq \mathbb{C}$ be a connected open set, let $M = \{M(z)\}_{z \in D}$ be a holomorphic family of subspaces of E, which is a Shubin family, and let $z_0 \in D$. Suppose the following condition is satisfied:*

(iii) *There exists a continuous function $C : D \to GL(E)$ such that $C(z)M(z_0) = M(z)$ for all $z \in D$.*

Then there exists a holomorphic function $A : D \to GL(E)$ *such that* $A(z)M(z_0) = M(z)$ *for all* $z \in D$.

Proof. This is a repetition of the proof of Theorem 6.7.3 until the moment where we use that one of conditions (i) or (ii) in Theorem 6.7.3 is satisfied. Instead here we use the function C from condition (iii), consider the continuous functions $C_j := A_j^{-1}C$ and observe that the values of these functions belong to $GL_{M(z_0)}(E)$. Then

$$C_j C_k^{-1} = A_j^{-1} C C^{-1} A_k = A_j^{-1} A_k = G_{jk} \quad \text{on } U_j \cap U_k$$

for all $j, k \in I$ with $U_j \cap U_k \neq \emptyset$. Hence the cocycle $\{G_{jk}\}$ is $\mathcal{C}^{GL_{M(z_0)}(E)}$-trivial. Now, again from Theorem 5.6.3 we get a family of holomorphic functions $B_j : U_j \to GL_{M(z_0)}(E)$ with

$$A_j^{-1} A_k = G_{jk} = B_j B_k^{-1} \qquad \text{on } U_j \cap U_k$$

for all $j, k \in I$ with $U_j \cap U_k \neq \emptyset$. It remains to set $A = A_j B_j$ on U_j, $j \in I$. □

6.8 Complemented families

In this section E is a Banach space.

6.8.1 Definition. Let $D \subseteq \mathbb{C}$ be an open set. A holomorphic family $M = \{M(z)\}_{z \in D}$ of subspaces of E will be called a holomorphic family $M = \{M(z)\}_{z \in D}$ of **complemented** subspaces of E if each of the spaces $M(z)$, $z \in D$, is a complemented subspace of E.

6.8.2 Lemma. *Let* $D \subseteq \mathbb{C}$ *be an open set, let* X *be a second Banach space, let* $A : D \to L(X, E)$ *be a holomorphic function such that, for all* $z \in D$, $A(z)$ *is right invertible, and let* $z_0 \in D$.

(i) *Then there exists a neighborhood* $U \subseteq D$ *of* z_0 *and a holomorphic function* $A^{(-1)} : U \to L(E, X)$ *such that* $AA^{(-1)} = I$ *on* U.

(ii) *If* U *and* $A^{(-1)}$ *are as in part (i), then, for all* $z \in U$, $P(z) := A^{(-1)}(z)A(z)$ *is a projector in* E *with* $\operatorname{Ker} P(z) = \operatorname{Ker} A(z)$.

Proof. (i) Let $B \in L(E, X)$ be a right inverse of $A(z_0)$. Then there is a neighborhood U of z_0 such that $A(z)B \in GL(E)$ for all $z \in U$. It remains to set $A^{(-1)}(z) = B\big(A(z)B\big)^{-1}$, $z \in U$.

(ii) Let $z \in U$. Then

$$P^2(z) = A^{(-1)}(z)\Big(A(z)A^{(-1)}(z)\Big)A(z) = A^{(-1)}(z)A(z) = P(z).$$

Hence, $P(z)$ is a projector. Further

$$P(z) \operatorname{Ker} A(z) = A^{(-1)}(z)A(z) \operatorname{Ker} A(z) = \{0\}.$$

Hence $\operatorname{Ker} A(z) \subseteq \operatorname{Ker} P(z)$. Finally,

$$\begin{aligned} A(z)(I - P(z)) &= A(z) - A(z)P(z) \\ &= A(z) - \Big(A(z)A^{(-1)}(z)\Big)A(z) = A(z) - A(z) = 0. \end{aligned}$$

Hence $\operatorname{Ker} P(z) = \operatorname{Im}(I - P(z)) \subseteq \operatorname{Ker} A(z)$. □

6.8.3 Lemma. *Let $D \subseteq \mathbb{C}$ be an open set, let $\{M(z)\}_{z\in D}$ be a holomorphic family of complemented subspaces of E, and let $z_0 \in D$.*

(i) *Then there exist a neighborhood U of z_0 and a closed subspace F of E such that*

$$E = F \dot{+} M(z) \qquad \text{for all } z \in U. \qquad (\text{direct sum}). \tag{6.8.1}$$

(ii) *Let U an F be as in part (i). For $z \in U$, we denote by $P(z)$ the projector in E with $\operatorname{Im} P(z) = M(z)$ and $\operatorname{Ker} P(z) = F$. Then the function $U \ni z \to P(z)$ is holomorphic on U.*

Proof. Assertion (i) follows from Proposition 6.1.5. We prove part (ii). It is sufficient to prove that, for each $v \in E$, the function Pv is holomorphic (Theorem 1.7.1). Let $v \in E$ be given.

Since M is a holomorphic family of subspaces of E, after shrinking U if necessary, we can find a Banach space X and a holomorphic function $A : U \to L(X, E)$ with $\operatorname{Im} A(z) = M(z)$ for all $z \in U$. Let $X \oplus F$ be the direct sum of X and F, and define a holomorphic function $\widetilde{A} : U \to L(X \oplus F, E)$, setting

$$\widetilde{A}(z)(x, f) = A(z)x + f \qquad \text{for } x \in X \text{ and } f \in F.$$

By (6.8.1), $\widetilde{A}(z)$ is surjective. By Lemma 6.4.6, after a further shrinking of U (or by Theorem 6.4.4 without shrinking of U), we can find holomorphic functions $x : U \to X$ and $f : U \to F$ such that $\widetilde{A}(z)\big(x(z), f(z)\big) = v$ for all $z \in U$. By definition of $\widetilde{A}$, this means that $A(z)x(z) + f(z) = v$ for all $z \in U$. Since $A(z)x(z) \in \operatorname{Im} A(z) = M(z) = \operatorname{Im} P(z)$, $f(z) \in F = \operatorname{Ker} P(z)$ and $P(z)$ is a projector, this implies that $A(z)x(z) = P(z)v$ for all $z \in U$. Hence the function Pv is holomorphic. □

6.8.4 Proposition. *Each holomorphic family of complemented subspaces of E is a Shubin family.*

Proof. Let $z_0 \in D$ be given. Then, by Lemma 6.8.3, there exist a neighborhood $U \subseteq D$ of z_0 and a holomorphic function $P : U \to L(E)$ such that the operators $P(z)$ are projectors with $\operatorname{Im} P(z) = M(z)$ for all $z \in U$. Set

$$A(z) = P(z) + I - P(z_0)\,, \qquad z \in U.$$

Since $A(z_0) = I$, after shrinking U, we may assume that $A(z) \in GL(E)$ for all $z \in U$. Moreover,

$$A(z)M(z_0) = P(z)M(z_0) \subseteq M(z) \qquad \text{for all } z \in U.$$

Since $\{A(z)M(z_0)\}_{z\in U}$ and $\{M(z)\}_{z\in U}$ are continuous families of subspaces and $A(z_0)M(z_0) = M(z_0)$, this implies, by Proposition 6.1.3, that $A(z)M(z_0) = M(z)$ for all $z \in U$ (assuming that U is connected). □

6.8.5 Theorem. *Let $D \subseteq \mathbb{C}$ be an open set, and let $M = \{M(z)\}_{z\in D}$ be a holomorphic family of complemented subspaces of E. Then there exists a holomorphic function $P : D \to L(E)$ such that, for all $z \in D$, the operator $P(z)$ is a projector with* $\operatorname{Im} P(z) = M(z)$.

Proof. By Lemma 6.8.3 (ii), there exist an open covering $\mathcal{U} = \{U_j\}_{j\in I}$ of D and holomorphic functions $P_j : U_j \to L(E)$, $j \in I$, such that $P_j(z)$ is a projector with $\operatorname{Im} P_j(z) = M(z)$ for all $z \in U_j$ and $j \in I$. Set

$$\mathcal{N}(z) = \Big\{T \in L(E) \;\Big|\; \operatorname{Im} T \subseteq M(z) \subseteq \operatorname{Ker} T\Big\} \qquad \text{for } z \in D,$$

and

$$\mathcal{P}_j(z)T = P_j(z)T\big(I - P_j(z)\big) \qquad \text{for } T \in L(E),\ z \in U_j,\ j \in I.$$

For each $j \in I$, then $\mathcal{P}_j$ is a holomorphic $L\big(L(E)\big)$-valued function on U_j, where the operators $\mathcal{P}_j(z)$ are projectors with $\operatorname{Im} \mathcal{P}_j(z) = \mathcal{N}(z)$ for all $z \in U_j$. By Proposition 6.1.4 this implies that $\mathcal{N} := \{\mathcal{N}(z)\}_{z\in D}$ is a holomorphic family of subspaces of $L(E)$.

Since $P_j(z) - P_k(z) \in \mathcal{N}(z)$ for $z \in U_j \cap U_k$, the differences $P_j - P_k$ define a $(\mathcal{U}, \mathcal{O}^{\mathcal{N}})$-cocycle (Def. 6.4.3). Therefore, from Theorem 6.4.5 (ii) we get sections $S_j \in \mathcal{O}^{\mathcal{N}}(U_j)$, $j \in I$, with

$$P_j - P_k = S_j - S_k \qquad \text{on } U_j \cap U_k$$

for all $j, k \in I$ with $U_j \cap U_k \neq \emptyset$, and, setting

$$P = P_j - S_j \qquad \text{on } U_j\,, \qquad j \in I,$$

we can define a global holomorphic function $P : D \to L(E)$. This function has the required properties.

Indeed, for each $j \in I$, we have

$$P^2 = \big(P_j - S_j\big)^2 = P_j^2 + S_j^2 - P_jS_j - S_jP_j = P_j - S_j = P \quad \text{on } U_j.$$

Hence, the values of P are projectors. Further, for each $j \in I$ and $z \in U_j$, we have

$$\operatorname{Im} P(z) \supseteq P(z)M(z) = \big(P_j(z) - S_j(z)\big)M(z) = P_j(z)M(z) = M(z)$$

and

$$\operatorname{Im} P(z) = \operatorname{Im}\big(P_j(z) - S_j(z)\big) \subseteq M(z).$$

Hence $\operatorname{Im} P(z) = M(z)$. □

6.8.6 Theorem. *Let $D \subseteq \mathbb{C}$ be a connected open set, let $M = \{M(z)\}_{z\in D}$ be a holomorphic family of complemented subspaces of E, and let $z_0 \in D$. Suppose at least one of the following two conditions is satisfied:*

(i) *D is simply connected.*

(ii) *The groups $GL\big(M(z_0)\big)$ and $GL\big(E/M(z_0)\big)$ are connected.*

Then there exists a holomorphic function $A : D \to GL(E)$ such that $A(z)M(z_0) = M(z)$ for all $z \in D$.

Proof. It is easy to see that from condition (ii) in Theorem 6.8.6 follows condition (ii) in Theorem 6.7.3. Since M is a Shubin family (Proposition 6.8.4), therefore Theorem 6.8.6 follows from Theorem 6.7.3. □

Since holomorphic families of complemented subspaces are Shubin families (Proposition 6.8.4), from Theorem 6.7.4 we get the following Oka-Grauert principle:

6.8.7 Corollary (to Theorem 6.7.4). *Let $D \subseteq \mathbb{C}$ be a connected open set, let $M = \{M(z)\}_{z\in D}$ be a holomorphic family of complemented subspaces of E, and let $z_0 \in D$. Suppose the following condition is satisfied:*

(iii) *There exists a continuous function $C : D \to GL(E)$ such that $C(z)M(z_0) = M(z)$ for all $z \in D$.*

Then there exists a holomorphic function $A : D \to GL(E)$ such that $A(z)M(z_0) = M(z)$ for all $z \in D$.

6.9 Finite dimensional and finite codimensional families

In this section E is a Banach space.

6.9.1 Theorem. *Let $D \subseteq \mathbb{C}$ be a connected open set, let $z_0 \in D$, and let $\big\{M(z)\big\}_{z\in D}$ be a holomorphic family of subspaces of E such that at least one of the following two conditions is satisfied:*

(i) *The spaces $M(z)$, $z \in D$, are finite dimensional;*

(ii) *The spaces $M(z)$, $z \in D$, are of finite codimension in E.*

Then there exists a holomorphic function $A : D \to \mathcal{G}^\infty(E)$ (Def. 5.12.1) with $A(z)M(z_0) = M(z)$ for all $z \in D$.

Proof. Since M is a holomorphic family of complemented subspaces, we can apply Theorem 6.8.5. Therefore we get a holomorphic function $P : D \to L(E)$ such that the operators $P(z)$ are projectors with $\operatorname{Im} P(z) = M(z)$ for all $z \in D$. Set $N(z) = \operatorname{Ker} P(z)$, $z \in D$, and $N = \{N(z)\}_{z\in D}$. Then, by Proposition 6.1.4, N is also a holomorphic family of subspaces of E.

For each $w \in D$, we choose a sufficiently small connected neighborhood $U_w \subseteq D$ of w and set

$$A_w^M(z) = P(z) + I - P(w)\,, \qquad A_w^N(z) = I - P(z) + P(w), \qquad z \in U_w.$$

Since the neighborhoods U_w are sufficiently small, then the values of each A_w^M and each A_w^N are invertible. Since at least one of the projectors $P(w)$ or $I - P(w)$ is finite dimensional, these values even belong to $\mathcal{G}^\infty(E)$. Moreover

$$A_w^M(z)M(w) = P(z)M(w) \subseteq M(z)$$

and

$$A_w^N(z)N(w) = \big(I - P(z)\big)N(w) \subseteq N(z)$$

for all $z \in U_w$ and $w \in D$. Since the neighborhoods U_w are connected, $A^M(w)M(w) = M(w)$ and $A^N(w)N(w) = N(w)$, by Proposition 6.1.3 this implies that

$$A_w^M(z)M(w) = M(z) \quad \text{and} \quad A_w^N(z)N(w) = N(z)\,, \quad z \in U_w\,,\ w \in D.$$

The spaces $M(w)$ and $N(w)$ are finite dimensional or closed and finite codimensional, and D is connected. Therefore we can find operators $B_w^M, B_w^N \in \mathcal{G}^\infty(E)$ such that $B_w^M M(z_0) = M(w)$ and $B_w^N N(z_0) = N(w)$, $w \in D$. Setting

$$\widetilde{A}_w^M(z) = A_w^M(z)B_w^M \quad \text{and} \quad \widetilde{A}_w^N(z) = A_w^N(z)B_w^N\,, \qquad z \in U_w,\ w \in D,$$

we obtain holomorphic functions $\widetilde{A}_w^M, \widetilde{A}_w^N : U_w \to \mathcal{G}^\infty(E)$ with

$$\widetilde{A}_w^M(z)M(z_0) = M(z)\,, \quad \widetilde{A}_w^N(z)N(z_0) = N(z)\,, \quad z \in U_w\,,\ w \in D. \tag{6.9.1}$$

Now we set

$$G_{vw}^M = \big(\widetilde{A}_v^M\big)^{-1}\widetilde{A}_w^M\Big|_{M(z_0)} \quad \text{and} \quad G_{vw}^N = \big(\widetilde{A}_v^N\big)^{-1}\widetilde{A}_w^N\Big|_{N(z_0)} \quad \text{on } U_v \cap U_w$$

for all $v, w \in D$ with $U_v \cap U_w \neq \emptyset$. Then, by (6.9.1), the family G_{vw}^M is a $\big(\{U_w\}_{w\in D}, \mathcal{O}^{GL(M(z_0))}\big)$-cocycle, and the family G_{vw}^N is a $\big(\{U_w\}_{w\in D}, \mathcal{O}^{GL(N(z_0))}\big)$-cocycle. Since the values of the functions G_{vw}^M belong to $\mathcal{G}^\infty\big(M(z_0)\big)$ and the values of the functions G_{vw}^N belong to $\mathcal{G}^\infty\big(N(z_0)\big)$, we can apply Theorem 5.12.5. Therefore we get holomorphic functions $G_w^M : U_w \to \mathcal{G}^\infty\big(M(z_0)\big)$ and $G_w^N : U_w \to \mathcal{G}^\infty\big(N(z_0)\big)$ such that

$$\big(\widetilde{A}_v^M\big)^{-1}\widetilde{A}_w^M\Big|_{M(z_0)} = G_{vw}^M = G_v^M(G_w^M)^{-1}$$

and

$$\big(\widetilde{A}_v^N\big)^{-1}\widetilde{A}_w^N\Big|_{N(z_0)} = G_{vw}^N = G_v^N(G_w^N)^{-1}$$

on $U_v \cap U_w$ for all $v, w \in D$ with $U_v \cap U_w \neq \emptyset$. It follows that

$$\widetilde{A}_w^M G_w^M P(z_0) = \widetilde{A}_v^M G_v^M P(z_0) \quad \text{on } U_v \cap U_w$$

and

$$\widetilde{A}_w^N G_w^N (I - P(z_0)) = \widetilde{A}_v^N G_v^N (I - P(z_0)) \quad \text{on } U_v \cap U_w$$

for all $v, w \in D$ with $U_v \cap U_w \neq \emptyset$. Therefore, we have a global holomorphic function $A : D \to \mathcal{G}^\infty(E)$ defined by

$$A = \widetilde{A}_w^M G_w^M P(z_0) + \widetilde{A}_w^N G_w^M (I - P(z_0)) \qquad \text{on } U_w,\ w \in D.$$

It follows from (6.9.1) that $A(z)M(z_0) = M(z)$ and $A(z)N(z_0) = N(z)$ for all $z \in D$. □

From Theorem 6.9.1 immediately follows:

6.9.2 Corollary. *Let $D \subseteq \mathbb{C}$ be a connected open set, and let $\{M(z)\}_{z\in D}$ be a holomorphic family of subspaces of E.*

(i) *If the spaces $M(z)$ are finite dimensional, then there exist holomorphic sections $f_1, \dots, f_r : E \to M$ such that $f_1(z), \dots, f_r(z)$ is a basis of $M(z)$, for all $z \in D$.*

(ii) *If the spaces $M(z)$ are of finite codimension in E, then there exist holomorphic functions $f_1, \dots, f_r : E \to E$ such that $f_1(z), \dots, f_r(z)$ induces a basis of $E/M(z)$, for all $z \in D$.*

6.10 One-sided and generalized invertible holomorphic operator functions

In this section E and F are two Banach spaces, and $D \subseteq \mathbb{C}$ is an open set. Then from Theorem 6.2.7 we obtain the following proposition:

6.10.1 Proposition. *Let let $A : D \to L(E, F)$ be holomorphic such that, for all $z \in D$, $\operatorname{Im} A(z)$ is closed. Then the following are equivalent:*

(i) *The function k_A (Def. 6.2.3) is continuous on D.*

(ii) *The family $\operatorname{Im} A$ (Def. 6.2.1) is continuous.*

(iii) *The family $\operatorname{Ker} A$ (Def. 6.2.1) is continuous.*

(iv) *For each compact set $K \subseteq D$, $\inf_{z\in K} k_A(z) > 0$.*

(v) *The family $\operatorname{Im} A$ is holomorphic.*

(vi) *The family $\operatorname{Ker} A$ is holomorphic.*

Proof. By Theorem 6.2.7, conditions (i)–(iv) are equivalent. Since here the function A is holomorphic, by definition of a holomorphic family of subspaces, (v) is equivalent to (ii) and (vi) is equivalent to (iii). □

6.10.2. An operator $A \in L(E, F)$ is called **generalized invertible** if there exists $B \in L(F, E)$ such that $ABA = A$ and $BAB = B$. The operator B then is called a **generalized inverse** of A.

6.10.3 Proposition. *An operator $A \in L(E, F)$ is generalized invertible, if and only if, there exist projectors $P \in L(E)$ and $Q \in L(F)$ such that $\operatorname{Im} P = \operatorname{Im} A$ and $\operatorname{Im} Q = \operatorname{Ker} A$, i.e., if the spaces $\operatorname{Im} A$ and $\operatorname{Ker} A$ are complemented.*

Proof. First assume that A has a generalized inverse B. Then

$$(AB)^2 = (ABA)B = AB \qquad \text{and} \qquad (BA)^2 = B(ABA) = BA.$$

Hence AB and BA are projectors. Obviously, $\operatorname{Im}(AB) \subseteq \operatorname{Im} A$, and from $A = (AB)A$ it follows that also $\operatorname{Im} A \subseteq \operatorname{Im}(AB)$, and thus $P := AB$ is a projector onto $\operatorname{Im} A$. Moreover it is clear that $\operatorname{Ker} A \subseteq \operatorname{Ker} BA$ and $A(BA) = A$ implies that $\operatorname{Ker} BA \subseteq \operatorname{Ker} A$. Thus BA is a projector whose kernel coincides with $\operatorname{Ker} A$. Hence $Q := I - BA$ is a projector onto $\operatorname{Ker} A$.

Conversely, if there exist such projectors P and Q, then the operator $\widehat{A} := A\big|_{\operatorname{Ker} Q}$ is an invertible operator from $\operatorname{Ker} Q$ onto $\operatorname{Im} A = \operatorname{Im} P$. Let $\widehat{A}^{-1} \in L(\operatorname{Im} P, \operatorname{Ker} Q)$ be the inverse of this operator, and set $B := \widehat{A}^{-1}P$. Then $ABA = (A\widehat{A}^{-1}P)A = PA = A$. Moreover, then $\widehat{A}^{-1}PA(I - Q) = I - Q$ and $\widehat{A}^{-1}P = (I - Q)\widehat{A}^{-1}P$, which implies that

$$\begin{aligned} BAB &= \left(\widehat{A}^{-1}PA\right)\left(\widehat{A}^{-1}P\right) = \left(\widehat{A}^{-1}PA\right)\left((I-Q)\widehat{A}^{-1}P\right) \\ &= \left(\widehat{A}^{-1}PA(I-Q)\right)\left(\widehat{A}^{-1}P\right) = (I-Q)\widehat{A}^{-1}P = \widehat{A}^{-1}P = B. \end{aligned}$$

□

6.10.4 Theorem. (i) *Let $A : D \to L(E, F)$ be holomorphic such that, for all $z \in D$, $A(z)$ is generalized invertible. Suppose the equivalent conditions* (i)–(vi) *in Proposition* 6.10.1 *are satisfied. Then there exists a holomorphic function $B : D \to L(F, E)$ with $ABA = A$ on D.*

(ii) *Let $A : D \to L(E, F)$ be holomorphic such that all values of A are left invertible. Then there exists a holomorphic function $B : D \to L(F, E)$ with $BA = I$ on D.*

(iii) *Let $A : D \to L(E, F)$ be holomorphic such that all values of A are right invertible. Then there exists a holomorphic function $B : D \to L(F, E)$ with $AB = I$ on D.*

Proof. (i) By Proposition 6.10.3, the spaces $\operatorname{Im} A(z)$ and $\operatorname{Ker} A(z)$ are complemented. Therefore, by Theorem 6.8.5, we have holomorphic functions $P : D \to L(F)$ and $Q : D \to L(E)$ such that, for all $z \in D$, the operators $P(z)$ and $Q(z)$

are projectors with $\operatorname{Im} P(z) = \operatorname{Im} A(z)$ and $\operatorname{Ker} Q(z) = \operatorname{Ker} A(z)$. Then, for each $z \in D$, $A(z)$ is invertible as an operator from $\operatorname{Im} Q(z)$ to $\operatorname{Im} A(z)$; let $A^{(-1)}(z)$ be the inverse of this operator. Set $B(z) = A^{(-1)}(z)P(z)$ for $z \in D$. Then it is clear that $A(z)B(z)A(z) = A(z)$ for all $z \in D$.

It remains to prove that the function B is holomorphic. Let $z_0 \in D$ be given. Since $\operatorname{Im} A$ and $\operatorname{Im} Q$ are Shubin families (Proposition 6.8.4), we can find a neighborhood U of z_0 and holomorphic functions $T : U \to GL(F)$ and $G : U \to GL(E)$ such that

$$T(z) \operatorname{Im} A(z_0) = \operatorname{Im} A(z) \quad \text{and} \quad G(z) \operatorname{Im} Q(z_0) = \operatorname{Im} Q(z), \quad z \in U.$$

Setting

$$S(z) = A^{(-1)}(z_0)T^{-1}(z)A(z)G(z)\Big|_{\operatorname{Im} Q(z_0)}, \qquad z \in U,$$

then we get a holomorphic function $S : U \to GL\big(\operatorname{Im} Q(z_0)\big)$. We claim that

$$B(z) = G(z)S^{-1}(z)A^{(-1)}(z_0)T^{-1}(z)P(z), \qquad z \in U. \tag{6.10.1}$$

Indeed, since $P(z)A(z) = A(z)$, we get

$$\begin{aligned}
&\Big(B(z) - G(z)S^{-1}(z)A^{(-1)}(z_0)T^{-1}(z)P(z)\Big)A(z)Q(z) \\
&= \Big(A^{(-1)}(z)P(z) - G(z)S^{-1}(z)A^{(-1)}(z_0)T^{-1}(z)P(z)\Big)A(z)Q(z) \\
&= Q(z) - G(z)S^{-1}(z)\Big(A^{(-1)}(z_0)T^{-1}(z)A(z)G(z)\Big)G^{-1}(z)Q(z) \\
&= Q(z) - G(z)S^{-1}(z)S(z)G^{-1}(z)Q(z) = 0, \quad z \in U.
\end{aligned}$$

This implies (6.10.1), as both sides of (6.10.1) vanish on $\operatorname{Ker} P(z)$ and $\operatorname{Im}\big(A(z)Q(z)\big) = \operatorname{Im} P(z)$,

$$B(z)P(z) = S^{-1}(z)A^{-1}(z_0)T^{-1}(z)P(z), \qquad z \in U.$$

Since the functions on the right-hand side of (6.10.1) are holomorphic, it follows that B is holomorphic.

(ii) Since the values of A are left invertible, $\operatorname{Ker} A(z) = \{0\}$ and $\operatorname{Im} A(z)$ is complemented, for all $z \in D$. Hence, by Proposition 6.10.1, $\operatorname{Im} A$ is a holomorphic family of complemented subspaces of F, and we can apply part (i) of the theorem. Therefore we have a holomorphic function $B : D \to L(F, E)$ with $ABA = A$ on D. Since the values of A are injective, this implies $BA = I$ on D.

(iii) Since the values of A are right invertible, $\operatorname{Im} A(z) = F$ and $\operatorname{Ker} A(z)$ is complemented, for all $z \in D$. Hence, by Proposition 6.10.1, $\operatorname{Ker} A$ is a holomorphic family of complemented subspaces of E, and we can apply part (i) of the theorem. Therefore we have a holomorphic function $B : D \to L(F, E)$ with $ABA = A$ on D. Since the values of A are surjective, this implies $AB = I$ on D. □

6.11 Example: A globally non-trivial complemented holomorphic family of subspaces

Here we construct a holomorphic family of complemented subspaces $\{M(z)\}_{z\in D}$ of a Banach space E which is not "globally continuously injective", by which we mean that there does not exist a Banach space M_0 and a continuous function $A : D \to L(M_0, E)$ with $\operatorname{Ker} A(z) = \{0\}$ and $\operatorname{Im} A(z) = M(z)$ for all $z \in D$.

6.11.1 Example. We use the notation from example 5.6.2. Additionally here we assume that

$$\frac{1}{\sqrt{8}} < r < \frac{1}{2}\,.$$

Now, by the block matrices

$$Z(z) := \begin{pmatrix} \left(\frac{1}{2} - z\right) I & -\left(\frac{1}{2} + z\right) I \\ \left(\frac{1}{2} + z\right) I & \left(\frac{1}{2} - z\right) I \end{pmatrix},$$

$$F_1(z) := \begin{pmatrix} A & 0 \\ 0 & I \end{pmatrix} Z \qquad \text{and} \qquad F_2(z) := Z \begin{pmatrix} I & 0 \\ 0 & A \end{pmatrix},$$

we define holomorphic operator functions $Z, F_1, F_2 : \mathbb{C} \to L(X \oplus X)$. Since

$$\det \begin{pmatrix} \frac{1}{2} - z & -\frac{1}{2} - z \\ \frac{1}{2} + z & \frac{1}{2} - z \end{pmatrix} = \frac{1}{4} + 2z^2 \neq 0 \qquad \text{for } z \neq \pm \frac{i}{\sqrt{8}},$$

these operator functions are invertible on

$$\mathbb{C} \setminus \left\{ \frac{i}{\sqrt{8}}\,,\ -\frac{i}{\sqrt{8}} \right\},$$

which is a neighborhood of $\overline{D}$. Setting

$$H(z) = \begin{cases} F_2(z) \begin{pmatrix} A & 0 \\ 0 & A^{-1} \end{pmatrix} F_1^{-1}(z) & \text{for } z \in V_1\,, \\ F_2(z) F_1^{-1}(z) & \text{for } z \in V_2\,, \end{cases}$$

we define a holomorphic operator function $H : U_1 \cap U_2 \to GL(X \oplus X)$. Since $-1/2 \in V_1$, $1/2 \in V_2$,

$$Z\left(-\frac{1}{2}\right) = \begin{pmatrix} I & 0 \\ 0 & I \end{pmatrix} \qquad \text{and} \qquad Z\left(\frac{1}{2}\right) = \begin{pmatrix} 0 & -I \\ I & 0 \end{pmatrix},$$

then we have

$$\begin{aligned} H\left(-\frac{1}{2}\right) &= Z\left(-\frac{1}{2}\right)\begin{pmatrix} I & 0 \\ 0 & A \end{pmatrix}\begin{pmatrix} A & 0 \\ 0 & A^{-1} \end{pmatrix} Z^{-1}\left(-\frac{1}{2}\right)\begin{pmatrix} A^{-1} & 0 \\ 0 & I \end{pmatrix} \\ &= \begin{pmatrix} I & 0 \\ 0 & I \end{pmatrix} \end{aligned}$$

and

$$\begin{aligned} H\left(\frac{1}{2}\right) &= Z\left(\frac{1}{2}\right)\begin{pmatrix} I & 0 \\ 0 & A \end{pmatrix} Z^{-1}\left(\frac{1}{2}\right)\begin{pmatrix} A^{-1} & 0 \\ 0 & I \end{pmatrix} \\ &= \begin{pmatrix} 0 & -I \\ I & 0 \end{pmatrix}\begin{pmatrix} I & 0 \\ 0 & A \end{pmatrix}\begin{pmatrix} 0 & I \\ -I & 0 \end{pmatrix}\begin{pmatrix} A^{-1} & 0 \\ 0 & I \end{pmatrix} \\ &= \begin{pmatrix} 0 & -A \\ I & 0 \end{pmatrix}\begin{pmatrix} 0 & I \\ -A^{-1} & 0 \end{pmatrix} = \begin{pmatrix} I & 0 \\ 0 & I \end{pmatrix}. \end{aligned}$$

Hence all values of H lie in the connected component of the unit operator in $GL(X \oplus X)$. Therefore, by Corollary 5.6.4, we can find holomorphic functions $H_j : U_j \to GL(X \oplus X)$ such that $H = H_2^{-1} H_1$ on $U_1 \cap U_2 = V_1 \cup V_2$, and therefore

$$H_2 F_2 \begin{pmatrix} A & 0 \\ 0 & A^{-1} \end{pmatrix} = H_1 F_1 \qquad \text{on } V_1 \tag{6.11.1}$$

and

$$H_2 F_2 = H_1 F_1 \qquad \text{on } V_2\,. \tag{6.11.2}$$

Since $H(-1/2)$ is the identical operator of $X \oplus X$, by multiplying by a constant operator, we may assume that also

$$H_1\left(-\frac{1}{2}\right) = H_2\left(-\frac{1}{2}\right) = \begin{pmatrix} I & 0 \\ 0 & I \end{pmatrix}. \tag{6.11.3}$$

Set $X_0 := X \oplus \{0\}$, and let $A_0\big((x,0)\big) := (Ax, 0)$ for $(x,0) \in X_0$. Since A_0 is an isomorphism of X_0, then it follows from (6.11.1) that

$$H_2 F_2\Big(X_0\Big) = H_2 F_2 \begin{pmatrix} A & 0 \\ 0 & A^{-1} \end{pmatrix}\Big(X_0\Big) = H_1 F_1\Big(X_0\Big) \qquad \text{on } V_1.$$

Together with (6.11.2) this implies that

$$H_2 F_2\Big(X_0\Big) = H_1 F_1\Big(X_0\Big) \qquad \text{on } U_1 \cap U_2 = V_1 \cup V_2.$$

Hence, there is a well-defined holomorphic family $\{M(z)\}_{z \in D}$ of complemented subspaces of $X \oplus X$ defined by

$$M(z) = H_j(z) F_j(z)\Big(X_0\Big) \qquad \text{for } z \in U_j\,,\ j = 1, 2. \tag{6.11.4}$$

By (6.11.3)

$$M\left(-\frac{1}{2}\right) = X_0. \tag{6.11.5}$$

We claim that there does not exist a continuous function $C : D \to L(X_0, X \oplus X)$ such that $\operatorname{Im} C(z) = M(z)$ and $\operatorname{Ker} C(z) = \{0\}$ for all $z \in D$.

Indeed, assume we have such a function C. Then, by (6.11.5), we may assume that

$$C\left(-\frac{1}{2}\right) = I. \tag{6.11.6}$$

For all $z \in D$, now we denote by $C^{(-1)}(z)$ the operator in $L(M(z), X_0)$ with $C^{(-1)}(z)C(z) = I$. Then, by setting

$$\gamma_j(z) = C^{(-1)}(z)H_j(z)F_j(z)\Big|_{X_0}, \qquad z \in U_j,$$

functions $\gamma_j : U_j \to GL(X_0)$, $j = 1, 2$, are well defined. By Corollary 6.2.5, these functions are continuous. From (6.11.2) we get

$$\gamma_1\left(\frac{1}{2}\right) = \gamma_2\left(\frac{1}{2}\right), \tag{6.11.7}$$

and from (6.11.3) and (6.11.6) it follows that

$$\gamma_1\left(-\frac{1}{2}\right) = F_1\left(-\frac{1}{2}\right)\Big|_{X_0} = \begin{pmatrix} A & 0 \\ 0 & I \end{pmatrix}\Big|_{X_0} = A_0 \tag{6.11.8}$$

and

$$\gamma_2\left(-\frac{1}{2}\right) = F_2\left(-\frac{1}{2}\right)\Big|_{X_0} = \begin{pmatrix} I & 0 \\ 0 & A \end{pmatrix}\Big|_{X_0} = I. \tag{6.11.9}$$

Since both U_1 and U_2 are connected, it follows from (6.11.7)–(6.11.9) that A_0 and I belong to the same connected component of $GL(X_0)$. This is a contradiction.

6.12 Comments

The gap metric was introduced and studied in [KKM, GK, GM]. The results on *continuous* (with respect to the gap metric) families of subspaces are not difficult and were observed by many authors.

The starting point of the theory of *holomorphic* families of subspaces was the paper of M.A. Shubin [Sh2], who gave the first definition and the first results, and the papers of the authors [GL1, GL2, GL3, Le5]. In those papers already most of the results for the more special families of subspaces studied in Sections 6.3, 6.4 and 6.6–6.11 can be found. Here this material is completed and extended in different directions.

Definition 6.4.1 and the local Lemma 6.4.6 are new in this generality. Using a terminology from Complex analysis in several variables, Lemma 6.4.6 states that the holomorphic families in the sense of Definition 6.4.1 are *Banach coherent sheaves* in the sense of [Le6], and the results then obtained in Section 6.4 could be viewed as special cases of that theory of Banach coherent sheaves. But we did not proceed in this way. Instead the more simple direct proofs are given which are possible in the case of one variable.

The example given in Section 6.5 is due to A. Douady [Do2].

Chapter 7

Plemelj-Muschelishvili factorization

This factorization for operator functions with respect to a contour is defined in the beginning of the first section. For scalar and matrix functions it was invented in the beginning of the last century as a tool for solving the linear Riemann-Hilbert boundary problem in complex analysis, singular integral equations and systems of such equations. It serves also as a tool for solving Wiener-Hopf equations and systems of Wiener-Hopf equations, both discrete and continuous. For details see [GK] and [F].

This type of factorization can be also used to solve transport equations, see [F] and [BGK]. For the plane symmetric case the mathematical equations describing the transport of energy through a medium is a linear integral-differential equation, which can be transformed into a Wiener-Hopf integral equation with an operator-valued kernel. The latter equation can be solved with the help of factorization of the mentioned earlier form of an operator-valued function acting in an infinite dimensional space, see [BGK]. The results on factorization which are mentioned here are also used in the spectral theory of operator polynomials, see [Ma] and [Ro] and the literature cited therein.

To transfer from the factorization problem of scalar functions to finite matrix functions already adds essential difficulties and appearance of the so-called partial indexes with a very complex behavior. The next step: factorization of operator functions is a much more difficult problem than the matrix function ones and in fact requires all techniques and tools which were developed in the previous chapters (except for Chapter 4). In fact these tools were started with this aim. As it turns out they are useful for many other purposes. This chapter is entirely dedicated to the factorization problem of operator functions [GL4, GL5].

7.1 Definitions and first remarks about factorization

Here $D_+ \subseteq \mathbb{C}$ is a bounded connected open set with piecewise $\mathcal{C}^1$-boundary Γ, $0 \in D_+$, and $D_- = \mathbb{C} \setminus \overline{D}_+$.

7.1.1 Definition. Let E be a Banach space, let G be one of the groups $GL(E)$, $\mathcal{G}^\infty(E)$ or $\mathcal{G}^\omega(E)$ (Def. 5.12.1), and let $A : \Gamma \to G$ be a continuous function. A representation of A in the form

$$A = A_- \Delta A_+ \tag{7.1.1}$$

will be called a **factorization** with respect to Γ and G of A if the factors A_-, A_+, Δ have the following properties:

- Either $\Delta \equiv I$ or Δ is of the form

$$\Delta(z) = P_0 + \sum_{j=1}^{n} z^{\kappa_j} P_j \tag{7.1.2}$$

  where $n \in \mathbb{N}^*$, $\kappa_1 > \ldots > \kappa_n$ are non-zero integers, $P_1, \ldots, P_n$ are mutually disjoint finite dimensional projectors[1] in E, and $P_0 = I - P_1 - \ldots - P_n$;
- A_+ is a continuous G-valued function on $\overline{D}_+$, which is holomorphic in D_+;
- A_- is a continuous G-valued function on $\overline{D}_- \cup \{\infty\}$, which is holomorphic in $D_- \cup \{\infty\}$ (Section 3.1.1).

If $\Delta \equiv I$, then this factorization will be called **canonical** (see also Definition 7.1.3).

By a factorization of A with respect to Γ (without mentioning a group G) we always mean a factorization of A with respect to Γ and $GL(E)$.

7.1.2. The integers $\kappa_1, \ldots, \kappa_n$ and the dimensions of the projectors $P_1, \ldots, P_n$ in this definition are uniquely determined by A. This will be established in Theorem 7.10.3 below. The numbers $\kappa_1, \ldots, \kappa_n$ will be called the **non-zero partial indices** of A (Def. 7.9.6), and $\dim P_j$ will be called the **multiplicity** of κ_j (Def. 7.9.8).

7.1.3 Definition. Let G be an open subgroup of the group of invertible elements of a Banach algebra with unit or let G be one of the groups $\mathcal{G}^\infty(E)$ or $\mathcal{G}^\omega(E)$, where E is a Banach space (Def. 5.12.1), and let $f : \Gamma \to G$ be continuous.

(i) For $z_0 \in \Gamma$, we say that f admits a **local factorization at** z_0 with respect to Γ and G if there exist a neighborhood U of z_0 and continuous functions $A_+ : U \cap \overline{D}_+ \to G$ and $A_- : U \cap \overline{D}_- \to G$, which are holomorphic in $U \cap D_-$ and $U \cap D_+$, respectively, such that

$$A = A_- A_+ \qquad \text{on } U \cap \Gamma\,. \tag{7.1.3}$$

We say that f admits **local factorizations** with respect to Γ and G if it admits at each point of Γ a local factorization with respect to Γ and G.

---

[1] By a projector in E we always mean a *bounded linear* projector in E. A family $P_1, \ldots, P_n$ of projectors in E is called **mutually disjoint** if $P_j P_k = 0$ for all $1 \le j, k \le n$ with $j \neq k$

(ii) If $g : \Gamma \to G$ is a second continuous function, then we say that f and g are **equivalent** with respect to Γ and G if there exist continuous functions $h_- : \overline{D}_- \cup \{\infty\} \to G$ and $h_+ : \overline{D}_+ \to G$, which are holomorphic in $D_- \cup \{\infty\}$ and D_+, respectively, such that

$$f = h_- g h_+ \qquad \text{on } \Gamma \,. \tag{7.1.4}$$

If, in this case, $g \equiv 1$, i.e., if (7.1.4) takes the form

$$f = h_- h_+ \qquad \text{on } \Gamma \,, \tag{7.1.5}$$

then this representation will be called a **canonical factorization** of f with respect to Γ and G.

As we already saw in the scalar case (Remark 3.11.4), not any continuous function admits local factorizations with respect to Γ. However already weak smoothness requirements ensure the existence of local factorizations with respect to Γ. In Section 7.2 we see that Wiener functions admit local factorizations. In Section 7.3 we prove that this is true also for Hölder functions.

7.1.4 Proposition. *Let G an open subgroup of the group of invertible elements of a Banach algebra with unit or let G be one of the groups $\mathcal{G}^\infty(E)$ or $\mathcal{G}^\omega(E)$, where E is a Banach space (Def. 5.12.1), and let $f : \Gamma \to G$ be continuous.*

Assume, at some point $z_0 \in \Gamma$, f admits two local factorizations with respect to Γ and G, i.e.,

$$f = h_- h_+ \qquad \textit{and} \qquad f = \widetilde{h}_- \widetilde{h}_+ \qquad \textit{on } \Gamma \cap U, \tag{7.1.6}$$

where U is a neighborhood of z_0 and $h_-, \widetilde{h}_- : U \cap \overline{D}_- \to GA$ and $h_+, \widetilde{h}_+ : U \cap \overline{D}_+ \to G$ are continuous functions, which are holomorphic in $U \cap D_-$ and $U \cap D_+$, respectively.

Then there is a holomorphic function $g : U \to G$ such that

$$\widetilde{h}_- = h_- g^{-1} \quad \textit{on} \quad \overline{D}_- \cap U \qquad \textit{and} \qquad \widetilde{h}_+ = g h_+ \quad \textit{on} \quad \overline{D}_+ \cap U. \tag{7.1.7}$$

Proof. By (7.1.6),

$$\widetilde{h}_-^{-1} h_- = \widetilde{h}_+ h_+^{-1} \qquad \text{on } \Gamma \cap U.$$

Hence, there is a continuous function $g : U \to G$ defined by

$$g = \begin{cases} \widetilde{h}_-^{-1} h_- & \text{on } \overline{D}_- \cap U, \\ \widetilde{h}_+ h_+^{-1} & \text{on } \overline{D}_+ \cap U, \end{cases}$$

which satisfies (7.1.7) by definition. As this function is holomorphic in $U \setminus \Gamma$, it follows from Theorem 1.5.4 that g is holomorphic on U. □

7.1.5 Proposition. *Let G an open subgroup of the group of invertible elements of a Banach algebra with unit or let G be one of the groups $\mathcal{G}^\infty(E)$ or $\mathcal{G}^\omega(E)$, where E is a Banach space (Def. 5.12.1), and let $f : \Gamma \to G$ be continuous.*

If f admits a canonical factorization $f = f_+ f_-$ with respect to Γ and G, then setting $\widetilde{f}_-(z) = f_-(z) f_-^{-1}(\infty)$ and $\widetilde{f}_+(z) = f_+(z) f_-(\infty)$ we get a canonical factorization $f = \widetilde{f}_- \widetilde{f}_+$ with the additional property $\widetilde{f}_-(\infty) = 1$.

With this additional property, the canonical factorization of f is uniquely determined by f.

Proof. Assume there are two canonical factorizations $f = f_- f_+$ and $f = \widetilde{f}_- \widetilde{f}_+$ of f with $f_-(\infty) = \widetilde{f}_-(\infty) = 1$. Then $f_- f_+ = \widetilde{f}_- \widetilde{f}_+$ and therefore

$$\widetilde{f}_-^{-1} f_- = \widetilde{f}_+ f_+^{-1} \qquad \text{on } \Gamma.$$

By Theorem 1.5.4, the two sides of this relation define a holomorphic function on $\mathbb{C} \cup \{\infty\}$ with value 1 at infinity. By Liouville's theorem this implies that $\widetilde{f}_- \equiv f_-$ and $\widetilde{f}_+ \equiv f_+$. □

7.1.6 Proposition. *Let E be a Banach space, and let $A : \Gamma \to GL(E)$ be a continuous function which admits a factorization with respect to Γ and GA.*

Then there exists always a factorization $A = A_- \Delta A_+$ with respect to Γ satisfying the additional condition $A_-(\infty) = I$.

Proof. If $A = \widetilde{A}_- \widetilde{\Delta} \widetilde{A}_+$ is an arbitrary factorization with respect to Γ and GA, then we obtain a factorization $A = A_- \Delta A_+$ with $A_-(\infty) = I$ by setting $A_-(z) = A_-(z)\big(A_-(\infty)\big)^{-1}$, $\Delta(z) = A_-(\infty)\widetilde{\Delta}(z)\big(A_-(\infty)\big)^{-1}$ and $A_+(z) = \widetilde{A}_-(\infty) A_+(\infty)$. □

7.1.7. Note however that the factorization with the extra condition $A_-(\infty) = I$ established in Proposition 7.1.6 is not uniquely determined, except for the case of a canonical factorization (Proposition 7.1.5).

7.2 The algebra of Wiener functions and other splitting $\mathcal{R}$-algebras

In this section $\mathbb{T}$ is the unit circle, D_+ is the open unit disc and $D_- = \mathbb{C} \setminus \overline{D}_+$.

7.2.1. Let A be a Banach algebra, and let $W(A)$ be the space of Wiener functions with values in A (Def. 3.1.5), i.e., the space of functions $f : \mathbb{T} \to A$ of the form

$$f(z) = \sum_{n=-\infty}^{\infty} z^n f_n \qquad \text{with} \qquad \|f\|_W := \sum_{n=-\infty}^{\infty} \|f_n\|_W < \infty.$$

Then it follows from Cauchy's product theorem that if $f, g \in W(A)$, then the pointwise defined product fg belongs again to $W(A)$, where

$$\|fg\|_W \leq \|f\|_W \|g\|_W.$$

Hence $W(A)$ is a Banach algebra.

In the present section we prove: If A is a Banach algebra with unit, then each function $f \in W(A)$ with invertible values is equivalent (Def. 7.1.3) to some holomorphic function on $\mathbb{T}$. If fact we prove the following stronger result:

7.2.2 Theorem. *Let A be a Banach algebra with unit, let GA be the group of invertible elements of A, and let $f \in W(A)$ such that $f(z) \in GA$ for all $z \in \mathbb{T}$. Then:*

(i) *The pointwise defined function f^{-1} again belongs to $W(A)$.*

(ii) *The function f can be written in the form $f = h_- h h_+$, where: $h : \mathbb{T} \to GA$ is holomorphic, $h_- : \overline{D}_- \cup \{\infty\} \to GA$ is continuous on $\overline{D}_- \cup \{\infty\}$ and holomorphic in $D_- \cup \{\infty\}$, $h_+ : \overline{D}_+ \to GA$ is continuous on $\overline{D}_+$ and holomorphic in D_+, and the functions h_-, h_-^{-1}, h_+ and h_+^{-1} belong to $W(A)$. (The latter statement follows also from* (i) *and* (iii).*)*

(iii) *Let $g : \mathbb{T} \to GA$ be a second function from $W(A)$ which is equivalent to f with respect to Γ and GA (Def. 7.1.3). Further, let $g_- : \overline{D}_- \cup \{\infty\} \to GA$ and $g_+ : \overline{D}_+ \to GA$ be any functions which are holomorphic in $D_- \cup \{\infty\}$ and $\overline{D}_+$, respectively, such that*

$$g = g_- f g_+ \qquad \text{on } \mathbb{T}. \tag{7.2.1}$$

(Such functions then exist by definition of equivalence.) Then the functions g_-, g_-^{-1}, g_+ and g_+^{-1} belong to $W(A)$.

The remainder of this section is devoted to the proof of this theorem. In fact we immediately prove a generalization of it (a second interesting example, covered by this generalization, we meet in Section 8.11).

7.2.3 Definition. Let E be a Banach space. By a **rational function with values in** E or by a **rational function** $f : \mathbb{C} \to E$ we mean an E-valued meromorphic function on $\mathbb{C}$ (Section 1.10.6) such that $f(1/z)$ is also meromorphic on $\mathbb{C}$. It is easy to see that any rational function with values in E can be written in the form

$$f = \frac{p}{\varphi},$$

where p is a holomorphic polynomial with coefficients in E, and φ is a scalar holomorphic polynomial. We shall say that ∞ is a **pole** of an E-valued rational function f if 0 is a pole of the function $f(1/z)$ (Section 1.10.6).

7.2.4. Let A be a Banach algebra with the norm $\|\cdot\|$. Then we are interested in Banach algebras $\mathcal{R}$ with the norm $\|\cdot\|_{\mathcal{R}}$, which are algebraic subalgebras of the algebra of continuous functions $f : \mathbb{T} \to A$ and which satisfy the following three conditions:

(A) $\max_{z\in\mathbb{T}} \|f(z)\| \le \|f\|_{\mathcal{R}}$ for all $f \in \mathcal{R}$.

(B) The algebra of all A-valued rational functions without poles on $\mathbb{T}$ is contained in $\mathcal{R}$ as a dense subset.

(C) Each $f \in \mathcal{R}$ admits a splitting $f = f_+ + f_-$ with respect to $\mathbb{T}$ (Def. 3.1.2), where $f_-, f_+ \in \mathcal{R}$.

Sometimes such algebras will be called **splitting $\mathcal{R}$-algebras.**

Obviously, the Wiener algebra $W(A)$ (Section 7.2.1) is a splitting $\mathcal{R}$-algebra.

7.2.5 Theorem. *Let A be a Banach algebra with unit, let GA be the group of invertible elements of A, and let $\mathcal{R}$ be a Banach algebra of continuous functions $f : \mathbb{T} \to A$ satisfying conditions* (A), (B) *and* (C) *from Section* 7.2.4. *Let $f \in \mathcal{R}$ such that $f(z) \in GA$ for all $z \in \mathbb{T}$. Then:*

(i) *The pointwise defined function f^{-1} again belongs to $\mathcal{R}$.*

(ii) *The function f can be written in the form*

$$f = h_- h h_+, \tag{7.2.2}$$

where $h_- : \overline{D}_- \cup \{\infty\} \to GA$ is continuous on $\overline{D}_- \cup \{\infty\}$ and holomorphic in $D_- \cup \{\infty\}$, $h_+ : \overline{D}_+ \to GA$ is continuous on $\overline{D}_+$ and holomorphic in D_+, h is an A-valued rational function without poles on $\mathbb{T}$, $h(z) \in GA$ for all $z \in \mathbb{T}$, and the functions h_-, h_-^{-1}, h_+ and h_+^{-1} belong to $\mathcal{R}$. (The latter statement follows also from (i) *and* (iii)*.)*

(iii) *Let $g : \mathbb{T} \to GA$ be a second function from $\mathcal{R}$ which is equivalent to f with respect to Γ and GA (Def.* 7.1.3*). Further, let $g_- : \overline{D}_- \cup \{\infty\} \to GA$ and $g_+ : \overline{D}_+ \to GA$ be any functions which are holomorphic in $D_- \cup \{\infty\}$ and $\overline{D}_+$, respectively, such that*

$$g = g_- f g_+ \qquad \text{on } \mathbb{T}. \tag{7.2.3}$$

(Such functions then exist by definition of equivalence.) Then the functions g_-, g_-^{-1}, g_+ and g_+^{-1} belong to $\mathcal{R}$.

We first prove the corresponding generalization of Proposition 3.10.1:

7.2.6 Proposition. *Let $\mathcal{R}$ be a Banach algebra of scalar continuous functions $f : \mathbb{T} \to \mathbb{C}$ satisfying conditions* (A), (B) *and* (C) *from Section* 7.2.4 *(with $A = \mathbb{C}$). If $f \in \mathcal{R}$ and $f(z) \neq 0$ for all $z \in \mathbb{T}$, then $f^{-1} \in \mathcal{R}$.*

Proof. Let $f \in \mathcal{R}$ be given, which is not an invertible element of $\mathcal{R}$. We have to find $\theta \in \mathbb{T}$ with $f(\theta) = 0$.

Since f is not invertible, by the theory of commutative Banach algebras, there exits a multiplicative functional Φ on $\mathcal{R}$ with

$$\Phi(f) = 0.$$

By condition (B), for all fixed complex numbers $\lambda \in \mathbb{C} \setminus \mathbb{T}$, the function $z - \lambda$ is an invertible element of $\mathcal{R}$. Hence, for each fixed $\lambda \in \mathbb{C} \setminus \mathbb{T}$, $\Phi(z - \lambda) \neq 0$ and, therefore,

$$\Phi(z) - \lambda = \Phi(z - \lambda) \neq 0.$$

Hence $\theta := \Phi(z) \in \mathbb{T}$. Since Φ is multiplicative and linear, this implies that, for each scalar rational function r without poles on $\mathbb{T}$ (by condition (B) such functions belong to $\mathcal{R}$), we have

$$r(\theta) = \Phi(r).$$

Since the rational functions are dense in $\mathcal{R}$ (condition (B)) and by condition (A), this further implies that $g(\theta) = \Phi(g)$ for all $g \in \mathcal{R}$. Hence $f(\theta) = \Phi(f) = 0$. □

Further, the proof of Theorem 7.2.5 is based on the following

7.2.7 Lemma. *Let B be a Banach algebra with unit, and let L be a maximal left ideal of B. Further, let Z be a closed subalgebra of B such that $1 \in Z$ and $zb = bz$ for all $b \in B$ and $z \in Z$ (i.e., Z is a subalgebra of the center of B with $1 \in Z$). Then $L \cap Z$ is a maximal ideal of Z.*

Proof. Here the unit element of B will be denoted by e. First we prove:

Proposition ♣: For each $a \in Z \setminus L$ we have:

(i) There exits $y \in B$ with $ya - e \in L$.

(ii) If $y_1, y_2 \in B$ with $y_1 a - e \in L$ and $y_2 a - e \in L$, then $y_1 - y_2 \in L$.

Part (i) follows from the hypotheses that L is maximal among the proper left ideals of B. (Otherwise the set of elements of the form $l + ya$, $l \in L, y \in B$, would be a bigger proper left ideal.)

For part (ii), we first prove that

$$L = \left\{ y \in B \;\middle|\; ya \in L \right\}. \tag{7.2.4}$$

Let K_a be the set on the right-hand side of (7.2.4). It is clear that K_a is a left ideal of B. Since $1 \notin K_a$, K_a is even a proper left ideal of B. If $l \in L$, then $la = al \in L$ (as a belongs to the center of B and L is a left ideal). Hence $L \subseteq K_a$. Since L is maximal among the proper left ideals, this means that $L = K_a$.

Now let $y_1, y_2 \in B$ with $y_1 a - e \in L$ and $y_2 a - e \in L$ be given. Then $(y_1 - y_2)a = y_1 a - e - (y_2 a - e) \in L$, which implies by (7.2.4) that $y_1 - y_2 \in L$.

Proposition ♣ is proved.

Now we prove that $Z \cap L$ is a maximal ideal of Z. It is clear that $Z \cap L$ is an ideal of Z. Assume it is not maximal. Then there exists $a \in Z$ with $a - ze \notin L$ for all $z \in \mathbb{C}$. Therefore, by part (i) of proposition ♣, for each $z \in \mathbb{C}$, we can choose an element $f(z) \in B$ with

$$f(z)(a - z) - e \in L. \tag{7.2.5}$$

To complete the proof, now it is sufficient to show that

$$f(z) \in L \qquad \text{for all } z \in \mathbb{C}, \tag{7.2.6}$$

because then from (7.2.5) we get the contradiction $e \in L$.

Let B/L be the factor space of B with respect to L, and let $\pi : B \to B/L$ be the canonical projector. (7.2.6) then is equivalent to the relation

$$(\pi \circ f)(z) = 0 \qquad \text{for all } z \in \mathbb{C}.$$

Therefore, by Liouville's theorem, it is sufficient to prove the folllowing

Proposition ♠*:* The function $\pi \circ f$ is holomorphic on $\mathbb{C}$ and $(\pi \circ f)(\infty) = 0$.

Setting

$$f^*(z) = \frac{1}{z}\left(\frac{a}{z} - e\right)^{-1} \qquad \text{for } |z| > \frac{1}{\|a\|}$$

we define a B-valued holomorphic function f^* such that

$$f^*(z)(a - ze) - e = 0 \in L \qquad \text{for all } |z| > \frac{1}{\|a\|}.$$

By (7.2.5) and part (ii) of proposition ♣, this implies that $f(z) - f^*(z) \in L$ for $|z| > 1/\|a\|$. Hence $\pi \circ f$ is equal to $\pi \circ f^*$ for $|z| > 1/\|a\|$. As f^* is holomorphic and $f^*(\infty) = 0$, this proves that $\pi \circ f$ is holomorphic for $|z| > 1/\|a\|$ and $(\pi \circ f)(\infty) = 0$.

It remains to prove that $\pi \circ f$ is holomorphic everywhere on $\mathbb{C}$. Let $\xi \in \mathbb{C}$ be given, and let U be the open disc centered at ξ and with radius $1/\|f(\xi)\|$. Then for all $z \in U$, $e - (z - \xi)f(\xi)$ is an invertible element of B, and, setting

$$f^\xi(z) = f(\xi)\Big(e - (z - \xi)f(\xi)\Big)^{-1}$$

we get a holomorphic function $f^\xi : U \to B$. Note that

$$f^\xi(z) = \sum_{k=0}^{\infty} (z - \xi)^k \big(f(\xi)\big)^{k+1}, \qquad z \in U.$$

Therefore

$$f^\xi(z)\Big((z - \xi)e - \big(f(\xi)\big)^{-1}\Big) = -e$$

and, further,

$$f^\xi(z)(a - ze) - e = f^\xi(z)\Big((a - ze) + (z - \xi e) - \big(f(\xi)\big)^{-1}\Big)$$
$$= f^\xi(z)\Big((a - \xi e) - \big(f(\xi)\big)^{-1}\Big) = f^\xi(z)\big(f(\xi)\big)^{-1}\Big(f(\xi)(a - \xi e) - e\Big)$$

for all $z \in U$. Since, by (7.2.5), $f(\xi)(a - \xi e) - e \in L$ and L is a left ideal, it follows that $f^\xi(z)(a - ze) - e \in L$ for all $z \in U$. Again by (7.2.5) and by part (ii) of proposition ♣, this implies that $f^\xi(z) - f(z) \in L$ for all $z \in U$. Hence $\pi \circ f = \pi \circ f^\xi$ on U. As f^ξ is holomorphic in U, this further implies that $\pi \circ f$ is holomorphic in U. Proposition ♠ is proved. □

7.2.8. Proof of Theorem 7.2.5 (i): Assume f is not an invertible element of the algebra $\mathcal{R}$. Then the set of elements of the form gf, $g \in \mathcal{R}$, is a proper left ideal of $\mathbb{R}$. Let L be a maximal left ideal in $\mathcal{R}$ with

$$\Big\{gf \;\Big|\; g \in \mathbb{R}\Big\} \subseteq L. \tag{7.2.7}$$

Further, let $\mathcal{R}_\mathbb{C}$ be the subalgebra of $\mathcal{R}$, which consists of the functions of the form ψe, where ψ is a scalar function and e is the unit element of A. By Lemma 7.2.7, $L \cap \mathbb{R}_\mathbb{C}$ is a maximal ideal of $\mathcal{R}_\mathbb{C}$, and it follows from Proposition 7.2.6 (and the theory of commutative Banach algebras) that there exists a point $\theta \in \mathbb{T}$ such that

$$L \cap \mathcal{R}_\mathbb{C} = \Big\{\varphi \in \mathcal{R}_\mathbb{C} \;\Big|\; \varphi(\theta) = 0\Big\}. \tag{7.2.8}$$

Therefore, for all $\varphi \in \mathbb{R}_\mathbb{C}$, the function

$$\varphi(z) - \varphi(\theta), \qquad z \in \mathbb{T},$$

belongs to $L \cap \mathcal{R}_\mathbb{C}$. Since L is a left ideal, it follows that, for all $a \in A$ and $\varphi \in L \cap \mathcal{R}_\mathbb{C}$, the function

$$a\varphi(z) - a\varphi(\theta), \qquad z \in \mathbb{T},$$

belongs to L. Hence, for all rational functions $r : \mathbb{C} \to A$ without poles on $\mathbb{T}$ (which belong to $\mathcal{R}$ by condition (B)), the function

$$r(z) - r(\theta), \qquad z \in \mathbb{T},$$

belongs to L.

Now we introduce the function $g \in \mathcal{R}$ defined by

$$g(z) := \big(f(\theta)\big)^{-1} f(z), \qquad z \in \mathbb{T}.$$

Then, by condition (B), there is a sequence $r_\nu : \mathbb{T} \to A$ of rational functions without poles on $\mathbb{T}$ such that

$$\lim_\nu \|g - r_\nu\|_\mathcal{R} = 0.$$

As $g(\theta) = e$ (the unit element of A), then it follows from condition (A), that also

$$\lim_{\nu} \|e - r_\nu(\theta)\| = \lim_{\nu} \|g(\theta) - r_\nu(\theta)\| = 0.$$

Hence, the sequence of functions

$$r_\nu(z) - r_\nu(\theta), \qquad z \in \mathbb{T},$$

converges to the function

$$g(z) - e, \qquad z \in \mathbb{T},$$

with respect to the norm of $\mathcal{R}$. Since L is closed with respect to this norm and each of the functions $r_\nu(z) - r_\nu(\theta)$ belongs to L, it follows that the function

$$g(z) - e, \qquad z \in \mathbb{T},$$

belongs to L. As, by definition of L, also $g \in L$, this implies that the constant function with value e, i.e., the unit element of $\mathcal{R}$, belongs to L. As L is a proper ideal, this is a contradiction. □

7.2.9. Let A be a Banach algebra with unit, and let $\mathcal{R}$ be a Banach algebra of continuous functions $f : \mathbb{T} \to A$ satisfying conditions (A), (B) and (C) from Section 7.2.4.

Then we denote by $\mathcal{R}_+$ the subalgebra of all $f \in \mathcal{R}$, which admit a continuous extension to $\overline{D}_+$, which is holomorphic in D_+, and by $\mathcal{R}_-$ we denote the subalgebra of all $f \in \mathcal{R}$, which admit a continuous extension to $\overline{D}_- \cup \{\infty\}$, which is holomorphic in $D_- \cup \{\infty\}$. Further, we denote by $\mathcal{R}_-^0$ the subalgebra of all $f \in \mathcal{R}_-(A)$ with $f(\infty) = 0$. Since all constant functions with value in A belong to $\mathcal{R}$ (condition (B)), it follows from condition (C) that $\mathcal{R}$ is the direct sum of $\mathcal{R}_+$ and $\mathcal{R}_-^0$. We denote by P_+ the linear projector from $\mathcal{R}$ to $\mathcal{R}_+$ parallel to $\mathcal{R}_-^0$, and we set $P_- = I - P_+$.

We denote by $\|P_\pm\|$ the operator norms of $P_\pm$ with respect to the norm $\|\cdot\|_{\mathcal{R}}$ of $\mathcal{R}$:

$$\|P_\pm\| := \sup_{f \in \mathcal{R}, \|f\|_{\mathcal{R}} = 1} \|P_\pm f\|_{\mathcal{R}}. \tag{7.2.9}$$

For $\mathcal{R} = W(A)$ (Section 7.2.1), obviously, $\|P_\pm\| = 1$.

Now, as a special case of the factorization Lemma 5.2.3, we immediately obtain:

7.2.10 Lemma. *Let A be a Banach algebra with unit, let $\mathcal{R}$ be a Banach algebra of continuous functions $f : \mathbb{T} \to A$ satisfying conditions* (A), (B) *and* (C) *from Section* 7.2.4*, and let*

$$C = max\{\|P_+\|, \|P_-\|\},$$

where P_+, P_- are the projectors defined in Section 7.2.9*. Then each $f \in \mathcal{R}$ with*

$$\|f - 1\|_{\mathcal{R}} < \frac{1}{C} \tag{7.2.10}$$

admits a canonical factorization $f = f_- f_+$ *with respect to* $\mathbb{T}$ *(Def. 7.1.3), where* $f_-, f_+ \in \mathcal{R}$ *and*

$$\|f_-^{-1} - 1\|_{\mathcal{R}} < \frac{C\|f-1\|_{\mathcal{R}}}{1 - C\|f-1\|_{\mathcal{R}}} \quad \textit{and} \quad \|f_+ - 1\|_{\mathcal{R}} < \frac{C\|f-1\|_{\mathcal{R}}}{1 - C\|f-1\|_{\mathcal{R}}}.$$

7.2.11. Proof of part (ii) of Theorem 7.2.5: Here we use without further references that part (i) of the theorem is already proved. Let $f \in \mathcal{R}$ be given. We use the notations introduced in Section 7.2.9, and set

$$C = max\{\|P_+\|, \|P_-\|\}.$$

By part (i) of Theorem 7.2.5 (proved in Section 7.2.8), f is an invertible element of $\mathcal{R}$. By condition (B), we can find a rational function $q : \mathbb{C} \to A$ without poles on $\mathbb{T}$ such that

$$\big\|f^{-1} - q\big\|_{\mathcal{R}} < \frac{1}{C\|f\|_{\mathcal{R}}}. \tag{7.2.11}$$

By condition (A) we can moreover assume that $q(z) \in GA$ for all $z \in \mathbb{T}$, where GA is the group of invertible elements of A. Then it follows that

$$\big\|qf - 1\big\|_{\mathcal{R}} \le \big\|f\big\|_{\mathcal{R}} \big\|f^{-1} - q\big\|_{\mathcal{R}} < \frac{1}{C}.$$

From (7.2.11) and Lemma 7.2.10 it follows that the function qf admits a canonical factorization

$$qf = g_- g_+ \tag{7.2.12}$$

with respect to $\mathbb{T}$, where $g_- \in \mathcal{R}_-$ and $g_+ \in \mathcal{R}_+$. By part (i) of Theorem 7.2.5 then also $g_-^{-1} \in \mathcal{R}_-$ and $g_+^{-1} \in \mathcal{R}_+$.

It follows that $g_-^{-1} q \in \mathcal{R}$, and from part (i) of Theorem 7.2.5 it follows that also $q^{-1} g_- \in \mathcal{R}$. By condition (B) we can find a rational function $p : \mathbb{C} \to A$ without poles on $\mathbb{T}$ with

$$\big\|g_-^{-1} q - p\big\|_{\mathcal{R}} < \frac{1}{C\|q^{-1} g_-\|_{\mathcal{R}}}. \tag{7.2.13}$$

Moreover, by condition (A), we may assume that $p(z) \in GA$ for all $z \in \mathbb{T}$. From (7.2.13) it follows that

$$\big\|q^{-1} g_- p - 1\big\|_{\mathcal{R}} \le \big\|q^{-1} g_-\big\|_{\mathcal{R}} \big\|p - g_-^{-1} q\big\|_{\mathcal{R}} < \frac{1}{C}.$$

Therefore, again by Lemma 7.2.10, $q^{-1} g_- p$ admits a canonical factorization

$$q^{-1} g_- p = h_- h_+ \tag{7.2.14}$$

with respect to $\mathbb{T}$, where $h_\pm \in \mathcal{R}$. Then

$$h_+ p^{-1} = h_-^{-1} q^{-1} g_-.$$

Let U be a neighborhood of $\mathbb{T}$ such that p and q are holomorphic and with values in GA on U. Then this equality shows (by Theorem 1.5.4) that there is an A-valued rational function h such that $h(z) \in GA$ for all $z \in \mathbb{T}$ which is defined by

$$h = \begin{cases} h_+ p^{-1} & \text{on } U \cap \overline{D}_+, \\ h_-^{-1} q^{-1} g_- & \text{on } U \cap \overline{D}_-. \end{cases}$$

For this function we have, by (7.2.12) and (7.2.14),

$$f = q^{-1} g_- g_+ = q^{-1} g_- p p^{-1} g_+ = h_- h_+ p^{-1} g_+ = h_- h g_+ \qquad \text{on } \mathbb{T}.$$

This completes the proof of part (ii) of Theorem 7.2.5. □

7.2.12. Proof of part (iii) of Theorem 7.2.5: By part (ii) of the theorem, the functions f and g can be written in the form

$$f = u_- p u_+ \qquad \text{and} \qquad g = v_- q v_+, \tag{7.2.15}$$

where p, q are A-valued rational functions, which are holomorphic and invertible in some neighborhood of $\mathbb{T}$, $u_-, v_- : \overline{D}_- \cup \{\infty\} \to GA$ are continuous on $\overline{D}_- \cup \{\infty\}$ and holomorphic in $D_- \cup \{\infty\}$, $u_+, v_+ : \overline{D}_+ \to GA$ are continuous on $\overline{D}_+$ and holomorphic in D_+, and moreover

$$u_\pm^{\pm 1}, v_\pm^{\pm 1} \in \mathcal{R}. \tag{7.2.16}$$

From (7.2.15) and (7.2.3) we get

$$v_- p v_+ = g_- u_- q u_+ g_+$$

and therefore

$$u_-^{-1} g_-^{-1} v_- = q u_+ g_+ v_+^{-1} p^{-1} \quad \text{and} \quad u_+ g_+ v_+^{-1} = q^{-1} u_-^{-1} g_-^{-1} v_- p^{-1} \qquad \text{on } \mathbb{T}.$$

Since the functions p and q are A-valued rational functions, which are holomorphic and invertible in a neighborhood of $\mathbb{T}$, these equalities imply (by Theorem 1.5.4 and the corresponding properties of the functions $u_\pm$ and $v_\pm$) that also $u_-^{-1} g_-^{-1} v_-$ and $u_+ g_+ v_+^{-1}$ are A-valued rational functions, which are holomorphic and invertible in a neighborhood of $\mathbb{T}$. Hence, these functions belong to $\mathcal{R}$ (by property (B)). In view of (7.2.16), this further implies that $g_+^{\pm 1}$ and $g_-^{\pm 1}$ belong to $\mathcal{R}$. □

7.3 Hölder continuous and differentiable functions

In this section $D_+ \subseteq \mathbb{C}$ is a bounded, connected, open set with piecewise $\mathcal{C}^1$-boundary Γ such that $0 \in D_+$, and $D_- := \mathbb{C} \setminus \overline{D}_+$. Further, A is a Banach algebra with unit 1, and GA is the group of invertible elements of A.

In this section we prove that each Hölder-α continuous function $f : \Gamma \to GA$, $0 < \alpha < 1$, is equivalent (Def. 7.1.3) with respect to Γ to a function, which is holomorphic in some neighborhood of Γ. If Γ is of class $\mathcal{C}^k$ (Def. 3.4.1) and f is of class $\mathcal{C}^{k+\alpha}$ (Def. 3.4.3), then the result is correspondingly stronger. We prove:

7.3.1 Theorem. *Let $0 < \alpha < 1$, $k \in \mathbb{N}$, and let $f : \Gamma \to GA$ be a function such that: If $k = 0$, then f is of class $\mathcal{C}^\alpha$ (Def. 2.1.6). If $k \geq 1$, then Γ is of class $\mathbb{C}^k$ (Def. 3.4.1) and f is of class $\mathcal{C}^{k+\alpha}$ (Def. 3.4.3). Then:*

(i) *The function f can be written in the form*

$$f = f_- h f_+ \qquad \text{on } \Gamma, \tag{7.3.1}$$

where $f_- : \overline{D}_- \cup \{\infty\} \to GA$ is continuous on $\overline{D}_- \cup \{\infty\}$ and holomorphic in $D_- \cup \{\infty\}$, $f_+ : \overline{D}_+ \to GA$ is continuous on $\overline{D}_+$ and holomorphic in D_+, and h is a holomorphic GA-valued function in some neighborhood of Γ.

(ii) *If two functions f_- and f_+ are as in part (i), then, automatically, f_- is of class $\mathcal{C}^{k+\alpha}$ on $\overline{D}_-$, and f_+ is of class $\mathcal{C}^{k+\alpha}$ on $\overline{D}_+$.*

The proof of this theorem will be given at the end of the section.

7.3.2 Definition. Let $0 \leq \alpha < 1$.

Then we denote by $(\mathcal{C}^\alpha)^A(\Gamma)$ the Banach algebra of all Hölder-α continuous functions $f : \Gamma \to A$, endowed with the norm $\|f\|_{\Gamma,\alpha}$ (Def. 2.1.6).

If Γ is of class $\mathcal{C}^k$, $k \in \mathbb{N}^*$, then we denote by $(\mathcal{C}^{k+\alpha})^A(\Gamma)$ the algebra of all functions $f : \Gamma \to A$ of class $\mathcal{C}^{k+\alpha}$ (Def. 3.4.3). In $(\mathcal{C}^{k+\alpha})^A(\Gamma)$ we introduce the following norm

$$\|fg\|_{\Gamma,k+\alpha} := \sum_{n=0}^{k} \frac{1}{n!} \|(fg)^{(n)}\|_{\Gamma,\alpha}. \tag{7.3.2}$$

7.3.3 Lemma. *Suppose Γ is of class $\mathcal{C}^k$, $k \in \mathbb{N}^*$, and let $0 \leq \alpha < 1$. Then $(\mathcal{C}^{k+\alpha})^A(\Gamma)$, endowed with the norm (7.3.2), is a Banach algebra.*

Proof. It is clear that $(\mathcal{C}^{k+\alpha})^A(\Gamma)$ is a Banach *space* with the norm (7.3.2). Let $f, g \in (\mathcal{C}^{k+\alpha})^A(\Gamma)$. It remains to prove that

$$\|fg\|_{\Gamma,k+\alpha} \leq \|f\|_{\Gamma,k+\alpha} \|g\|_{\Gamma,k+\alpha}. \tag{7.3.3}$$

As $(\mathcal{C}^\alpha)^A(\Gamma)$ is a Banach algebra with respect to $\|\cdot\|_{\Gamma,\alpha}$, we have

$$\begin{aligned}
\|fg\|_{\Gamma,k+\alpha} &= \sum_{n=0}^{k} \frac{1}{n!} \|(fg)^{(n)}\|_{\Gamma,\alpha} \\
&= \sum_{n=0}^{k} \frac{1}{n!} \left\| \sum_{j=1}^{n} \binom{n}{j} f^{(j)} f^{(n-j)} \right\|_{\Gamma,\alpha} = \sum_{n=0}^{k} \left\| \sum_{j=1}^{n} \frac{1}{j!(n-j)!} f^{(j)} f^{(n-j)} \right\|_{\Gamma,\alpha} \\
&\leq \sum_{n=0}^{k} \sum_{j=1}^{n} \frac{1}{j!(n-j)!} \|f^{(j)}\|_{\Gamma,\alpha} \|f^{(n-j)}\|_{\Gamma,\alpha} \\
&= \sum_{0 \leq j \leq n \leq k} \frac{1}{j!(n-j)!} \|f^{(j)}\|_{\Gamma,\alpha} \|f^{(n-j)}\|_{\Gamma,\alpha}.
\end{aligned} \tag{7.3.4}$$

On the other hand

$$\begin{aligned}\|f\|_{\Gamma,k+\alpha}\|g\|_{\Gamma,k+\alpha} &= \left(\sum_{n=0}^{k}\frac{1}{n!}\|f^{(n)}\|_{\Gamma,\alpha}\right)\left(\sum_{n=0}^{k}\frac{1}{n!}\|g^{(n)}\|_{\Gamma,\alpha}\right)\\ &= \sum_{n,m=0}^{k}\frac{1}{n!m!}\|f^{(n)}\|_{\Gamma,\alpha}\|f^{(m)}\|_{\Gamma,\alpha}\end{aligned} \tag{7.3.5}$$

Comparing (7.3.4) and (7.3.5) we get (7.3.3). □

7.3.4 Lemma. *Let $0 \le \alpha < 1$ and $k \in \mathbb{N}$. If $k \ge 1$, then we additionally assume that Γ is of class $\mathcal{C}^k$. Let $f, g \in (\mathcal{C}^{k+\alpha})^A(\Gamma)$. Then*

$$\|fg\|_{\Gamma,k+\alpha} \le \|f\|_{\Gamma,k}\|g\|_{\Gamma,k+\alpha} + \|f\|_{\Gamma,k+\alpha}\|g\|_{\Gamma,k}. \tag{7.3.6}$$

Proof. For all $z, w \in \Gamma$ with $z \ne w$, we have

$$\begin{aligned}\frac{\|f(z)g(z)-f(w)g(w)\|}{|z-w|^\alpha} &= \frac{\|f(z)g(z)-f(z)g(w)+f(z)g(w)-f(w)g(w)\|}{|z-w|^\alpha}\\ &\le \|f(z)\|\frac{\|g(z)-g(w)\|}{|z-w|^\alpha} + \frac{\|f(z)-f(w)\|}{|z-w|^\alpha}\|g(w)\|\\ &\le \|f\|_{\Gamma,0}\frac{\|g(z)-g(w)\|}{|z-w|^\alpha} + \frac{\|f(z)-f(w)\|}{|z-w|^\alpha}\|g\|_{\Gamma,0}.\end{aligned}$$

Hence

$$\begin{aligned}\|fg\|_{\Gamma,\alpha} &= \|fg\|_{\Gamma,0} + \sup_{z,w\in\Gamma,z\ne w}\frac{\|f(z)g(z)-f(w)g(w)\|}{|z-w|^\alpha}\\ &\le \|fg\|_{\Gamma,0} + \|f\|_{\Gamma,0}\sup_{z,w\in\Gamma,z\ne w}\frac{\|g(z)-g(w)\|}{|z-w|^\alpha} + \sup_{z,w\in\Gamma,z\ne w}\frac{\|f(z)-f(w)\|}{|z-w|^\alpha}\|g\|_{\Gamma,0}\\ &= \|f\|_{\Gamma,0}\left(\|g\|_{\Gamma,0} + \sup_{z,w\in\Gamma,z\ne w}\frac{\|g(z)-g(w)\|}{|z-w|^\alpha}\right) + \sup_{z,w\in\Gamma,z\ne w}\frac{\|f(z)-f(w)\|}{|z-w|^\alpha}\|g\|_{\Gamma,0}\\ &\le \|f\|_{\Gamma,0}\|g\|_{\Gamma,\alpha} + \|f\|_{\Gamma,\alpha}\|g\|_{\Gamma,0},\end{aligned}$$

which is the assertion of the lemma for $k = 0$. If $k \ge 1$, then this further implies

$$\begin{aligned}\|fg\|_{\Gamma,k+\alpha} &= \sum_{n=0}^{k}\frac{1}{n!}\|(fg)^{(n)}\|_{\Gamma,\alpha} = \sum_{n=0}^{k}\left\|\sum_{j=1}^{n}\frac{1}{j!(n-j)!}f^{(j)}f^{(n-j)}\right\|_{\Gamma,\alpha}\\ &\le \sum_{0\le j\le n\le k}\frac{1}{j!(n-j)!}\left(\|f^{(j)}\|_{\Gamma,\alpha}\|f^{(n-j)}\|_{\Gamma,\alpha} + \|f^{(j)}\|_{\Gamma,0}\|f^{(n-j)}\|_{\Gamma,0}\right).\end{aligned} \tag{7.3.7}$$

On the other hand

$$\begin{aligned}
&\|f\|_{\Gamma,k}\|g\|_{\Gamma,k+\alpha} + \|f\|_{\Gamma,k+\alpha}\|\|g\|_{\Gamma,k} \\
&\quad = \Big(\sum_{n=0}^{k}\frac{1}{n!}\|f^{(n)}\|_{\Gamma,k}\Big)\Big(\sum_{n=0}^{k}\frac{1}{n!}\|g^{(n)}\|_{\Gamma,\alpha}\Big) \\
&\quad + \Big(\sum_{n=0}^{k}\frac{1}{n!}\|f^{(n)}\|_{\Gamma,\alpha}\Big)\Big(\sum_{n=0}^{k}\frac{1}{n!}\|g^{(n)}\|_{\Gamma,0}\Big) \\
&\quad = \sum_{n,m=0}^{k}\frac{1}{n!m!}\Big(\|f^{(n)}\|_{\Gamma,k}\|f^{(m)}\|_{\Gamma,\alpha} + \|f^{(n)}\|_{\Gamma,\alpha}\|f^{(m)}\|_{\Gamma,k}\Big).
\end{aligned} \tag{7.3.8}$$

Comparing (7.3.7) and (7.3.8) we get (7.3.6). □

7.3.5 Lemma. *Let $0 < \beta < \alpha < 1$ and $k \in \mathbb{N}$, where, for $k \geq 1$, we assume that Γ is of class $\mathcal{C}^k$. Then there exist constants $\delta > 0$ and $C < \infty$ such that each $f : \Gamma \to GA$ which belongs to $(\mathcal{C}^{k+\alpha})^A(\Gamma)$ and which satisfies the estimate*

$$\|f - 1\|_{\Gamma,k+\beta} < \delta, \tag{7.3.9}$$

admits a canonical factorization $f = f_-f_+$ with respect to Γ such that f_- is of class $\mathcal{C}^{k+\alpha}$ on $\overline{D}_-$, and f_+ is of class $\mathcal{C}^{k+\alpha}$ on $\overline{D}_+$. Moreover

$$\|f_\pm - 1\|_{\Gamma,k+\beta} \leq C\|f - 1\|_{\Gamma,k+\beta} < 1. \tag{7.3.10}$$

Proof. Let $\gamma = \alpha, \beta$. Then we denote by $(\mathcal{C}_+^{k+\gamma})^A(\Gamma)$ the algebra of all holomorphic functions $f : D_+ \to A$, which are of class $\mathcal{C}^{k+\gamma}$ on $\overline{D}_+$ (Def. 3.4.4), and by $(\mathcal{C}_-^{k+\gamma})_0^A$ we denote the algebra of all holomorphic functions $f : D_- \cup \{\infty\} \to A$ with $f(\infty) = 0$, which are of class $\mathcal{C}^{k+\gamma}$ on $\overline{D}_-$. It follows from the maximum principle for holomorphic functions and from theorems 3.3.2 and 3.4.5 that these algebras are closed subalgebras of the Banach algebra $(\mathcal{C}^{k+\gamma})^A(\Gamma)$, endowed with the norm $\|\cdot\|_{\Gamma,k+\gamma}$. By Liouville's theorem, $(\mathcal{C}_+^{k+\gamma})^A(\Gamma)\cap(\mathcal{C}_-^{k+\gamma})_0^A = \{0\}$ (Proposition 3.1.3), and again by theorems 3.3.2 and 3.4.5, each $f \in (C^{k+\gamma})^A(\Gamma)$ can be written in the form $f = f_+ + f_-$ with $f_+ \in (\mathcal{C}_+^{k+\gamma})^A(\Gamma)$ and $f_- \in (\mathcal{C}_-^{k+\gamma})_0^A$. Hence $(\mathcal{C}^{k+\gamma})^A(\Gamma)$ is the direct sum of $(\mathcal{C}_+^{k+\gamma})^A(\Gamma)$ and $(\mathcal{C}_-^{k+\gamma})_0^A$.

Now we could apply Lemma 5.2.1 or Lemma 5.2.3. This would give constants $\delta > 0$ and $C < \infty$ with the following property: If $f \in (\mathcal{C}^{k+\alpha})^A(\Gamma)$ satisfies (7.3.9), then f admits a canonical factorization $f = f_+f_-$ with $f_+ \in (\mathcal{C}_+^{k+\beta})^A(\Gamma)$ and $f_- - 1 \in (\mathcal{C}_-^{k+\beta})_0^A(\Gamma)$. But we want to prove that the factors $f_\pm$ also of class $\mathcal{C}^{k+\alpha}$. Although this does not follow from lemmas 5.2.1 and 5.2.3, the main idea of the proof of these lemmas will be used also here, however with more care to the estimates.

Let P_+ be the linear projector from $(\mathcal{C}^{k+\beta})^A(\Gamma)$ to $(\mathcal{C}_+^{k+\beta})^A(\Gamma)$ parallel to $(\mathcal{C}_-^{k+\beta})_0^A(\Gamma)$, and let $P_- := I - P_+$. As $(\mathcal{C}^{k+\beta})^A(\Gamma)$ is the direct sum of $(\mathcal{C}_+^{k+\beta})^A(\Gamma)$

and $(\mathcal{C}_-^{k+\beta})_0^A$, these projectors are continuous. Moreover, the restriction of these projectors to $(\mathcal{C}^{k+\alpha})^A(\Gamma)$ are continuous with respect to the norm $\|\cdot\|_{\Gamma,k+\alpha}$. Therefore, we can find a constant $C < \infty$ with

$$\begin{aligned} \|P_\pm f\|_{\Gamma,k+\beta} &\le C\|f\|_{\Gamma,k+\beta} && \text{if } f \in (\mathcal{C}^{k+\beta})^A(\Gamma), \\ \|P_\pm f\|_{\Gamma,k+\alpha} &\le C\|f\|_{\Gamma,k+\alpha} && \text{if } f \in (\mathcal{C}^{k+\alpha})^A(\Gamma). \end{aligned} \tag{7.3.11}$$

We set

$$\delta = \frac{1}{4C}$$

and prove that these constants δ and C have the required property.

Let $f \in (\mathcal{C}^{k+\alpha})^A(\Gamma)$ with (7.3.9) be given. Set $a = 1 - f$. Then

$$\|a\|_{\Gamma,k+\beta} < \delta = \frac{1}{4C}. \tag{7.3.12}$$

We define a sequence $a_n^- \in (\mathcal{C}_-^{k+\alpha})_0^A(\Gamma)$, $n \in \mathbb{N}^*$, setting

$$a_1^- = P_- a \qquad \text{and} \qquad a_n^- = P_-(a a_{n-1}^-) \quad \text{for } n \ge 2.$$

We claim that then, for all $n \in \mathbb{N}^*$,

$$\|a_n^-\|_{\Gamma,k+\beta} \le \frac{C\|a\|_{\Gamma,k+\beta}}{2^{n-1}} \quad \text{and} \quad \|a_n^-\|_{\Gamma,k+\alpha} \le \frac{C\|a\|_{\Gamma,k+\alpha}}{2^{n-1}}. \tag{7.3.13}$$

Indeed, from (7.3.11) we get

$$\|a_1^-\|_{\Gamma,k+\beta} = \|P_- a\|_{\Gamma,k+\beta} \le C\|a\|_{\Gamma,k+\beta},$$

and

$$\|a_1^-\|_{\Gamma,k+\alpha} = \|P_- a\|_{\Gamma,k+\alpha} \le C\|a\|_{\Gamma,k+\alpha},$$

which proves (7.3.13) for $n = 1$. Now we assume that $m \in \mathbb{N}^*$ and (7.3.13) is already proved for $n = m$. Together with (7.3.11) and (7.3.12) this yields

$$\begin{aligned} \|a_{m+1}^-\|_{\Gamma,k+\beta} &= \|P_-(a a_m^-)\|_{\Gamma,k+\beta} \\ &\le C\|a\|_{\Gamma,k+\beta}\|a_m^-\|_{\Gamma,k+\beta} \le \frac{C^2\|a\|_{\Gamma,k+\beta}^2}{2^{m-1}} \le \frac{C\|a\|_{\Gamma,k+\beta}}{2^m}, \end{aligned}$$

and, taking into account also Lemma 7.3.4,

$$\begin{aligned} \|a_{m+1}^-\|_{\Gamma,k+\alpha} &= \|P_-(a a_m^-)\|_{\Gamma,k+\alpha} \le C\|a a_m^-\|_{\Gamma,k+\alpha} \\ &\le C\Big(\|a\|_{\Gamma,k+\beta}\|a_m^-\|_{\Gamma,k+\alpha} + \|a\|_{\Gamma,k+\alpha}\|a_m^-\|_{\Gamma,k+\beta}\Big) \\ &\le C^2\Big(\frac{\|a\|_{\Gamma,k+\beta}\|a\|_{\Gamma,k+\alpha}}{2^{m-1}} + \frac{\|a\|_{\Gamma,k+\alpha}\|a\|_{\Gamma,k+\beta}}{2^{m-1}}\Big) \\ &= C^2\frac{\|a\|_{\Gamma,k+\beta}\|a\|_{\Gamma,k+\alpha}}{2^{m-2}} \le \frac{C\|a\|_{\Gamma,k+\alpha}}{2^m}, \end{aligned}$$

i.e., (7.3.13) holds for $n = m + 1$.

From (7.3.13) it follows that the series

$$a_- := \sum_{n=1}^{\infty} a_n^-$$

converges absolutely both in $(\mathcal{C}_-^{k+\beta})_0^A(\Gamma)$ and in $(\mathcal{C}_-^{k+\alpha})_0^A(\Gamma)$, where (taking into account also (7.3.12))

$$\|a_-\|_{\Gamma,k+\beta} \le 2C\|a\|_{\Gamma,k+\beta} < \frac{1}{2}. \tag{7.3.14}$$

Define a second sequence $a_n^+ \in (\mathcal{C}_+^{k+\alpha})^A(\Gamma)$, $n \in \mathbb{N}^*$, setting

$$a_1^+ = P_+ a \quad \text{and} \quad a_n^+ = P_+(a a_{n-1}^-) = \quad \text{if } n \ge 2.$$

Then it follows from (7.3.11), (7.3.12) and (7.3.13) that

$$\|a_n^+\|_{\Gamma,k+\beta} \le \frac{C\|a\|_{\Gamma,k+\beta}}{2^{n-1}}, \qquad n \in \mathbb{N}^*,$$

and, taking into account also Lemma 7.3.4,

$$\begin{aligned}\|a_n^+\|_{\Gamma,k+\alpha} &\le C\|a a_{n-1}^-\|_{\Gamma,k+\alpha} \le C\Big(\|a\|_{\Gamma,k+\alpha}\|a_{n-1}^-\|_{\Gamma,k+\beta} + \|a\|_{\Gamma,k+\beta}\|a_{n-1}^-\|_{\Gamma,k+\alpha}\Big)\\ &\le C^2\left(\frac{\|a\|_{\Gamma,k+\alpha}\|a\|_{\Gamma,k+\beta}}{2^{n-2}} + \frac{\|a\|_{\Gamma,k+\beta}\|a\|_{\Gamma,k+\alpha}}{2^{n-2}}\right)\\ &= \frac{C^2\|a\|_{\Gamma,k+\beta}\|a\|_{\Gamma,k+\alpha}}{2^{n-3}} \le \frac{C\|a\|_{\Gamma,k+\alpha}}{2^{n-1}}.\end{aligned}$$

Hence the series

$$a_+ := \sum_{n=1}^{\infty} a_n^+$$

converges absolutely both in $(\mathcal{C}_+^{k+\beta})^A(\Gamma)$ and in $(\mathcal{C}_+^{k+\alpha})^A(\Gamma)$, where (taking into account also (7.3.12))

$$\|a_+\|_{\Gamma,k+\beta} \le 2C\|a\|_{\Gamma,k+\beta} < \frac{1}{2}. \tag{7.3.15}$$

From (7.3.14) and (7.3.15) it follows in particular that $1 - a_-(z) \in GA$ for all $z \in \overline{D}_- \cup \{\infty\}$ and that $1 - a_+(z) \in GA$ for all $z \in \overline{D}_+$. Therefore, setting $f_- = (1 + a_-)^{-1}$, we get a function $f_- : \overline{D}_- \cup \{\infty\} \to GA$, which is holomorphic in $D_- \cup \{\infty\}$ and of class $\mathcal{C}^{k+\alpha}$ on $\overline{D}_-$, and setting $f_+ = 1 - a_+$ we get a function

$f_+ : \overline{D}_+ \to GA$, which is holomorphic in D_+ and of class $\mathcal{C}^{k+\alpha}$ on $\overline{D}_+$. By (7.3.14) and (7.3.15) these functions satisfy (7.3.14). Moreover

$$f_-^{-1} f = (1 + a_-)(1 - a) = 1 + \sum_{n=1}^{\infty} a_n^- - a - \sum_{n=1}^{\infty} a a_n^-$$
$$= 1 + a_1^- + \sum_{n=2}^{\infty} a_n^- - a - \sum_{n=2}^{\infty} a a_{n-1}^- = 1 + (a_1^- - a) + \sum_{n=2}^{\infty} (a_n^- - a a_{n-1}^-).$$

Since

$$a_1^- - a = P_- a - a = -P_+ a = a_1^+$$

and

$$a_n^- - a a_{n-1}^- = P_-(a a_{n-1}^-) - a a_{n-1}^- = -P_+(a a_{n-1}^-) = -a_n^+ \qquad \text{for } n \geq 2,$$

this implies that

$$f_-^{-1} f = 1 - \sum_{n=1}^{\infty} a_n^+ = 1 - a_+ = f_+,$$

i.e., $f = f_- f_+$. □

Proof of Theorem 7.3.1. Let $\delta > 0$ be the constant from Lemma 7.3.5. Then, by the approximation Theorem 3.5.1, we can find a neighborhood U of Γ and a holomorphic function $q : U \to GA$ such that

$$\left\| q^{-1} f - 1 \right\|_{\Gamma, k+\beta} < \delta.$$

Then, by Lemma 7.3.5, $q^{-1} f$ admits a canonical factorization $q^{-1} f = g_- g_+$ with respect to Γ such that $g_\pm$ is of class $\mathcal{C}^{k+\alpha}$ on $\overline{D}_\pm$. Now, again by Theorem 3.5.1 and by Lemma 7.3.5, we can find a neighborhood V of Γ and a holomorphic function $p : V \to GA$ such that

$$\left\| q g_- p^{-1} - 1 \right\|_{\Gamma, k+\beta} < \delta$$

and, hence, $q g_- p^{-1}$ admits a canonical factorization $q g_- p^{-1} = h_- h_+$ with respect to Γ such that $h_\pm$ is of class $\mathcal{C}^{k+\alpha}$ on $\overline{D}_\pm$. Then

$$h_+ p = h_-^{-1} q g_- \qquad \text{on } \Gamma.$$

This implies (by Theorem 1.5.4) that there is a holomorphic function h on $U \cap V$ defined by

$$h = \begin{cases} h_+ p & \text{on } U \cap V \cap \overline{D}_+ \\ h_-^{-1} q g_- & \text{on } U \cap V \cap \overline{D}_-. \end{cases}$$

Then

$$f = q g_- g_+ = q g_- p^{-1} p g_+ = h_- h_+ p g_+ = h_- h g_+,$$

which is a representation of f as required in part (i) of the theorem, where, additionally, we already know that the factors h_- and g_+ are of class $\mathcal{C}^{k\alpha}$ on $\overline{D}_-$ and $\overline{D}_+$ respectively.

To prove part (ii), we assume that we have some other representation of f in the form $f = \widetilde{h}_-\widetilde{h}\widetilde{g}_+$, where $\widetilde{h}_- : \overline{D}_- \cup \{\infty\} \to GA$ is continuous on $\overline{D}_- \cup \{\infty\}$ and holomorphic in $D_- \cup \{\infty\}$, $\widetilde{g}_+ : \overline{D}_+ \to GA$ is continuous on $\overline{D}_+$ and holomorphic in D_+, and $\widetilde{h}$ is holomorphic in some neighborhood of Γ. Then

$$\widetilde{h}_-\widetilde{h}\widetilde{g}_+ = h_-hg_+$$

and therefore

$$h_-^{-1}\widetilde{h}_-\widetilde{h} = hg_+\widetilde{g}_+^{-1} \qquad \text{on } \Gamma.$$

The latter equality shows (by Theorem 1.5.4) that the two sides of this equality define holomorphic function r in some neighborhood U of Γ such that

$$\widetilde{h}_- = h_-r\widetilde{h}^{-1} \quad \text{and} \quad \widetilde{g}_+ = r^{-1}hg_+.$$

Since the functions on the right-hand side of these relations are of class $\mathcal{C}^{k\alpha}$ on $\overline{D}_- \cap U$ and $\overline{D}_+ \cap U$ respectively, it follows that $\widetilde{h}_-$ and $\widetilde{g}_+$ and of class $\mathcal{C}^{k\alpha}$ on $\overline{D}_-$ and $\overline{D}_+$, respectively. □

7.4 Reduction of the factorization problem to functions, holomorphic and invertible on $\mathbb{C}^*$

In this section, $D_+ \subseteq \mathbb{C}$ is a bounded, connected, open set with piecewise $\mathcal{C}^1$-boundary Γ such that $0 \in D_+$, and $D_- := \mathbb{C} \setminus D_+$. Further, throughout this section, A is a Banach algebra with unit 1, and G is a (possibly not open) subgroup of the group of invertible elements of A.

7.4.1 Lemma. *Let $\mathbb{P}^1 = \mathbb{C} \cup \{\infty\}$ be the Riemann sphere (see the beginning of Section 5.10), let p be an arbitrary point in $\mathbb{P}^1 \setminus \Gamma$, and let*

$$\begin{aligned} D'_+ &:= D_+ \quad \text{and} \quad D'_- := \big(D_- \cup \{\infty\}\big) \setminus \{p\} \qquad \text{if } p \in D_- \cup \{\infty\}, \\ D'_+ &:= D_+ \setminus \{p\} \quad \text{and} \quad D'_- := D_- \cup \{\infty\} \qquad \text{if } p \in D_+ . \end{aligned}$$

Denote by D''_- and D''_+ the closures in $\mathbb{P}^1 \setminus \{p\}$ of D'_- and D'_+ , respectively.[2]

Let $f : \Gamma \to G$ be a continuous function which admits local factorizations with respect to Γ and G (Def. 7.1.3), and suppose that at least one of the following conditions is satisfied:

(I) *G is open in A.*

[2]If $p = \infty$, then $D''_- = \overline{D}_-$ and $D''_+ = \overline{D}_+$. If $p \in D_-$, then $D''_- = (\overline{D}_- \cup \{\infty\}) \setminus \{p\}$ and $D''_+ = \overline{D}_+$. If $p \in D_+$, then $D''_- = \overline{D}_- \cup \{\infty\}$ and $D''_+ = \overline{D}_+ \setminus \{p\}$.

(II) *$A = L(E)$ for some Banach space E, and $G = \mathcal{G}^\infty(E)$ (Def.* 5.12.1).

(III) *$A = L(E)$ for some Banach space E, $G = \mathcal{G}^\omega(E)$ (Def.* 5.12.1).

Then there exist continuous functions $f_- : D''_- \to G$ and $f_+ : D''_+ \to G$, which are holomorphic in D'_- and D'_+, respectively, such that

$$f = f_- f_+ \qquad \text{on } \Gamma\,. \tag{7.4.1}$$

Proof. By hypothesis, we can find open sets $U_1, \ldots, U_m \subseteq \mathbb{P}^1 \setminus \{p\}$, continuous functions $f_j^+ : U_j \cap D''_+ \to G$, which are holomorphic in $U_j \cap D'_+$, and continuous functions $f_j^- : U_j \cap D''_- \to G$, which are holomorphic in $U_j \cap D'_-$, such that $\Gamma \subseteq U_1 \cup \ldots \cup U_m$ and

$$f = f_j^- f_j^+ \qquad \text{on } U_j \cap \Gamma\,, \qquad 1 \le j \le m.$$

Set

$$U_0 = \big(\mathbb{P}^1 \setminus \{p\}\big) \setminus \Gamma \qquad \text{and} \qquad f_0^+ = f_0^- = 1 \quad \text{on } U_0\,.$$

Then, for all $0 \le j, k \le m$ with $U_j \cap U_k \cap \Gamma \neq \emptyset$,

$$f_j^+ \big(f_k^+\big)^{-1} = \big(f_j^-\big)^{-1} f_k^- \qquad \text{on } U_j \cap U_k \cap \Gamma. \tag{7.4.2}$$

Indeed, for $1 \le j, k \le m$ this is clear, as $f_j^- f_j^+ = A = f_k^- f_k^+$ on $U_j \cap U_k \cap \Gamma$, and, for $j = 0$ or $k = 0$ this is trivial, because $U_0 \cap \Gamma = \emptyset$. Now, by (7.4.2), there is a well-defined family $g_{jk} \in \mathcal{C}^G(U_j \cap U_k)$, $0 \le j, k \le m$, such that, for all $0 \le j, k \le m$,

$$g_{jk} = \begin{cases} f_j^+ \big(f_k^+\big)^{-1} & \text{on } U_j \cap U_k \cap D''_+ \ \text{ if } \ U_j \cap U_k \cap \overline{D}'_+ \neq \emptyset, \\ \big(f_j^-\big)^{-1} f_k^- & \text{on } U_j \cap U_k \cap D''_- \ \text{ if } \ U_j \cap U_k \cap D''_- \neq \emptyset. \end{cases} \tag{7.4.3}$$

On $(U_j \cap U_k) \setminus \Gamma$, these functions are holomorphic, since the functions $f_j^\pm$ and $\big(f_j^\pm\big)^{-1}$ are holomorphic on $U_j \cap D_j^\pm$. Hence, by Theorem 1.5.4,

$$g_{jk} \in \mathcal{O}^G(U_j \cap U_k)\,, \qquad 0 \le j, k \le m.$$

Moreover it is clear from (7.4.3) that, for all $0 \le j, k, l \le m$ with $U_j \cap U_k \cap U_l \neq \emptyset$,

$$g_{jk} g_{kl} = g_{jl} \qquad \text{on } U_j \cap U_k \cap U_l\,,$$

i.e.,, the family $\{g_{jk}\}_{0 \le j,k \le m}$ is a $\big(\{U_0, \ldots, U_m\}, \mathcal{O}^G\big)$-cocycle (Def. 5.6.1). Since $U_0 \cup \ldots \cup U_m = \mathbb{P}^1 \setminus \{p\}$ is simply connected, by theorems 5.6.3 and 5.12.5, this cocycle is $\mathcal{O}^G$-trivial. Hence we have a family $h_j \in \mathcal{O}^G(U_j)$, $0 \le j \le m$, with

$$g_{jk} = h_j h_k^{-1} \qquad \text{on } U_j \cap U_k$$

for all $0 \le j,k \le m$ with $U_j \cap U_k \neq \emptyset$. Then it follows from (7.4.3) that, for all $0 \le j,k \le m$,

$$h_j h_k^{-1} = \begin{cases} \left(f_j^-\right)^{-1} f_k^- & \text{on } U_j \cap U_k \cap D''_- \text{ if } U_j \cap U_k \cap D''_- \neq \emptyset, \\ f_j^+ \left(f_k^+\right)^{-1} & \text{on } U_j \cap U_k \cap D''_+ \text{ if } U_j \cap U_k \cap D''_+ \neq \emptyset, \end{cases}$$

and therefore

$$f_j^- h_j = f_k^- h_k \quad \text{on } U_j \cap U_k \cap D''_- \quad \text{if} \quad U_j \cap U_k \cap D''_- \neq \emptyset$$

and

$$h_j^{-1} f_j^+ = h_k^{-1} f_k^+ \quad \text{on } U_j \cap U_k \cap D''_+ \quad \text{if} \quad U_j \cap U_k \cap D''_+ \neq \emptyset.$$

Hence, there is a well-defined continuous function $f_- : D''_- \to G$, which is holomorphic in D''_-, and a well-defined continuous function $f_+ : D''_+ \to G$, which is holomorphic in D'_+, such that

$$f_-\big|_{U_j \cap D''_-} = f_j^- h_j\,, \qquad 0 \le j \le m\,, \tag{7.4.4}$$

$$f_+\big|_{U_j \cap D''_+} = h_j^{-1} f_j^+\,, \qquad 0 \le j \le m\,. \tag{7.4.5}$$

Since $f = f_j^- f_j^+$ on $U_j \cap \Gamma$, $1 \le j \le m$, it follows from (7.4.4) and (7.4.5) that (7.4.1) holds. □

7.4.2 Theorem. *Let $f : \Gamma \to G$ be a continuous function which admits local factorizations with respect to Γ and G (Def. 7.1.3), and suppose that at least one of the following conditions is satisfied:*

(I) *G is open in A.*

(II) *$A = L(E)$ for some Banach space E, and $G = \mathcal{G}^\infty(E)$ (Def. 5.12.1).*

(III) *$A = L(E)$ for some Banach space E, and $G = \mathcal{G}^\omega(E)$ (Def. 5.12.1).*

Then there exists a holomorphic function $h : \mathbb{C}^* \to G$ such that f and h are equivalent with respect to Γ and G.

Proof. Setting $p = \infty$ in Lemma 7.4.1, we get continuous functions $h_- : \overline{D}_- \to G$ and $f_+ : \overline{D}_+ \to G$, which are holomorphic in D_- and D_+, respectively, such that

$$f = h_- f_+ \qquad \text{on } \Gamma. \tag{7.4.6}$$

Applying Lemma 7.4.1 with $p = 0$ to h_-, we get continuous functions $f_- : \overline{D}_- \cup \{\infty\} \to G$ and $h : \overline{D}_+ \setminus \{0\} \to G$, which are holomorphic in $D_- \cup \{\infty\}$ and $D_+ \setminus \{0\}$, respectively, such that

$$h_- = f_- h \qquad \text{on } \Gamma \tag{7.4.7}$$

and therefore

$$h = f_-^{-1} h_- \qquad \text{on } \Gamma\,.$$

Since f_-^{-1} and h_- are continuous on $\overline{D}_-$ and holomorphic in D_-, from the latter relation it follows that h admits a continuous G-valued extension to $\mathbb{C}\setminus\{0\}$, which is holomorphic in D_-, and which we also denote by h. Since h is also holomorphic on $D_+\setminus\{0\}$, it follows from Theorem 1.5.4 that h is holomorphic on $\mathbb{C}\setminus\{0\}$. From (7.4.6) and (7.4.7) it follows that f and h are equivalent. □

7.5 Factorization of holomorphic functions close to the unit

Let $D_+ \subseteq \mathbb{C}$ be a bounded, connected, open set with piecewise $\mathcal{C}^1$-boundary Γ such that $0 \in D_+$. Set $D_- = \mathbb{C} \setminus D_+$. Further, let A be a Banach algebra with unit 1, and let G be an open subgroup of the group of invertible elements of A. Here we prove the following

7.5.1 Theorem. *Let U be a bounded neighborhood of Γ, and let $\varepsilon > 0$. Then there exists $\delta > 0$ such that, for each $g \in \overline{\mathcal{O}}^A(\overline{U})$ (Def. 5.4.5) with*

$$\max_{z\in\overline{U}} \|g(z)\| < \delta\,, \tag{7.5.1}$$

there exist $g_- \in \overline{\mathcal{O}}^A(D_- \cup \overline{U} \cup \{\infty\})$ and $g_+ \in \overline{\mathcal{O}}^A(D_+ \cup \overline{U})$ such that

$$\max_{z\in D_+\cup\overline{U}} \|g_+(z)\| < \varepsilon\,, \qquad \max_{z\in D_-\cup\overline{U}\cup\{\infty\}} \|g_-(z)\| < \varepsilon$$

and

$$(1+g) = (1+g_-)(1+g_+) \qquad \text{on } \overline{U}.$$

Proof. Set $U_+ = D_+ \cup U$ and $U_- = D_- \cup U \cup \{\infty\}$. We consider $\overline{\mathcal{O}}^A(\overline{U}_-)$, $\overline{\mathcal{O}}^A(\overline{U}_+)$ and $\overline{\mathcal{O}}^A(\overline{U})$ as Banach algebras endowed with the maximum norm. By Theorem 3.7.3, each function $g \in \overline{\mathcal{O}}^A(\overline{U})$ can be written in the form $g = g_+ + g_-$ with $g_\pm \in \overline{\mathcal{O}}^A(\overline{U}_\pm)$. Therefore the assertion follows from the factorization Lemma 5.2.1. □

7.6 Reduction of the factorization problem to polynomials in z and $1/z$

In this section, $D_+ \subseteq \mathbb{C}$ is a bounded, connected, open set with piecewise $\mathcal{C}^1$-boundary Γ such that $0 \in D_+$, and $D_- = \mathbb{C} \setminus D_+$.

7.6.1 Theorem. *Let A be a Banach algebra with unit, and let GA be the group of invertible elements of A. Let $0 < r < R < \infty$ such that*

$$\overline{D}_+ \subseteq W := \left\{ z \in \mathbb{C} \;\middle|\; r < |z| < R \right\}, \tag{7.6.1}$$

and let $f : \Gamma \to GA$ be a continuous function which admits local factorizations with respect to Γ and GA (Def. 7.1.3). Then:

(i) *The function f is equivalent with respect to Γ and GA (Def. 7.1.3) to a function h of the form*

$$h(z) = \sum_{n=N}^{M} z^n h_n \,, \qquad N, M \in \mathbb{Z}, \tag{7.6.2}$$

such that $h(z) \in GA$ for all $z \in \overline{W}$.

(ii) *The function f is equivalent with respect to Γ and GA to a function h such that h^{-1} is of the form*

$$h^{-1}(z) = \sum_{n=N}^{M} z^n h_n \,, \qquad N, M \in \mathbb{Z}, \tag{7.6.3}$$

and $h^{-1}(z) \in GA$ for all $z \in \overline{W}$.

Proof. It is sufficient to prove part (i). Part (ii) then follows by changing the order of multiplication in A and applying part (i) to f^{-1}. Set

$$W_- = \Big\{ z \in \mathbb{C} \;\Big|\; r < |z| \Big\} \quad \text{and} \quad W_+ = \Big\{ z \in \mathbb{C} \;\Big|\; |z| < R \Big\}.$$

Then, by Theorem 7.5.1, for some $\varepsilon > 0$, the following holds:

(*) For each $\varphi \in \overline{\mathcal{O}}^{GA}(\overline{W})$ with $\max_{z \in \overline{W}} \|\varphi(z) - 1\| < \varepsilon$, there exist $\varphi_+ \in \overline{\mathcal{O}}^{G}(\overline{W}_+)$ and $\varphi_- \in \overline{\mathcal{O}}^{GA}(\overline{W}_- \cup \{\infty\})$ with $\varphi = \varphi_- \varphi_+$ on $\overline{W}$.

By Theorem 7.4.2 we may assume that f is a GA-valued holomorphic function on $\mathbb{C} \setminus \{0\}$. Let

$$f(z) = \sum_{n=-\infty}^{\infty} z^n f_n$$

be the Laurent expansion of f at zero. Choose integers $K < M$ such that the function

$$u(z) := \sum_{n=K}^{M} z^n f_n$$

is so close to f over $\overline{W}$ that $u(z) \in GA$ for all $z \in \overline{W}$ and

$$\max_{z \in \overline{W}} \left\| u^{-1}(z) f(z) - 1 \right\| < \varepsilon.$$

Then, by statement (*), we can find $f_+ \in \overline{\mathcal{O}}^{G}(\overline{W}_+)$ and $g_- \in \overline{\mathcal{O}}^{G}\left(\overline{W}_- \cup \{\infty\}\right)$ such that $u^{-1} f = g_- f_+$ on $\overline{W}$. Hence the function $f f_+^{-1} = u g_-$ has a Laurent expansion of the form

$$f(z) f_+^{-1}(z) = \sum_{n=-\infty}^{M} z^n a_n \,. \tag{7.6.4}$$

Choose an integer N such that the function

$$v(z) := \sum_{n=N}^{M} z^n a_n$$

is so close to ff_+^{-1} over $\overline{W}$ that $v(z) \in GA$ for all $z \in \overline{W}$ and

$$\max_{z\in\overline{W}} \left\| f(z)f_+^{-1}(z)v^{-1}(z) - 1 \right\| < \varepsilon.$$

Then, again by statement (*), we can find $g_+ \in \overline{\mathcal{O}}^{GA}(\overline{W}_+)$ and $f_- \in \overline{\mathcal{O}}^{GA}\left(\overline{W}_- \cup \{\infty\}\right)$ such that $ff_+^{-1}v^{-1} = f_-g_+$ on $\overline{W}$. Set

$$h = f_-^{-1}ff_+^{-1}.$$

Then $h \in \overline{\mathcal{O}}^{GA}(\overline{W})$, as f is holomorphic and GA-valued on $\mathbb{C} \setminus \{0\}$. Moreover, since $h = g_+v$, the Laurent expansion of h is of the form

$$h(z) = \sum_{n=N}^{\infty} z^n h_n \,.$$

On the other hand, as $h = f_-^{-1}ff_+^{-1}$, it follows from (7.6.4) that the Laurent expansion of h is of the form

$$h(z) = \sum_{n=-\infty}^{M} z^n h_n \,.$$

Hence, the Laurent expansion of h is of the form

$$h(z) = \sum_{n=N}^{M} z^n h_n \,.$$

As $f = f_-hf_+$, this completes the proof. □

7.7 The finite dimensional case

In this section, $L(n, \mathbb{C})$, $n \in \mathbb{N}^*$, is the algebra of complex $n\times n$-matrices, $GL(n, \mathbb{C})$ is the group of invertible complex $n\times n$-matrices, $D_+ \subseteq \mathbb{C}$ is a bounded, connected, open set with piecewise $\mathcal{C}^1$-boundary Γ such that $0 \in D_+$, and $D_- = \mathbb{C} \setminus D_+$. Here we prove the following theorem:

7.7.1 Theorem. *Let $A : \Gamma \to GL(n, \mathbb{C})$ be a continuous function which admits local factorizations with respect to Γ. Then A admits a factorization with respect to Γ.*

Proof. By Theorem 7.6.1 (i) we may assume that A is of the form

$$A(z) = \sum_{j=N}^{M} z^j A_j \,, \qquad A_j \in L(n, \mathbb{C}) \,, \tag{7.7.1}$$

where $-\infty < N \leq M < \infty$. We need here the following definition: For $\kappa \in \mathbb{Z}$ and $x \in \mathbb{C}^n \setminus \{0\}$, a pair (φ^-, φ^+) is called a **κ-section of** x if $\varphi^+ : \overline{D}_+ \to \mathbb{C}^n$ and $\varphi^- : \overline{D}_- \cup \{\infty\} \to \mathbb{C}^n$ are holomorphic functions such that

$$z^\kappa \varphi^-(z) = A(z)\varphi^+(z) \quad \text{for all } z \in \Gamma \qquad \text{and} \qquad \varphi^+(0) = x.$$

If $\kappa \geq M$, then, for each $x \in \mathbb{C}^n \setminus \{0\}$, there exists a κ-section of x, namely

$$\varphi^+(z) := x \qquad \text{and} \qquad \varphi^-(z) := z^{-\kappa} A(z)x = \sum_{j=N-\kappa}^{M-\kappa} z^j A_{j+\kappa} x \,.$$

On the other hand, if $\kappa \leq N - 1$, then no $x \in \mathbb{C}^n \setminus \{0\}$ has a κ-section. Indeed, let $\kappa \leq N - 1$, and let (φ^-, φ^+) be a κ-section of a vector $x \in \mathbb{C}^n \setminus \{0\}$. Then

$$\varphi^-(z) = z^{-\kappa} \Big(\sum_{j=N}^{M} z^j A_j \Big) \varphi^+(z) \quad \text{for all } z \in \Gamma \,.$$

Then the left-hand side of this equation is holomorphic on $\overline{D}_- \cup \{\infty\}$ and, as $j - \kappa \geq 1$ for $N \leq j \leq M$, the right-hand side is holomorphic on $\overline{D}_+$. Hence, by Liouville's theorem, both sides vanish identically. It follows that

$$A(z)\varphi^+(z) = \Big(\sum_{j=N}^{M} z^j A_j \Big) \varphi^+(z) = 0 \qquad \text{for } z \in \Gamma \,.$$

Since the values of A on Γ are invertible, this implies that $\varphi^+ = 0$ on Γ. Hence, by uniqueness of holomorphic functions, $\varphi^+ \equiv 0$ on $\overline{D}_+$. In particular $x = \varphi^+(0) = 0$.

Hence, for each $x \in \mathbb{C}^n \setminus \{0\}$, there exists a smallest integer κ such that x has a κ-section. We denote this integer by $\kappa(x)$. Note that, for each $\kappa \in \mathbb{Z}$, the set

$$\{0\} \cup \big\{ x \in \mathbb{C}^n \setminus \{0\} \;\big|\; \kappa(x) \leq \kappa \big\}$$

is a linear subspace of $\mathbb{C}^n$. Therefore, we can find a basis $e_1, \ldots, e_n$ of $\mathbb{C}^n$ such that

$$\begin{aligned} \kappa(e_n) &= \min_{x \in \mathbb{C}^n \setminus \{0\}} \kappa(x), \\ \kappa(e_j) &= \min_{x \in \mathbb{C}^n \setminus \operatorname{span}(e_{j+1}, \ldots, e_n)} \kappa(x) \qquad \text{for } 1 \leq j \leq n-1. \end{aligned} \tag{7.7.2}$$

Fix a $\kappa(e_j)$-section $(\varphi_j^-, \varphi_j^+)$ of e_j for each j. Denote by A_+ and A_- the matrices formed by the columns φ_j^+ and φ_j^-, respectively, and let Δ be the diagonal matrix with diagonal $z^{\kappa(e_1)}, \dots, z^{\kappa(e_n)}$. Then

$$\Delta A_- = AA_+ , \tag{7.7.3}$$

and $A_+(0)$ is the matrix with the columns $e_1, \dots, e_n$. It remains to prove that $A_+(z)$ is invertible for all $z \in \overline{D}_+$, and $A_-(z)$ is invertible for all $z \in \overline{D}_- \cup \{\infty\}$. To prove this we assume the contrary.

First assume that the matrix $A_+(z_0)$ is not invertible for some $z_0 \in \overline{D}_+$. Since $A_+(0)$ is invertible, $z_0 \neq 0$. Then there exist $1 \leq k \leq n$ and numbers $\lambda_k, \dots, \lambda_n$ with

$$\sum_{j=k}^{n} \lambda_j \varphi_j^+(z_0) = 0 \qquad \text{and } \lambda_k \neq 0,$$

and there is a holomorphic function ψ^+ on $\overline{D}_+$ with

$$\psi^+(z) = \frac{1}{z - z_0} \sum_{j=k}^{n} \lambda_j \varphi_j^+(z) \quad \text{for } z \in \overline{D}_+ \setminus \{z_0\} .$$

Moreover, since $\kappa(e_1) \geq \dots \geq \kappa(e_n)$, there is a holomorphic function ψ^- on $\overline{D}_- \cup \{\infty\}$ with

$$\psi^-(z) = \frac{z}{z - z_0} \sum_{j=k}^{n} \lambda_j z^{\kappa(e_j) - \kappa(e_k)} \varphi_j^-(z) \quad \text{for } z \in \overline{D}_- .$$

Since $(\varphi_j^-, \varphi_j^+)$ is a $\kappa(e_j)$-section of e_j, we get

$$\begin{aligned} A(z)\psi^+(z) &= \frac{1}{z - z_0} \sum_{j=k}^{n} \lambda_j A(z) \varphi_j^+(z) = \frac{1}{z - z_0} \sum_{j=k}^{n} \lambda_j z^{\kappa(e_j)} \varphi_j^-(z) \\ &= z^{\kappa(e_k)-1} \frac{z}{z - z_0} \sum_{j=k}^{n} \lambda_j z^{\kappa(e_j) - \kappa(e_k)} \varphi_j^-(z) = z^{\kappa(e_k)-1} \psi^-(z) \quad \text{for } z \in \Gamma. \end{aligned}$$

Hence (ψ^-, ψ^+) is a $(\kappa(e_k) - 1)$-section of the vector

$$\psi^+(0) = -\frac{1}{z_0} \sum_{j=k}^{n} \lambda_j e_j .$$

Since $\lambda_k \neq 0$, this vector belongs to $\mathbb{C}^n \setminus \operatorname{span}(e_{k+1}, \dots, e_n)$, which is a contradiction to (7.7.2).

Now we assume that the matrix $A_-(z_0)$ is not invertible for some $z_0 \in \overline{D}_- \cup \{\infty\}$. Then there exist $1 \leq k \leq n$ and numbers $\lambda_k, \dots, \lambda_n$ with

$$\sum_{j=k}^{n} \lambda_j \varphi_j^-(z_0) = 0 \qquad \text{and } \lambda_k \neq 0,$$

and there is a holomorphic function ψ^- on $\overline{D}_- \cup \{\infty\}$ with

$$\psi^-(z) = \begin{cases} \frac{z}{z-z_0} \sum_{j=k}^{n} \lambda_j \varphi_j^-(z) & \text{if } z_0 \in \overline{D}_- \text{ and } z \in \overline{D}_- \setminus \{z_0\}, \\ z \sum_{j=k}^{n} \lambda_j \varphi_j^-(z) & \text{if } z_0 = \infty \text{ and } z \in \overline{D}_-. \end{cases}$$

Moreover, since $\kappa(e_1) \geq \ldots \geq \kappa(e_n)$, there is a holomorphic function ψ^+ on $\overline{D}_+$ with

$$\psi^+(z) = \begin{cases} \frac{1}{z-z_0} \sum_{j=k}^{n} \lambda_j z^{\kappa(e_k)-\kappa(e_j)} \varphi_j^+(z) & \text{if } z_0 \in \overline{D}_- \text{ and } z \in \overline{D}_+, \\ \sum_{j=k}^{n} \lambda_j z^{\kappa(e_k)-\kappa(e_j)} \varphi_j^+(z) & \text{if } z_0 = \infty \text{ and } z \in \overline{D}_+. \end{cases}$$

Using again that $(\varphi_j^-, \varphi_j^+)$ is a $\kappa(e_j)$-section of e_j, for all $z \in \Gamma$ we get

$$A(z)\psi^+(z) = \begin{cases} \frac{1}{z-z_0} \sum_{j=k}^{n} \lambda_j z^{\kappa(e_k)} \varphi_j^-(z) = z^{\kappa(e_k)-1}\psi^-(z) & \text{if } z_0 \in \overline{D}_-, \\ \sum_{j=k}^{n} \lambda_j z^{\kappa(e_k)} \varphi_j^-(z) = z^{\kappa(e_k)-1}\psi^-(z) & \text{if } z_0 = \infty. \end{cases}$$

Hence (ψ^-, ψ^+) is a $(\kappa(e_k)-1)$-section of the vector $\psi^+(0)$. Let m be the index with $k \leq n \leq m$ such that $\kappa(e_j) = \kappa(e_k)$ for $k \leq j \leq m$ and, if $m < n$, $\kappa(e_{m+1}) < \kappa(e_k)$. Then

$$\psi^+(0) = \begin{cases} -\frac{1}{z_0} \sum_{j=k}^{m} \lambda_j e_j & \text{if } z_0 \in \overline{D}_-, \\ \sum_{j=k}^{m} \lambda_j e_j & \text{if } z_0 = \infty. \end{cases}$$

Since $\lambda_k \neq 0$, in both cases $\psi^+(0)$ belongs to $\mathbb{C}^n \setminus \operatorname{span}(e_{k+1}, \ldots, e_n)$, which is again a contradiction to (7.7.2). □

7.8 Factorization of $\mathcal{G}^\infty(E)$-valued functions

In this section E is a Banach space with $\dim E = \infty$, $D_+ \subseteq \mathbb{C}$ is a bounded, connected, open set with piecewise $\mathcal{C}^1$-boundary Γ such that $0 \in D_+$, and $D_- = \mathbb{C} \setminus D_+$.

Here we study the factorization problem for functions with values in $\mathcal{G}^\infty(E)$ or $\mathcal{G}^\omega(E)$ (Def. 5.12.1). For $\mathcal{G}^\infty(E)$ we obtain the complete solution (Theorem 7.8.6) at the end of this section. For $\mathcal{G}^\omega(E)$ this is more difficult and can be done only later (Theorem 8.2.2). But a part of the argument works for both $\mathcal{G}^\infty(E)$ and $\mathcal{G}^\omega(E)$. We start with these points of the argument.

7.8.1 Proposition. *Let $\aleph = \infty, \omega$, and let $A : \Gamma \to \mathcal{G}^{\aleph}(E)$ be a continuous function which admits a factorization with respect to Γ. Let $A = A_- \Delta A_+$ be a factorization of A with respect to Γ and $GL(E)$. Assume moreover that $A_-(\infty) = I$ (see Proposition* 7.1.6*). Then, automatically, the values of A_- and A_+ belong to $\mathcal{G}^{\aleph}(E)$.*

Proof. Denote by 1 the unit element in the factor algebra $L(E)/\mathcal{F}^{\aleph}(E)$, and let $\pi : L(E) \to L(E)/\mathcal{F}^{\aleph}(E)$ be the canonical map. Then $\pi(A) = \pi(\Delta) = 1$ and therefore

$$1 = \pi(A_-)\pi(A_+) \quad \text{on } \Gamma \qquad \text{and} \qquad \pi\big(A_-(\infty)\big) = 1\,.$$

Hence

$$\pi\big(A_-^{-1}\big) = \pi\big(A_+\big) \qquad \text{on } \Gamma \tag{7.8.1}$$

and

$$\pi\big(A_-^{-1}(\infty)\big) = 1\,. \tag{7.8.2}$$

By Theorem 1.5.4 the two sides of (7.8.1) define a holomorphic function $f : \mathbb{C} \cup \{\infty\} \to L(E)/\mathcal{F}^{\aleph}(E)$. By (7.8.2), $f(\infty) = 1$. It follows from Liouville's theorem that $f \equiv 1$, i.e.,

$$\pi\big(A_-^{-1}\big) = 1 \text{ on } \overline{D}_- \cup \{\infty\} \qquad \text{and} \qquad \pi\big(A_+\big) = 1 \text{ on } \overline{D}_+\,.$$

This means that the values of A_- and A_+ belong to $\mathcal{G}^{\aleph}(E)$. □

7.8.2 Theorem. *Let $\aleph = \infty, \omega$, and let $G\big(L(E)/\mathcal{F}^{\aleph}(E)\big)$ be the group of invertible elements of the factor algebra $L(E)/\mathcal{F}^{\aleph}(E)$, and let*

$$\pi : L(E) \to L(E)/\mathcal{F}^{\aleph}(E)$$

be the canonical map. Further, let $W \subseteq \mathbb{C}$ be an open set, and let $f : D \to G\big(L(E)/\mathcal{F}^{\aleph}(E)\big)$ be holomorphic. Then there exists a holomorphic function $A : D \to GL(E)$ such that

$$f(z) = \pi\big(A(z)\big) \qquad \text{for all } z \in W. \tag{7.8.3}$$

Proof. We first prove this locally. Consider an arbitrary point $z_0 \in W$. Let

$$f(z) = \sum_{n=0}^{\infty} (z - z_0)^n f_n$$

be the Taylor expansion of f at z_0. Since π is bounded and surjective, then, by the Banach open mapping theorem, we can find operators $F_n \in L(E)$ with $\pi(F_n) = f_n$ and $\|F_n\| < 2\|f_n\|$. Let $U \subseteq W$ be an open disc around z_0. Then, setting

$$F(z) = \sum_{n=0}^{\infty} (z - z_0)^n F_n\,, \qquad z \in U,$$

we get a holomorphic function $F : U \to L(E)$ with $\pi\big(F(z)\big) = f(z)$ for all $z \in U$. Since $f(z_0)$ is invertible, then F_0 is a Fredholm operator with index zero. Therefore we can find a finite dimensional operator K in E such that $F_0 + K$ is invertible. Choose a neighborhood $V \subseteq U$ so small that $F(z) + K$ is invertible for all $z \in V$. Setting $A_0(z) = F(z) + K$, $z \in V$, we obtain a holomorphic function $A_0 : V \to GL(E)$ with $f(z) = \pi\big(A_0(z)\big)$ for $z \in V$.

By the local statement just proved, there exist an open covering $\mathcal{U} = \{U_j\}_{j \in I}$ of D and holomorphic functions $A_j : U_j \to GL(E)$ with $\pi\big(A_j(z)\big) = f(z)$ for $z \in U_j$. Then

$$\pi\big(A_j A_k^{-1}\big) = \pi\big(f f^{-1}\big) = 1 \qquad \text{on } U_j \cap U_k$$

for all $j, k \in I$ with $U_j \cap U_k \neq \emptyset$. Therefore, the values of the functions $A_j A_k^{-1}$ lie in $\mathcal{G}^\aleph(E)$. Now, from Theorem 5.12.5 we get a family of functions $V_j : U_j \to \mathcal{G}^\aleph(E)$ with

$$A_j A_k^{-1} = V_j V_k^{-1} \qquad \text{on } U_j \cap U_k$$

for all $j, k \in I$ with $U_j \cap U_k \neq \emptyset$. Setting $A = V_j^{-1} A_j$ on U_j we complete the proof. □

7.8.3 Lemma. *Let $\aleph = \infty, \omega$, and let $A : \Gamma \to \mathcal{G}^\aleph(E)$ be a continuous function which admits local factorizations with respect to Γ and $GL(E)$ (Def. 7.1.3). Then A admits local factorizations with respect to Γ and $\mathcal{G}^\aleph(E)$.*

Proof. Let $w \in \Gamma$ be given. By hypothesis there exist a neighborhood U of w and continuous functions $\widetilde{A}_- : U \cap \overline{D}_- \to GL(E)$ and $\widetilde{A}_+ : U \cap \overline{D}_+ \to GL(E)$, which are holomorphic in $U \cap D_-$ and $U \cap D_+$, respectively, such that

$$A = \widetilde{A}_- \widetilde{A}_+ \qquad \text{on } U \cap \Gamma . \tag{7.8.4}$$

Let $G\big(L(E)/\mathcal{F}^\aleph(E)\big)$ be the group of invertible elements of the factor algebra $L(E)/\mathcal{F}^\aleph(E)$, let 1 be its unit element, and let $\pi : L(E) \to L(E)/\mathcal{F}^\aleph(E)$ be the canonical map. Since $\pi(A) = 1$, then it follows from (7.8.4) that

$$\pi\big(\widetilde{A}_-^{-1}\big) = \pi\big(\widetilde{A}_+\big) \qquad \text{on } U \cap \Gamma . \tag{7.8.5}$$

Hence, by Theorem 1.5.4, there is a holomorphic function $f : U \to G\big(L(E)/\mathcal{F}^\infty(E)\big)$ with

$$f\big|_{U \cap \overline{D}_-} = \pi\big(\widetilde{A}_-^{-1}\big) \qquad \text{and} \qquad f\big|_{U \cap \overline{D}_+} = \pi\big(\widetilde{A}_+\big) . \tag{7.8.6}$$

By Theorem 7.8.2 we can find a holomorphic function $A : U \to GL(E)$ such that

$$\pi(A) = f . \tag{7.8.7}$$

Set $A_- = \widetilde{A}_- A$ on $U \cap \overline{D}_-$ and $A_+ = A^{-1} \widetilde{A}_+$ on $U \cap \overline{D}_+$. Then it follows from (7.8.4) that

$$A = A_- A_+ \qquad \text{on } \Gamma,$$

and from (7.8.7) and (7.8.6) it follows that $\pi(A_-) = 1$ on $U \cap \overline{D}_-$ and $\pi(A_+) = 1$ on $U \cap \overline{D}_+$, i.e., the values of A_- and A_+ belong to $\mathcal{G}^\aleph(E)$. □

7.8.4 Proposition. *Let $\aleph = \infty, \omega$, let $0 < r < R < \infty$ such that*

$$\overline{D}_+ \subseteq W := \left\{ z \in \mathbb{C} \;\middle|\; r < |z| < R \right\}, \tag{7.8.8}$$

and let $A : \Gamma \to \mathcal{G}^{\aleph}(E)$ be a continuous function which admits local factorizations with respect to Γ and $GL(E)$ (Def. 7.1.3). Then:

(i) *The function A can be written in the form*

$$A = A_- H A_+, \tag{7.8.9}$$

where the functions $A_\pm$ and H have the following properties:

$A_- : \overline{D}_- \cup \{\infty\} \to \mathcal{G}^{\aleph}(E)$ is continuous on $\overline{D}_- \cup \{\infty\}$ and holomorphic in $D_- \cup \{\infty\}$;

$A_+ : \overline{D}_+ \to \mathcal{G}^{\aleph}(E)$ is continuous on $\overline{D}_+$ and holomorphic in D_+;

H is of the form

$$H(z) = I + \sum_{n=N}^{M} z^n H_n, \qquad N, M \in \mathbb{Z}, \quad H_n \in \mathcal{F}^{\aleph}(E), \tag{7.8.10}$$

and $H(z) \in \mathcal{G}^{\aleph}(E)$ for all $z \in \overline{W}$;

(ii) *The function A can be written in the form*

$$A = A_- H A_+, \tag{7.8.11}$$

where the functions $A_\pm$ and H have the following properties:

$A_- : \overline{D}_- \cup \{\infty\} \to \mathcal{G}^{\aleph}(E)$ is continuous on $\overline{D}_- \cup \{\infty\}$ and holomorphic in $D_- \cup \{\infty\}$;

$A_+ : \overline{D}_+ \to \mathcal{G}^{\aleph}(E)$ is continuous on $\overline{D}_+$ and holomorphic in D_+;

H^{-1} is of the form

$$H^{-1}(z) = I + \sum_{n=N}^{M} z^n H_n, \qquad N, M \in \mathbb{Z}, \quad H_n \in \mathcal{F}^{\aleph}(E), \tag{7.8.12}$$

and $H^{-1}(z) \in \mathcal{G}^{\aleph}(E)$ for all $z \in \overline{W}$.

Proof. The proofs of parts (i) and (ii) are similar (for (i) we use part (i) of Theorem 7.6.1, and for (ii) we use part (ii) of that theorem). We restrict ourselves to part (i).

Let $\mathcal{F}_I^{\aleph}(E)$ be the Banach algebra of Definition 5.12.1. By Lemma 7.8.3, A admits also local factorizations with respect to Γ and $G\mathcal{F}_I^{\aleph}(E)$. Therefore, we can apply part (i) of Theorem 7.6.4 (with $A = \mathcal{F}_I^{\aleph}(E)$), and we obtain a representation of A in the form $A = a_- h a_+$, where

$a_- : \overline{D}_- \cup \{\infty\} \to G\mathcal{F}_I^{\aleph}(E)$ is continuous on $\overline{D}_- \cup \{\infty\}$ and holomorphic in $D_- \cup \{\infty\}$;

$a_+ : \overline{D}_+ \to G\mathcal{F}_I^{\aleph}(E)$ is continuous on $\overline{D}_+$ and holomorphic in D_+;

h is of the form

$$h(z) = \sum_{n=N}^{M} z^n h_n\,, \qquad N, M \in \mathbb{Z},$$

and $h(z) \in G\mathcal{F}_I^{\aleph}(E)$ for all $z \in \overline{W}$.

Let $V : \Gamma \to \mathcal{F}^{\aleph}(E)$, $V_- : \overline{D}_- \cup \{\infty\} \to \mathcal{F}^{\aleph}(E)$, $V_+ : \overline{D}_+ \to \mathcal{F}^{\aleph}(E)$, $U : \overline{W} \to \mathcal{F}^{\aleph}(E)$, $\lambda_- : \overline{D}_- \cup \{\infty\} \to \mathbb{C}^*$, $\lambda_+ : \overline{D}_+ \cup \{\infty\} \to \mathbb{C}^*$ and $\lambda : \overline{W} \to \mathbb{C}^*$ be the functions with $A = I + V$, $a_\pm = \lambda_\pm I + V_\pm$ and $h = \lambda I + U$. Then

$$I + V = \big(\lambda_- I + V_-\big)\big(\lambda I + U\big)\big(\lambda_+ I + V_+\big),$$

which implies (as $\dim E = \infty$) that $\lambda^{-1} = \lambda_- \lambda_+$. Hence

$$A = \left(I + \frac{V_-}{\lambda_-}\right)\left(I + \lambda^{-1} U\right)\left(I + \frac{V_+}{\lambda_+}\right).$$

Setting $A_\pm = \lambda_\pm^{-1} a_\pm$ and $H = \lambda^{-1} h$, we conclude the proof. □

The following Lemma 7.8.5 as well as the subsequent theorem here will be proved only for the group $\mathcal{G}^\infty(E)$. Below we obtain these results also for $\mathcal{G}^\omega(E)$ (Theorem 8.2.2). But then the proof is more difficult, because then the approximation argument used in the proof of Lemma 7.8.5 does not work.

7.8.5 Lemma. *Let $A : \Gamma \to \mathcal{G}^\infty(E)$ be a continuous function which admits local factorizations with respect to Γ and $GL(E)$ (Def. 7.1.3). Moreover, let $0 < r < R < \infty$ be given such that*

$$\overline{D}_+ \subseteq W := \Big\{ z \in \mathbb{C} \,\Big|\, r < |z| < R \Big\}. \tag{7.8.13}$$

Then A is equivalent with respect to Γ and $\mathcal{G}^\infty(E)$ to a function of the form $Q + PA_P P$, where:

- *P is a finite dimensional projector in E and $Q = I - P$;*
- *A_P is a rational function with values in $L(\operatorname{Im} P)$ such that $A_P(z)$ is invertible for all $z \in \overline{W}$.*

Proof. By part (i) of Proposition 7.8.4, we may assume that A is already of the form

$$A(z) = I + \sum_{n=N}^{M} z^n A_n\,, \qquad N, M \in \mathbb{Z}, \quad A_n \in \mathcal{F}^\infty(E),$$

and that $A(z)$ is invertible for all $z \in \overline{W}$. Set

$$W_- = \left\{ z \in \mathbb{C} \;\middle|\; r < |z| \right\} \quad \text{and} \quad W_+ = \left\{ z \in \mathbb{C} \;\middle|\; |z| < R \right\}.$$

Then, by Theorem 7.5.1, for some $\varepsilon > 0$, the following holds:

(*) For each $B \in \overline{\mathcal{O}}^{GL(E)}(\overline{W})$ with $\max_{z \in \overline{W}} \|B(z) - 1\| < \varepsilon$, there exist $B_+ \in \overline{\mathcal{O}}^{GL(E)}(\overline{W}_+)$ and $B_- \in \overline{\mathcal{O}}^{Gl(E)}(\overline{W}_- \cup \{\infty\})$ with $B = B_- B_+$ on $\overline{W}$.

As $A_n \in \mathcal{F}^\infty(E)$, we can chose finite dimensional operators F_n so close to A_n that, for the function

$$F(z) := I + \sum_{n=N}^{M} z^n F_n$$

we have:

$$F(z) \in GL(E) \quad \text{for all } z \in \overline{W}, \qquad \text{and} \qquad \max_{z \in \overline{W}} \left\| AF^{-1}(z) - I \right\| < \varepsilon.$$

Then, by (*), AF^{-1} can be written in the form

$$AF^{-1} = B_- B_+ \tag{7.8.14}$$

with $B_+ \in \overline{\mathcal{O}}^{GL(E)}(\overline{W}_+)$ and $B_- \in \overline{\mathcal{O}}^{GL(E)}(\overline{W}_- \cup \{\infty\})$. Choose a finite dimensional projector in E with $\operatorname{Im} F_n \subseteq \operatorname{Im} P$ and $\operatorname{Ker} P \subseteq \operatorname{Ker} F_n$ for $N \le n \le M$, and set $Q = I - P$. Then $F = PFP + Q$ on $\overline{W}$. Setting $A_P = PF\big|_{\operatorname{Im} P}$, we conclude the proof. □

7.8.6 Theorem. *Let $A : \Gamma \to \mathcal{G}^\infty(E)$ be a continuous function which admits local factorizations with respect to Γ and $GL(E)$. Then A admits a global factorization with respect to Γ and $\mathcal{G}^\infty(E)$.*

Proof. This follows immediately from the preceding Lemma 7.8.5 and Theorem 7.7.1. □

Under certain additional assumptions about the function A in Theorem 7.8.6, we can say correspondingly more about the factors of the factorizations of A. We have:

7.8.7 Corollary. *Let $0 < \alpha < 1$ and $k \in \mathbb{N}$, where, for $k \ge 1$, we additionally assume that Γ is of class $\mathcal{C}^k$ (Def. 3.4.1). Let $A : \Gamma \to \mathcal{G}^\infty(E)$ be a function of class $\mathbb{C}^{k+\alpha}$ (Def. 3.4.3). Then:*

(i) *A admits a factorization with respect to Γ and $\mathcal{G}^\infty(E)$.*

(ii) *If $A = A_- \Delta A_+$ is an arbitrary factorization of A with respect to Γ, then automatically, the factors A_- and A_+ are of class $\mathbb{C}^{k+\alpha}$ on $\overline{D}_-$ and $\overline{D}_+$, respectively.*

Proof. Part (i) of Theorem 7.3.1 in particular states that A admits local factorizations with respect to Γ. Therefore part (i) of the corollary follows from Theorem 7.8.6. Part (ii) of the corollary follows from part (ii) of Theorem 7.3.1 □

7.8.8 Corollary. *Let $\Gamma = \mathbb{T}$ be the unit circle, and let $\mathcal{R}$ be a Banach algebra of continuous $L(E)$-valued functions satisfying conditions* (A), (B) *and* (C) *in Section* 7.2.4. *For example, let $\mathcal{R} = W\big(L(E)\big)$ be the Wiener algebra* (*see Section* 7.2.1). *Let $A : \Gamma \to \mathcal{G}^\infty(E)$ be a function which belongs to $\mathcal{R}$. Then:*

(i) *A admits a factorization with respect to Γ and $\mathcal{G}^\infty(E)$.*

(ii) *If $A = A_-\Delta A_+$ is an arbitrary factorization of A with respect to Γ, then automatically, the factors A_- and A_+ belong to the algebra $\mathcal{R}$.*

Proof. Part (ii) of Theorem 7.2.5 in particular states that A admits local factorizations with respect to Γ. Therefore part (i) of the corollary follows from Theorem 7.8.6. Part (ii) of the corollary follows from part (iii) of Theorem 7.2.5. □

7.9 The filtration of an operator function with respect to a contour

In this section E is a Banach space, $D_+ \subseteq \mathbb{C}$ is a bounded connected open set with piecewise C^1-boundary Γ such that $0 \in D_+$. We set $D_- = \mathbb{C} \setminus D_+$.

7.9.1 Definition. Let $A : \Gamma \to GL(E)$ be a continuous function, and let $\kappa \in \mathbb{Z}$.

(i) A pair $\varphi = (\varphi_-, \varphi_+)$ will be called a (Γ, κ)**-section** or simply a κ**-section** of A if $\varphi_- : \overline{D}_- \cup \{\infty\} \to E$ and $\varphi_+ : \overline{D}_+ \to E$ are continuous E-valued functions which are holomorphic in $D_- \cup \{\infty\}$ and D_+, respectively, such that

$$z^\kappa \varphi_-(z) = A(z)\varphi_+(z) \qquad \text{for } z \in \Gamma. \tag{7.9.1}$$

(ii) We denote by $M(\kappa, A) = M(\kappa, \Gamma, A)$ the space of all κ-sections of A. We consider $M(\kappa, A)$ as a Banach space endowed with the norm defined by

$$\|\varphi\| := \max_{z \in \overline{D}_- \cup \{\infty\}} \|\varphi_-(z)\| + \max_{z \in \overline{D}_+} \|\varphi_+(z)\| \tag{7.9.2}$$

for $\varphi = (\varphi_-, \varphi_+) \in M(\kappa, A)$.

(iii) We define

$$\begin{aligned} M_-(z, \kappa, A) &= M_-(z, \kappa, \Gamma, A) \\ &= \Big\{\varphi_-(z) \,\Big|\, (\varphi_-, \varphi_+) \in M(\kappa, A)\Big\} \quad \text{for } z \in \overline{D}_- \cup \{\infty\} \end{aligned}$$

and

$$\begin{aligned} M_+(z,\kappa,A) &= M_+(z,\kappa,\Gamma,A) \\ &= \Big\{\varphi_+(z) \;\Big|\; (\varphi_-,\varphi_+) \in M(\kappa,A)\Big\} \quad \text{for } z \in \overline{D}_+ . \end{aligned}$$

Since $0 \notin \Gamma$, it follows from (7.9.1) that

$$M_-(z,\kappa,\Gamma,A) = A(z)M_+(z,\kappa,\Gamma,A) \quad \text{for all } z \in \Gamma \text{ and } \kappa \in \mathbb{Z}. \tag{7.9.3}$$

7.9.2. Let $A : \Gamma \to E$ be a continuous function, let $\kappa, \mu \in \mathbb{Z}$, and let (φ_-, φ_-) be a κ-section of A. Since Γ is the boundary both of D_- and D_+, then it follows by uniqueness of holomorphic functions from (7.9.1) that each of the two components φ_+ and φ_- is uniquely determined by the other one.

7.9.3. Let $A : \Gamma \to E$ be a continuous function, and let $\kappa, \mu \in \mathbb{Z}$ with $\kappa \geq \mu$. Then

$$M_-(z,\kappa,\Gamma,A) \supseteq M_-(z,\mu,\Gamma,A) \qquad \text{for } z \in \overline{D}_- \cup \{\infty\}, \tag{7.9.4}$$

$$M_+(z,\kappa,\Gamma,A) \supseteq M_+(z,\mu,\Gamma,A) \qquad \text{for } z \in \overline{D}_+ . \tag{7.9.5}$$

This follows from two obvious statements:

(i) If $(\omega_-, \omega_+) \in M(\mu,\Gamma,A)$ and $z \in \overline{D}_- \cup \{\infty\}$, then $(\phi_-, \phi_+) := (\omega_-, \widetilde{\omega}_+)$ with

$$\widetilde{\omega}_+(\zeta) := \zeta^{\kappa-\mu}\omega_+(\zeta), \qquad \zeta \in \overline{D}_+ ,$$

is a κ-section of A with $\phi_-(z) = \omega_-(z)$.

(ii) If $(\omega_-, \omega_+) \in M(\mu,\Gamma,A)$ and $z \in \overline{D}_+$, then $(\phi_-, \phi_+) := (\widetilde{\omega}_-, \omega_+)$ with

$$\widetilde{\omega}_-(\zeta) := \zeta^{\mu-\kappa}\omega_-(\zeta), \qquad \zeta \in \overline{D}_- \cup \{\infty\},$$

is a κ-section of A with $\phi_+(z) = \omega_+(z)$.

7.9.4 Lemma. *Let $A : \Gamma \to GL(E)$ be a continuous function which admits local factorizations with respect to Γ. Then there exist integers $\kappa_- \leq \kappa_+$ such that:*

(i) *If $\kappa \in \mathbb{Z}$ with $\kappa < \kappa_-$, then*

$$M(\kappa,\Gamma,A) = 0, \tag{7.9.6}$$

i.e.,

$$\left.\begin{aligned} M_-(z,\kappa,\Gamma,A) &= 0 \qquad \textit{for all } z \in \overline{D}_- \cup \{\infty\}, \\ M_+(z,\kappa,\Gamma,A) &= 0 \qquad \textit{for all } z \in \overline{D}_+ . \end{aligned}\right. \tag{7.9.7}$$

(ii) *If $\kappa \in \mathbb{Z}$ with $\kappa \geq \kappa_+$, then*

$$\left.\begin{aligned} M_-(z,\kappa,\Gamma,A) &= E \qquad \textit{for all } z \in \overline{D}_- \cup \{0\}, \\ M_+(z,\kappa,\Gamma,A) &= E \qquad \textit{for all } z \in \overline{D}_+ . \end{aligned}\right. \tag{7.9.8}$$

Proof. By Theorem 7.6.1, there exist integers $N \leq M$, functions $H, T : \Gamma \to GL(E)$ of the form

$$S(z) = \sum_{n=N}^{M} z^n S_n \qquad \text{and} \qquad H(z) = \sum_{n=N}^{M} z^n H_n\,,$$

and functions $A_-, T_- \in \overline{\mathcal{O}}^{GL(E)}(\overline{D}_- \cup \{\infty\})$ (Def. 5.4.5) and $A_+, T_+ \in \overline{\mathcal{O}}^{GL(E)}(\overline{D}_+)$ such that

$$A = A_- H A_+ \qquad \text{on } \Gamma \tag{7.9.9}$$

and

$$A = T_- S^{-1} T_+ \qquad \text{on } \Gamma\,. \tag{7.9.10}$$

Set

$$\kappa_- = N \qquad \text{and} \qquad \kappa_+ = \max(-N, M)\,.$$

We now first prove part (i). Let $\kappa \leq \kappa_- - 1 = N - 1$ be given, and let $(\varphi_-, \varphi_+) \in M(\kappa, A)$. Then it follows from (7.9.9) that

$$z^\kappa \varphi_-(z) = A_-(z) \sum_{n=N}^{M} z^n H_n A_+(z) \varphi_+(z)\,,$$

and therefore

$$A_-^{-1}(z)\varphi_-(z) = \sum_{n=N}^{M} \zeta^{n-\kappa} H_n A_+(z)\varphi_+(z)\,, \qquad z \in \Gamma\,.$$

Since $n - \kappa \geq n - N + 1 \geq 1$ for $n \geq N$, the two sides of this relation define a continuous function on $\mathbb{C} \cup \{\infty\}$ which is equal to zero for $z = 0$. This function is holomorphic outside Γ and, hence, by Theorem 1.5.4, holomorphic on $\mathbb{C} \cup \{\infty\}$. Therefore, by Liouville's theorem, it is identically zero. In particular $A_-^{-1}\varphi_- \equiv 0$. Since the values of A_- are invertible, it follows that $\varphi_- \equiv 0$ and, consequently, $\varphi_+ \equiv 0$.

We prove part (ii). First let $\kappa \geq \kappa_+$ and $z \in \overline{D}_- \cup \{\infty\}$ be given.

For $n \geq N$, then $n + \kappa \geq n - N \geq 0$. Therefore, setting

$$\varphi_+(\zeta) = \zeta^\kappa T_+^{-1}(\zeta) S(\zeta) T_-^{-1}(z) v\,, \qquad \zeta \in \overline{D}_+\,,$$

we get a function $\varphi_+ \in \overline{\mathcal{O}}^E(\overline{D}_+)$. Moreover, setting

$$\varphi_-(\zeta) = T_-(\zeta) T_-^{-1}(z) v\,, \qquad \zeta \in \overline{D}_- \cup \{\infty\},$$

we get a function $\varphi_- \in \overline{\mathcal{O}}^E(\overline{D}_- \cup \{\infty\})$. From (7.9.10) it follows that

$$A(\zeta)\varphi_+(\zeta) = T_-(\zeta) S^{-1}(\zeta) T_+(\zeta) \zeta^\kappa T_+^{-1}(\zeta) S(\zeta) T_-^{-1}(z) v = \zeta^\kappa \varphi_-(\zeta)$$

for $\zeta \in \Gamma$. Hence $(\varphi_-, \varphi_+) \in M_A(\kappa, A)$. It remains to observe that $\varphi_-(z) = T_-(z)T_-^{-1}(z)v = v$.

Now let $\kappa \geq \kappa_+$ and $z \in D_+$. For $n \leq M$, then $n - \kappa \leq M - \kappa_+ \leq 0$. Therefore, setting

$$\varphi_-(\zeta) = \zeta^{-\kappa} A_-(\zeta) H(\zeta) A_+(z) v\,, \qquad \zeta \in \overline{D}_-\,,$$

we get a function $\varphi_- \in \overline{\mathcal{O}}^E\big(\overline{D}_- \cup \{\infty\}\big)$. Moreover, setting

$$\varphi_+(\zeta) = A_+^{-1}(\zeta) A_+(z) v\,, \qquad \zeta \in \overline{D}_+,$$

we get a function $\varphi_+ \in \overline{\mathcal{O}}^E\big(\overline{D}_+\big)$. From (7.9.9) it follows that

$$A(\zeta)\varphi_+(\zeta) = A_-(\zeta)H(\zeta)A_+(\zeta)A_+^{-1}(\zeta)A_+(z)v = \zeta^\kappa \varphi_-(\zeta)$$

for $\zeta \in \Gamma$. Hence $(\varphi_-, \varphi_+) \in M_A(\kappa, A)$. It remains to observe that $\varphi_+(z) = A_+^{-1}(z)A_+(z)v = v$. □

7.9.5. Let $A, \widetilde{A} : \Gamma \to GL(E)$ be two continuous functions, which are equivalent with respect to Γ and $GL(E)$ (Def. 7.1.3) and which admit local factorizations with respect to Γ (Def. 7.1.3), and let $A_- \in \overline{\mathcal{O}}^{GL(E)}\big(\overline{D}_- \cup \{\infty\}\big)$ and $A_+ \in \overline{\mathcal{O}}^{GL(E)}\big(\overline{D}_+\big)$ be functions (which then exist) such that $\widetilde{A} = A_- A A_+$ on Γ. Then it is clear that:

> (φ_-, φ_+) is a (Γ, κ)-section of A, if and only if, $\big(A_-\varphi_-, A_+^{-1}\varphi_+\big)$ is a (Γ, κ)-section of $\widetilde{A}$.

Hence

$$M_-(z, \kappa, \Gamma\widetilde{A}) = A_-(z) M_-(z, \kappa, \Gamma, A) \text{ for all } z \in \overline{D}_- \cup \{\infty\} \text{ and } \kappa \in \mathbb{Z},$$

$$M_+(z, \kappa, \Gamma, \widetilde{A}) = A_+^{-1}(z) M_+(z, \kappa, \Gamma, A) \text{ for all } z \in \overline{D}_+ \text{ and } \kappa \in \mathbb{Z}.$$

7.9.6 Definition. Let $A : \Gamma \to GL(E)$ be a continuous function. An integer κ will be called a **partial index of** A **relative to** Γ or simply a **partial index of** A if there exists a point $z \in \mathbb{C} \cup \{\infty\}$ such that:

$$\text{if } z \in \overline{D}_- \cup \{\infty\}, \text{ then } M_-(z, \kappa - 1, \Gamma, A) \underset{\neq}{\subset} M_-(z, \kappa, \Gamma, A)\,,$$

$$\text{if } z \in \overline{D}_+, \text{ then } M_+(z, \kappa - 1, \Gamma, A) \underset{\neq}{\subset} M_+(z, \kappa, \Gamma, A)\,.$$

Note that if A admits local factorizations relative to Γ, then, by Lemma 7.9.4, the set of partial indices of A is not empty and finite.

7.9.7 Theorem. *Let $A : \Gamma \to GL(E)$ be a continuous function which admits local factorizations with respect to Γ (Def. 7.1.3), and let $k_1 > \ldots > k_n$ be the partial indices of A with respect to Γ. Then:*

(i) *For all* $z \in \overline{D}_- \cup \{\infty\}$,

$$M_-(z, k_1, \Gamma, A) = E \quad \textit{and} \quad M_-(z, k_n - 1, \Gamma, A) = \{0\},$$

and, for all $z \in \overline{D}_+$,

$$M_+(z, k_1, \Gamma, A) = E \quad \textit{and} \quad M_+(z, k_n - 1, \Gamma, A) = \{0\}.$$

(ii) *For each partial index* k_j, $1 \le j \le n$, *there exists* $d_j \in \mathbb{N}^* \cup \{\infty\}$ *such that*

$$d_j = \begin{cases} \dim\big(M_-(z, k_j, \Gamma, A)/M_-(z, k_j - 1, \Gamma, A)\big) & \textit{if } z \in \overline{D}_- \cup \{\infty\}, \\ \dim\big(M_+(z, k_j, \Gamma, A)/M_+(z, k_j - 1, \Gamma, A)\big) & \textit{if } z \in \overline{D}_+ . \end{cases}$$

(iii) *Let* $1 \le j \le n$, *let* $d \in \mathbb{N}^*$, *and let* $(\varphi_\nu^-, \varphi_\nu^+) \in M(k_j, \Gamma, A)$, $1 \le \nu \le d$. *Assume, for at least one point* $z \in \mathbb{C} \cup \{\infty\}$, *the following condition* $C(z)$ *is satisfied:*

$$C(z): \begin{cases} \textit{If } z \in \overline{D}_- \cup \{\infty\}, \textit{ then the classes in the factor space} \\ M_-(z, k_j, \Gamma, A)/M_-(z, k_j - 1, \Gamma, A), \textit{ defined by the vectors} \\ \varphi_1^-(z), \dots, \varphi_d^-(z), \textit{ are linearly independent.} \\ \textit{If } z \in \overline{D}_+, \textit{ then the classes in the factor space} \\ M_+(z, k_j, \Gamma, A)/M_+(z, k_j - 1, \Gamma, A), \textit{ defined by the vectors} \\ \varphi_1^+(z), \dots, \varphi_d^+(z), \textit{ are linearly independent.} \end{cases}$$

Then condition $C(z)$ *is satisfied for all* $z \in \mathbb{C} \cup \{\infty\}$.

Before proving this theorem, we use it for the following

7.9.8 Definition. With the notations from the preceding theorem we define: d_j will be called the **multiplicity** of k_j (as a partial index of A with respect to Γ), $1 \le j \le n$. The family of families of subspaces

$$\Big\{M_+(z, k_j, \Gamma, A)\Big\}_{z \in \overline{D}_+} \quad \text{and} \quad \Big\{M_-(z, k_j, \Gamma, A)\Big\}_{z \in \overline{D}_- \cup \{\infty\}}, \qquad 1 \le j \le n,$$

will be called the **filtration of** A **relative to** Γ or simply the **filtration of** A.

7.9.9 Remark. Let $A, \widetilde{A} : \Gamma \to GL(E)$ be two continuous functions which are equivalent relative to Γ and $GL(E)$ (Def. 7.1.3), and which admit local factorizations with respect to Γ (Def. 7.1.3). Then it follows from Section 7.9.5 that A and $\widetilde{A}$ have the same partial indices with the same multiplicities relative to Γ.

Proof of Theorem 7.9.7. Clearly, (ii) follows from (iii). Moreover, taking into account Lemma 7.9.4, assertion (i) follows from (ii). Therefore, it is sufficient to prove (iii). By Theorem 7.4.2, A is equivalent with respect to Γ and $GL(E)$ to a holomorphic function in a neighborhood of Γ. Therefore and by Section 7.9.5, we may assume that A is holomorphic in a neighborhood of Γ.

To prove (iii), we assume that (iii) is not true, i.e., we assume that there are two points $z, w \in \mathbb{C} \cup \{\infty\}$ such that $C(z)$ is satisfied, but $C(w)$ is not satisfied.

Since $C(w)$ is not true, we can find a non-zero vector $(\lambda_1, \ldots, \lambda_d)$ of complex numbers such that:

$$\sum_{\nu=1}^{d} \lambda_\nu \varphi_\nu^-(w) \in M_-(w, k_j - 1, A) \qquad \text{if } w \in \overline{D}_- \cup \{\infty\},$$

$$\sum_{\nu=1}^{d} \lambda_\nu \varphi_\nu^+(w) \in M_+(w, k_j - 1, A) \qquad \text{if } w \in \overline{D}_+ .$$

By definition of $M_\pm(w, k_j - 1, A)$, then there exists a $(k_j - 1)$-section (ω^-, ω^+) of A such that

$$\omega^-(w) = \sum_{\nu=1}^{d} \lambda_\nu \varphi_\nu^-(w) \qquad \text{if } w \in \overline{D}_- ,$$

$$\omega^+(w) = \sum_{\nu=1}^{d} \lambda_\nu \varphi_\nu^+(w) \qquad \text{if } w \in \overline{D}_+ .$$

As observed in 7.9.3, this $(k_j - 1)$-section can be modified so that we get a k_j-section $(\widetilde{\omega}^-, \widetilde{\omega}^+)$ of A such that still

$$\begin{aligned} \widetilde{\omega}^-(w) &= \sum_{\nu=1}^{d} \lambda_\nu \varphi_\nu^-(w) \qquad \text{if } w \in \overline{D}_- \cup \{\infty\}, \\ \widetilde{\omega}^+(w) &= \sum_{\nu=1}^{d} \lambda_\nu \varphi_\nu^+(w) \qquad \text{if } w \in \overline{D}_+ , \end{aligned} \tag{7.9.11}$$

but

$$\begin{aligned} \widetilde{\omega}^-(\zeta) &\in M_-(\zeta, k_j - 1, A) \quad \text{for all } \zeta \in \overline{D}_- \cup \{\infty\}, \\ \widetilde{\omega}^+(\zeta) &\in M_-(\zeta, k_j - 1, A) \quad \text{for all } \zeta \in \overline{D}_+ . \end{aligned} \tag{7.9.12}$$

If $w \neq \infty$, then we set

$$\psi_-(\zeta) = \frac{\zeta}{\zeta - w}\Big(\widetilde{\omega}^-(\zeta) - \sum_{\nu=1}^{d} \lambda_\nu \varphi_\nu^-(\zeta)\Big) \quad \text{for } \zeta \in \overline{D}_- \setminus \{w\},$$

and

$$\psi_+(\zeta) = \frac{1}{\zeta - w}\Big(\widetilde{\omega}^+(\zeta) - \sum_{\nu=1}^{d} \lambda_\nu \varphi_\nu^+(\zeta)\Big) \quad \text{for } \zeta \in \overline{D}_+ \setminus \{w\}.$$

If $w = \infty$, then we set

$$\psi^-(\zeta) = \zeta\Big(\widetilde{\omega}^-(\zeta) - \sum_{\nu=1}^{d} \lambda_\nu \varphi_\nu^-(\zeta)\Big) \quad \text{for } \zeta \in \overline{D}_-,$$

and

$$\psi^+(\zeta) = \widetilde{\omega}^+(\zeta) - \sum_{\nu=1}^{d} \lambda_\nu \varphi_\nu^+(\zeta) \quad \text{for } \zeta \in \overline{D}_+ .$$

Since A is holomorphic in a neighborhood of Γ and therefore the functions $\widetilde{\omega}^\pm$ and $\widetilde{\varphi}_\nu^\pm$ are holomorphic in a neighborhood of Γ, then, by (7.9.11), we obtain a pair of holomorphic functions $\psi^- : \overline{D}_- \cup \{\infty\} \to E$ and $\psi^+ : \overline{D}_+ \to E$ with

$$A(\zeta)\psi^+(\zeta) = \zeta^{k_j - 1}\psi^-(\zeta) \qquad \text{for } \zeta \in \Gamma.$$

Hence (ψ^-, ψ^+) is a $(k_j - 1)$-section of A, and therefore

$$\begin{aligned} &\psi^-(z) \in M_-(z, k_j - 1, A) \quad \text{if } z \in \overline{D}_- \cup \{\infty\}, \\ &\psi^+(z) \in M_+(z, k_j - 1, A) \quad \text{if } z \in \overline{D}_+ . \end{aligned} \tag{7.9.13}$$

On the other hand,

$$\psi^-(z) = \frac{z}{z-w}\Big(\widetilde{\omega}^-(z) - \sum_{\mu=1}^{d} \lambda_\mu \varphi^-(z)\Big) \qquad \text{if } w \neq \infty \text{ and } z \in \overline{D}_-,$$

$$\psi^-(z) = \widetilde{\omega}^-(z) - \sum_{\mu=1}^{d} \lambda_\mu \varphi^-(z) \qquad \text{if } w \neq \infty \text{ and } z = \infty,$$

$$\psi^+(z) = \frac{z}{z-w}\Big(\widetilde{\omega}^+(z) - \sum_{\mu=1}^{d} \lambda_\mu \varphi^+(z)\Big) \qquad \text{if } w \neq \infty \text{ and } z \in \overline{D}_+,$$

$$\psi^-(z) = z\Big(\widetilde{\omega}^-(z) - \sum_{\mu=1}^{d} \lambda_\mu \varphi^-(z)\Big) \qquad \text{if } w = \infty \text{ and } z \in \overline{D}_-,$$

$$\psi^+(z) = \widetilde{\omega}^+(z) - \sum_{\mu=1}^{d} \lambda_\mu \varphi^+(z) \qquad \text{if } w = \infty \text{ and } z \in \overline{D}_+ .$$

By (7.9.12) and condition $C(z)$, this implies that

$$\begin{aligned} &\psi^-(z) \in M_-(z, k_j, A) \setminus M_-(z, k_j - 1, A) \quad \text{if } z \in \overline{D}_- \cup \{\infty\}, \\ &\psi^+(z) \in M_+(z, k_j, A) \setminus M_+(z, k_j - 1, A) \quad \text{if } z \in \overline{D}_+, \end{aligned}$$

which contradicts (7.9.13). □

Theorem 7.9.7 contains interesting information concerning the Riemann-Hilbert boundary problem. To state it, we first give the following

7.9.10 Definition. Let $A : \Gamma \to GL(E)$ be a continuous function, let $z \in \mathbb{C} \cup \{\infty\}$, let $v \in E$, let $\kappa \in \mathbb{Z}$, and let (φ_+, φ_-) be a κ-section of A (Def. 7.9.1). We shall say that (φ_+, φ_-) is a (Γ, κ)-section of simply a κ-section of A **through** (z, v) if [3]

$$\varphi_-(z) = v \text{ if } z \in \overline{D}_- \cup \{\infty\} \qquad \text{and} \qquad \varphi_+(z) = v \text{ if } z \in D_+ .$$

With this definition, from Theorem 7.9.7 we immediately obtain:

7.9.11 Corollary (to Theorem 7.9.7). *Let $A : \Gamma \to GL(E)$ be a continuous function which admits local factorizations with respect to Γ, let $k_1 > \ldots > k_n$ be the partial indices of A (Def. 7.9.6), and let d_j be the multiplicity of k_j as a partial index of A (Def. 7.9.8), $1 \le j \le n$. Then, for each point $z \in \mathbb{C} \cup \{\infty\}$, there exist (possibly not closed) linear subspaces $E_1, \ldots, E_n$ of E such that E is the algebraically direct sum of $E_1, \ldots, E_n$,*

$$\dim E_j = d_j \qquad \text{for } 1 \le j \le n$$

and such that, for each $1 \le j \le n$ and each $v \in E_j$ with $v \ne 0$ the following hold:

(i) *There exists a k_j-section of A through (z, v).*

(ii) *If $\mu < k_j$, then there exists no μ-section of A through (z, v).*

(iii) *If (φ_-, φ_+) is a k_j-section of A through (z, v), then*

$$\varphi_-(\zeta) \in M_-(\zeta, k_j, A) \setminus M_-(z, k_j - 1, A) \quad \text{for all } \zeta \in \overline{D}_- \cup \{\infty\},$$
$$\varphi_+(\zeta) \in M_+(\zeta, k_j, A) \setminus M_+(z, k_j - 1, A) \quad \text{for all } \zeta \in \overline{D}_+ .$$

(iv) *If (φ_-, φ_+) and (ψ_-, ψ_+) are two k_j-section of A through (z, v), then*

$$\varphi_-(\zeta) - \psi_-(\zeta) \in M_-(\zeta, k_j - 1, A) \quad \text{for all } \zeta \in \overline{D}_- \cup \{\infty\},$$
$$\varphi_+(\zeta) - \psi_-(\zeta) \in M_+(z, k_j - 1, A) \quad \text{for all } \zeta \in \overline{D}_+ .$$

Proof. Let $z \in \mathbb{C} \cup \{\infty\}$ be given. Put

$$M(z, \kappa, A) = \begin{cases} M_-(z, \kappa, A) & \text{if } z \in \overline{D}_- \cup \{\infty\}, \\ M_+(z, \kappa, A) & \text{if } z \in \overline{D}_+ , \end{cases} \qquad \kappa \in \mathbb{Z}.$$

The space E_n is uniquely determined. We have to set $E_n = M(z, k_n, A)$. If $1 \le j \le n-1$, then for E_j we can (and have to) choose an arbitrary algebraic complement of $M(z, k_j - 1, A)$ in $M_j(z, k_j, A)$. □

[3] For $z \in \Gamma$ we have to make a choice. Just as well we could require that $\varphi_+(z) = v$.

7.10 A general criterion for the existence of factorizations

In this section, E is a Banach space, $D_+ \subseteq \mathbb{C}$ is a bounded, connected, open set with piecewise $\mathcal{C}^1$-boundary Γ such that $0 \in D_+$, and $D_- = \mathbb{C} \setminus D_+$.

7.10.1 Theorem. *Let $A : \Gamma \to GL(E)$ be a continuous function which admits local factorizations with respect to Γ. Then the following are equivalent:*

(i) *A admits a canonical factorization with respect to Γ.*

(ii) *Zero is the only partial index of A with respect to Γ (Def. 7.9.6).*

Proof. (i)⇒(ii): Assume that $A = A_- A_+$ is a canonical factorization of A with respect to Γ.

Let (φ_-, φ_+) be a -1-section of A. Then

$$\frac{1}{z} A_-^{-1}(z)\varphi_-(z) = A_+(z)\varphi_+(z) \qquad \text{for all } z \in \Gamma.$$

Then, by Theorem 1.5.4, the two sides of this equation define a holomorphic function on $\mathbb{C} \cup \{\infty\}$ which vanishes at $z = \infty$. Hence, by Liouville's theorem, this function is identically zero. It follows that $A_-^{-1}\varphi_- \equiv 0$ on $\overline{D}_- \cup \{\infty\}$ and $A_+\varphi_+ \equiv 0$ on $\overline{D}_+$. Since functions A_-^{-1} and A_+ are invertible, this implies that $\varphi_- \equiv 0$ on $\overline{D}_- \cup \{\infty\}$ and $\varphi_+ \equiv 0$ on $\overline{D}_+$. Hence

$$M(-1, A) = \{0\}. \tag{7.10.1}$$

If $z \in \overline{D}_- \cup \{\infty\}$ and $v \in E$, then, setting

$$\begin{aligned} \varphi_-(\zeta) &= A_-(\zeta)A_-^{-1}(z)v \qquad && \text{for } \zeta \in \overline{D}_- \cup \{\infty\}, \\ \varphi_+(\zeta) &= A_+^{-1}(\zeta)A_-^{-1}(z)v \qquad && \text{for } \zeta \in \overline{D}_+, \end{aligned}$$

we get a 0-section (φ_-, φ_+) of A with $\varphi_-(z) = v$. Hence

$$M(z, 0, A) = E \qquad \text{for all } z \in \overline{D}_- \cup \{\infty\}. \tag{7.10.2}$$

If $z \in \overline{D}_+$ and $v \in E$, then, setting

$$\begin{aligned} \varphi_-(\zeta) &= A_-(\zeta)A_+(z)v \qquad && \text{for } \zeta \in \overline{D}_- \cup \{\infty\} \\ \varphi_+(\zeta) &= A_+^{-1}(\zeta)A_+(z)v \qquad && \text{for } \zeta \in \overline{D}_+, \end{aligned}$$

we get a 0-section (φ_-, φ_+) of A with $\varphi_+(z) = v$. Hence

$$M(z, 0, A) = \{0\} = E \qquad \text{for all } z \in \overline{D}_+. \tag{7.10.3}$$

From (7.10.1)–(7.10.3) it follows that zero is the only partial index of A with respect to Γ.

(ii)⇒(i): Assume that zero is the only partial index of A. By Theorem 7.9.6 this means that

$$M(-1, A) = \{0\} \tag{7.10.4}$$

and

$$\begin{aligned} M_-(z, 0, A) &= E \qquad \text{for all } z \in \overline{D}_- \cup \{\infty\}, \\ M_+(z, 0, A) &= E \qquad \text{for all } z \in \overline{D}_+ . \end{aligned} \tag{7.10.5}$$

For each $\varphi = (\varphi_-, \varphi_+) \in M(0, A)$, we define

$$\begin{aligned} \Phi_-(z)\varphi &= \varphi_-(z) \qquad \text{for } z \in \overline{D}_- \cup \{\infty\}, \\ \Phi_+(z)\varphi &= \varphi_+(z) \qquad \text{for } z \in \overline{D}_+ . \end{aligned}$$

Recall that we consider $M(0, A)$ as a Banach space (cf. Def. 7.9.1). Then it follows from the maximum principle that

$$\begin{aligned} \Phi_-(z) &\in L\Big(M(0, \Gamma, A), E\Big) \quad \text{for all } z \in \overline{D}_- \cup \{\infty\}, \\ \Phi_+(z) &\in L\Big(M(0, \Gamma, A), E\Big) \quad \text{for all } z \in \overline{D}_+ . \end{aligned}$$

By Theorem 7.9.7, the operators $\Phi_-(z)$, $z \in \overline{D}_- \cup \{\infty\}$, and $\Phi_+(z)$, $z \in \overline{D}_+$, are injective. Moreover, by (7.10.5), these operators are surjective. Hence, by the Banach open mapping theorem, they are invertible from $M(0, A)$ onto E.

By Theorem 7.4.2 and Remark 7.9.9, in this proof, we may assume that A is holomorphic in a neighborhood of Γ. Then, for each $\varphi = (\varphi_-, \varphi_+) \in M(0, A)$, it follows from the relation

$$\varphi_-(z) = A(z)\varphi_+(z), \qquad z \in \Gamma,$$

that $\Phi_-\varphi = \varphi_-$ is holomorphic on $\overline{D}_- \cup \{\infty\}$, and $\Phi_+\varphi = \varphi_+$ is holomorphic on $\overline{D}_+$. Hence the functions

$$\begin{aligned} \Phi_- &: \overline{D}_- \cup \{\infty\} \to L\Big(M(0, A), E\Big), \\ \Phi_+ &: \overline{D}_+ \to L\Big(M(0, A), E\Big) \end{aligned}$$

are holomorphic (Theorem 1.7.1).

Now we fix some point $z_0 \in \Gamma$, and set

$$\begin{aligned} A_-(z) &= \Phi_-(z)\Phi_-^{-1}(z_0) \qquad \text{for } z \in \overline{D}_- \cup \{\infty\}, \\ A_+(z) &= \Phi_+(z)\Phi_-^{-1}(z_0) \qquad \text{for } z \in \overline{D}_+ . \end{aligned}$$

Then $A = A_- A_+^{-1}$ on Γ. Indeed, let $v \in E$ and $z \in \Gamma$ be given. Set

$$\varphi = (\varphi_-, \varphi_+) = \Phi_+^{-1}(z)v.$$

Then $\varphi_-(z) = A(z)\varphi_+(z)$ and $\Phi_+(z)\varphi = v$, and therefore

$$A_-(z)A_+^{-1}(z)v = \Phi_-(z)\Phi_-^{-1}(z_0)\Phi_-(z_0)\Phi_+^{-1}(z)v = \Phi_-(z)\varphi$$
$$= \varphi_-(z) = A(z)\varphi_+(z) = A(z)\Phi_+(z)\varphi = A(z)v\,. \qquad \square$$

7.10.2 Lemma. *Let $A : \Gamma \to GL(E)$ be a continuous function which admits local factorizations with respect to Γ. Suppose A has negative partial indices with respect to Γ (Def. 7.9.6). Denote these negative partial indices by $\kappa_1 > \ldots > \kappa_m$. Let N be the subspace of $M(0, A)$ (Def. 7.9.1) which consist of the sections $(\varphi_-, \varphi_+) \in M(0, A)$ with*

$$\varphi_-(z) \in M_-(z, \kappa_1, A) \qquad \text{for all } z \in \overline{D}_- \cup \{\infty\}\,,$$
$$\varphi_+(z) \in M_+(z, \kappa_1, A) \qquad \text{for all } z \in \overline{D}_+\,.$$

If each of the negative partial indices $\kappa_1, \ldots, \kappa_m$ has finite multiplicity (Def. 7.9.8), then N is finite dimensional.

Proof. Let d_j be the multiplicity of κ_j, $1 \le j \le m$. Set $\kappa_0 = 0$ and $\kappa_{m+1} = \kappa_m - 1$. Suppose $d_j < \infty$ for $1 \le j \le m$. Then we can find vectors $v_{j,1}, \ldots, v_{j,d_j} \in M(0, \kappa_j, A)$, $1 \le j \le m$, which define a basis in the factor space

$$M(0, \kappa_j, A)\Big/M(0, \kappa_{j+1}, A)\,.$$

Further, for $1 \le j \le m$ and $1 \le \nu \le d_j$, we choose a section

$$\varphi_{j,\nu} = \left(\varphi_{j,\nu}^-, \varphi_{j,\nu}^+\right) \in M(\kappa_j, A)$$

with $\varphi_{j,\nu}^+(0) = v_{j,\nu}$. By Theorem 7.9.7, then, for $1 \le j \le m$,

$$\begin{aligned}&\text{for all } z \in \overline{D}_- \cup \{\infty\}\,, \text{ the vectors } \varphi_{j,1}^-(z), \ldots, \varphi_{j,d_j}^-(z)\\ &\text{define a basis in the factor space } M_-(z, \kappa_j, A)/M_-(z, \kappa_{j+1}, A)\,,\\ &\text{and, for all } z \in \overline{D}_+\,, \text{ the vectors } \varphi_{j,1}^+(z), \ldots, \varphi_{j,d_j}^+(z)\\ &\text{define a basis in the factor space } M_+(z, \kappa_j, A)/M_+(z, \kappa_{j+1}, A)\,.\end{aligned} \tag{7.10.6}$$

Note that this in particular implies that

$$\begin{aligned}&\text{for all } z \in \overline{D}_- \cup \{\infty\}\,, \text{ the vectors } \varphi_{j,\nu}^-(z)\,, 1 \le j \le m,\\ &1 \le \nu \le d_j, \text{ form a basis of the space } M_-(z, \kappa_1, A),\\ &\text{and, for all } z \in \overline{D}_+\,, \text{ the vectors } \varphi_{j,\nu}^+(z), 1 \le j \le m,\\ &1 \le \nu \le d_j, \text{ form a basis in the space } M_+(z, \kappa_1, A)\,.\end{aligned} \tag{7.10.7}$$

Moreover, for $1 \le j \le m$, $1 \le \nu \le d_j$ and $0 \le s \le -\kappa_j$, we define

$$\varphi_{s,j,\nu}^-(z) = z^{\kappa_j+s}\varphi_{j,\nu}^-(z) \qquad \text{for } z \in \overline{D}_-\,,$$
$$\varphi_{s,j,\nu}^+(z) = z^s\varphi_{j,\nu}^+(z) \qquad \text{for } z \in \overline{D}_+\,.$$

Then each $(\varphi^-_{s,j,\nu}\varphi^+_{s,j,\nu})$ belongs to the space N. To prove that $\dim N < \infty$, now it is sufficient to prove that the finite system

$$\{\varphi_{s,j,\nu}\}_{1\le j\le m, 1\le \nu\le d_j, 0\le s\le -\kappa_j}$$

generates N.

For that we consider an arbitrary $(\phi_-, \phi_+) \in N$. By (7.10.7), there are uniquely determined continuous functions $\lambda^-_{j,\nu} : \overline{D}_- \cup \{\infty\} \to \mathbb{C}$ and $\lambda^+_{j,\nu} : \overline{D}_+ \to \mathbb{C}$, which are holomorphic in $D_- \cup \{\infty\}$ and D_+, respectively, such that

$$\begin{aligned} \phi_-(z) &= \sum_{1\le j\le m, 1\le \nu\le d_j} \lambda^-_{j,\nu}(z)\varphi^-_{j,\nu}(z) \quad \text{for } z \in \overline{D}_- \cup \{\infty\}, \\ \phi_+(z) &= \sum_{1\le j\le m, 1\le \nu\le d_j} \lambda^+_{j,\nu}(z)\varphi^+_{j,\nu}(z) \quad \text{for } z \in \overline{D}_+ . \end{aligned} \tag{7.10.8}$$

Since $\phi_-(z) = A(z)\phi_+(z)$ and $z^{\kappa_j}\varphi^-_{j,\nu}(z) = A(z)\varphi^+_{j,\nu}(z)$ for $z \in \Gamma$, this implies that

$$\sum_{1\le j\le m, 1\le \nu\le d_j} \lambda^-_{j,\nu}(z)\varphi^-_{j,\nu}(z) = \sum_{1\le j\le m, 1\le \nu\le d_j} z^{\kappa_j}\lambda^+_{j,\nu}(z)\varphi^-_{j,\nu}(z) \quad \text{for } z \in \Gamma .$$

By (7.10.7) this implies that

$$\lambda^-_{j,\nu}(z) = z^{\kappa_j}\lambda^+_{j,\nu}(z) \qquad \text{for } z \in \Gamma, \ 1 \le j \le m, 1 \le \nu \le d_j.$$

It follows by Liouville's theorem that, for some numbers $\alpha_{s,j,\nu}$,

$$\lambda^-_{j,\nu}(z) = \sum_{s=0}^{-\kappa_j} \alpha_{s,j,\nu} z^{\kappa_j+s} \quad \text{and} \quad \lambda^+_{j,\nu}(z) = \sum_{s=0}^{-\kappa_j} \alpha_{s,j,\nu} z^s .$$

Together with (7.10.8) this implies

$$\begin{aligned} \phi_- &= \sum_{1\le j\le m, 1\le \nu\le d_j, 0\le s\le -\kappa_j} \alpha_{s,j,\nu}\varphi^-_{s,j,\nu}, \\ \phi_+ &= \sum_{1\le j\le m, 1\le \nu\le d_j, 0\le s\le -\kappa_j} \alpha_{s,j,\nu}\varphi^+_{s,j,\nu} . \end{aligned}$$

□

7.10.3 Theorem. *Let $A : \Gamma \to GL(E)$ be a continuous function which admits local factorizations with respect to Γ. Then the following are equivalent:*

(i) *A admits a factorization with respect to Γ.*

(ii) *If A has non-zero partial indices (Def. 7.9.6), then each of them has finite multiplicity (Def. 7.9.8).*

Moreover: If these equivalent conditions are satisfied and if $A = A_-\Delta A_+$ is a factorization of A with respect to Γ, then:

(a) *If zero is the only partial index of A with respect to Γ, then $\Delta \equiv I$.*

(b) *If there exist non-zero partial indices of A with respect to Γ, then:*

- *The numbers $\kappa_1 > \ldots > \kappa_n$ from Definition 7.1.1 are the non-zero partial indices of A.*
- *If $P_1, \ldots, P_n$ are the projections from Definition 7.1.1, the $\dim P_j$ is the multiplicity of κ_j (as a partial index of A), $1 \le j \le n$.*
- *If zero is a partial index of A (which is always the case for $\dim E = \infty$), then $\dim P_0$ is the multiplicity of zero (as a partial index of A).*

Proof. First assume that there exists a factorization $A = A_-\Delta A_+$ of A with respect to Γ. Then, by Section 7.9.5, the diagonal factor Δ and A have the same partial indices with the same multiplicities. If zero is the only non-zero partial index of Δ, then it is clear that $\Delta \equiv I$. If there exist non-zero partial indices of Δ, then it is also clear that, with the notation from Definition 7.1.1, the numbers $\kappa_1 > \ldots > \kappa_n$ are these non-zero partial indices with the multiplicities $\dim P_1, \ldots, \dim P_n$, respectively. Further it is clear that zero is a partial index of Δ, if and only if, $P_0 \neq 0$ and that then $\dim P_0$ is the multiplicity of zero.

Now we assume that condition (ii) is satisfied. It remains to prove that then A admits a factorization. If zero is the only partial index of A, we know this from Theorem 7.10.1.

Assume there exist non-zero partial indices of A. For simplicity we consider only the case when there exist both positive and negative partial indices. (It is clear how to modify the proof in the other cases.) Let $\kappa_1 > \ldots > \kappa_m$ be the positive partial indices of A, and let $\kappa_{m+1} > \ldots > \kappa_n$ be the negative partial indices of A. Let N be the subspace of $M(0, A)$ defined in Lemma 7.10.2. By this lemma, N is finite dimensional.

If $(\varphi_-, \varphi_+) \in M(0, A)$, then we define

$$\Phi_-(z)\Big((\varphi_-, \varphi_+)\Big) = \varphi_-(z) \qquad \text{for } z \in \overline{D}_- \cup \{\infty\},$$
$$\Phi_+(z)\Big((\varphi_-, \varphi_+)\Big) = \varphi_+(z) \qquad \text{for } z \in \overline{D}_+ .$$

It follows from the maximum principle for holomorphic functions that in this way functions

$$\begin{aligned} \Phi_- &: \overline{D}_- \cup \{\infty\} \to L\Big(M(0, A), E\Big), \\ \Phi_+ &: \overline{D}_+ \to L\Big(M(0, A), E\Big) \end{aligned} \tag{7.10.9}$$

are defined.

By Theorem 7.4.2 and Remark 7.9.9 we may assume that A is holomorphic in a neighborhood of Γ. Then, for each $\varphi = (\varphi_-, \varphi_+) \in M(0, A)$, it follows from

$$\varphi_-(z) = A(z)\varphi_+(z)\,, \qquad z \in \Gamma\,,$$

that $\Phi_-\varphi = \varphi_-$ is holomorphic on $\overline{D}_- \cup \{\infty\}$, and $\Phi_+\varphi = \varphi_+$ is holomorphic on $\overline{D}_+$. Hence the functions (7.10.9) are holomorphic (Theorem 1.7.1).

Note that, by definition,

$$\begin{aligned} M_-(z,0,A) &= \operatorname{Im}\Phi_-(z) \qquad &&\text{for } z \in \overline{D}_- \cup \{\infty\}\,, \\ M_+(z,0,A) &= \operatorname{Im}\Phi_+(z) \qquad &&\text{for } z \in \overline{D}_+ \end{aligned} \tag{7.10.10}$$

and

$$\Phi_-(z) = A(z)\Phi_+(z) \qquad \text{for } z \in \Gamma\,. \tag{7.10.11}$$

Since the positive partial indices of A have finite multiplicities, it follows from Theorem 7.9.7 that the spaces $M_-(z, 0, A)$, $z \in \overline{D}_- \cup \{\infty\}$, and $M_+(z, 0, A)$, $z \in \overline{D}_+$, are of finite codimension in E. Since the operators $\Phi_-(z)$ and $\Phi_+(z)$ are linear and bounded, this implies together with (7.10.10) that these spaces are closed in E.

Since N is finite dimensional, N is complemented in $M(0, A)$. Therefore, $M(0, A)$ can be written as a direct sum

$$M(0,A) = N \dotplus N^\perp$$

with some closed subspace $N^\perp$ of $M(0, A)$. Set

$$\begin{aligned} E_0^-(z) &= \Phi_-(z)N^\perp \qquad &&\text{for } z \in \overline{D}_- \cup \{\infty\}\,, \\ E_0^+(z) &= \Phi_+(z)N^\perp \qquad &&\text{for } z \in \overline{D}_+\,. \end{aligned} \tag{7.10.12}$$

Then, by definition of N and by Theorem 7.9.7,

$$\begin{aligned} &\Phi_-(z)N \subseteq M_-(z,\kappa_{m+1},A)\,, \qquad && z \in \overline{D}_- \cup \{\infty\}\,, \\ &\Phi_+(z)N \subseteq M_+(z,\kappa_{m+1},A)\,, \qquad && z \in \overline{D}_+\,, \\ &E_0^-(z) \cap M_-(z,\kappa_{m+1},A) = \{0\}\,, \qquad && z \in \overline{D}_- \cup \{\infty\}\,, \\ &E_0^+(z) \cap M_+(z,\kappa_{m+1},A) = \{0\}\,, \qquad && z \in \overline{D}_+\,. \end{aligned}$$

Since the spaces $M_\pm(z, 0, A)$ are closed and the spaces $M_\pm(z, \kappa_{m+1}, A)$ are finite dimensional, this together with (7.10.10) implies that also the spaces $E_0^-(z)$, $z \in \overline{D}_- \cup \{\infty\}$, and $E_0^+(z)$, $z \in \overline{D}_+$, are closed and that

$$\begin{aligned} M_-(z,0,A) &= E_0^-(z) \dotplus M_-(z,\kappa_{m+1},A)\,, \quad && z \in \overline{D}_- \cup \{\infty\}\,, \\ M_+(z,0,A) &= E_0^+(z) \dotplus M_+(z,\kappa_{m+1},A)\,, \quad && z \in \overline{D}_+\,. \end{aligned} \tag{7.10.13}$$

Moreover, by Theorem 7.9.7 and the definition of N,

$$\begin{aligned} &\operatorname{Ker}\Phi_-(z)\cap N^\perp = \{0\}\,, \qquad z\in \overline{D}_-\cup\{\infty\}\,,\\ &\operatorname{Ker}\Phi_+(z)\cap N^\perp = \{0\}\,, \qquad z\in \overline{D}_+\,. \end{aligned} \tag{7.10.14}$$

Hence, by the Banach open mapping theorem, for $z\in\overline{D}_-\cup\{\infty\}$, the operator $\Phi_-(z)\big|_{N^\perp}$ is invertible as an operator from $N^\perp$ to $E_0^-(z)$, and, for $z\in\overline{D}_+$, the operator $\Phi_+(z)\big|_{N^\perp}$ is invertible as an operator from $N^\perp$ to $E_0^+(z)$.

Now we fix some point $z_0\in\Gamma$. Let $\Phi_-^{(-1)}(z_0): E_0^-(z_0)\to N^\perp$ be the inverse of $\Phi_-(z_0)$ as an operator from $N^\perp$ to $E_0^-(z_0)$, and let $\Phi_+^{(-1)}(z_0): E_0^+(z_0)\to N^\perp$ be the inverse of $\Phi_+(z_0)$ as an operator from $N^\perp$ to $E_0^+(z_0)$. Then, setting

$$\begin{aligned} &A_0^-(z) = \Phi_-(z)\Phi_-^{(-1)}(z_0) \qquad \text{for } z\in\overline{D}_-\cup\{\infty\}\,,\\ &A_0^+(z) = \Phi_+(z)\Phi_+^{(-1)}(z_0)A^{-1}(z_0) \qquad \text{for } z\in\overline{D}_+\,, \end{aligned}$$

we obtain holomorphic functions

$$\begin{aligned} &A_0^- : \overline{D}_-\cup\{\infty\} \longrightarrow L\big(E_0^-(z_0),E\big)\,,\\ &A_0^+ : \overline{D}_+ \longrightarrow L\big(E_0^-(z_0),E\big)\,, \end{aligned}$$

with injective values and such that

$$\begin{aligned} &A_0^-(z)E_0^-(z_0) = E_0^-(z) \qquad \text{for} z\in\overline{D}_-\cup\{\infty\}\,,\\ &A_0^+(z)E_0^-(z_0) = E_0^+(z) \qquad \text{for } z\in\overline{D}_+\,. \end{aligned} \tag{7.10.15}$$

Note that, by (7.10.11), for $z\in\Gamma$,

$$A(z)A_0^+(z) = A(z)\Phi_+(z)\Phi_+^{(-1)}(z_0)A^{-1}(z_0) = \Phi_-(z)\Phi_+^{(-1)}(z_0)A^{-1}(z_0).$$

Since, again by (7.10.11), $\Phi_+^{(-1)}(z_0)A^{-1}(z_0) = \Phi_-^{(-1)}(z_0)$, this implies that

$$A(z)A_0^+(z) = A_0^-(z) \qquad \text{for } z\in\Gamma\,. \tag{7.10.16}$$

Set

$$E_n^-(z_0) = M_-(z_0,\kappa_n,A) \tag{7.10.17}$$

and choose, for $1\le j\le n-1$, a subspace $E_j^-(z_0)$ of $M(z_0,\kappa_j,A)$ such that

$$\begin{aligned} &M_-(z_0,\kappa_j,A) = E_j^-(z_0)\dot{+}M_-(z_0,\kappa_{j+1},A) \quad \text{if } j\neq m\,,\\ &M_-(z_0,\kappa_m,A) = E_m^-(z_0)\dot{+}M_-(z_0,0,A)\,. \end{aligned} \tag{7.10.18}$$

From (7.10.13), (7.10.17) and (7.10.18) it follows that

$$E = E_0^-(z_0)\dot{+}E_1^-(z_0)\dot{+}\ldots\dot{+}E_n^-(z_0)\,. \tag{7.10.19}$$

By (7.10.17) and (7.10.18), for $1 \le j \le n$, we can find sections

$$\left(\varphi_{j1}^{-}, \varphi_{j1}^{+}\right), \ldots, \left(\varphi_{jd_j}^{-}, \varphi_{jd_j}^{+}\right) \in M(z_0, \kappa_j, A), \qquad 1 \le j \le n,$$

such that $\varphi_{j1}^{-}(z_0), \ldots, \varphi_{jd_j}^{-}(z_0)$ is a basis of $E_j^{-}(z_0)$. Let $E_j^{-}(z)$, $z \in \overline{D}_- \cup \{\infty\}$, be the space spanned by $\varphi_1^{j-}(z), \ldots, \varphi_{d_j}^{j-}(z)$, and let $E_j^{+}(z)$, $z \in \overline{D}_+$, be the space spanned by $\varphi_{1j}^{+}(z), \ldots, \varphi_{jd_j}^{+}(z)$. Then, from Theorem 7.9.7 and (7.10.19) it follows that

$$E = E_0^{-}(z) \dot{+} E_1^{-}(z) \dot{+} \ldots \dot{+} E_n^{-}(z), \quad z \in \overline{D}_- \cup \{\infty\},$$
$$E = E_0^{+}(z) \dot{+} E_1^{+}(z) \dot{+} \ldots \dot{+} E_n^{+}(z), \quad z \in \overline{D}_+ .$$

Now, for $1 \le j \le n$, let

$$A_j^{-} : \overline{D}_- \cup \{\infty\} \longrightarrow L\left(E_j^{-}(z_0), E\right)$$

be the holomorphic function defined by

$$A_j^{-}(z)\varphi_{j\nu}^{-}(z_0) = \varphi_{j\nu}^{-}(z), \qquad z \in \overline{D}_- \cup \{\infty\}, \quad 1 \le \nu \le d_j,$$

and let

$$A_j^{+} : \overline{D}_+ \longrightarrow L\left(E_j^{-}(z_0), E\right)$$

be the holomorphic function defined by

$$A_j^{+}(z)\varphi_{j\nu}^{-}(z_0) = \varphi_{j\nu}^{+}(z), \qquad z \in \overline{D}_+, \quad 1 \le \nu \le d_j .$$

Note that then

$$\begin{aligned} A_j^{-}(z)E_j^{-}(z_0) &= E_j^{-}(z), \qquad z \in \overline{D}_- \cup \{\infty\}, \quad 1 \le j \le n, \\ A_j^{+}(z)E_j^{-}(z_0) &= E_j^{+}(z), \qquad z \in \overline{D}_+, \quad 1 \le j \le n. \end{aligned} \tag{7.10.20}$$

Moreover, then, for $z \in \Gamma$, $1 \le j \le n$ and $1 \le \nu \le d_j$,

$$A(z)A_j^{+}(z)\varphi_{j\nu}^{-}(z_0) = A(z)\varphi_{j\nu}^{+}(z) = z^{\kappa_j}\varphi_{j\nu}^{-}(z) = z^{\kappa_j}A_j^{-}(z)\varphi_{j\nu}^{-}(z_0).$$

Hence

$$z^{\kappa_j}A_j^{-}(z) = A(z)A_j^{+}(z), \qquad z \in \Gamma, \quad 1 \le j \le n. \tag{7.10.21}$$

By (7.10.19) there are mutually disjoint projectors P_j, $0 \le j \le n$, in E with

$$I = P_0 + P_1 + \ldots + P_n \quad \text{and} \quad \operatorname{Im} P_j = E_j^{-}(z_0), \quad 0 \le j \le n.$$

Set

$$A_-(z) = \sum_{j=0}^{n} A_j^{-}(z)P_j \quad \text{for } z \in \overline{D}_- \cup \{\infty\},$$

$$A_+(z) = \sum_{j=0}^{n} A_j^{+}(z)P_j \quad \text{for } z \in \overline{D}_+,$$

$$\Delta(z) = P_0 + \sum_{j=1}^{n} z^{\kappa_j}P_j .$$

By (7.10.19), (7.10.15) and (7.10.20), in this way holomorphic functions

$$A_- : \overline{D}_- \cup \{\infty\} \longrightarrow GL(E) \quad \text{and} \quad A_+ : \overline{D}_- \longrightarrow GL(E)$$

are defined. It follows from (7.10.21) and (7.10.16) that

$$A(z)A_+(z) = A_-(z)\Delta(z) \qquad \text{for } z \in \Gamma\,.$$

Hence $A = A_- \Delta A_+^{-1}$ is a factorization of A with respect to Γ. □

7.11 Comments

This type of factorization written without the diagonal factor for matrix functions was considered for the first time in the pioneering work of Plemelj [Pl]. The complete proofs were given by N.I. Muschelishvili [Mu]. This was the motivation behind the term *Plemelj-Muschelishvili factorisation.* This form of the factorization which is considered here was first proposed for matrix functions in [GK]. In some sources, this form of factorization is called *Gohberg-Krein factorization* or *Wiener-Hopf factorization* or *factorization along a contour.* For connections with Wiener-Hopf and Töplitz operators, see the next chapter. The example (8.13.2) was given in [GK]. A general review for factorizations in algebras and applications can be found in [Go4]. The first factorization theorem for operator-valued functions (the result of Section 7.8) was proved in [Go2]. Another proof of this result was given in [Le1, Le3] (see also [CG]). The local principle for matrix functions was proposed by M.A. Shubin [Sh1]. The extension of this principle to operator functions as well as applications one can find in [GL4]. For other directions of developments in matrix and operator functions, see [BGK, GKS, GLR, GGK1, CG]. Factorization theory of operator-valued functions represents also an important tool in spectral theory of operator functions and operator polynomials (see, for instance, [Ma] and [Ro]).

Chapter 8

Wiener-Hopf operators, Toeplitz operators and factorization

In this chapter we continue to study the factorization problem, where now the emphasis is on the connection with Wiener-Hopf and Toeplitz operators. This chapter also contains applications to operator-valued Wiener-Hopf equations and equations with infinite Töplitz matrices. The local principle continues to play an important role.

8.1 Holomorphic operator functions

In this section E is a Banach space, $D_+ \subseteq \mathbb{C}$ is a bounded connected open set with piecewise $\mathcal{C}^1$-boundary Γ such that $0 \in D_+$, and $D_- := \mathbb{C} \setminus D_+$. Further, we assume that a bounded neighborhood W of Γ is fixed such that each connected component of W contains at least one connected component of Γ.[1] We set $W_+ = D_+ \cup W$ and $W_- = D_- \cup W$.

8.1.1 Definition. We denote by $\overline{\mathcal{O}}^E(\overline{W})$ the Banach space of continuous E-valued functions, which are holomorphic in W, endowed with the norm

$$\|f\|_{\overline{W}} := \max_{z \in \overline{W}} \|f(z)\|\,, \qquad f \in \overline{\mathcal{O}}^E(\overline{W})\,.$$

Further, let $\overline{\mathcal{O}}^E(\overline{W}_+)$ be the subspace of all functions in $\overline{\mathcal{O}}^E(\overline{W})$ which admit a holomorphic extension to W_+, and let $\overline{\mathcal{O}}^E_0(\overline{W}_- \cup \{\infty\})$ be the subspace of all

[1] It is possible but not necessary that W is a *small* neighborhood of Γ. An important example is also $W = \{z \in \mathbb{C} \mid r < |z| < R\}$ with $0 < r < R < \infty$ such that $\Gamma \subseteq W$.

functions $f \in \overline{\mathcal{O}}^E(\overline{W})$ which admit a holomorphic extension to $W_- \cup (\infty\}$ with $f(\infty) = 0$.

Since each connected component of W contains at least one connected component of Γ, it follows from Theorem 3.7.3 and Proposition 3.1.3 that $\overline{\mathcal{O}}^E(\overline{W})$ is the direct sum of $\overline{\mathcal{O}}^E(\overline{W}_+)$ and $\overline{\mathcal{O}}_0^E(\overline{W}_- \cup \{\infty\})$.

We denote by $\mathcal{P}$ the projector from $\overline{\mathcal{O}}^E(\overline{W})$ onto $\overline{\mathcal{O}}^E(\overline{W}_+)$ parallel to $\overline{\mathcal{O}}_0^E(\overline{W}_- \cup\{\infty\})$. As $\overline{\mathcal{O}}^E(\overline{W}_+)$ and $\overline{\mathcal{O}}_0^E(\overline{W}_-\cup\{\infty\})$ are closed subspaces of $\overline{\mathcal{O}}^E(\overline{W})$, this projector is continuous with respect to the topology of $\overline{\mathcal{O}}^E(\overline{W})$ (by the Banach open mapping theorem). We set $\mathcal{Q} = I - \mathcal{P}$.

8.1.2 Definition. Let $A : \overline{W} \to L(E)$ be a continuous function which is holomorphic in W, and let $\mathcal{M}_A : \overline{\mathcal{O}}^E(\overline{W}) \longrightarrow \overline{\mathcal{O}}^E(\overline{W})$ be the operator of multiplication by A:

$$(\mathcal{M}_A f)(z) = A(z)f(z)\,, \qquad f \in \overline{\mathcal{O}}^E(\overline{W})\,, \qquad z \in \overline{W}.$$

Then we denote by $\mathcal{W}_A$ the bounded linear operator on $\overline{\mathcal{O}}^E(\overline{W}_+)$ defined by

$$(\mathcal{W}_A)f = \mathcal{P}\big(\mathcal{M}_A f\big)\,, \qquad f \in \overline{\mathcal{O}}^E(\overline{W}_+)\,.$$

This operator $\mathcal{W}_A$ will be called the **Wiener-Hopf operator** defined by A on $\overline{\mathcal{O}}^E(\overline{W}_+)$.

8.1.3 Lemma. *Let $A, B : \overline{W} \to GL(E)$ be continuous functions which are holomorphic in W, and let $\mathcal{W}_A$ and $\mathcal{W}_B$ be the Wiener-Hopf operators defined in $\overline{\mathcal{O}}^E(\overline{W}_+)$ by A and B respectively. Suppose A and B are equivalent with respect to Γ (Def. 7.1.3). Then $\mathcal{W}_A$ and $\mathcal{W}_B$ are equivalent.*[2]

Proof. By hypothesis there are continuous functions $T_- : \overline{D}_- \cup \{\infty\} \to GL(E)$ and $T_+ : \overline{D}_+ \to GL(E)$ which are holomorphic in $D_- \cup \{\infty\}$ and D_+, respectively, such that

$$A(z) = T_-(z)B(z)T_+(z) \tag{8.1.1}$$

for $z \in \Gamma$. Since A and B are continuous on $\overline{W}$ and holomorphic in W, and since each connected component of W contains at least one connected component of Γ, from this relation it follows that T_- admits a continuous and invertible extension to $\overline{W}_- \cup \{\infty\}$ which is holomorphic in W_-, and T_+ admits a continuous and invertible extension to $\overline{W}_+$ which is holomorphic in W_+. We denote these extensions also by T_- and T_+. Then (8.1.1) holds for all $z \in \overline{W}$. It is clear that

$$\mathcal{P}\mathcal{M}_{T_+}\mathcal{P} = \mathcal{M}_{T_+}\mathcal{P} \qquad \text{and} \qquad M_{T_+^{-1}}\mathcal{P} = \mathcal{M}_{T_+^{-1}}\mathcal{P}\,. \tag{8.1.2}$$

Also it is clear that $\mathcal{P}\mathcal{M}_{T_-}\mathcal{Q} = \mathcal{P}\mathcal{M}_{T_-^{-1}}\mathcal{Q} = 0$, which implies that

$$\mathcal{P}\mathcal{M}_{T_-}\mathcal{P} = \mathcal{P}\mathcal{M}_{T_-} \qquad \text{and} \qquad \mathcal{P}\mathcal{M}_{T_-^{-1}}\mathcal{P} = \mathcal{P}\mathcal{M}_{T_-^{-1}}\,. \tag{8.1.3}$$

---

[2]Two operators T and S in a Banach space X are called **equivalent** if there exist invertible operators V, W in X such that $T = VSW$.

From (8.1.2) and (8.1.3) it follows that $\mathcal{W}_{T_-}$ and $\mathcal{W}_{T_+}$ are invertible, where

$$\mathcal{W}_{T_-}^{-1} = \mathcal{W}_{T_-^{-1}} \qquad \text{and} \qquad \mathcal{W}_{T_+}^{-1} = \mathcal{W}_{T_+^{-1}} .$$

From (8.1.1)–(8.1.3) it follows that

$$\begin{aligned}\mathcal{W}_{T_-}\mathcal{W}_B\mathcal{W}_{T_+} &= \mathcal{P}\mathcal{M}_{T_-}\mathcal{P}\mathcal{M}_B\mathcal{P}\mathcal{M}_{T_+}\big|_{\overline{\mathcal{O}}^E(\overline{W}_+)} \\ &= \mathcal{P}\mathcal{M}_{T_-}\mathcal{M}_B\mathcal{M}_{T_+}\big|_{\overline{\mathcal{O}}^E(\overline{W}_+)} = \mathcal{P}\mathcal{M}_A\big|_{\overline{\mathcal{O}}^E(\overline{W}_+)} = \mathcal{W}_A .\end{aligned}$$

Hence $\mathcal{W}_A$ and $\mathcal{W}_B$ are equivalent. □

8.1.4 Theorem. *Let $A : \overline{W} \to GL(E)$ be a continuous function which is holomorphic in W, and let $\mathcal{W}_A$ be the Wiener-Hopf operator defined by A in $\overline{\mathcal{O}}^E(\overline{W}_+)$. Then:*

(i) *The function A admits a canonical factorization with respect to Γ, if and only if, $\mathcal{W}_A$ is invertible.*

(ii) *The function A admits a factorization with respect to Γ, if and only if, $\mathcal{W}_A$ is a Fredholm operator. If this is the case, with the notations from Definition 7.1.1,*

$$\begin{aligned}\dim \operatorname{Coker} \mathcal{W}_A &= \sum_{1\le j\le n\,,\,\kappa_j>0} \kappa_j \dim P_j\,, \\ \dim \operatorname{Ker} \mathcal{W}_A &= -\sum_{1\le j\le n\,,\,\kappa_j<0} \kappa_j \dim P_j\,,\end{aligned} \tag{8.1.4}$$

where the term on the right means zero if there is no j with $\kappa_j > 0$ resp. if there is no j with $\kappa_j < 0$.

Proof. First assume that A admits a factorization $A = A_-\Delta A_+$. Then it is easy to see (using Theorem 1.5.4) that $\mathcal{W}_\Delta$ is a Fredholm operator, where, with the notations from Definition 7.1.1,

$$\begin{aligned}\dim \operatorname{Coker} \mathcal{W}_\Delta &= \sum_{1\le j\le n\,,\,\kappa_j>0} \kappa_j \dim P_j\,, \\ \dim \operatorname{Ker} \mathcal{W}_\Delta &= -\sum_{1\le j\le n\,,\,\kappa_j<0} \kappa_j \dim P_j\,.\end{aligned} \tag{8.1.5}$$

Since by Lemma 8.1.3 the operators $\mathcal{W}_A$ and $\mathcal{W}_\Delta$ are equivalent, this implies that $\mathcal{W}_A$ is invertible if $\Delta \equiv I$ and that $\mathcal{W}_A$ is a Fredholm operator satisfying (8.1.4) if $\Delta \not\equiv I$.

It remains to prove that A admits a factorization if $\mathcal{W}_A$ is a Fredholm operator, and that this factorization is canonical if $\mathcal{W}_A$ is invertible. By theorems 7.10.1 and 7.9.9 for that it is sufficient to prove the following two statements (I) and (II):

(I) If there exists a negative partial index κ of A and if d is its multiplicity, then $\dim \operatorname{Ker} \mathcal{W}_A \geq d$.

(II) If there exists a positive partial index κ of A and if d is its multiplicity, then $\dim \operatorname{Coker} \mathcal{W}_A \geq d$.

Proof of (I): Let $\kappa < 0$ be a partial index of A with multiplicity d. By Corollary 7.9.11, then we can find a d-dimensional linear subspace V of E such that, for each vector $v \in V$ there exists a κ-section of A through $(0, v)$. Let B be an algebraic basis of V. For each $b \in B$, we fix a κ-section $(\varphi_b^-, \varphi_b^+)$ of A with $\varphi_+(0) = b$. Since A is continuous on $\overline{W}$, holomorphic in W and

$$z^\kappa \varphi_b^- = A(z)\varphi_b^+(z) \tag{8.1.6}$$

for $z \in \Gamma$, it follows that φ_b^- is continuous on $\overline{W}_- \cup \{\infty\}$ and holomorphic in $W_- \cup \{\infty\}$, the function φ_b^+ is continuous on $\overline{W}_+$ and holomorphic in W_+, and (8.1.6) holds for all $z \in \overline{W}$. Hence

$$\varphi_b^+ \in \overline{\mathcal{O}}^E(\overline{W}_+), \qquad b \in B, \tag{8.1.7}$$

$$z^\kappa \varphi_b^-(z) \in \overline{\mathcal{O}}_0^E\big(\overline{W}_- \cup \{\infty\}\big), \qquad b \in B, \tag{8.1.8}$$

and

$$\mathcal{W}_A \varphi_b^+ = \mathcal{P}\Big(z^\kappa \varphi_b^-(z)\Big) = 0, \qquad b \in B.$$

Hence $\varphi_b^+ \in \operatorname{Ker} \mathcal{W}_A$ for all $b \in B$. Since the set B is linearly independent and $\varphi_b^+(0) = b$ for all $b \in B$, also the set $\{\varphi_b^+\}_{b \in B}$ is linearly independent. Hence $\dim \operatorname{Ker} \mathcal{W}_A \geq \dim V = d$.

Proof of (II): Let $\kappa > 0$ be a partial index of A with multiplicity d. By Corollary 7.9.11, there is a d-dimensional linear subspace V of E such that, for each vector $v \in V$ with $v \neq 0$, there exists no $(\kappa - 1)$-section of A through (∞, v). Since $\kappa - 1 \geq 0$, it follows (cf. Remark 7.9.3) that there exist no 0-section through (∞, v).

Now let $\mathcal{V}$ be the linear subspace of $\overline{\mathcal{O}}^E(\overline{W}_+)$ which consists of all constant functions with value in V. Then

$$\dim \mathcal{V} = d. \tag{8.1.9}$$

Moreover

$$\mathcal{V} \cap \operatorname{Im} \mathcal{W}_A = \{0\}. \tag{8.1.10}$$

Indeed, assume for some $v_0 \in V$ the constant function with value v_0 belongs to $\operatorname{Im} \mathcal{W}_A$. Then there exist $\varphi_+ \in \overline{\mathcal{O}}^E(\overline{W}_+)$ and $\varphi_- \in \overline{\mathcal{O}}_0^E\big(\overline{W}_- \cup \{\infty\}\big)$ with

$$A(z)\varphi_+(z) = v_0 + \varphi_-(z) \qquad \text{for } z \in \overline{W}.$$

Since $\varphi_-(\infty) = 0$ (by definition of $\overline{\mathcal{O}}_0^E(\overline{W}_- \cup \{\infty\})$), then $(\varphi_+, v_0 + \varphi_-)$ is a 0-section of A through (∞, v). Since there do not exist 0-sections of A through

(∞, v) if $v \in V$ and $v \neq 0$, it follows that $v_0 = 0$. From (8.1.10) and (8.1.9) it follows that $\operatorname{Coker} \mathcal{W}_A = d$. $\square$

8.2 Factorization of $\mathcal{G}^\omega(E)$-valued functions

Here we use the same notation as in Section 7.8 and prove the result on factorization of $\mathcal{G}^\omega(E)$-valued functions announced already there.

8.2.1 Lemma. *Let $0 < r < R < \infty$ such that*

$$\Gamma \subseteq W := \left\{ z \in \mathbb{C} \,\middle|\, r < |z| < R \right\} ,$$

and let $V : \overline{W} \to \mathcal{F}^\omega(E)$ be holomorphic. Then, with the notations introduced at the beginning of Section 8.1 *and in Definition* 8.1.2, *the operators $\mathcal{P}\mathcal{M}_V\mathcal{Q}$ and $\mathcal{Q}\mathcal{M}_V\mathcal{P}$ are compact as operators on $\overline{\mathcal{O}}^E(\overline{W})$.*

Proof. Let

$$V(z) = \sum_{n=-\infty}^{\infty} z^n V_n$$

be the Laurent expansion of V. Since V is holomorphic on $\overline{W}$ (i.e., in a neighborhood of $\overline{W}$), this series converges uniformly on $\overline{W}$. It follows that

$$\mathcal{P}\mathcal{M}_V\mathcal{Q} = \sum_{n=-\infty}^{\infty} \mathcal{P}\mathcal{M}_{z^n V_n}\mathcal{Q} \quad \text{and} \quad \mathcal{Q}\mathcal{M}_V\mathcal{P} = \sum_{n=-\infty}^{\infty} \mathcal{Q}\mathcal{M}_{z^n V_n}\mathcal{P}$$

where the sums converge with respect to the operator norm of $L(\overline{\mathcal{O}}^E(\overline{W}))$. Therefore it is sufficient to prove that for all $n \in \mathbb{Z}$ the operators $\mathcal{P}\mathcal{M}_{z^n V_n}\mathcal{Q}$ and $\mathcal{Q}\mathcal{M}_{z^n V_n}\mathcal{P}$ are compact as operators on $\overline{\mathcal{O}}^E(\overline{W})$.

Let $n \in \mathbb{Z}$ be given, and let $(\varphi_\nu)_{\nu \in \mathbb{N}}$ be a bounded sequence in $\overline{\mathcal{O}}^E(\overline{W})$. We have to find a subsequence φ_{ν_j} of it such that the sequences $\left(\mathcal{P}\mathcal{M}_{z^n V_n}\mathcal{Q}\right)\varphi_{\nu_j}$ and $\left(\mathcal{Q}\mathcal{M}_{z^n V_n}\mathcal{P}\right)\varphi_{\nu_j}$ converge in $\overline{\mathcal{O}}^E(\overline{W})$. Let

$$\varphi_\nu(z) = \sum_{k=-\infty}^{\infty} z^k \varphi_{\nu k} , \qquad z \in W$$

be the Laurent expansion of φ_ν. Then by Cauchy's inequality, for each $k \in \mathbb{Z}$, the sequence $(\varphi_{\nu k})_{\nu \in \mathbb{N}}$ is bounded in E. Since the operator V_n is compact, then there exists a sequence $\nu_1 \leq \nu_2 \leq \ldots$ of numbers $\nu_j \in \mathbb{N}$ such that, for all $k \in \mathbb{Z}$ with

$$-n \leq k \leq -1 \quad \text{if } n \geq 1 \qquad \text{and} \qquad 0 \leq k \leq -n-1 \quad \text{if } n \leq -1 ,$$

the subsequence $(V_n\varphi_{\nu_j,k})_{j\in\mathbb{N}^*}$ of $(V_n\varphi_{\nu,k})_{\nu\in\mathbb{N}}$ converges in E. Since

$$\begin{aligned}
&\left(\mathcal{P}\mathcal{M}_{z^nV_n}\mathcal{Q}\right)\varphi_{\nu_j} = 0 \quad \text{if } n \le 0\,,\\
&\left(\mathcal{P}\mathcal{M}_{z^nV_n}\mathcal{Q}\right)\varphi_{\nu_j} = \mathcal{P}z^nV_n\sum_{k=-\infty}^{-1} z^k\varphi_{\nu_j,k} = \sum_{k=-n}^{-1} z^{k+n}V_n\varphi_{\nu_j,k} \quad \text{if } n \ge 1\,,\\
&\left(\mathcal{Q}\mathcal{M}_{z^nV_n}\mathcal{P}\right)\varphi_{\nu_j} = 0 \quad \text{if } n \ge 0\,,\\
&\left(\mathcal{Q}\mathcal{M}_{z^nV_n}\mathcal{P}\right)\varphi_{\nu_j} = \mathcal{Q}z^nV_n\sum_{k=0}^{\infty} z^k\varphi_{\nu_j,k} = \sum_{k=0}^{-n-1} z^{k+n}V_n\varphi_{\nu_j,k} \quad \text{if } n \le -1\,,
\end{aligned}$$

this implies that sequences $\left(\mathcal{P}\mathcal{M}_{z^nV_n}\mathcal{Q}\right)\varphi_{\nu_j}$ and $\left(\mathcal{Q}\mathcal{M}_{z^nV_n}\mathcal{P}\right)\varphi_{\nu_j}$ converge in $\overline{\mathcal{O}}^E(\overline{W})$. □

8.2.2 Theorem. *Let $A:\Gamma\to\mathcal{G}^\omega(E)$ be a continuous function which admits local factorizations with respect to Γ and $GL(E)$. Then A admits a factorization with respect to Γ and $\mathcal{G}^\omega(E)$.*

Proof. By Theorem 7.4.2, we may assume that A is defined, holomorphic and with values in $\mathcal{G}^\omega$ on $\mathbb{C}^*$. Choose $0<r<R<\infty$ such that

$$\Gamma \subseteq W := \left\{ z\in\mathbb{C} \,\middle|\, r<|z|<R \right\}.$$

Let $V:\overline{W}\to\mathcal{F}^\omega(E)$ be the function with $A=I+V$. Let $\mathcal{M}_A$ and $\mathcal{M}_V$ be the operators of multiplication by A and V, respectively. By Lemma 8.2.1, $\mathcal{P}\mathcal{M}_V\mathcal{Q}$ and $\mathcal{Q}\mathcal{M}_V\mathcal{P}$ are compact as operators on $\overline{\mathcal{O}}^E(\overline{W})$. Since $\mathcal{M}_A$ is invertible as an operator on $\overline{\mathcal{O}}^E(\overline{W})$ and

$$\mathcal{M}_A = \mathcal{P}\mathcal{M}_A\mathcal{P} + \mathcal{Q}\mathcal{M}_A\mathcal{Q} + \mathcal{P}\mathcal{M}_V\mathcal{Q} + \mathcal{P}\mathcal{M}_V\mathcal{Q}\,,$$

this implies that

$$\mathcal{P}\mathcal{M}_A\mathcal{P} + \mathcal{Q}\mathcal{M}_A\mathcal{Q}$$

is Fredholm as an operator on $\overline{\mathcal{O}}^E(\overline{W})$. In particular, the Wiener-Hopf operator $\mathcal{W}_A = \mathcal{P}\mathcal{M}_A\big|_{\overline{\mathcal{O}}^E(\overline{W}_+)}$ defined by A in $\overline{\mathcal{O}}^E(\overline{W}_+)$ is a Fredholm operator. By Theorem 8.1.4 this implies that A admits a factorization with respect to Γ and $GL(E)$. Finally, it follows from Proposition 7.8.1 that A admits a factorization with respect to Γ and $\mathcal{G}^\omega(E)$. □

Under certain additional assumptions about the function A in Theorem 8.2.2, we can say correspondingly more about the factors of the factorizations of A. We have:

8.2.3 Corollary (to Theorem 8.2.2). *Let $0<\alpha<1$ and $k\in\mathbb{N}$, where, for $k\ge 1$, we additionally assume that Γ is of class $\mathcal{C}^k$ (Def. 3.4.1). Let $A:\Gamma\to\mathcal{G}^\omega(E)$ be a function of class $\mathbb{C}^{k+\alpha}$ (Def. 3.4.3). Then:*

(i) *A admits a factorization with respect to* Γ *and* $\mathcal{G}^\omega(E)$.

(ii) *If* $A = A_-\Delta A_+$ *is an arbitrary factorization of* A *with respect to* Γ, *then automatically, the factors* A_- *and* A_+ *are of class* $\mathbb{C}^{k+\alpha}$ *on* $\overline{D}_-$ *and* $\overline{D}_+$, *respectively.*

Proof. Part (i) of Theorem 7.3.1 in particular states that A admits local factorizations with respect to Γ. Therefore part (i) of the corollary follows from Theorem 8.2.2. Part (ii) of the corollary follows from part (ii) of Theorem 7.3.1 □

8.2.4 Corollary (to Theorem 8.2.2). *Let* $\Gamma = \mathbb{T}$ *be the unit circle, and let* $\mathcal{R}$ *be a Banach algebra of continuous* $L(E)$*-valued functions satisfying conditions* (A), (B) *and* (C) *in Section* 7.2.4. *For example, let* $\mathcal{R} = W\big(L(E)\big)$ *be the Wiener algebra (see Section* 7.2.1). *Let* $A : \Gamma \to \mathcal{G}^\omega(E)$ *be a function which belongs to* $\mathcal{R}$. *Then:*

(i) *The function* A *admits a factorization with respect to* Γ *and* $\mathcal{G}^\omega(E)$.

(ii) *If* $A = A_-\Delta A_+$ *is an arbitrary factorization of* A *with respect to* Γ, *then automatically, the factors* A_- *and* A_+ *belong to the algebra* $\mathcal{R}$.

Proof. Part (ii) of Theorem 7.2.5 in particular states that A admits local factorizations with respect to Γ. Therefore part (i) of the corollary follows from Theorem 8.2.2. Part (ii) of the corollary follows from part (iii) of Theorem 7.2.5. □

Theorem 8.2.2 admits the following generalization:

8.2.5 Theorem. *Let* $A : \Gamma \to GL(E)$ *be a continuous function which admits local factorizations with respect to* Γ *and* $GL(E)$. *Let* $G\big(L(E)/\mathcal{F}^\omega(E)\big)$ *be the group of invertible elements of the factor algebra* $L(E)/\mathcal{F}^\omega(E)$, *and let*

$$\pi : L(E) \to L(E)/\mathcal{F}^\omega(E)$$

be the canonical map. Then the following are equivalent:

(i) *The function* A *admits a factorization with respect to* Γ *and* $GL(E)$.

(ii) *The function* $\pi(A)$ *admits a canonical factorization with respect to* Γ *and* $G\big(L(E)/\mathcal{F}^\omega(E)\big)$.

Proof. (i)⇒(ii) is clear. We prove (ii)⇒(i). Assume there is a canonical factorization $\pi(A) = f_-f_+$ of $\pi(A)$ with respect to Γ and $G\big(L(E)/\mathcal{F}^\omega(E)\big)$. By Theorem 7.4.2 we may assume that A is holomorphic in a neighborhood of Γ. Then also $\pi(A)$ is holomorphic in a neighborhood of Γ, and it follows from the relations

$$f_- = \pi(A)f_+^{-1} \qquad \text{and} \qquad f_+ = f_-^{-1}\pi(A)$$

that f_- is holomorphic in a neighborhood of $\overline{D}_-\cup\{\infty\}$ and $f+$ is holomorphic in a neighborhood of $\overline{D}_+$. Then, by Theorem 7.8.2, we can find holomorphic functions $A_- : \overline{D}_- \cup \{\infty\} \to GL(E)$ and $A_+ : \overline{D}_+ \to GL(E)$ such that $\pi(A_-) = f_-$ and $\pi(A_+) = f_+$. Setting

$$\widetilde{A} = A_-^{-1}AA_+^{-1},$$

we define a holomorphic function $\widetilde{A} : \Gamma \to GL(E)$. Since $\pi(\widetilde{A}) = f_-^{-1}\pi(A)f_+^{-1} = 1$ on Γ, where 1 is the unit element in $L(E)/\mathcal{F}^\omega(E)$, we see that the values of $\widetilde{A}$ belong to $\mathcal{G}^\omega(E)$. By Theorem 7.7, $\widetilde{A}$ admits a factorization with respect to Γ and $GL(E)$. Since, by definition, $\widetilde{A}$ is equivalent to A with respect to Γ and $GL(E)$, this completes the proof. □

Note the following obvious corollary to Theorem 8.2.5 (which is not obvious at all without this theorem):

8.2.6 Corollary. *Let $A : \Gamma \to GL(E)$ be a continuous function which admits local factorizations with respect to Γ. Suppose, there is a finite number of points $p_1, \dots, p_m \in D_+$ such that A admits a continuous extension to $\overline{D}_+ \setminus \{p_1, \dots, p_m\}$ (also denoted by A) which is holomorphic in $D_+ \setminus \{p_1, \dots, p_m\}$ and has the following property: For each $z \in D_+ \setminus \{p_1, \dots, p_m\}$, the value $A(z)$ is a Fredholm operator, and, for all $1 \le j \le m$, the Laurent expansion of A at p_j is of the form*

$$A(z) = \sum_{n=-\infty}^{\infty} (z - p_j)^n A_n,$$

where A_0 is a Fredholm operator and each of the coefficients A_n with $n \le 0$ is compact.

Then A admits a factorization with respect to Γ.

Clearly, there is also a corresponding corollary with respect to $D_- \cup \{\infty\}$.

8.3 The space $\mathcal{L}^2(\Gamma, H)$

The results on factorization obtained up to now can be essentially improved if we restrict ourselves to Hilbert spaces. In the present section, we introduce a technical tool for this: the Hilbert space of Hilbert space-valued functions on a contour. Here we use the following notations:

- H is a separable Hilbert space with the scalar product $\langle \cdot, \cdot \rangle_H$ and the norm $\| \cdot \|_H$.
- $D_+ \subseteq \mathbb{C}$ is a bounded connected open set with piecewise $\mathcal{C}^1$-boundary Γ. We assume that $0 \in D_+$, and we set $D_- := \mathbb{C} \setminus \overline{D}_+$.
- We give Γ the orientation as the boundary of D_+, and we denote by $|dz|$ the Euclidean volume form of Γ.
- If $U \subseteq \mathbb{C}$ is an open set, then we denote by $\overline{\mathcal{O}}^H(\overline{U})$ the space of continuous functions on $\overline{U}$ which are holomorphic in U, and by $\mathcal{O}^H(\overline{U})$ we denote the subspace of $\overline{\mathcal{O}}^H(\overline{U})$ which consists of the functions which admit a holomorphic extension to some neighborhood of $\overline{U}$.
- $\mathcal{C}^0(\Gamma, H)$ is the space of continuous H-valued functions on Γ.

- $\mathcal{O}^H(\Gamma)$ is the subspace of $\mathcal{C}^0(\Gamma, H)$ which consists of the functions which admit a holomorphic extension to some neighborhood of Γ.
- $\mathcal{O}_0^H(\overline{D}_-)$ is the subspace of $\mathcal{O}^H(\overline{D}_-)$ which consists of the functions with $f(\infty) = 0$.

8.3.1. In $\mathcal{C}^0(\Gamma, H)$ we introduce a scalar product $\langle \cdot, \cdot \rangle_{\mathcal{L}^2(\Gamma,H)}$ and the corresponding norm $\| \cdot \|_{\mathcal{L}^2(\Gamma,H)}$, setting

$$\langle f, g \rangle_{\mathcal{L}^2(\Gamma,H)} = \int_\Gamma \langle f(z), g(z) \rangle_H \, |dz| \qquad \text{for } f, g \in \mathcal{C}^0(\Gamma, H)$$

and

$$\|f\|_{\mathcal{L}^2(\Gamma,H)} = \sqrt{\langle f, f \rangle_{\mathcal{L}^2(\Gamma,H)}} \qquad \text{for } f \in \mathcal{C}^0(\Gamma, H)\,.$$

We define $\mathcal{L}^2(\Gamma, H)$ to be the completion of $\mathcal{C}^0(\Gamma, H)$ with respect to the norm $\| \cdot \|_{\mathcal{L}^2(\Gamma,H)}$.

8.3.2. For the scalar case $H = \mathbb{C}$, the space $\mathcal{L}^2(\Gamma, \mathbb{C})$ is usually interpreted as the space of square integrable complex-valued functions on Γ. We avoid such an interpretation in the case of an arbitrary separable Hilbert space H (although this is possible). By our definition, the elements of $\mathcal{L}^2(\Gamma, H)$ are equivalence classes of Cauchy sequences of continuous functions. Nevertheless, many operations which can be defined pointwise in the case of a true function can be applied also to the elements of $\mathcal{L}^2(\Gamma, H)$.

For example, let $\varphi : \Gamma \to \mathbb{C}$ be a scalar continuous function and let $f \in L^2(\Gamma, H)$ be represented by the Cauchy sequence $\{f_n\}_{n\in\mathbb{N}}$ of functions $f_n \in \mathcal{C}^0(\Gamma, H)$. Then φf is defined to be the element in $L^2(\Gamma, H)$ represented by the Cauchy sequence $\{\varphi f_n\}_{n\in\mathbb{N}}$. Note that then

$$\big\|\varphi f\big\|_{\mathcal{L}^2(\Gamma,H)} = \max_{z\in\Gamma} |\varphi(z)| \big\|\varphi f\big\|_{\mathcal{L}^2(\Gamma,H)}$$

In the same way, for each continuous operator function $A : \Gamma \to L(H)$ and each $f \in \mathcal{L}^2(\Gamma, H)$, the product Af is well defined. Then

$$\big\|Af\big\|_{\mathcal{L}^2(\Gamma,H)} = \max_{z\in\Gamma} \big\|A(z)\big\|_{L(H)} \big\|\varphi f\big\|_{\mathcal{L}^2(\Gamma,H)}\,.$$

Moreover, denote by $\mathcal{L}^1(\Gamma, \mathbb{C})$ the Banach space of integrable complex-valued functions on Γ with the norm

$$\|f\|_{\mathcal{L}^1(\Gamma,\mathbb{C})} = \int_\Gamma |f||dz|\,, \qquad f \in \mathcal{L}^1(\Gamma, \mathbb{C})\,.$$

Then, for any two elements $f, g \in \mathcal{L}^2(\Gamma, H)$, a function $\langle f, g \rangle_H \in \mathcal{L}^1(\Gamma, \mathbb{C})$ is well defined such that

$$\langle f, g \rangle_{\mathcal{L}^2(\Gamma,H)} = \int_\Gamma \langle f, g \rangle_H |dz|\,.$$

Indeed, let $\{f_n\}_{n\in\mathbb{N}}$ and $\{g_n\}_{n\in\mathbb{N}}$ be two sequences of functions from $\mathcal{C}^0(\Gamma, H)$ which represent f and g, respectively. Then, by the Cauchy-Schwarz inequality in H, pointwise on Γ we have

$$\Big|\big\langle f_n, g_n\big\rangle_H - \big\langle f_m, g_m\big\rangle_H\Big| \le \big\|f_n - f_m\big\|_H\big\|g_n\big\|_H + \big\|f_m\big\|_H\big\|g_n - g_m\big\|_H\,,$$

which implies, by the Cauchy-Schwarz inequality in $L^2(\Gamma, \mathbb{C})$, that

$$\begin{aligned}&\Big\|\big\langle f_n, g_n\big\rangle_H - \big\langle f_m, g_m\big\rangle_H\Big\|_{\mathcal{L}^1(\Gamma,\mathbb{C})}\\ &\quad\le \big\|f_n - f_m\big\|_{\mathcal{L}^2(\Gamma,H)}\big\|g_n\big\|_{\mathcal{L}^2(\Gamma,H)} + \big\|f_m\big\|_{\mathcal{L}^2(\Gamma,H)}\big\|g_n - g_m\big\|_{\mathcal{L}^2(\Gamma,H)}\,.\end{aligned}$$

Hence $\big\{\langle f_n, g_n\rangle_H\big\}_{n\in\mathbb{N}}$ is a Cauchy sequence in $\mathcal{L}^1(\Gamma, \mathbb{C})$, and the function $\langle f, g\rangle_H$ can be defined to be the limit of this sequence in $\mathcal{L}^1(\Gamma, \mathbb{C})$.

Finally we note that, for each $f \in \mathcal{L}^2(\Gamma, H)$, the function

$$\|f\|_H := \sqrt{\langle f, f\rangle_H}$$

belongs to $\mathcal{L}^2(\Gamma, \mathbb{C})$ and

$$\|f\|^2_{\mathcal{L}^2(\Gamma,H)} = \int\limits_\Gamma \|f\|^2_H |dz|\,.$$

8.3.3. Note that, by Corollary 3.3.3, $\mathcal{O}^H(\Gamma)$ is dense in $\mathcal{C}^0(\Gamma, H)$ with respect to uniform convergence on Γ. Hence $\mathcal{O}^H(\Gamma)$ is dense in $\mathcal{L}^2(\Gamma, H)$ with respect to the norm $\|\cdot\|_{\mathcal{L}^2(\Gamma,H)}$.

8.3.4. The linear map I_Γ from $\mathcal{C}^0(\Gamma, H)$ to H defined by

$$I_\Gamma(f) := \int\limits_\Gamma f(z)dz\,, \qquad f \in \mathcal{C}^0(\Gamma, H)\,, \tag{8.3.1}$$

is bounded with respect to the norm $\|\cdot\|_{\mathcal{L}^2(\Gamma,H)}$. As (by definition) $\mathcal{C}^0(\Gamma, H)$ is dense in $\mathcal{L}^2(\Gamma, H)$, this implies that I_Γ admits a uniquely determined continuous linear extension to $\mathcal{L}^2(\Gamma, H)$, which we denote also by I_Γ. We define

$$\int\limits_\Gamma f(z)dz := I_\Gamma(f) \qquad \text{for } f \in \mathcal{L}^2(\Gamma, H)\,.$$

Indeed, let $|\Gamma|$ be the Euclidean length of Γ. Then, for each $f \in \mathcal{L}^2(\Gamma, H)$, we get from the Cauchy-Schwarz inequality:

$$\begin{aligned}\|I_\Gamma(f)\| &\le \int_\Gamma \|f(z)\|_H|dz| \le \Big(\int\limits_\Gamma |dz|\Big)^{1/2}\Big(\int\limits_\Gamma \|f(z)\|^2_H|dz|\Big)^{1/2}\\ &= \sqrt{|\Gamma|}\,\|f\|_{\mathcal{L}^2(\Gamma,H)}\,.\end{aligned}$$

8.3.5 Definition. Let $f \in \mathcal{L}^2(\Gamma, H)$, and let $U \subseteq \mathbb{C} \cup \{\infty\}$ be a neighborhood of Γ.

(i) Let $f^- \in \mathcal{O}^H\big(U \cap (D_- \cup \{\infty\})\big)$. We shall say that f^- is a **holomorphic extension** of f if there exist a neighborhood $V \subseteq U$ of Γ and a sequence $f_n^- \in \mathcal{O}^H(\Gamma) \cap \mathcal{O}^H(V \cap D_-)$, $n \in \mathbb{N}$, such that

$$\lim_{n\to\infty} \|f - f_n^-\|_{\mathcal{L}^2(\Gamma,H)} = 0 \tag{8.3.2}$$

and

$$\lim_{n\to\infty} \max_{z\in K} \|f^-(z) - f_n^-(z)\|_H = 0 \quad if K \subseteq V \cap D_- \text{ is compact}. \tag{8.3.3}$$

(ii) Let $f^+ \in \mathcal{O}^H(U \cap D_+)$. We shall say that f^+ is a **holomorphic extension** of f if there exists a neighborhood $V \subseteq U$ of Γ and a sequence $f_n^+ \in \mathcal{O}^H(\Gamma) \cap \mathcal{O}^H(V \cap D_+)$, $n \in \mathbb{N}$, such that

$$\lim_{n\to\infty} \|f - f_n^+\|_{\mathcal{L}^2(\Gamma,H)} = 0 \tag{8.3.4}$$

and

$$\lim_{n\to\infty} \max_{z\in K} \|f^+(z) - f_n^+(z)\|_H = 0 \quad if K \subseteq V \cap D_+ \text{ is compact}. \tag{8.3.5}$$

8.3.6 Proposition. *Let $f \in \mathcal{L}^2(\Gamma, H)$, let $U \subseteq \mathbb{C} \cup \{\infty\}$ be a neighborhood of Γ, and let W be a neighborhood of Γ with piecewise $\mathcal{C}^1$-boundary ∂W, oriented by W, such that $\overline{W} \subseteq U \cap \mathbb{C}$.*

(i) *Assume $f^- \in \mathcal{O}^H\big(U \cap (D_- \cup \{\infty\})\big)$ is a holomorphic extension of f. Then*

$$\int_{D_-\cap\partial W} f^-(z)\,dz = \int_\Gamma f(z)\,dz. \tag{8.3.6}$$

Moreover, if $w \in W \cap D_-$, then the Cauchy formula holds:

$$f^-(w) = \frac{1}{2\pi i}\int_{D_-\cap\partial W} \frac{f^-(z)}{z-w}\,dz - \frac{1}{2\pi i}\int_\Gamma \frac{f(z)}{z-w}\,dz. \tag{8.3.7}$$

(ii) *Assume $f^+ \in \mathcal{O}^H\big(U \cap D_+\big)$ is a holomorphic extension of f. Then*

$$\int_{D_+\cap\partial W} f^+(z)\,dz = -\int_\Gamma f(z)\,dz. \tag{8.3.8}$$

Moreover, if $w \in W \cap D_+$, then the Cauchy formula holds:

$$f^+(w) = \frac{1}{2\pi i}\int_{D_+\cap\partial W} \frac{f^+(z)}{z-w}\,dz + \frac{1}{2\pi i}\int_\Gamma \frac{f(z)}{z-w}\,dz. \tag{8.3.9}$$

Proof. The proofs of parts (i) and (ii) are similar. We may restrict ourselves to part (i).

By hypothesis, there exist a neighborhood $V \subseteq U$ of Γ and a sequence $f_n^- \in \mathcal{O}^H(\Gamma) \cap \mathcal{O}^H(V \cap D_-)$, $n \in \mathbb{N}$, with (8.3.2) and (8.3.3). After shrinking V we may assume that $\overline{V} \subseteq W$ and the boundary ∂V of V is also piecewise of class $\mathcal{C}^1$. We orient ∂V by V. Then, by Cauchy's theorem,

$$\int_{D_- \cap \partial W} f^-(\zeta)\, d\zeta = \int_{D_- \cap \partial V} f^-(\zeta)\, d\zeta . \tag{8.3.10}$$

Moreover, by Cauchy's theorem,

$$\int_{D_- \cap \partial V} f_n^-(\zeta)\, d\zeta = \int_\Gamma f_n^-(\zeta)\, d\zeta \qquad \text{for all } n .$$

By (8.3.3) and (8.3.2), this implies

$$\int_{D_- \cap \partial V} f^-(\zeta)\, d\zeta = \int_\Gamma f^-(\zeta)\, d\zeta .$$

Together with (8.3.10) this proves (8.3.6).

Now we consider some point $w \in W \cap D_-$. Choose a neighborhood W' of Γ with piecewise $\mathcal{C}^1$-boundary $\partial W'$, oriented by W', so small that $\overline{W}' \subseteq W$ and $w \in W \setminus \overline{W}'$. Then, by Cauchy's formula,

$$f^-(w) = \frac{1}{2\pi i} \int_{D_- \cap \partial W} \frac{f^-(z)}{z-w}\, dz - \frac{1}{2\pi i} \int_{D_- \cap \partial W'} \frac{f^-(z)}{z-w}\, dz . \tag{8.3.11}$$

Since f^- is a holomorphic extension of f, it follows easily that the function

$$W' \cap D_- \ni z \to f^-(z)/(z-w)$$

is a holomorphic extension of the function

$$\Gamma \ni z \to f(z)/(z-w) .$$

Therefore, to the second integral in (8.3.11) we can apply (8.3.6) with W' instead of W, $f(z)/(z-w)$ instead of $f(z)$ and $f^-(z)/(z-w)$ instead of $f^-(z)$. This gives (8.3.7). □

8.3.7. Since holomorphic functions in a neighborhood of Γ are uniquely determined by their values on Γ, from (8.3.7) and (8.3.9) we get: If $f \in \mathcal{L}^2(\Gamma, H)$ admits a holomorphic extension to some open set of the form $W \cap D_-$ or $W \cap D_+$, where W is a neighborhood of Γ such that each connected component of W intersects Γ, then this extension is uniquely determined.

The opposite is also true:

8.3.8 Proposition. (i) *Let $f, g \in \mathcal{L}^2(\Gamma, H)$, let U be a neighborhood of Γ, and let $h \in \mathcal{O}^H(U \cap D_-)$ such that h is both a holomorphic extension of f and a holomorphic extension of g. Then $f = g$.*

(ii) *Let $f, g \in \mathcal{L}^2(\Gamma, H)$, let U be a neighborhood of Γ, and let $h \in \mathcal{O}^H(U \cap D_+)$ such that h is both a holomorphic extension of f and a holomorphic extension of g. Then $f = g$.*

Proof. The proofs of (i) and (ii) are similar. We restrict ourselves to the proof of part (i).

Set $h = f - g$. We have to prove that $h = 0$.

By hypothesis, the zero function on D_- is a holomorphic extension of h. This means, by definition, that there exist a neighborhood V of Γ and a sequence $h_n \in \mathcal{O}^H(\Gamma) \cap \mathcal{O}^H(V \cap D_-)$, $n \in \mathbb{N}$, such that

$$\lim_{n\to\infty} \|h - h_n\|_{\mathcal{L}^2(\Gamma,H)} = 0 \tag{8.3.12}$$

and

$$\lim_{n\to\infty} \max_{z\in K} \|h_n(z)\|_H = 0 \quad if K \subseteq V \cap D_- \text{ is compact}. \tag{8.3.13}$$

Let $\{e_j\}_{j\in I}$ be an orthonormal basis of H, and let $h_j \in \mathcal{L}^2(\Gamma, \mathbb{C})$ and $h_{nj} \in \mathcal{O}^{\mathbb{C}}(\Gamma)$ be the functions with

$$h = \sum_{j\in I} h_j e_j \qquad \text{and} \qquad h_n = \sum_{j\in I} h_{nj} e_j\,. \tag{8.3.14}$$

For each $j \in I$, it follows from (8.3.12) and (8.3.13) that

$$\lim_{n\to\infty} \|h_j - h_{nj}\|_{\mathcal{L}^2(\Gamma,\mathbb{C})} = 0$$

and

$$\lim_{n\to\infty} \max_{z\in K} |h_{nj}(z)| = 0 \quad if K \subseteq V \cap D_- \text{ is compact.}$$

If $W \subseteq V$ is a further (arbitrarily small) neighborhood of Γ and $\varphi \in \mathcal{O}^{\mathbb{C}}(W)$, then this yields, for all $j \in I$,

$$\lim_{n\to\infty} \|h_j\varphi - h_{nj}\varphi\|_{\mathcal{L}^2(\Gamma,\mathbb{C})} = 0 \tag{8.3.15}$$

and

$$\lim_{n\to\infty} \max_{z\in K} |h_{nj}(z)\varphi(z)| = 0 \quad if K \subseteq W \cap D_- \text{ is compact.} \tag{8.3.16}$$

Hence, for each $\varphi \in \mathcal{O}^{\mathbb{C}}(\Gamma)$ and $j \in I$, the (scalar) zero function on D_- is a holomorphic extension of $h_j\varphi$. By (8.3.6) this implies that

$$\int_\Gamma h_j(z)\,\varphi(z)\,dz = 0 \qquad \text{for all } \varphi \in \mathcal{O}^{\mathbb{C}}(\Gamma) \text{ and } j \in I\,. \tag{8.3.17}$$

Let $\theta : \Gamma \to \mathbb{C}$ be the function with $dz = \theta|dz|$. As Γ is piecewise $\mathcal{C}^1$, this function is piecewise continuous and $|\theta| = 1$. Then (8.3.17) takes the form

$$\left\langle h_j \,,\, \overline{\varphi\theta} \right\rangle_{\mathcal{L}^2(\Gamma,\mathbb{C})} = 0 \qquad \text{for all } \varphi \in \mathcal{O}^{\mathbb{C}}(\Gamma) \text{ and } j \in I\,. \tag{8.3.18}$$

Since $\mathcal{O}^{\mathbb{C}}(\Gamma)$ is dense in $\mathcal{L}^2(\Gamma,\mathbb{C})$, also the functions of the form $\overline{\varphi\theta}$, $\varphi \in \mathcal{O}^{\mathbb{C}}(\Gamma)$, are dense in $\mathcal{L}^2(\Gamma,\mathbb{C})$. Therefore (8.3.18) means that $h_j = 0$ for all $j \in I$. By (8.3.14) this yields $h = 0$. □

8.3.9 Proposition. *Let $f \in \mathcal{L}^2(\Gamma, H)$, let $U \subseteq \mathbb{C} \cup \{\infty\}$ be a neighborhood of Γ, and let $h \in \mathcal{O}^H(U \setminus \Gamma)$ such that both $h\big|_{U\cap(D_-\cup\{\infty\})}$ and $h\big|_{U\cap D_+}$ are holomorphic extensions of f. Then h admits a holomorphic extension $\widetilde{h}$ to U and $\widetilde{h}\big|_\Gamma = f$.*

Proof. It is sufficient to prove that h admits a holomorphic extension $\widetilde{h}$ to U. The relation $\widetilde{h}\big|_\Gamma = f$ then follows from Proposition 8.3.8. Choose a neighborhood W of Γ with piecewise $\mathcal{C}^1$-boundary, oriented by W, such that $\overline{W} \subseteq U \cap \mathbb{C}$. We define a holomorphic function $F : W \to \mathbb{C}$, setting

$$\widetilde{h}(w) = \frac{1}{2\pi i} \int\limits_{\partial W} \frac{h(z)}{z-w} dz\,, \qquad w \in W\,.$$

We have to prove that $\widetilde{h}(w) = h(w)$ for $w \in W \setminus \Gamma$. The proofs are similar for $w \in W \cap D_-$ and $w \in W \cap D_+$. Let $w \in D_- \cap W$ be given.

Since $h\big|_{U\cap D_+}$ is a holomorphic extension of f, it follows that the function

$$D_+ \cap W \ni z \longrightarrow \frac{1}{2\pi i}\frac{h(z)}{z-w} \tag{8.3.19}$$

is a holomorphic extension of the function

$$\Gamma \ni z \longrightarrow \frac{1}{2\pi i}\frac{f(z)}{z-w}. \tag{8.3.20}$$

Therefore, by (8.3.8) (with (8.3.19) instead of f^+ and (8.3.20) instead of f),

$$\frac{1}{2\pi i} \int\limits_{D_+\cap\,\partial W} \frac{h(z)}{z-w}\, dz = -\frac{1}{2\pi i} \int\limits_{\Gamma} \frac{f(z)}{z-w}\, dz\,.$$

Moreover, from the Cauchy formula (8.3.7) we get

$$h(w) = \frac{1}{2\pi i} \int\limits_{D_-\cap\,\partial W} \frac{h(z)}{z-w}\, dz - \frac{1}{2\pi i} \int\limits_{\Gamma} \frac{f(z)}{z-w}\, d\zeta\,.$$

Together this proves that $h(w) = \widetilde{h}(w)$. □

8.3.10 Theorem and Definition. Recall that, by Theorem 3.7.3 and Proposition 3.1.3, the space $\mathcal{O}^H(\Gamma)$ is the (algebraic) direct sum of $\mathcal{O}^H(\overline{D}_+)$ and $\mathcal{O}_0^H(\overline{D}_- \cup \{\infty\})$. We denote by $\mathcal{P} = \mathcal{P}_\Gamma$ the linear projector from $\mathcal{O}^H(\Gamma)$ to $\mathcal{O}^H(\overline{D}_+)$ parallel to $\mathcal{O}_0^H(\overline{D}_- \cup \{\infty\})$, and we set $\mathcal{Q} = I - \mathcal{P}$.

(i) *The projector $\mathcal{P}$ is continuous with respect to the norm $\|\cdot\|_{L_2(\Gamma,H)}$.*

Since $\mathcal{O}^H(\Gamma)$ is dense in $\mathcal{L}^2(\Gamma, H)$ (Section 8.3.3), it follows that $\mathcal{P}$ admits a uniquely determined continuous linear extension to $L_2(\Gamma, H)$. This extension will be also denoted by $\mathcal{P}$, and we set $\mathcal{Q} = I - \mathcal{P}$, $\mathcal{L}_+^2(\Gamma, H) = \mathcal{P}\mathcal{L}^2(\Gamma, H)$ and $\mathcal{L}_-^2(\Gamma, H) = \mathcal{Q}\mathcal{L}^2(\Gamma, H)$.

(ii) *For each $f \in L^2(\Gamma, H)$, there exist uniquely determined functions $f^- \in \mathcal{O}_0^H(D_- \cup \{\infty\})$ and $f^+ \in \mathcal{O}^H(D_+)$ such that f^+ is the holomorphic extension of $\mathcal{P}f$* [3]*, and f^- is the holomorphic extension of $\mathcal{Q}f$. These functions are given by*

$$f_+(z) = \frac{1}{2\pi i}\int\limits_\Gamma \frac{f(\zeta)}{\zeta - z}d\zeta = \frac{1}{2\pi i}\int\limits_\Gamma \frac{(\mathcal{P}f)(\zeta)}{\zeta - z}d\zeta\,, \quad z \in D_+\,, \tag{8.3.21}$$

and

$$f_-(z) = -\frac{1}{2\pi i}\int\limits_\Gamma \frac{f(\zeta)}{\zeta - z}d\zeta = -\frac{1}{2\pi i}\int\limits_\Gamma \frac{(\mathcal{Q}f)(\zeta)}{\zeta - z}d\zeta\,, \quad z \in D_-\,. \tag{8.3.22}$$

(iii) *Let Γ be the unit circle, and let $\{e_j\}_{j\in\mathbb{N}}$ be an orthonormal basis of H. Then the functions*

$$\frac{z^n}{\sqrt{2\pi}}e_j\,, \qquad n \in \mathbb{Z},\ j \in \mathbb{N}, \tag{8.3.23}$$

form an orthonormal basis of $\mathcal{L}^2(\Gamma, \mathbb{C})$, and, hence, by Laurent decomposition, the family

$$\frac{z^n}{\sqrt{2\pi}}e_j\,, \qquad n \in \mathbb{N},\ j \in \mathbb{N}, \tag{8.3.24}$$

forms an orthonormal basis of $\mathcal{L}_+^2(\Gamma, \mathbb{C})$, and the family

$$\frac{z^n}{\sqrt{2\pi}}\,, \qquad n \in \mathbb{Z},\ n \leq -1,\ j \in \mathbb{N}, \tag{8.3.25}$$

forms an orthonormal basis of $\mathcal{L}_-^2(\Gamma, \mathbb{C})$. In particular, then $\mathcal{P}$ is an orthogonal projector.

---

[3] By the observation in Section 8.3.7, here we may speak about <u>the</u> holomorphic extension.

Proof. (iii) We have

$$\int\limits_{|z|=1} \left|\frac{z^n}{\sqrt{2\pi}}\right|^2 |dz| = \frac{1}{2\pi} \int\limits_{|z|=1} |dz| = 1 \qquad \text{for all } n \in \mathbb{Z}\,,$$

and

$$\begin{aligned}\int\limits_{|z|=1} z^n\overline{z^m}\,|dz| &= \int\limits_{|z|=1} z^{n-m}|dz| = \int\limits_0^{2\pi} e^{i(n-m)t}\,dt \\ &= \frac{1}{i(n-m)}e^{i(n-m)t}\Big|_{t=0}^{t=2\pi} = 0 \qquad \text{for all } n, m \in \mathbb{Z} \text{ with } n \neq m\,.\end{aligned}$$

If Γ is the unit circle, this shows that

$$\frac{z^n}{\sqrt{2\pi}}, \qquad n \in \mathbb{Z},$$

is an orthonormal system in $\mathcal{L}^2(\Gamma, \mathbb{C})$, which further implies that (8.3.23) is orthonormal in $\mathcal{L}^2(\Gamma, H)$. Moreover if Γ is the unit circle and

$$f(z) = \sum_{n=-\infty}^{\infty} z^n f_n$$

is the Laurent expansion of a function $f \in \mathcal{O}^H(\Gamma)$, then

$$f(z) = \sum_{n=-\infty}^{\infty} \sum_{j=0}^{\infty} \langle f_n, e_j \rangle_H\, z^n e_j\,.$$

This implies that f belongs to the closed linear hull of (8.3.23). As $\mathcal{O}^H(\Gamma)$ is dense in $\mathcal{L}^2(\Gamma, H)$, this completes the proof of part (iii).

(i) We first discuss the scalar case $H = \mathbb{C}$, where the main difficulties already appear. In this discussion, we may restrict ourselves to the case when Γ is connected and, hence, D_+ is simply connected. The general case then easily follows applying this special case to each of the connected components of Γ.

If Γ is of class $\mathcal{C}^1$ (and not only *piecewise* $\mathcal{C}^1$, as usually in this book), for a proof we can refer to the book [GKru][4], where the proof is reduced to the case of the unit circle using a conformal mapping from D_+ to the unit disc and its boundary properties. For the general case (when Γ is only piecewise $\mathcal{C}^1$) we can refer only to the original paper [CMM], where the much more general case of Lipschitz contours is considered.[5]

[4] In [GKru] even the more general case of Ljapunov contours is considered.

[5] We do not know whether there exists in the literature a more simple direct proof for the case of piecewise $\mathcal{C}^1$-contours.

We now give a proof of part (i), using the fact that this is already known for $H = \mathbb{C}$.

Let $\{e_j\}_{j\in I}$ be an orthonormal basis of H. Denote by $\mathcal{F}$ the subspace of $\mathcal{L}^2(\Gamma, H)$ which consists of the functions of the form

$$f = \sum_{j\in J} \varphi_j(z) e_j \,, \qquad z \in \Gamma \,,$$

where $J \subseteq I$ is finite and $\varphi_j \in \mathcal{O}^{\mathbb{C}}(\Gamma)$, $j \in J$. For such functions it is easy to see that

$$\|f\|^2_{\mathcal{L}^2(\Gamma,H)} = \sum_{j\in J} \|\varphi_j\|^2_{\mathcal{L}^2(\Gamma,\mathbb{C})} \,, \qquad \mathcal{P}f = \sum_{j\in J} (\mathcal{P}\varphi_j) e_j$$

and

$$\|\mathcal{P}f\|^2_{\mathcal{L}^2(\Gamma,H)} = \sum_{j\in J} \|\mathcal{P}\varphi_j\|^2_{\mathcal{L}^2(\Gamma,\mathbb{C})} \le \|\mathcal{P}\|^2_{\mathcal{L}^2(\Gamma,\mathbb{C})} \sum_{j\in J} \|\varphi_j\|^2_{\mathcal{L}^2(\Gamma,\mathbb{C})} \,,$$

where $\|\mathcal{P}\|_{\mathcal{L}^2(\Gamma,\mathbb{C})}$ denotes the norm of $\mathcal{P}$ as an operator in $\mathcal{L}^2(\Gamma,\mathbb{C})$. If $\|\mathcal{P}\|_{\mathcal{L}^2(\Gamma,H)}$ is the norm of $\mathcal{P}$ as an operator in $\mathcal{L}^2(\Gamma, H)$, this implies that

$$\|\mathcal{P}f\|_{\mathcal{L}^2(\Gamma,H)} \le \|\mathcal{P}\|_{\mathcal{L}^2(\Gamma,\mathbb{C})} \|f\|_{\mathcal{L}^2(\Gamma,H)} \quad \text{for all } f \in \mathcal{F} \,. \tag{8.3.26}$$

Set

$$f_j(z) = \langle f(z), e_j\rangle_H \qquad \text{for } z \in \Gamma \text{ and } f \in \mathcal{O}^H(\Gamma) \,.$$

Then

$$\|f\|^2_{\mathcal{L}^2(\Gamma,H)} = \sum_{j\in I} \|f_j\|^2_{\mathcal{L}^2(\Gamma,\mathbb{C})} \,.$$

for each $f \in \mathcal{O}^H(\Gamma)$. Therefore, for each $f \in \mathcal{O}^H(\Gamma)$ and each $\varepsilon > 0$, there exists a finite set $J \subseteq I$ with

$$\Big\| f - \sum_{j\in J} f_j e_j \Big\|_{\mathcal{L}^2(\Gamma,H)} < \varepsilon \,.$$

Since $\mathcal{O}^H(\Gamma)$ is dense in $\mathcal{L}^2(\Gamma, H)$, this implies that $\mathcal{F}$ is dense in $\mathcal{L}^2(\Gamma, H)$. Together with (8.3.26) this completes the proof of part (i).

(ii) Let $f \in \mathcal{L}^2(\Gamma, H)$ be given. We define f^+ by the first equality in (8.3.21), and we define f^- by the first equality in and (8.3.22). It is clear that in this way functions $f^+ \in \mathcal{O}^H(D_+)$ and $f^- \in \mathcal{O}^H_0(D_- \cup \{\infty\})$ are well defined. It remains to prove the second equality in (8.3.21), the second equality in (8.3.22) and the facts that f^+ is a holomorphic extension of $\mathcal{P}f$ and f^- is a holomorphic extension of $\mathcal{Q}f$.

Since $\mathcal{O}^H(\Gamma)$ is dense in $\mathcal{L}^2(\Gamma, H)$, we can choose a sequence $h_n \in \mathcal{O}^H(\Gamma)$, $n \in \mathbb{N}$, with

$$\lim_{n\to\infty} \|h_n - f\|_{\mathcal{L}^2(\Gamma,H)} = 0 \,. \tag{8.3.27}$$

Since, by part (i) of the theorem, $\mathcal{P}$ and $\mathcal{Q}$ are continuous, this implies that

$$\lim_{n\to\infty} \|\mathcal{P}h_n - \mathcal{P}f\|_{\mathcal{L}^2(\Gamma,H)} = \lim_{n\to\infty} \|\mathcal{Q}h_n - \mathcal{Q}f\|_{\mathcal{L}^2(\Gamma,H)} = 0\,. \tag{8.3.28}$$

Moreover, by Cauchy's formula and Cauchy's theorem,

$$(\mathcal{P}h_n)(z) = \frac{1}{2\pi i}\int_\Gamma \frac{(\mathcal{P}h_n)(\zeta)}{\zeta - z}d\zeta = \frac{1}{2\pi i}\int_\Gamma \frac{h_n(\zeta)}{\zeta - z}d\zeta \tag{8.3.29}$$

for $z \in D_+$, and

$$(\mathcal{Q}h_n)(z) = -\frac{1}{2\pi i}\int_\Gamma \frac{(\mathcal{Q}h_n)(\zeta)}{\zeta - z}d\zeta = -\frac{1}{2\pi i}\int_\Gamma \frac{h_n(\zeta)}{\zeta - z}d\zeta \tag{8.3.30}$$

for $z \in D_-$. The second equality in (8.3.21) follows passing to the limit in (8.3.29), and the second equality in (8.3.22) follows passing to the limit in (8.3.30).

Now let K be a compact subset of D_+. Then, by (8.3.29) and the definition of f^+,

$$\max_{z\in K} \big\|(\mathcal{P}h_n)(z) - f^+(z)\big\|_H \le \frac{1}{2\pi}\int_\Gamma \frac{\|h_n(\zeta) - f(\zeta)\|_H}{|\zeta - z|}|d\zeta|\,.$$

If $|\Gamma|$ is the length of Γ and d is the distance between K and Γ, using the Cauchy-Schwarz inequality, this further implies

$$\max_{z\in K} \big\|(\mathcal{P}h_n)(z) - f^+(z)\big\|_H \le \frac{1}{2\pi d}|\Gamma|^{1/2}\|h_n - f\|_{\mathcal{L}^2(\Gamma,H)}\,.$$

Together with (8.3.28) this implies that

$$\lim_{n\to\infty}\max_{z\in K} \big\|(\mathcal{P}h_n)(z) - f^+(z)\big\|_H = 0\,.$$

Hence f^+ is a holomorphic extension of $\mathcal{P}f$.

In the same way it follows from (8.3.30) and (8.3.28) that f^- is a holomorphic extension of $\mathcal{Q}f$. □

8.3.11 Corollary (to the preceding theorem and definition). *Since $\mathcal{O}^H(\Gamma)$ is dense in $\mathcal{L}^2(\Gamma,H)$ and $\mathcal{P}$ and $\mathcal{Q}$ are continuous with respect to the topology of $\mathcal{L}^2(\Gamma,H)$, it follows immediately from the definitions of $\mathcal{L}^2_+(\Gamma,H)$ and $\mathcal{L}^2_-(\Gamma,H)$ that the space $\mathcal{O}^H(\overline{D}_+)$ is dense in $\mathcal{L}^2_+(\Gamma,H)$ and the space $\mathcal{O}^H_0\big(\overline{D}_- \cup \{\infty\}\big)$ is dense in $\mathcal{L}^2_-(\Gamma,H)$.*

8.4 Operator functions with values acting in a Hilbert space

In this section, H is a separable Hilbert space[6], $D_+ \subseteq \mathbb{C}$ is a bounded connected open set with piecewise $\mathcal{C}^1$-boundary Γ such that $0 \in D_+$, and $D_- := \mathbb{C} \setminus \overline{D}_+$. Further, throughout this section $\mathcal{L}^2(\Gamma, H)$, $\mathcal{L}^2_+(\Gamma, H)$ and $\mathcal{L}^2_-(\Gamma, H)$ denote the Hilbert spaces introduced in Section 8.3.1 and Theorem and Definition 8.3.10, and $\mathcal{P}$ denotes the projector from $\mathcal{L}^2(\Gamma, H)$ onto $\mathcal{L}^2_+(\Gamma, H)$ parallel to $\mathcal{L}^2_-(\Gamma, H)$.

8.4.1 Definition. Let $A : \Gamma \to L(H)$ be a continuous function. Then we denote by $\mathcal{W}_A$ (or by $\mathcal{W}_A^\Gamma$) the bounded linear operator acting in $\mathcal{L}^2_+(\Gamma, H)$ by

$$\mathcal{W}_A f = \mathcal{P}(Af)\,, \qquad f \in \mathcal{L}^2_+(\Gamma, H)\,.$$

This operator $\mathcal{W}_A$ will be called the **Wiener-Hopf operator** defined by A on $\mathcal{L}^2_+(\Gamma, H)$. Sometimes we use also the notation $\mathcal{M}_A$ to denote the operator acting in $\mathcal{L}^2(\Gamma, H)$ by multiplication by A. Then

$$\mathcal{W}_A = \mathcal{P}\mathcal{M}_A\big|_{\mathcal{L}^2_+(\Gamma,H)}.$$

8.4.2 Theorem. *Let $A : \Gamma \to GL(H)$ be a continuous function which admits local factorizations with respect to Γ and $GL(H)$ (Def. 7.1.3), and let $\mathcal{W}_A$ be the Wiener-Hopf operator defined by A on $\mathcal{L}^2_+(\Gamma, H)$. Then:*

(i) *The function A admits a canonical factorization with respect to Γ and $GL(H)$ (Def. 7.1.1), if and only if, $\mathcal{W}_A$ is invertible.*

(ii) *The function A admits a factorization with respect to Γ and $GL(H)$ (Def. 7.1.1), if and only if, $\mathcal{W}_A$ is a Fredholm operator. If this is the case, with the notations from Definition* 7.1.1,

$$\begin{aligned} \dim \operatorname{Coker} \mathcal{W}_A &= \sum_{1\le j\le n\,,\,\kappa_j>0} \kappa_j \dim P_j\,, \\ \dim \operatorname{Ker} \mathcal{W}_A &= -\sum_{1\le j\le n\,,\,\kappa_j<0} \kappa_j \dim P_j\,, \end{aligned} \tag{8.4.1}$$

where the term on the right means zero if there is no j with $\kappa_j > 0$ resp. if there is no j with $\kappa_j < 0$.

The remainder of this section is devoted to the proof of this theorem. We will deduce it from Theorem 8.1.4. The first step is the following lemma.

8.4.3 Lemma. *Let W be a bounded neighborhood of Γ such that each connected component of W contains at least one connected component of Γ. We set $W_+ =$*

[6] For simplicity we consider only *separable* Hilbert spaces, although the results can be generalized to the non-separable case.

$D_+ \cup W$ and $W_- = D_- \cup W$. Let $A : \overline{W} \to GL(H)$ be a continuous function which is holomorphic in W. Let $\mathcal{W}_A$ be the Wiener-Hopf operator defined by A on $\mathcal{L}^2_+(\Gamma, H)$ (Def. 8.4.1), and let $\widetilde{\mathcal{W}}_A$ be the Wiener-Hopf operator defined by A on $\overline{\mathcal{O}}^H(\overline{W}_+)$ (Def. 8.1.2). Then

$$\operatorname{Ker} \mathcal{W}_A = \operatorname{Ker} \widetilde{\mathcal{W}}_A \tag{8.4.2}$$

and

$$\overline{\mathcal{O}}^H(\overline{W}_+) \cap \operatorname{Im} \mathcal{W}_A = \operatorname{Im} \widetilde{\mathcal{W}}_A \,. \tag{8.4.3}$$

Proof. (See Def. 8.1.1 for the notations.) We first prove (8.4.2). The relation

$$\operatorname{Ker} \mathcal{W}_A \supseteq \operatorname{Ker} \widetilde{\mathcal{W}}_A$$

is obvious. To prove the opposite relation, let $f \in \operatorname{Ker} \mathcal{W}_A$ be given. Then $g := Af \in \mathcal{L}^2_-(\Gamma, H)$. By statement (ii) in Theorem and Definition 8.3.10, g admits a holomorphic extension g_- to D_-. Since A^{-1} is holomorphic on W and continuous on $\overline{W}$, this implies that $f = A^{-1}g$ admits the holomorphic extension $A^{-1}g_-$ to $W \cap D_-$, which further extends continuously to $\overline{W} \cap D_-$. On the other hand, by the same statement (ii) in Theorem and Definition 8.3.10, f admits a holomorphic extension to D_+. By Proposition 8.3.9, together this implies that

$$f \in \overline{\mathcal{O}}^H(\overline{W}_+) \,. \tag{8.4.4}$$

Hence $Af \in \overline{\mathcal{O}}^H(\overline{W})$. Since, moreover, $Af \in \mathcal{L}^2_-(\Gamma, H)$, it follows that $Af \in \overline{\mathcal{O}}^H_0(\overline{W}_- \cup \{\infty\})$. Together with (8.4.4) this means that $f \in \operatorname{Ker} \overline{\mathcal{W}}_A$.

Now we prove (8.4.3). The relation

$$\overline{\mathcal{O}}^H(\overline{W}_+) \cap \operatorname{Im} \mathcal{W}_A \supseteq \operatorname{Im} \overline{\mathcal{W}}_A$$

is obvious. To prove the opposite relation, let

$$f_+ \in \overline{\mathcal{O}}^H(\overline{W}_+) \cap \operatorname{Im} \mathcal{W}_A$$

be given. Then there exist $u_+ \in \mathcal{L}^2_+(\Gamma, H)$ and $g_- \in \mathcal{L}^2_-(\Gamma, H)$ such that

$$f_+ + g_- = Au_+ \,. \tag{8.4.5}$$

Hence $g_- = Au_+ - f_+$. Since A is continuous on $\overline{W}$ and holomorphic in W, by Proposition 8.3.9 this implies that $g_- \in \overline{\mathcal{O}}^H(\overline{W})$ and hence (as $g_- \in \mathcal{L}^2_-(\Gamma, H)$)

$$g_- \in \overline{\mathcal{O}}^H_0\left(\overline{W}_- \cup \{\infty\}\right) . \tag{8.4.6}$$

Since $f_+ \in \overline{\mathcal{O}}^H\left(\overline{W}_+\right)$, in view of (8.4.5) this further implies that $Au_+ \in \overline{\mathcal{O}}^H\left(\overline{W}\right)$. As A^{-1} is continuous on $\overline{W}$ and holomorphic in W, so we get $u_+ \in \overline{\mathcal{O}}^H\left(\overline{W}\right)$ and hence (as $u_+ \in \mathcal{L}^2_+(\Gamma, H)$)

$$u_+ \in \overline{\mathcal{O}}^H\left(\overline{W}_+\right) . \tag{8.4.7}$$

From (8.4.5)–(8.4.7) it follows that $f \in \operatorname{Im} \widetilde{\mathcal{W}}_A$. □

8.4.4 Proposition. *(see the beginning of Section* 8.3 *for the notations) The space* $\overline{\mathcal{O}}^H(\overline{D}_+)$ *is contained in* $\mathcal{L}^2_+(\Gamma, H)$ *as a dense subspace, and* $\overline{\mathcal{O}}^H_0(\overline{D}_- \cup \{\infty\})$ *is contained in* $\mathcal{L}^2_-(\Gamma, H)$ *as a dense subspace.*

Proof. By the Mergelyan approximation Theorem 2.2.1, $\mathcal{O}^H(\overline{D}_+)$ is dense in $\overline{\mathcal{O}}^H(\overline{D}_+)$ and $\mathcal{O}^H(\overline{D}_- \cup \{\infty\})$ is dense in $\overline{\mathcal{O}}^H(\overline{D}_- \cup \{\infty\})$ with respect to uniform convergence and, hence, with respect to the topology of $\mathcal{L}^2(\Gamma, H)$. Since, by definition of $\mathcal{P}$ and Q, $\mathcal{P}f = f$ for $f \in \mathcal{O}^H(\overline{D}_+)$ and $\mathcal{Q}f = f$ for $f \in \mathcal{O}^H(\overline{D}_- \cup \{\infty\})$ and since $\mathcal{P}$ and $\mathcal{Q}$ are continuous on $\mathcal{L}^2(\Gamma, H)$, this implies that $\overline{\mathcal{O}}^H(\overline{D}_+)$ is contained in $\mathcal{L}^2_+(\Gamma, H)$ and $\overline{\mathcal{O}}^H(\overline{D}_- \cup \{\infty\})$ is contained in $\mathcal{L}^2_-(\Gamma, H)$.

To prove the density, let $f^+ \in \mathcal{L}^2_+(\Gamma, H)$ and $f^- \in \mathcal{L}^2_-(\Gamma, H)$ be given. Since $\mathcal{O}^H(\Gamma)$ is dense in $\mathcal{L}^2(\Gamma, H)$ (cf. Section 8.3.3), then there are sequences h_n^+ and h_n^- in $\mathcal{O}^H(\Gamma)$ such that

$$\lim_{n \to \infty} \|h_n^+ - f^+\|_{\mathcal{L}^2(\Gamma,H)} = 0 \quad \text{and} \quad \lim_{n \to \infty} \|h_n^- - f^-\|_{\mathcal{L}^2(\Gamma,H)} = 0$$

and hence, as $\mathcal{P}$ and $\mathcal{O}$ are continuous,

$$\lim_{n \to \infty} \|\mathcal{P}h_n^+ - f^+\|_{\mathcal{L}^2(\Gamma,H)} = 0 \quad \text{and} \quad \lim_{n \to \infty} \|\mathcal{Q}h_n^- - f^-\|_{\mathcal{L}^2(\Gamma,H)} = 0.$$

Since $\mathcal{P}h_n^+ \in \mathcal{O}^H(\overline{D}_+)$ and $\mathcal{Q}h_n^- \in \mathcal{O}^H(\overline{D}_- \cup \{\infty\})$, this completes the proof of the density. □

8.4.5 Lemma. *Let* $A, B : \Gamma \to L(E)$ *be continuous functions, and let* $T_- : \overline{D}_- \cup \{\infty\} \to GL(E)$ *and* $T_+ : \overline{D}_+ \to GL(E)$ *be continuous functions, which are holomorphic in* $D_- \cup \{\infty\}$ *and* D_+*, respectively, such that*

$$A = T_- B T_+ \qquad \text{on } \Gamma. \tag{8.4.8}$$

Then:

(i) $\mathcal{W}_A = \mathcal{W}_{T_-}\mathcal{W}_B\mathcal{W}_{T_+}$.

(ii) *If* $T_-(z)$ *is invertible for all* $z \in \overline{D}_- \cup \{\infty\}$*, then* $\mathcal{W}_{T_-}$ *is invertible and*

$$\mathcal{W}_{T_-}^{-1} = \mathcal{W}_{T_-^{-1}}.$$

(iii) *If* $T_+(\lambda)$ *is invertible for all* $z \in \overline{D}_+$*, then* $\mathcal{W}_{T_+}$ *is invertible and*

$$\mathcal{W}_{T_+}^{-1} = \mathcal{W}_{T_+^{-1}}.$$

In particular, if A *and* B *are equivalent relative to* Γ *and* $GL(E)$ *(Def.* 7.1.3*), then* $\mathcal{W}_A$ *and* $\mathcal{W}_B$ *are equivalent.*[7]

[7] Two operators T and S in a Banach space X are called **equivalent** if there exist invertible operators V, W in X such that $T = VSW$.

Proof. It is sufficient to prove part (i), because (ii) and (iii) then follow.

It follows from Proposition 8.4.4 that $\mathcal{M}_{T_+}\mathcal{L}^2_+(\Gamma,H) \subseteq \mathcal{L}^2_+(\Gamma,H)$ and $\mathcal{M}_{T_-}\mathcal{L}^2_-(\Gamma,H) \subseteq \mathcal{L}^2_-(\Gamma,H)$. Therefore

$$\mathcal{P}\mathcal{M}_{T_+}\mathcal{P} = \mathcal{M}_{T_+}\mathcal{P} \tag{8.4.9}$$

and

$$\mathcal{P}\mathcal{M}_{T_-}(I-\mathcal{P}) = 0\,.$$

From the second relation it follows that

$$\mathcal{P}\mathcal{M}_{T_-}\mathcal{P} = \mathcal{P}\mathcal{M}_{T_-}\,. \tag{8.4.10}$$

From (8.4.8)–(8.4.10) together we obtain the assertion:

$$\begin{aligned}\mathcal{W}_{T_-}\mathcal{W}_B\mathcal{W}_{T_+} &= \mathcal{P}\mathcal{M}_{T_-}\mathcal{P}\mathcal{M}_B\mathcal{P}\mathcal{M}_{T_+}\Big|_{\mathcal{L}^2_+(\Gamma,H)} \\ &= \mathcal{P}\mathcal{M}_{T_-}\mathcal{M}_B\mathcal{M}_{T_+}\Big|_{\mathcal{L}^2_+(\Gamma,H)} = \mathcal{P}\mathcal{M}_A\Big|_{\mathcal{L}^2_+(\Gamma,H)} = \mathcal{W}_A\,.\end{aligned}$$

□

Proof of Theorem 8.4.2. First assume that A admits a factorization $A = A_-\Delta A_+$ with respect to Γ and $GL(H)$. Using Proposition 8.3.9, then it is easy to prove that $\mathcal{W}_\Delta$ is a Fredholm operator, where, with the notations from Definition 7.1.1,

$$\begin{aligned}\dim \operatorname{Coker} \mathcal{W}_\Delta &= \sum_{1\le j\le n\,,\,\kappa_j>0} \kappa_j \dim P_j\,,\\ \dim \operatorname{Ker} \mathcal{W}_\Delta &= -\sum_{1\le j\le n\,,\,\kappa_j<0} \kappa_j \dim P_j\,.\end{aligned} \tag{8.4.11}$$

Since, by Lemma 8.4.5, the operators $\mathcal{W}_A$ and $\mathcal{W}_\Delta$ are equivalent, this implies that $\mathcal{W}_A$ is invertible if $\Delta \equiv I$ and that $\mathcal{W}_A$ is a Fredholm operator satisfying (8.4.1) if $\Delta \not\equiv 0$.

Now we assume that $\mathcal{W}_A$ is a Fredholm operator on $\mathcal{L}^2_+(\Gamma,H)$. By Theorem 7.4.2, A is equivalent with respect to Γ and $GL(H)$ to some holomorphic function $B : \mathbb{C}^* \to GL(H)$. Let $\mathcal{W}_B$ be the Wiener-Hopf operator defined by B in $\mathcal{L}^2_+(\Gamma,H)$. Then, by Lemma 8.4.5, the operators $\mathcal{W}_A$ and $\mathcal{W}_B$ are equivalent. Hence $\mathcal{W}_B$ is a Fredholm operator, and if $\mathcal{W}_A$ is invertible, then $\mathcal{W}_B$ is invertible.

Now we choose a neighborhood W as in Lemma 8.4.3 and set $W_+ = D_+ \cup W$. Let, as in this lemma, $\widetilde{\mathcal{W}_B}$ be the Wiener-Hopf operator defined by B in $\overline{\mathcal{O}}^H(\overline{W}_+)$. Since $\mathcal{W}_A$ is a Fredholm operator, it follows from relations (8.4.2) and (8.4.3) in Lemma 8.4.3 that also $\widetilde{\mathcal{W}_B}$ is a Fredholm operator, which is invertible if $\mathcal{W}_A$ is invertible. Now the two assertions (i) and (ii) of the theorem under proof follow from Theorem 8.1.4. □

8.5 Functions close to the unit operator or with positive real part

In this section, H is again a separable Hilbert space, $D_+ \subseteq \mathbb{C}$ is a bounded connected open set with piecewise $\mathcal{C}^1$-boundary Γ such that $0 \in D_+$, and $D_- := \mathbb{C} \setminus \overline{D}_+$. For many results, we have to assume that $\Gamma = \mathbb{T}$ is the unit circle. If this is the case, this will be mentioned explicitly.

The following theorem is true for general contours.

8.5.1 Theorem. *There exists a constant $\varepsilon > 0$ such that the following holds: Any continuous operator function $A : \Gamma \to GL(H)$, which admits local factorizations with respect to Γ (Def. 7.1.3) and which satisfies*

$$\max_{z\in\Gamma} \|A(z) - I\|_{L(H)} < \varepsilon \,, \tag{8.5.1}$$

admits a canonical factorization with respect to Γ (Def. 7.1.1).

Proof. Let $\mathcal{L}^2(\Gamma, H)$, $\mathcal{L}^2_+(\Gamma, H)$ and $\mathcal{L}^2_-(\Gamma, H)$ be the Hilbert spaces introduced in Section 8.3.1 and Theorem and Definition 8.3.10, and let $\mathcal{P}$ be the projector from $\mathcal{L}^2(\Gamma, H)$ onto $\mathcal{L}^2_+(\Gamma, H)$ parallel to $\mathcal{L}^2_-(\Gamma, H)$. Recall that (by Theorem and Definition 8.3.10) $\mathcal{P}$ is a bounded linear operator on $\mathcal{L}^2_-(\Gamma, H)$, and set

$$\varepsilon = \frac{1}{\|\mathcal{P}\|_{L(\mathcal{L}^2(\Gamma,H))}} \,.$$

Now let a continuous function $A : \Gamma \to GL(H)$ be given which satisfies (8.5.1), and let $\mathcal{M}_{A-I} : \mathcal{L}^2(\Gamma, H) \to \mathcal{L}^2(\Gamma, H)$ be the operator of multiplication by $A - I$, and let $\mathcal{W}_A$ be the Wiener-Hopf operator defined by A on $\mathcal{L}^2_+(\Gamma, H)$ (Def. 8.4.1). Then

$$\|\mathcal{M}_{A-I}\|_{L(\mathcal{L}^2(\Gamma,H))} < \varepsilon = \frac{1}{\|\mathcal{P}\|_{L(\mathcal{L}^2(\Gamma,H))}} \,.$$

Since $\mathcal{W}_A - I = \mathcal{P}\mathcal{M}_{A-I}\big|_{\mathcal{L}^2_+(\Gamma,H)}$, this implies

$$\|\mathcal{W}_A - I\|_{L\left(\mathcal{L}^2_+(\Gamma,H)\right)} \le \|\mathcal{P}\|_{L(\mathcal{L}^2(\Gamma,H))} \|\mathcal{M}_{A-I}\|_{L(\mathcal{L}^2(\Gamma,H))} < 1 \,.$$

Hence $\mathcal{W}_A$ is invertible. If, in addition, A admits local factorizations with respect to Γ and $GL(H)$, it follows from Theorem 8.4.2 that A admits a canonical factorization with respect to Γ. □

8.5.2 Corollary (to Theorem 8.5.1). *Let $0 < \alpha < 1$ and $k \in \mathbb{N}$, where, for $k \ge 1$, we additionally assume that Γ is of class $\mathcal{C}^k$ (Def. 3.4.1). Then there exists $\varepsilon > 0$ such that, for any $\mathbb{C}^{k+\alpha}$-function $A : \Gamma \to GL(H)$ (Def. 3.4.3), which satisfies*

$$\max_{z\in\Gamma} \|A(z) - I\|_{L(H)} < \varepsilon \,,$$

the following holds:

(i) *A admits a canonical factorization with respect to* Γ.

(ii) *If* $A = A_- A_+$ *is an arbitrary canonical factorization of* A *with respect to* Γ, *then automatically, the factors* A_- *and* A_+ *are of class* $\mathcal{C}^{k+\alpha}$ *on* $\overline{D}_-$ *and* $\overline{D}_+$, *respectively.*

Proof. Part (i) of Theorem 7.3.1 in particular yields that A admits local factorizations with respect to Γ. Therefore part (i) of the corollary follows from Theorem 8.5.1. Part (ii) of the corollary follows from part (ii) of Theorem 7.3.1 □

8.5.3 Theorem. *Let* $\Gamma = \mathbb{T}$ *be the unit circle. Then any continuous function* $A : \mathbb{T} \to GL(H)$, *which admits local factorizations with respect to* $\mathbb{T}$ *(Def. 7.1.3) and which satisfies*

$$\max_{z\in\Gamma} \|A(z) - I\|_{L(H)} < 1\,, \tag{8.5.2}$$

admits a canonical factorization with respect to $\mathbb{T}$ *(Def. 7.1.1).*

Proof. This is a repetition of the proof of Theorem 8.5.1, taking into account that now (as $\Gamma = \mathbb{T}$), by statement (iii) in definition and Theorem 8.3.10,

$$\|\mathcal{P}\|_{L(\mathcal{L}^2(\Gamma,H))} = 1\,.$$

□

8.5.4 Corollary (to Theorem 8.5.3). *Let* $\Gamma = \mathbb{T}$ *be the unit circle, and let* $A : \Gamma \to GL(H)$ *be a* $\mathbb{C}^{k+\alpha}$*-function,* $0 < \alpha < 1$, $k \in \mathbb{N}$ *(Def. 3.4.3), which satisfies*

$$\max_{z\in\mathbb{T}} \|A(z) - I\|_{L(H)} < 1\,.$$

Then:

(i) *A admits a canonical factorization with respect to* $\mathbb{T}$.

(ii) *If* $A = A_- A_+$ *is an arbitrary canonical factorization of* A *with respect to* $\mathbb{T}$, *then automatically, the factors* A_- *and* A_+ *are of class* $\mathbb{C}^{k+\alpha}$ *on* $\overline{D}_-$ *and* $\overline{D}_+$, *respectively.*

Proof. Part (i) of Theorem 7.3.1 in particular yields that A admits local factorizations with respect to $\mathbb{T}$. Therefore part (i) of the corollary follows from Theorem 8.5.3. Part (ii) of the corollary follows from part (ii) of Theorem 7.3.1 □

8.5.5 Corollary (to Theorem 8.5.3). *Let* $\Gamma = \mathbb{T}$ *be the unit circle, and let* $\mathcal{R}$ *be a Banach algebra of continuous* $L(H)$*-valued functions satisfying conditions* (A), (B) *and* (C) *in Section* 7.2.4. *For example, let* $\mathcal{R} = W\big(L(H)\big)$ *be the Wiener algebra (see Section* 7.2.1). *Let* $A : \mathbb{T} \to GL(H)$ *be a function which belongs to* $\mathcal{R}$ *and satisfies*

$$\max_{z\in\mathbb{T}} \|A(z) - I\|_{L(H)} < 1\,.$$

Then:

(i) *A admits a canonical factorization with respect to* $\mathbb{T}$.

(ii) *If* $A = A_- A_+$ *is an arbitrary canonical factorization of* A *with respect to* Γ, *then automatically, the factors* A_- *and* A_+ *belong to the algebra* $\mathcal{R}$.

Proof of Theorem 8.5.1. Part (ii) of Theorem 7.2.5 yields that A admits local factorizations with respect to Γ. Therefore part (i) of the corollary follows from Theorem 8.5.3. Part (ii) of the corollary follows from part (iii) of Theorem 7.2.5. □

8.5.6. It is clear that the assertion of Theorem 8.5.3 remains valid if we replace the unit circle $\mathbb{T}$ by an arbitrary circle in $\mathbb{C}$. However this is not true for more general contours. V.I. Macaev and A.I. Virozub [ViMa] even proved the following: If the assertion of Theorem 8.5.3 is valid for any finite dimensional Hilbert space H with $\mathbb{T}$ replaced by an arbitrary Jordan curve, then this Jordan curve is a circle.

However there is the following result for operator functions of a special form.

8.5.7 Theorem. *Assume that the contour* Γ *(of the generality as described at the beginning of this section) is connected, i.e., we assume that* D_+ *is simply connected (Section* 2.5.1*).*

Let $A : \Gamma \to GL(H)$ *be a function of the form*

$$A(z) = \frac{T}{z - z_0} + B_+(z)\,, \qquad z \in \Gamma\,, \tag{8.5.3}$$

where $z_0 \in D_+$, $T \in L(H)$ *and* B_+ *is a continuous* $L(H)$*-valued function on* $\overline{D}_+$, *which is holomorphic in* D_+, *and assume that*

$$\max_{z\in\Gamma} \|A(z) - I\|_{L(H)} < 1\,. \tag{8.5.4}$$

Then A *admits a canonical factorization with respect to* Γ *and* $GL(H)$.

8.5.8. Of course, also for this theorem there is a corollary for functions of class $\mathcal{C}^{k+\alpha}$ corresponding to Corollary 8.5.2 of Theorem 8.5.1.

Proof of Theorem 8.5.7. As A is of the form (8.5.3), it is continuous on $\overline{D}_+ \setminus \{z_0\}$ and holomorphic in $D_+ \setminus \{z_0\}$. Choose a neighborhood $U \subseteq \mathbb{C} \setminus \{z_0\}$ of Γ so small that

$$q := \sup_{z\in U\cap\overline{D}_+} \|A(z) - I\|_{L(H)} < 1\,. \tag{8.5.5}$$

Set

$$\mathbb{T} = \left\{ z \in \mathbb{C} \,\middle|\, |z| = 1 \right\} \qquad \text{and} \qquad \mathbb{D} = \left\{ z \in \mathbb{C} \,\middle|\, |z| < 1 \right\}.$$

Since D_+ is simply connected, we can find a biholomorphic map φ from a neighborhood of $\overline{\mathbb{D}}$ onto D_+ such that

$$\Gamma' := \varphi(\mathbb{T}) \subseteq U\,, \qquad z_0 \in D_+ := \varphi(\mathbb{D}) \qquad \text{and} \qquad \varphi(0) = z_0\,.$$

It is easy to see that then any canonical factorization of A with respect to Γ' and $GL(H)$ is also a canonical factorization with respect to Γ and $GL(H)$. Therefore it is sufficient to prove that A admits a canonical factorization with respect to Γ' and $GL(H)$.

For that we want to apply theroem 8.4.2. Since A is holomorphic in a neighborhood of Γ', it is trivial that A admits local factorizations with respect to Γ' and $GL(H)$. Therefore, to apply this theroem, we only have to prove that the Wiener-Hopf operator defined by A in $\mathcal{L}^2_+(\Gamma', H)$ is invertible.

Since the derivative of a biholomorphic map does not vanish,

$$\rho(z) := \frac{|z - z_0|^2}{|\varphi'(\varphi^{-1}(z))|}, \qquad z \in \Gamma',$$

is a well-defined continuous function on Γ'. Moreover, $h(z) \neq 0$ for all $z \in \Gamma'$ (as $z_0 \notin \Gamma'$). Therefore, setting

$$\langle f, g \rangle_\rho = \int_{\Gamma'} \langle f(z), g(z) \rangle_H \, \rho(z) \, |dz|,$$

we get a scalar product $\langle \cdot, \cdot \rangle_\rho$ on $\mathcal{L}^2(\Gamma', H)$, such that the corresponding norm $\|\cdot\|_\rho$ is equivalent to the norm $\|\cdot\|_{\mathcal{L}^2(\Gamma', H)}$ introduced in Definition 8.3.1.

Therefore, to prove the invertibility of the above mentioned Wiener-Hopf operator, now it is sufficient to show that

$$\|\mathcal{P}(Af) - f\|_\rho \leq q \|f\|_\rho \tag{8.5.6}$$

for all $f \in \mathcal{L}^2_+(\Gamma', H)$, where $\mathcal{P}$ is the projector from $\mathcal{L}^2(\Gamma', H)$ onto $\mathcal{L}^2_+(\Gamma', H)$ parallel to $\mathcal{L}^2_-(\Gamma', H)$. Since $\mathcal{O}^H(\overline{D}'_+)$ is dense in $\mathcal{L}^2_+(\Gamma', H)$ (Corollary 8.3.11), it is sufficient to prove this for all $f \in \mathcal{O}^H(\overline{D}'_+)$. Let such f be given. Let T be the operator from (8.5.3), set

$$v = Tf(z_0) \qquad \text{and} \qquad h(z) = \frac{v}{z - z_0}, \qquad z \in \mathbb{C} \setminus D'_+ .$$

Then

$$\begin{aligned}
\langle \mathcal{P}(Af) - f, h \rangle_\rho &= \int_{\Gamma'} \left\langle \big(\mathcal{P}(Af) - f\big)(z), \frac{v}{z - z_0} \right\rangle_H \frac{|z - z_0|^2}{|\varphi'(\varphi^{-1}(z))|} |dz| \\
&= \int_{\mathbb{T}} \left\langle \big(\mathcal{P}(Af) - f\big)(\varphi(z)), \frac{v}{\varphi(z) - z_0} \right\rangle_H \frac{|\varphi(z) - z_0|^2}{|\varphi'(\varphi^{-1}(\varphi(z)))|} |\varphi'(z)||dz| \\
&= \int_{\mathbb{T}} \left\langle \big(\mathcal{P}(Af) - f\big)(\varphi(z)), \frac{v}{\varphi(z) - z_0} \right\rangle_H |\varphi(z) - z_0|^2 |dz| \\
&= \int_{\mathbb{T}} \left\langle \big(\mathcal{P}(Af) - f\big)(\varphi(z)), v \right\rangle_H (\varphi(z) - z_0) |dz| .
\end{aligned}$$

Since f is holomorphic in a neighborhood of $\overline{D}'_+$ and A is holomorphic in a neighborhood of Γ', the function under the last integral is holomorphic in a neighborhood of $\overline{\mathbb{D}}$. Moreover, as $\varphi(0) = z_0$, this function vanishes at zero. Therefore the Taylor expansion at zero of this function is of the form

$$\Big\langle \big(\mathcal{P}(Af) - f\big)(\varphi(z)) \, , v \Big\rangle_H (\varphi(z) - z_0) = \sum_{n=1}^{\infty} a_n z^n \, .$$

Since this series converges uniformly on $\mathbb{T}$ and

$$\int_{\mathbb{T}} z^n |dz| = \int_0^{2\pi} e^{int} dt = \frac{1}{in} e^{int} \Big|_{t=0}^{t=2\pi} = 0 \quad \text{for } n \geq 1 \, ,$$

it follows that $\big\langle \mathcal{P}(Af) - f, h \big\rangle_\rho = 0$. Hence

$$\|\mathcal{P}(Af) - f\|_\rho \leq \|\mathcal{P}(Af) - f + h\|_\rho \, .$$

Since, by (8.5.3), $\mathcal{P}(Af) + h = Af$ and therefore $\mathcal{P}(Af) - f + h = Af - f$, it follows that

$$\|\mathcal{P}(Af) - f\|_\rho \leq \|Af - f\|_\rho \, .$$

As, by (8.5.5), $\|Af - f\|_\rho \leq q\|f\|_\rho$, this proves (8.5.6). □

8.5.9 Theorem. *Let $\Gamma = \mathbb{T}$ be the unit circle. Assume $A : \mathbb{T} \to GL(H)$ is a continuous function which admits local factorizations with respect to $\mathbb{T}$ (Def.* 7.1.3, *such that, for some $c > 0$,*

$$\operatorname{Re} \big\langle A(z)v, v \big\rangle_H \geq c\|v\|_H \qquad \textit{for all } v \in H \textit{ and } z \in \mathbb{T} \, , \tag{8.5.7}$$

where $\langle \cdot, \cdot \rangle_H$ denotes the scalar product of H. Then A admits a canonical factorization with respect to $\mathbb{T}$ (Def. 7.1.1).

Proof. By Theorem 8.4.2 it is sufficient to prove that the Wiener-Hopf operator $\mathcal{W}_A$ defined by A on $\mathcal{L}^2(\mathbb{T}, H)$ is invertible. It follows from (8.5.7) that, for each continuous function $f : \mathbb{T} \to H$,

$$\begin{aligned} \operatorname{Re} \big\langle Af, f \big\rangle_{\mathcal{L}^2(\mathbb{T},H)} &= \int_{\mathbb{T}} \operatorname{Re} \big\langle A(z)f(z), f(z) \big\rangle_H |dz| \\ &\geq c \int_{\mathbb{T}} \|f(z)\|_H |dz| = c\|f\|^2_{\mathcal{L}^2(\mathbb{T},H)} . \end{aligned}$$

Since the continuous functions are dense in $\mathcal{L}^2(\mathbb{T}, H)$, this implies that

$$\operatorname{Re} \big\langle Af, f \big\rangle_{\mathcal{L}^2(\mathbb{T},H)} \geq c\|f\|^2_{\mathcal{L}^2(\mathbb{T},H)} \quad \text{for all } f \in \mathcal{L}^2(\mathbb{T}, H) \, .$$

Since, by statement (iii) in Theorem and Definition 8.3.10, the projector $\mathcal{P}$ is orthogonal and $\mathcal{P}f = f$ for $f \in \mathcal{L}^2_+(\mathbb{T}, H)$, this implies that

$$\begin{aligned}\operatorname{Re}\left\langle \mathcal{W}_A f, f\right\rangle_{\mathcal{L}^2(\mathbb{T},H)} &= \operatorname{Re}\left\langle \mathcal{P}(Af), f\right\rangle_{\mathcal{L}^2(\mathbb{T},H)} = \operatorname{Re}\left\langle Af, f\right\rangle_{\mathcal{L}^2(\mathbb{T},H)} \\ &\geq c\left\|f\right\|^2_{\mathcal{L}^2(\mathbb{T},H)} \qquad \text{for all } f \in \mathcal{L}^2_+(\mathbb{T}, H)\,.\end{aligned}$$

Hence the real part of $\mathcal{W}_A$ is positive, which implies that $\mathcal{W}_A$ is invertible. □

8.5.10 Theorem. *Let $\Gamma = \mathbb{T}$ be the unit circle. Assume $A : \Gamma \to GL(H)$ is a continuous function which admits local factorizations with respect to Γ (Def. 7.1.3), such that at least one of the following two conditions is satisfied:*

$$\max_{z\in\mathbb{T}} \|A^{-1}(z) - I\|_{L(H)} < 1\,, \tag{8.5.8}$$

or, for some $c > 0$,

$$\operatorname{Re}\langle A^{-1}(z)v, v\rangle_H \geq c\|v\|_H \qquad \textit{for all } v \in H \textit{ and } z \in \mathbb{T}\,. \tag{8.5.9}$$

Then A admits a canonical factorization with respect to $\mathbb{T}$ (Def. 7.1.1).

Proof. Set

$$B(z) = A^{-1}\left(\frac{1}{z}\right).$$

Since A admits local factorizations with respect to $\mathbb{T}$, then B admits local factorizations with respect to $\mathbb{T}$. Moreover, since A satisfies at least one of the conditions (8.5.8) or (8.5.9), B satisfies at least one of the conditions

$$\max_{z\in\mathbb{T}} \|B(z) - I\|_{L(H)} < 1\,,$$

or

$$\operatorname{Re}\langle B(z)v, v\rangle_H \geq c\|v\|_H \qquad \text{for all } v \in H\,.$$

Therefore, by theorems 8.5.3 and 8.5.9, B admits a canonical factorization $B = B_- B_+$ with respect to $\mathbb{T}$. Setting

$$A_-(z) = B_+^{-1}\left(\frac{1}{z}\right) \qquad \text{and} \qquad A_+(z) = B_-^{-1}\left(\frac{1}{z}\right),$$

we obtain a required canonical factorization $A = A_- A_+$ of A with respect to $\mathbb{T}$. □

8.5.11. Finally we note that also theorems 8.5.9 and 8.5.10 have corollaries corresponding to corollaries 8.5.4 and 8.5.4 of Theorem 8.5.3.

8.6 Block Töplitz operators

In this section H is a separable Hilbert space and

$$l^2(H) = H \oplus H \oplus H \oplus \dots$$

is defined to be the Hilbert space of square integrable sequences $v = (v_n)_{n\in\mathbb{N}}$ of vectors $v_n \in H$, endowed with the scalar product

$$\langle v, w\rangle = \sum_{n=0}^{\infty} \langle v_n, w_n\rangle, \qquad v = (v_n)_{n\in\mathbb{N}} \in l^2(H), \quad w = (w_n)_{n\in\mathbb{N}} \in l^2(H),$$

and the norm

$$\|v\| = \sqrt{\sum_{n=0}^{\infty} \|v_n\|^2}, \qquad v = (v_n)_{n\in\mathbb{N}} \in l^2(H).$$

8.6.1. Let $\pi_n : l^2(H) \to H$, $n \in \mathbb{N}$, be the projectors defined by

$$\pi_n v = v_n \qquad \text{for } v = (v_j)_{j=0}^{\infty} \in l^2(H),$$

and let $\tau_n : H \to l^2(H)$, $n \in \mathbb{N}$, be the injections defined by

$$\tau_n h = (\delta_{nj} h)_{j=0}^{\infty}, \qquad \text{for } h \in H,$$

where δ_{nj} is the Kronecker symbol. Then, with each operator $T \in L\big(l^2(H)\big)$, we associate the infinite matrix with elements from $L(H)$ given by

$$(T_{jk})_{j,k=0}^{\infty} = \begin{pmatrix} T_{00} & T_{01} & T_{02} & \dots \\ T_{10} & T_{11} & T_{12} & \dots \\ T_{20} & T_{21} & T_{22} & \dots \\ \vdots & \vdots & \vdots & \ddots \end{pmatrix} \qquad \text{with} \qquad T_{jk} = \pi_j T \tau_k.$$

This matrix will be called the **block matrix** of T. An operator $T \in L\big(l^2(H)\big)$ will be called a **block Töplitz operator** if the elements T_{jk} of its block matrix depend only on the difference $j-k$, i.e., if the block matrix of T is of the form

$$\big(T_{j-k}\big)_{j,k=0}^{\infty} = \begin{pmatrix} T_0 & T_{-1} & T_{-2} & \dots \\ T_1 & T_0 & T_{-1} & \dots \\ T_2 & T_1 & T_0 & \dots \\ \vdots & \vdots & \vdots & \ddots \end{pmatrix} \quad \text{with} \quad T_n = \begin{cases} \pi_n T \tau_0 & \text{if } n \geq 0, \\ \pi_0 T \tau_{-n} & \text{if } n \leq 0. \end{cases}$$

8.6.2. Let $\mathbb{T}$ be the unit circle, and let $\mathcal{L}^2(\mathbb{T}, H)$, $\mathcal{L}^2_+(\mathbb{T}, H)$, $\mathcal{L}^2_-(\mathbb{T}, H)$ be the Hilbert spaces introduced in Section 8.3.1 and in Theorem and Definition 8.3.10.

Recall that, by part (iii) of Theorem and Definition 8.3.10, we have the orthogonal decomposition

$$\mathcal{L}^2(\mathbb{T},H) = \mathcal{L}^2_+(\mathbb{T},H) \oplus \mathcal{L}^2_-(\mathbb{T},H). \tag{8.6.1}$$

Let $\{e_\nu\}_{\nu=1}^\infty$ be an orthonormal basis of H. Recall that then, by part (iii) of Theorem and Definition 8.3.10, the family $\{\Psi_{n\nu}\}_{n\in\mathbb{N},\nu\in\mathbb{N}^*}$ of functions $\Psi_{n\nu} \in \mathcal{L}^2_+(\mathbb{T},H)$ defined by

$$\Psi_{n\nu}(z) = \frac{z^n}{\sqrt{2\pi}} e_\nu, \qquad z \in \mathbb{T}, \quad n \in \mathbb{N},\ \nu \in \mathbb{N}^*, \tag{8.6.2}$$

forms an orthonormal basis of $\mathcal{L}^2_+(\mathbb{T},H)$. On the other hand, by definition of $l^2(H)$, the family $\{\psi_{n\nu}\}_{n\in\mathbb{N},\nu\in\mathbb{N}^*}$ of sequences $\psi_{n\nu} \in l^2(H)$ defined by

$$\psi_{n\nu} = \left(\delta_{nj} e_\nu\right)_{j=0}^\infty, \qquad n \in \mathbb{N},\ \nu \in \mathbb{N}^*, \tag{8.6.3}$$

forms an orthonormal basis of $l^2(H)$. We denote by $\mathbf{M}$ the linear isometry from $\mathcal{L}^2_+(\mathbb{T},H)$ onto $l^2(H)$ defined by

$$\mathbf{M}\Psi_{n\nu} = \psi_{n\nu}\,, \qquad n \in \mathbb{N},\ \nu \in \mathbb{N}^*. \tag{8.6.4}$$

8.6.3 Lemma and Definition. Let $A : \mathbb{T} \to L(H)$ be a continuous function, and let $\mathcal{W}_A$ be the Wiener-Hopf operator defined by A in $\mathcal{L}^2_+(\mathbb{T},H)$ (cf. Section 8.4.1). Then the operator T_A defined by

$$T_A = \mathbf{M}\mathcal{W}_A\mathbf{M}^{-1} \tag{8.6.5}$$

is a block Töplitz operator, where

$$\pi_j T_A \tau_k = \frac{1}{2\pi i}\int_{\mathbb{T}} \frac{A(z)}{z^{j-k+1}}dz, \qquad j,k \in \mathbb{N}. \tag{8.6.6}$$

The operator T_A defined by (8.6.5) will be called the **block Töplitz operator defined by** A.

Proof. We only have to prove (8.6.6). Let $j,k \in \mathbb{N}$ be given. First note that, for all $k \in \mathbb{N}^*$ and $v \in H$,

$$\left(\mathbf{M}^{-1}\tau_k v\right)(z) = \frac{z^k}{\sqrt{2\pi}} v, \qquad z \in \mathbb{T}. \tag{8.6.7}$$

Indeed, for $v = e_\nu$ this holds by definition (8.6.4). Since $\{e_\nu\}_{\nu\in\mathbb{N}^*}$, is an orthonormal basis of H, it follows for all $v \in H$.

Next we prove that, for all $f_+ \in \mathcal{L}^2_+(\mathbb{T},H)$ and $j \in \mathbb{N}$,

$$\pi_j \mathbf{M} f_+ = \frac{1}{i\sqrt{2\pi}}\int_{\mathbb{T}} \frac{f_+(z)}{z^{j+1}}dz. \tag{8.6.8}$$

Since the family $\{\Psi_{m\nu}\}_{m\in\mathbb{N},\nu\in\mathbb{N}^*}$ is an orthonormal basis of $\mathcal{L}^2_+(\mathbb{T},H)$ and the two sides of (8.6.8) depend continuously on f_+ with respect to the norm of $\mathcal{L}^2_+(\mathbb{T},H)$, it is sufficient to prove that this holds for $f_+ = \Psi_{m\nu}$, $m \in \mathbb{N}$, $\nu \in \mathbb{N}^*$, which is the case, namely:

$$\pi_j \mathbf{M}\Psi_{m\nu} = \pi_j \psi_{m\nu} = \delta_{mj} e_\nu = \frac{\delta_{mj}}{2\pi i}\left(\int_{\mathbb{T}} \frac{1}{z} dz\right) e_\nu = \frac{1}{2\pi i}\left(\int_{\mathbb{T}} \frac{z^m}{z^{j+1}} dz\right) e_\nu$$
$$= \frac{1}{i\sqrt{2\pi}} \int_{\mathbb{T}} \frac{\Psi_{m\nu}(z)}{z^{j+1}} dz.$$

Now let $\mathcal{P}$ be the orthogonal projector from $\mathcal{L}^2(\mathbb{T},H)$ onto $\mathcal{L}^2_+(\mathbb{T},H)$. Then, for all $f \in \mathcal{L}^2(\mathbb{T},H)$,

$$\int_{\mathbb{T}} \frac{(\mathcal{P}f)(z)}{z^{j+1}} dz = \int_{\mathbb{T}} \frac{f(z)}{z^{j+1}} dz, \qquad j \in \mathbb{N}. \tag{8.6.9}$$

Indeed, if f is a function of the form

$$f(z) = \frac{v}{z^m}, \qquad z \in \mathbb{T},$$

where $v \in H$ and $m \in \mathbb{N}^*$, this is obviously the case. As the functions of this form are dense in $\mathcal{L}^2_-(\mathbb{T},H)$ (part (iii) of Theorem and Definition 8.3.10), this implies (8.6.9) for all $f \in \mathcal{L}^2(\mathbb{T},H)$. [8]

From (8.6.9) and (8.6.8) it follows that, for all $f \in \mathcal{L}^2(\mathbb{T},H)$,

$$\pi_j \mathbf{M}\mathcal{P} f = \frac{1}{i\sqrt{2\pi}} \int_{\mathbb{T}} \frac{f(z)}{z^{j+1}} dz.$$

Hence, for all $f_+ \in \mathcal{L}^2_+(\mathbb{T},H)$,

$$\pi_j \mathbf{M}\mathcal{W}_A f_+ = (\pi_j \mathbf{M}\mathcal{P})(A f_+) = \frac{1}{i\sqrt{2\pi}} \int_{\mathbb{T}} \frac{A(z) f_+(z)}{z^{j+1}} dz.$$

Together with (8.6.7) this further implies that, for all $v \in H$,

$$\pi_j T_A \tau_k h = \pi_j \mathbf{M}\mathcal{W}_A \mathbf{M}^{-1} \tau_k v = \frac{1}{i\sqrt{2\pi}} \int_{\mathbb{T}} \frac{A(z) z^k v/\sqrt{2\pi}}{z^{j+1}} dz = \frac{1}{2\pi i} \int_{\mathbb{T}} \frac{A(z) v}{z^{j-k+1}} dz,$$

i.e., we have (8.6.6). □

8.6.4. If $P \in L(H)$ is a projector, then we denote by $S_{\kappa,P}$, $\kappa \in \mathbb{Z}$, the Töplitz operator defined by the operator function $z^\kappa P$, $z \in \mathbb{T}$.

[8] Of course, (8.6.9) follows also from the more general formula 8.3.21 in part (ii) of Theorem and Definition 8.3.10.

Then it follows from (8.6.6) that, for all $j, k \in \mathbb{N}$ and $\kappa \in \mathbb{Z}$,

$$\pi_j S_{\kappa,P} \tau_k := \frac{1}{2\pi i} \int_{\mathbb{T}} \frac{P}{z^{j-k+1-\kappa}} dz = \begin{cases} P & \text{if } j = k + \kappa, \\ 0 & \text{if } j \neq k + \kappa. \end{cases} \tag{8.6.10}$$

Hence, for each projector $P \in L(H)$ and each sequence $v = (v_0, v_1, \ldots) \in l^2(H)$,

$$\begin{aligned} S_{0,P} v &= (P v_0, P v_1, \ldots), \\ S_{1,P} v &= (0, P v_1, P v_2, \ldots), \\ S_{-1,P} v &= (P v_2, P v_3, \ldots), \end{aligned} \tag{8.6.11}$$

which implies that, for each projector $P \in L(H)$ and for all $\kappa \in \mathbb{N}$,

$$\begin{aligned} S_{\kappa,P} &= S_{1,P}^{\kappa}, \\ S_{-\kappa,P} &= S_{-1,P}^{\kappa}, \\ S_{-\kappa,P} S_{\kappa,P} &= S_{-1,P}^{\kappa} S_{1,P}^{\kappa} = S_{0,P}, \end{aligned} \tag{8.6.12}$$

and, for all projectors $P, P' \in L(H)$ with $PP' = P'P = 0$ and all $\kappa, \kappa' \in \mathbb{Z}$,

$$S_{\kappa,P} S_{\kappa',P'} = 0. \tag{8.6.13}$$

From Lemma 8.4.5 we immediately obtain the following

8.6.5 Proposition. *Let D_+ be the open unit disc, let $D_- := \mathbb{C} \setminus \overline{D}_+$, let $A : \mathbb{T} \to L(H)$ be a continuous function, let $A_- : \overline{D}_- \cup \{\infty\} \to L(H)$ be a continuous function which is holomorphic in $D_- \cup \{\infty\}$, and let $A_+ : \overline{D}_+ \cup \{\infty\} \to L(H)$ be a continuous function which is holomorphic in D_+. Then*

(i) $T_{A_- A A_+} = T_{A_-} T_A T_{A_+}$.

(ii) *If $A_-(z)$ is invertible for all $z \in \overline{D}_- \cup \{\infty\}$, then T_{A_-} is invertible and*

$$T_{A_-}^{-1} = T_{A_-^{-1}}.$$

(iii) *If $A_+(z)$ is invertible for all $z \in \overline{D}_+$, then T_{A_+} is invertible and*

$$T_{A_+}^{-1} = T_{A_+^{-1}}.$$

We also have the following

8.6.6 Proposition. *Let $A : \mathbb{T} \to L(H)$ be an arbitrary continuous function such that $\mathcal{W}_A$ and T_A are Fredholm operators. (By definition of T_A it is clear that if one of them is Fredholm, then the other one is also Fredholm.) Then $A(z)$ is invertible for all $z \in \mathbb{T}$.*

Proof. Assume the contrary, i.e., assume that, for some $z_0 \in \mathbb{T}$, $A(z_0)$ is not invertible. This is equivalent to the statement that at least one of the operators $A(z_0)$ and $A^*(z_0)$ is not left invertible, where $A^*(z_0)$ is the Hilbert space adjoint $A(z_0)$.

Assume first that $A(z_0)$ is not left invertible.

By part (iii) of Theorem and Definition 8.3.10, the functions $z^n/\sqrt{2\pi}$, $n \in \mathbb{Z}$, form an orthonormal basis of $\mathcal{L}^2(\mathbb{T}, H)$. Let P_m, $m \in \mathbb{N}$, be the orthogonal projector from $\mathcal{L}^2_+(\mathbb{T}, H)$ onto the closed subspace of $\mathcal{L}^2_+(\mathbb{T}, H)$ spanned by the functions $z^n/\sqrt{2\pi}$, $n \geq m$.

As $\mathcal{W}_A$ is a Fredholm operator, we can find $k \in \mathbb{N}$ so large that

$$\alpha := \inf_{u \in \operatorname{Im} P_k,\, \|u\|_{\mathcal{L}^2(\mathbb{T},H)}=1} \|\mathcal{W}_A u\|_{\mathcal{L}^2(\mathbb{T},H)} > 0. \tag{8.6.14}$$

On the other hand, as $A(z_0)$ is not left invertible, we can find $v_0 \in H$ with

$$\|v_0\| = 1 \qquad \text{and} \qquad \big\|\big(A(z_0)\big)v_0\big\| < \frac{\alpha}{4}.$$

Choose $\delta > 0$ so small that

$$\frac{\delta}{2} < \int_{z\in\mathbb{T}, |z-z_0|<\delta} |dz| < 2\delta$$

and

$$\big\|\big(A(z)\big)v_0\big\| < \frac{\alpha}{4} \qquad \text{if } |z - z_0| < \delta,$$

and set

$$f(z) := \begin{cases} \frac{v_0}{\sqrt{\delta}} & \text{if } |\Phi^{-1}(z) - \lambda_0| < \delta, \\ 0 & \text{if } |\lambda - \lambda_0| \geq \delta. \end{cases}$$

Then

$$\|f\|_{\mathcal{L}^2(\mathbb{T},H)} > 1, \tag{8.6.15}$$

but

$$\|Af\|_{\mathcal{L}^2(\mathbb{T},H)} < \frac{\alpha}{2}. \tag{8.6.16}$$

As the functions $z^n/\sqrt{2\pi}$, $n \in \mathbb{Z}$, form an orthonormal basis of $\mathcal{L}^2(\mathbb{T}, H)$, we have

$$\lim_{n\to\infty} \big\|P_k(z^n f) - z^n f\big\|_{\mathcal{L}^2(\mathbb{T},H)} = 0. \tag{8.6.17}$$

By (8.6.15) this implies that

$$\lim_{n\to\infty} \|P_k(z^n f)\|_{\mathcal{L}^2(\mathbb{T},H)} = \lim_{n\to\infty} \|z^n f\|_{\mathcal{L}^2(\mathbb{T},H)} = \|f\|_{\mathcal{L}^2(\mathbb{T},H)} > 1. \tag{8.6.18}$$

On the other hand, by (8.6.17),

$$\begin{aligned}\lim_{n\to\infty}\left\|\mathcal{W}_A\Big(P_k(z^n f)\Big)\right\|_{\mathcal{L}^2(\mathbb{T},H)} &= \lim_{n\to\infty}\left\|\mathcal{W}_A(z^n f)\right\|_{\mathcal{L}^2(\mathbb{T},H)}\\ &= \lim_{n\to\infty}\left\|\mathcal{P}\Big(Az^n f\Big)\right\|_{\mathcal{L}^2(\mathbb{T},H)}.\end{aligned}$$

As the projector $\mathcal{P}$ is orthogonal and in view of (8.6.16), this further implies that

$$\lim_{n\to\infty}\left\|\mathcal{W}_A\Big(P_k(z^n f)\Big)\right\|_{\mathcal{L}^2(\mathbb{T},H)} \le \lim_{n\to\infty}\left\|Az^n f\Big)\right\|_{\mathcal{L}^2(\mathbb{T},H)}$$
$$= \left\|Af\right\|_{\mathcal{L}^2(\mathbb{T},H)} < \frac{\alpha}{2}.$$

Together with (8.6.18) this is a contradiction to (8.6.14).

Now we assume that $A^*(z_0)$ is not left invertible. Since $\mathcal{W}_{A^*}$ is the Hilbert space adjoint of $\mathcal{W}_A$, which is also a Fredholm operator (as $\mathcal{W}_A$ is a Fredholm operator), then, with A replaced by A^*, we get the same contradiction. □

We now can prove the following two theorems:

8.6.7 Theorem. *Let $A : \mathbb{T} \to GL(H)$ be a continuous function which admits local factorizations relative to $\mathbb{T}$ and $GL(H)$ (Def. 7.1.3). Then the following two conditions are equivalent:*

(i) *The Töplitz operator T_A defined by A is invertible.*

(ii) *A admits a canonical factorization relative to $\mathbb{T}$ and $GL(H)$ (Def. 7.1.1).*

If these two equivalent conditions are satisfied and $A = A_- A_+$ is a canonical factorization of A relative to $\mathbb{T}$ and $GL(H)$, then the inverse of T_A is given by

$$T_A^{-1} = T_{A_+^{-1}} T_{A_+^{-1}}. \tag{8.6.19}$$

Moreover, the block matrix of the operator $T_{A_+^{-1}}$ has the lower triangular form

$$\begin{pmatrix} \Gamma_0^+ & 0 & 0 & \cdots\\ \Gamma_1^+ & \Gamma_0^+ & 0 & \cdots\\ \Gamma_2^+ & \Gamma_1^+ & \Gamma_0^+ & \cdots\\ \vdots & \vdots & \vdots & \ddots \end{pmatrix} \quad \textit{with} \quad \Gamma_n^+ := \frac{1}{2\pi i}\int_{\mathbb{T}} \frac{A_+^{-1}(z)}{z^{n+1}}dz, \quad n\in\mathbb{N}, \tag{8.6.20}$$

the block matrix of the operator $T_{A_-^{-1}}$ has the upper triangular form

$$\begin{pmatrix} \Gamma_0^- & \Gamma_{-1}^- & \Gamma_{-2}^- & \cdots\\ 0 & \Gamma_0^- & \Gamma_{-1}^- & \cdots\\ 0 & 0 & \Gamma_0^- & \cdots\\ \vdots & \vdots & \vdots & \ddots \end{pmatrix} \quad \textit{with} \quad \Gamma_{-n}^- := \frac{1}{2\pi i}\int_{\mathbb{T}} \frac{A_-^{-1}(z)}{z^{-n+1}}dz, \quad n\in\mathbb{N}, \tag{8.6.21}$$

and, hence, the block matrix $\left(\Gamma_{jk}\right)_{j,k=0}^{\infty}$ *of* T_A^{-1} *can be computed by the finite sums*

$$\Gamma_{jk} := \begin{cases} \sum_{\nu=0}^{k} \Gamma_{j-\nu}^{+}\Gamma_{\nu-k}^{-}, & j \geq k, \\ \sum_{\nu=0}^{j} \Gamma_{j-\nu}^{+}\Gamma_{\nu-k}^{-}, & j \leq k. \end{cases} \tag{8.6.22}$$

8.6.8 Theorem. *Let* $A : \mathbb{T} \to GL(H)$ *be a continuous function which admits local factorizations relative to* $\mathbb{T}$ *and* $GL(H)$ *(Def. 7.1.3). Then the following two conditions are equivalent:*

(i) *The Töplitz operator* T_A *defined by* A *is a Fredholm operator.*

(ii) A *admits a factorization relative to* $\mathbb{T}$ *and* $GL(H)$ *(Def. 7.1.1).*

If these two equivalent conditions are satisfied, if

$$A(z) = A_-(z)\Big(P_0 + \sum_{j=1}^{n} z^{\kappa_j} P_j\Big)A_+(z), \qquad z \in \mathbb{T},$$

is a factorization of A *relative to* $\mathbb{T}$ *and* $GL(H)$*, and* r *is the index with* $\kappa_1 > \ldots > \kappa_r > 0 > \kappa_{r+1} > \ldots > \kappa_n$*, then*

$$\dim \operatorname{Ker} T_A = -\sum_{j=r+1}^{n} \kappa_j \dim P_j \quad \textit{and} \quad \dim \operatorname{Coker} T_A = \sum_{j=1}^{r} \kappa_j \dim P_j. \tag{8.6.23}$$

Moreover, if $\Delta(z) := P_0 + \sum_{j=1}^{n} z^{\kappa_j} P_j$*, then*

$$T_{A_+^{-1}} T_{\Delta^{-1}} T_{A_-^{-1}} \tag{8.6.24}$$

is a generalized inverse (Section 6.10.2*) of* T_A*, where, with the notations introduced in Section* 8.6.4,

$$T_{\Delta^{-1}} = S_{0,P_0} + \sum_{j=1}^{n} S_{-\kappa_j,P_j} = S_{0,P_0} + \sum_{j=1}^{r} S_{-1,j}^{\kappa_j} + \sum_{j=r+1}^{n} S_{1,P_j}^{-\kappa_j}. \tag{8.6.25}$$

8.6.9. Proof of Theorem 8.6.7. By (8.6.6) it is clear that the statement on the equivalence of conditions (i) and (ii) coincides with part (i) of Theorem 8.3.7.

Now let $A = A_-A_+$ be a canonical factorization of A relative to $\mathbb{T}$ and $GL(H)$. Then, by parts (ii) and (iii) of Proposition 8.6.5, the operators T_{A_+} and T_{A_-} are invertible, where $T_{A_+}^{-1} = T_{A_+^{-1}}$ and $T_{A_-}^{-1} = T_{A_-^{-1}}$. Since, by part (i) of this proposition, $T_A = T_{A_-}T_{A_+}$, this further implies that

$$T_A^{-1} = T_{A_+}^{-1} T_{A_-}^{-1} = T_{A_+^{-1}} T_{A_-^{-1}}.$$

Moreover, by (8.6.6), the block matrix of the operator $T_{A_+^{-1}}$ is given by

$$\begin{pmatrix} \Gamma_0^+ & \Gamma_{-1}^+ & \Gamma_{-2}^+ & \cdots \\ \Gamma_1^+ & \Gamma_0^+ & \Gamma_{-1}^+ & \cdots \\ \Gamma_2^+ & \Gamma_1^+ & \Gamma_0^+ & \cdots \\ \vdots & \vdots & \vdots & \ddots \end{pmatrix} \quad \text{with} \quad \Gamma_n^+ := \frac{1}{2\pi i}\int_{\mathbb{T}} \frac{A_+^{-1}(z)}{z^{n+1}}dz\,, \quad n \in \mathbb{Z},$$

and the block matrix of the operator $T_{A_-^{-1}}$ is given by

$$\begin{pmatrix} \Gamma_0^- & \Gamma_{-1}^- & \Gamma_{-2}^- & \cdots \\ \Gamma_1^- & \Gamma_0^- & \Gamma_{-1}^- & \cdots \\ \Gamma_2^- & \Gamma_1^- & \Gamma_0^- & \cdots \\ \vdots & \vdots & \vdots & \ddots \end{pmatrix} \quad \text{with} \quad \Gamma_n^- := \frac{1}{2\pi i}\int_{\mathbb{T}} \frac{A_-^{-1}(z)}{z^{n+1}}dz\,, \quad n \in \mathbb{Z}.$$

As

$$\Gamma_n^+ = \frac{1}{2\pi i}\int_{\mathbb{T}} \frac{A_+^{-1}(z)}{z^{n+1}}dz = 0 \qquad \text{if } n < 0,$$

and

$$\Gamma_n^- = \frac{1}{2\pi i}\int_{\mathbb{T}} \frac{A_-^{-1}(z)}{z^{n+1}}dz = 0 \qquad \text{if } n > 0,$$

this implies (8.6.20) and (8.6.21). □

8.6.10. Proof of Theorem 8.6.8. Again in view of (8.6.6), the statement on the equivalence of conditions (i) and (ii) coincides with part (ii) of Theorem 8.3.7.

Now let these two equivalent conditions be satisfied, let

$$A(z) = A_-(z)\Big(P_0 + \sum_{j=1}^{n} z^{\kappa_j} P_j\Big)A_+(z), \qquad z \in \mathbb{T},$$

be a factorization of A relative to $\mathbb{T}$ and $GL(H)$, and let r be the index with $\kappa_1 > \ldots > \kappa_r > 0 > \kappa_{r+1} > \ldots > \kappa_n$.

Then (8.6.23) follows from part (ii) of Theorem 8.3.7.

From part (i) of proposition (8.6.5) it follows that

$$T_A = T_{A_-} T_\Delta T_{A_+}. \tag{8.6.26}$$

By definition (see Section 8.6.4), we have

$$T_\Delta = S_{0,P_0} + \sum_{j=1}^{n} S_{\kappa_j,P_j} \quad \text{and} \quad T_{\Delta^{-1}} = S_{0,P_0} + \sum_{j=1}^{n} S_{-\kappa_j,P_j}. \tag{8.6.27}$$

By (8.6.13) and (8.6.12) this yields

$$T_\Delta T_{\Delta^{-1}} T_\Delta = S_{0,P_0} + \sum_{j=1}^{n} S_{\kappa_j,P_j} S_{-\kappa_j,P_j} S_{\kappa_j,P_j} = S_{0,P_0} + \sum_{j=1}^{n} S_{\kappa_j,P_j} = T_\Delta.$$

In the same way we get the relation

$$T_{\Delta^{-1}} T_{\Delta} T_{\Delta^{-1}} = T_{\Delta^{-1}}.$$

Hence $T_{\Delta^{-1}}$ is a generalized inverse of T_{Δ}. Taking into account (8.6.26) and the fact that, by parts (ii) and (iii) of proposition (8.6.5), $T_{A_+}^{-1} = T_{A_+^{-1}}$ and $T_{A_-}^{-1} = T_{A_-^{-1}}$, this implies that the operator (8.6.24) is a generalized inverse of T_A.

The first equality in (8.6.25) holds by (8.6.27) and the second follows from (8.6.12). □

8.7 The Fourier transform of $\mathcal{L}^1(\mathbb{R}, E)$

8.7.1. The space $\mathcal{L}^p(\mathbb{R}, E)$. Let E be a Banach space, and let $1 \le p < \infty$. We denote by $C_0^0(\mathbb{R}, E)$ the complex linear space of continuous functions $f : \mathbb{R} \to E$ with compact support. We introduce a norm $\| \cdot \|_{\mathcal{L}^p(\mathbb{R},E)}$ on $C_0^0(\mathbb{R}, E)$, setting

$$\|f\|^p_{\mathcal{L}^p(\mathbb{R},A)} = \int_{-\infty}^{\infty} \|f(x)\|^p \, dx, \qquad f \in C_0^0(\mathbb{R}, E),$$

and we denote by $\mathcal{L}^p(\mathbb{R}, E)$ the completion of $C_0^0(\mathbb{R}, E)$ with respect to this norm. For $E = \mathbb{C}$ this is the usual space of scalar $\mathcal{L}^p$-functions on $\mathbb{R}$. In simple cases, also for general Banach spaces E, we identify the elements of $\mathcal{L}^p(\mathbb{R}, E)$ with E-valued functions. For example, each piecewise continuous function $f : \mathbb{R} \to E$ with $\int_{-\infty}^{\infty} \|F(x)\|^p dx < \infty$ will be viewed (in the obvious way) as an element of $\mathcal{L}^p(\mathbb{R}, E)$. Note that then

$$\|f\|^p_{\mathcal{L}^p(\mathbb{R},A)} = \int_{-\infty}^{\infty} \|f(x)\|^p \, dx.$$

Also the functions of the form $\varphi \cdot v$, where v is a fixed vector in E and $\varphi \in \mathcal{L}^p(\mathbb{R}, \mathbb{C})$, will be identified with the corresponding element in $\mathcal{L}^p(\mathbb{R}, E)$.

In general however, we view the elements of $\mathcal{L}^p(\mathbb{R}, E)$ as equivalence classes of Cauchy sequences rather than true functions (although the latter is also possible).

Nevertheless, for each element $f \in \mathcal{L}^1(\mathbb{R}, E)$, we define the integral

$$\int_{-\infty}^{\infty} f(x) \, dx \tag{8.7.1}$$

as follows: Take a Cauchy sequence $f_n \in C_0^0(\mathbb{R}, E)$ defining f. Then

$$\lim_{n \to \infty} \int_{-\infty}^{\infty} f_n(x) \, dx$$

exists and is independent of the choice of the Cauchy sequence. We define the integral (8.7.1) to be this limit.

8.7.2. Convolution. Let $\mathfrak{A}$ be a Banach algebra. For two functions $K, L \in \mathcal{L}^1(\mathbb{R}, \mathfrak{A})$, we introduce a product $K * L \in \mathcal{L}^1(\mathbb{R}, \mathfrak{A})$, called the **convolution** of K and L. If K and L are piecewise continuous, this is defined (as usual) by

$$(K * L)(y) := \int_{-\infty}^{\infty} K(x)L(y-x)dx = \int_{-\infty}^{\infty} K(y-x)L(x)dx, \qquad y \in \mathbb{R}.$$

By Fubini's theorem, for such functions we have

$$\begin{aligned}
\|K * L\|_{\mathcal{L}^1(\mathbb{R},\mathfrak{A})} &= \int_{-\infty}^{\infty} \left\| \int_{-\infty}^{\infty} K(y-x)L(x)dx \right\| dy \\
&\leq \int_{-\infty}^{\infty} \int_{-\infty}^{\infty} \|K(y-x)\| \, \|L(x)\| dx\, dy \\
&= \int_{-\infty}^{\infty} \left(\int_{-\infty}^{\infty} \|K(y-x)\| \, dy \right) \|L(x)\| \, dx \\
&= \int_{-\infty}^{\infty} \left(\int_{-\infty}^{\infty} \|K(y)\| \, dy \right) \|L(x)\| \, dx = \|K\|_{\mathcal{L}^1(\mathbb{R},\mathfrak{A})} \|L\|_{\mathcal{L}^1(\mathbb{R},\mathfrak{A})}.
\end{aligned}$$

Therefore, we can extend the convolution by continuity to all of $\mathcal{L}^1(\mathbb{R}, \mathfrak{A})$, and then

$$\|K * L\|_{\mathcal{L}^1(\mathbb{R},\mathfrak{A})} \leq \|K\|_{\mathcal{L}^1(\mathbb{R},\mathfrak{A})} \|L\|_{\mathcal{L}^1(\mathbb{R},\mathfrak{A})} \tag{8.7.2}$$

for all $K, L \in \mathcal{L}^1(\mathbb{R}, \mathfrak{A})$. Hence $\mathcal{L}^1(\mathbb{R}, \mathfrak{A})$ is an algebra with respect to convolution.

8.7.3. The spaces $\mathcal{L}^p_\pm(\mathbb{R}, E)$. Let E be a Banach space, let $1 \leq p < \infty$, and let $(\mathcal{C}^0_0)_-(\mathbb{R}, E)$ be the subspace of $\mathcal{C}^0_0(\mathbb{R}, E)$ which consists of the functions with support in $]-\infty, 0]$, and let $(\mathcal{C}^0_0)_+(\mathbb{R}, E)$ be the subspace of $\mathcal{C}^0_0(\mathbb{R}, E)$ which consists of the functions with support in $[0, \infty[$.

Then we denote by $\mathcal{L}^p_-(\mathbb{R}, E)$ and $\mathcal{L}^p_+(\mathbb{R}, E)$ the closures of $(\mathcal{C}^0_0)_-(\mathbb{R}, E)$ and $(\mathcal{C}^0_0)_+(\mathbb{R}, E)$ in $\mathcal{L}^p(\mathbb{R}, E)$, respectively. Since $\mathcal{L}^p_-(\mathbb{R}, E) \cap \mathcal{L}^1_+(\mathbb{R}, E) = \{0\}$ and the functions of the form $f_+ + f_-$ with $f_\pm \in (\mathcal{C}^0_0)_\pm(\mathbb{R}, E)$ are dense in $\mathcal{C}^0_0(\mathbb{R}, E)$ with respect to the norm $\|\cdot\|_{\mathcal{L}^p(\mathbb{R},E)}$, we see that $\mathcal{L}^p(\mathbb{R}, E)$ splits into the direct sum

$$\mathcal{L}^p(\mathbb{R}, E) = \mathcal{L}^p_-(\mathbb{R}, E) \oplus \mathcal{L}^p_+(\mathbb{R}, E). \tag{8.7.3}$$

If $\mathfrak{A}$ is a Banach algebra, then, obviously,

$$\begin{aligned}
f * g &\in \mathcal{L}^1_+(\mathbb{R}, \mathfrak{A}) \qquad \text{if } f, g \in \mathcal{L}^1_+(\mathbb{R}, \mathfrak{A}), \\
f * g &\in \mathcal{L}^1_-(\mathbb{R}, \mathfrak{A}) \qquad \text{if } f, g \in \mathcal{L}^1_-(\mathbb{R}, \mathfrak{A}),
\end{aligned} \tag{8.7.4}$$

i.e., $\mathcal{L}^1_+(\mathbb{R}, \mathfrak{A})$ and $\mathcal{L}^1_-(\mathbb{R}, \mathfrak{A})$ are subalgebras of the algebra $\mathcal{L}^1(\mathbb{R}, \mathfrak{A})$ with respect to covolution.

8.7.4. The Fourier transform of $\mathcal{L}^1(\mathbb{R}, E)$. Let E be a Banach space. For functions $f \in C_0^0(\mathbb{R}, E)$, we define by

$$\widehat{f}(\lambda) = \int_{-\infty}^{\infty} e^{i\lambda x} f(x)\, dx\,, \qquad \lambda \in \mathbb{C}, \tag{8.7.5}$$

the Fourier transform of f. Differentiating under the sign of integration, we see that $\widehat{f}$ is holomorphic on $\mathbb{C}$ for all $f \in C_0^0(\mathbb{R}, E)$. Moreover, if $f \in C_0^0(\mathbb{R}, E)$ and $\lambda \in \mathbb{R}$, then

$$\|\widehat{f}(\lambda)\| \le \int_{-\infty}^{\infty} |e^{i\lambda x}| \|f(x)\|\, dx = \|f\|_{\mathcal{L}^1(\mathbb{R},E)}.$$

Therefore, the Fourier transformation extends by continuity to all of $\mathcal{L}^1(\mathbb{R}, E)$, where, for all elements $f \in \mathcal{L}^1(\mathbb{R}, E)$, $\widehat{f}$ is a continuous E-valued function on $\mathbb{R}$, and

$$\sup_{\lambda \in \mathbb{R}} \|\widehat{f}(\lambda)\| \le \|f\|_{\mathcal{L}^1(\mathbb{R},E)} \qquad \text{for all } f \in f \in \mathcal{L}^1(\mathbb{R}, E). \tag{8.7.6}$$

We denote by $\widehat{\mathcal{L}}^1(\mathbb{R}, E)$, $\widehat{\mathcal{L}}^1_+(\mathbb{R}, E)$ and $\widehat{\mathcal{L}}^1_-(\mathbb{R}, E)$ the spaces of all continuous functions $\varphi : \mathbb{R} \to E$ of the form $\varphi = \widehat{f}$ with $f \in \mathcal{L}^1(\mathbb{R}, E)$, $f \in \mathcal{L}^1_+(\mathbb{R}, E)$ and $f \in \mathcal{L}^1_-(\mathbb{R}, E)$, respectively. Note that

$$\widehat{f} \not\equiv 0 \qquad \text{if } f \in \mathcal{L}^1(\mathbb{R}, E) \text{ and } f \not\equiv 0. \tag{8.7.7}$$

This can be deduced by the Hahn-Banach theorem from the scalar case. Therefore the Fourier transform is a linear isomorphism from $\mathcal{L}^1(\mathbb{R}, E)$ onto $\widehat{\mathcal{L}}^1(\mathbb{R}, E)$, and we can introduce a norm in $\widehat{\mathcal{L}}^1(\mathbb{R}, E)$, setting

$$\|\widehat{f}\|_{\widehat{\mathcal{L}}^1(\mathbb{R},E)} = \|f\|_{\mathcal{L}^1(\mathbb{R},E)} \qquad \text{for } f \in \mathcal{L}^1(\mathbb{R}, E). \tag{8.7.8}$$

Note that then (8.7.6) takes the form

$$\sup_{\lambda \in \mathbb{R}} \|\varphi(\lambda)\| \le \|\varphi\|_{\widehat{\mathcal{L}}^1(\mathbb{R},E)} \qquad \text{for all } \varphi \in \widehat{\mathcal{L}}^1(\mathbb{R}, E). \tag{8.7.9}$$

As $\mathcal{L}^1(\mathbb{R}, E) = \mathcal{L}^1_+(\mathbb{R}, E) \oplus \mathcal{L}^1_-(\mathbb{R}, E)$, the space $\mathcal{L}^1(\mathbb{R}, E)$ splits into the direct sum

$$\widehat{\mathcal{L}}^1(\mathbb{R}, E) = \widehat{\mathcal{L}}^1_+(\mathbb{R}, E) \oplus \mathcal{L}^1_-(\mathbb{R}, E). \tag{8.7.10}$$

8.7.5. Now we consider again a Banach algebra $\mathfrak{A}$.

Then for $K, L \in C_0^0(\mathbb{R}, \mathfrak{A})$, it follows from Fubini's theorem that

$$\begin{aligned}
(\widehat{K * L})(\lambda) &= \int_{-\infty}^{\infty} e^{i\lambda y} \Big(\int_{-\infty}^{\infty} K(y - x) L(x) dx \Big) dy \\
&= \int_{-\infty}^{\infty} \Big(\int_{-\infty}^{\infty} e^{i\lambda(y-x)} K(y - x)\, dy \Big) e^{i\lambda x} L(x) dx \qquad (8.7.11) \\
&= \int_{-\infty}^{\infty} \Big(\int_{-\infty}^{\infty} e^{i\lambda(y)} K(y)\, dy \Big) e^{i\lambda x} L(x) dx = \widehat{K}(\lambda) \widehat{L}(\lambda)
\end{aligned}$$

for all $\lambda \in \mathbb{C}$. Since the convolution is continuous with respect to the $\mathcal{L}^1$-norm (inequality (8.7.2)) and in view of estimate (8.7.6), this implies that

$$(\widehat{K * L})(\lambda) = \widehat{K}(\lambda)\widehat{L}(\lambda) \tag{8.7.12}$$

for all $K, L \in \mathcal{L}^1(\mathbb{R}, \mathfrak{A})$ and $\lambda \in \mathbb{R}$. Hence, for all $K, L \in \widehat{\mathcal{L}}^1(\mathbb{R}, \mathfrak{A})$, the pointwise defined product AL belongs again to $\widehat{\mathcal{L}}^1(\mathbb{R}, \mathfrak{A})$, and it follows from (8.7.2) and the definition of the norm in $\widehat{\mathcal{L}}^1(\mathbb{R}, \mathfrak{A})$ that

$$\|KL\|_{\widehat{\mathcal{L}}^1(\mathbb{R},\mathfrak{A})} \leq \|K\|_{\widehat{\mathcal{L}}^1(\mathbb{R},\mathfrak{A})}\|L\|_{\widehat{\mathcal{L}}^1(\mathbb{R},\mathfrak{A})} \tag{8.7.13}$$

for all $K, L \in \widehat{\mathcal{L}}^1(\mathbb{R}, \mathfrak{A})$. Hence $\widehat{\mathcal{L}}^1(\mathbb{R}, \mathfrak{A})$ is a Banach algebra, and, as $\mathcal{L}^1_+(\mathbb{R}, \mathfrak{A})$ and $\mathcal{L}^1_-(\mathbb{R}, \mathfrak{A})$ are subalgebras of $\mathcal{L}^1(\mathbb{R}, \mathfrak{A})$ with respect to convolution, $\widehat{\mathcal{L}}^1_+(\mathbb{R}, \mathfrak{A})$ and $\widehat{\mathcal{L}}^1_-(\mathbb{R}, \mathfrak{A})$ are subalgebras of $\widehat{\mathcal{L}}^1(\mathbb{R}, \mathfrak{A})$.

8.7.6. We denote by $\mathbb{H}_+$ the upper open half plane, and by $\mathbb{H}_-$ the lower open half plane:

$$\mathbb{H}_+ := \left\{ z \in \mathbb{C} \;\middle|\; \operatorname{Im} z > 0 \right\}, \qquad \mathbb{H}_- := \left\{ z \in \mathbb{C} \;\middle|\; \operatorname{Im} z < 0 \right\}.$$

The closures of $\mathbb{H}_+$ and $\mathbb{H}_-$ in $\mathbb{C}$ (and not in the Riemann sphere) will be denoted by $\overline{\mathbb{H}}_+$ and $\overline{\mathbb{H}}_-$. If E is a Banach space, then a function

$$f : \overline{\mathbb{H}}_+ \cup \{\infty\} \to E$$

will be called **continuous** on $\overline{\mathbb{H}}_+ \cup \{\infty\}$ if f is continuous on $\overline{\mathbb{H}}_+$ and $f(1/z)$ is also continuous on $\overline{\mathbb{H}}_+$. In the same way we define what it means that a function defined on $\overline{\mathbb{H}}_- \cup \{\infty\}$ or on $\mathbb{R} \cup \{\infty\}$ is continuous.

8.7.7 Theorem. *Let E be a Banach space.*

(i) *A function $\varphi \in \widehat{\mathcal{L}}^1(\mathbb{R}, E)$ belongs to $\widehat{\mathcal{L}}^1_+(\mathbb{R}, E)$, if and only if it admits a continuous extension to $\overline{\mathbb{H}}_+ \cup \{\infty\}$ which is holomorphic in $\mathbb{H}_+$ and vanishes at infinity. Moreover*

$$\max_{\lambda \in \overline{\mathbb{H}}_+ \cup \{\infty\}} \|\varphi(\lambda)\| \leq \|\varphi\|_{\widehat{\mathcal{L}}^1(\mathbb{R},E)} \qquad \textit{for each } \varphi \in \widehat{\mathcal{L}}^1_+(\mathbb{R}, E).$$

(ii) *A function $\varphi \in \widehat{\mathcal{L}}^1(\mathbb{R}, E)$ belongs to $\widehat{\mathcal{L}}^1_-(\mathbb{R}, E)$, if and only if it admits a continuous extension to $\overline{\mathbb{H}}_- \cup \{\infty\}$ which is holomorphic in $\mathbb{H}_-$ and vanishes at infinity. Moreover*

$$\max_{\lambda \in \overline{\mathbb{H}}_- \cup \{\infty\}} \|\varphi(\lambda)\| \leq \|\varphi\|_{\widehat{\mathcal{L}}^1(\mathbb{R},E)} \qquad \textit{for each } \varphi \in \widehat{\mathcal{L}}^1_-(\mathbb{R}, E). \tag{8.7.14}$$

(iii) *Each function* $\varphi \in \widehat{\mathcal{L}}^1(\mathbb{R}, E)$ *is continuous on* $\mathbb{R} \cup \{\infty\}$ *and* $\varphi(\infty) = 0$. *Moreover*

$$\max_{\lambda \in \mathbb{R}\cup\{\infty\}} \|\varphi(\lambda)\| \leq \|\varphi\|_{\widehat{\mathcal{L}}^1(\mathbb{R},E)} \qquad \text{for each } \varphi \in \widehat{\mathcal{L}}^1(\mathbb{R}, E).$$

Proof. As $\mathcal{L}^1(\mathbb{R}, E) = \mathcal{L}^1_-(\mathbb{R}, E) \oplus \mathcal{L}^1_+(\mathbb{R}, E)$, part (iii) follows from parts (i) and (ii) and estimate (8.7.6). By the same reason, for the proof of (i) and (ii), it is sufficient to prove the "only if" parts of (i) and (ii).

Indeed, assume that this is proved.

Let $\varphi_+ \in \widehat{\mathcal{L}}^1(\mathbb{R}, E)$ be a function which admits a continuous extension to $\overline{\mathbb{H}}_+ \cup \{\infty\}$, which is holomorphic in $\mathbb{H}_+$ and vanishes at infinity. Then it follows from the decomposition $\mathcal{L}^1(\mathbb{R}, E) = \mathcal{L}^1_-(\mathbb{R}, E) \oplus \mathcal{L}^1_+(\mathbb{R}, E)$ that

$$\varphi_+ = \widetilde{\varphi}_- + \widetilde{\varphi}_+ \qquad \text{on } \mathbb{R} \cup \{\infty\}$$

with $\widetilde{\varphi}_\pm \in \widehat{\mathcal{L}}^1_\pm(\mathbb{R}, E)$. Then

$$\varphi_+ - \widetilde{\varphi}_+ = \widetilde{\varphi}_- \qquad \text{on } \mathbb{R} \cup \{\infty\}.$$

The two sides of this equation define a holomorphic function on $\mathbb{C} \cup \{\infty\}$ vanishing at infinity (see Theorem 1.5.4). By Liouville's theorem this implies that $\varphi_+ = \widetilde{\varphi}_+ \in \widehat{\mathcal{L}}^1_+(\mathbb{R}, E)$.

In the same way, one proves that each $\varphi_- \in \widehat{\mathcal{L}}^1(\mathbb{R}, E)$, which admits a continuous extension to $\overline{\mathbb{H}}_- \cup \{\infty\}$, which is holomorphic in $\mathbb{H}_-$ and vanishes at infinity, belongs to $\mathcal{L}^1_-(\mathbb{R}, E)$.

The proofs of the "only if" parts of (i) and (ii) are similar. We therefore restrict ourselves to part (i). First we prove the following weaker statement:

(i′) Each function $\varphi \in \widehat{\mathcal{L}}^1_+(\mathbb{R}, E)$ admits a continuous extension to $\overline{\mathbb{H}}_+ \cup \{\infty\}$ which is holomorphic in $\mathbb{H}_+$ and which satisfies the estimate

$$\sup_{\lambda \in \overline{\mathbb{H}}_+} \|\varphi(\lambda)\| \leq \|\varphi\|_{\widehat{\mathcal{L}}^1(\mathbb{R},E)}. \tag{8.7.15}$$

To prove (i′), we first note that, for each $u \in \left(\mathcal{C}^0_0\right)_+(\mathbb{R}, E)$ and $\lambda \in \overline{\mathbb{H}}_+$,

$$\|\widehat{u}(\lambda)\| \leq \int_0^\infty |e^{i\lambda x}| \|u(x)\| \, dx = \int_0^\infty |e^{-x \operatorname{Im} \lambda}| \|u(x)\| \, dx \leq \|u\|_{\mathcal{L}^1(\mathbb{R},A)}. \tag{8.7.16}$$

Now let an arbitrary function $\varphi \in \widehat{\mathcal{L}}^1_+(\mathbb{R}, E)$ be given, and let $f \in \mathcal{L}^1_+(\mathbb{R}, E)$ be the element with $\varphi = \widehat{f}$. Choose a sequence $u_\nu \in \left(\mathcal{C}^0_0\right)_+(\mathbb{R}, E)$ with

$$\lim_{\mu \to \infty} \|f - u_\mu\|_{\mathcal{L}^1(\mathbb{R},E)} = 0. \tag{8.7.17}$$

Then

$$\lim_{\mu,\nu\to\infty} \|u_\mu - u_\nu\|_{\mathcal{L}^1(\mathbb{R},E)} = 0. \tag{8.7.18}$$

Recall that $\|\varphi - \widehat{u}_\mu\|_{\widehat{\mathcal{L}}^1(\mathbb{R},E)} = \|f - u_\mu\|_{\mathcal{L}^1(\mathbb{R},E)}$, by definition of the norm in $\widehat{\mathcal{L}}^1(\mathbb{R},E)$. Therefore it follows from (8.7.17) and (8.7.9) that

$$\lim_{\mu\to\infty} \sup_{\lambda\in\mathbb{R}} \|\varphi(\lambda) - \widehat{u}_\mu(\lambda)\| = 0. \tag{8.7.19}$$

From (8.7.18) and (8.7.16) it follows that

$$\lim_{\mu,\nu\to\infty} \sup_{\lambda\in\overline{\mathbb{H}}_+} \|\widehat{u}_\mu(\lambda) - \widehat{u}_\nu(\lambda)\| = 0.$$

Since the functions $\widehat{u}_\mu$ are holomorphic on $\mathbb{C}$, this implies that the sequence $\widehat{u}_\nu$ converges uniformly on $\overline{\mathbb{H}}_+$ to some continuous function which is holomorphic in $\mathbb{H}_+$. By (8.7.19), on $\mathbb{R}$, this function coincides with φ. So it is proved that φ admits a continuous extension to $\overline{\mathbb{H}}_+$ which is holomorphic in $\mathbb{H}_+$. We denote this extension also by φ. Moreover, since, again by (8.7.16)

$$\sup_{\lambda\in\overline{\mathbb{H}}_+} \|\widehat{u}_\mu(\lambda)\| \le \|u_\mu\|_{\mathcal{L}^1(\mathbb{R},A)},$$

we get also (8.7.17). The proof of statement (i′) is complete.

Now we consider a function u of the form

$$u(x) = \begin{cases} a & \text{if } x \in [\alpha,\beta], \\ 0 & \text{if } x \notin [\alpha,\beta], \end{cases} \tag{8.7.20}$$

where $0 \le \alpha < \beta < \infty$ and $a \in E$. Then, it follows immediately from the definition of the Fourier transformation that, for all $\lambda \in \mathbb{R}$ with $\lambda \ne 0$,

$$\widehat{u}(\lambda) = a\int_\alpha^\beta e^{i\lambda x}dx = \frac{e^{i\lambda x}}{i\lambda}\Big|_{x=\alpha}^{x=\beta} = a\frac{e^{i\lambda\beta} - e^{i\lambda\alpha}}{i\lambda}.$$

Since $\alpha \ge 0$, this formula shows that $\widehat{u}$ admits a continuous extension to $\mathbb{C}$, which can be expressed by the same formula for all $\lambda \in \mathbb{C}$ with $\lambda \ne 0$. For $\lambda \in \overline{\mathbb{H}}_+$ with $\lambda \ne 0$, this implies that

$$\|\widehat{u}_+(\lambda)\| \le \|a\|\frac{e^{-\beta\,\mathrm{Im}\,\lambda} + e^{-\alpha\,\mathrm{Im}\,\lambda}}{|\lambda|} \le \|a\|\frac{2}{|\lambda|}.$$

Hence

$$\lim_{R\to\infty} \sup_{\lambda\in\overline{\mathbb{H}}_+,|\lambda|\ge R} \|\widehat{u}_+(\lambda)\| = 0 \tag{8.7.21}$$

for each function u of the form (8.7.20). Now let $\mathcal{M}$ be the space of all finite sums of functions of the form (8.7.20). Then, for each $u \in \mathcal{M}$, $\widehat{u}$ is holomorphic on $\mathbb{C}$, and it follows from (8.7.21) that

$$\lim_{R\to\infty} \sup_{\lambda\in\overline{\mathbb{H}}_+, |\lambda|\geq R} \|\widehat{u}(\lambda)\| = 0 \qquad \text{for each } u \in \mathcal{M}. \tag{8.7.22}$$

Finally let an arbitrary function $\varphi \in \widehat{\mathcal{L}}^1_+(\mathbb{R}, E)$ be given. By statement (i′) we already know that φ admits a continuous extension to $\overline{\mathbb{H}}_+$ which is holomorphic in $\mathbb{H}_+$ and which satisfies the estimate (8.7.15). It remains to prove that

$$\lim_{R\to\infty} \sup_{\lambda\in\overline{\mathbb{H}}_+, |\lambda|\geq R} \|\varphi(\lambda)\| = 0. \tag{8.7.23}$$

Let $f \in \mathcal{L}^1_+(\mathbb{R}, E)$ be the function with $\varphi = \widehat{f}$. Since $\mathcal{M}$ is dense in $\mathcal{L}^1_+(\mathbb{R}, E)$, then we can find a sequence $u_\nu \in \mathcal{M}$ with $\lim\limits_{\nu\to\infty} \|u_\nu - f\|_{\mathcal{L}^1(\mathbb{R},A)} = 0$. As

$$\|\widehat{u}_\nu - \varphi\|_{\widehat{\mathcal{L}}^1(\mathbb{R},E)} = \|u_\nu - f\|_{\mathcal{L}^1(\mathbb{R},E)},$$

then it follows from (8.7.15) that

$$\lim_{\nu\to\infty} \sup_{\lambda\in\overline{\mathbb{H}}_+} \|\widehat{u}_\nu(\lambda) - \varphi(\lambda)\| = 0,$$

and from (8.7.22) we get

$$\lim_{R\to\infty} \sup_{\lambda\in\overline{\mathbb{H}}_+, |\lambda|\geq R} \|\widehat{u}_\nu(\lambda)\| = 0 \quad \text{for all } \nu.$$

Together this implies (8.7.23). □

We conclude this section by computing the Fourier transform of some special functions, which we need in sections 8.9 and 8.10 below.

8.7.8. For all $n \in \mathbb{N}$ we define

$$\theta_n^+(x) = \begin{cases} x^n e^{-x} & \text{for } x \geq 0, \\ 0 & \text{for } x < 0, \end{cases} \quad \text{and} \quad \theta_n^-(x) = \begin{cases} -x^n e^{x} & \text{for } x \leq 0, \\ 0 & \text{for } x > 0. \end{cases} \tag{8.7.24}$$

Note that $\theta_n^+ \in \mathcal{L}^1_+(\mathbb{R}, \mathbb{C}) \cap L^2(\mathbb{R}, \mathbb{C})$ and $\theta_n^- \in \mathcal{L}^1_-(\mathbb{R}, \mathbb{C}) \cap L^2(\mathbb{R}, \mathbb{C})$.

8.7.9 Lemma.

$$\theta_n^+ = (n+1)!\, \underbrace{\theta_0^+ * \ldots * \theta_0^+}_{n+1 \text{ times}} \ \textit{and}\ \theta_n^- = (n+1)!\, \underbrace{\theta_0^- * \ldots * \theta_0^-}_{n+1 \text{ times}}, \quad n \in \mathbb{N}^*.$$

Proof. We have for all $n \in \mathbb{N}$,

$$\begin{aligned}(\theta_n^+ * \theta_0^+)(y) &= \int_{-\infty}^{\infty} \theta_0^+(y-x)\theta_n^+(x)dx = \int_0^y e^{x-y}x^n e^{-x}dx \\ &= e^{-y}\int_0^y x^n dx = e^{-y}\frac{y^{n+1}}{n+1} = \frac{1}{n+1}\theta_{n+1}^+(y),\end{aligned}$$

and

$$\begin{aligned}(\theta_n^- * \theta_0^-)(y) &= \int_{-\infty}^{\infty} \theta_0^-(y-x)\theta_n^-(x)dx = \int_y^0 e^{y-x}x^n e^{x}dx \\ &= e^{y}\int_y^0 x^n dx = -e^{y}\frac{y^{n+1}}{n+1} = \frac{1}{n+1}\theta_{n+1}^-(y),\end{aligned}$$

i.e., $\theta_{n+1}^+ = (n+1)\theta_n^+ * \theta_0^+$ and $\theta_{n+1}^- = (n+1)\theta_n^- * \theta_0^-$ which implies the assertion by induction. □

8.7.10. If $\xi \in \mathbb{H}_+$, then $\operatorname{Re} i\xi < 0$ and therefore, setting

$$\vartheta_\xi^-(x) = \begin{cases} ie^{-i\xi x}, & \text{if } x \le 0, \\ 0 & \text{if } x > 0, \end{cases}$$

then we get a function $\vartheta_\xi^- \in \mathcal{L}_-^1(\mathbb{R}, \mathbb{C})$. If $\xi \in \mathbb{H}_-$, then $\operatorname{Re} i\xi > 0$ and therefore, setting

$$\vartheta_\xi^+(x) = \begin{cases} -ie^{-i\xi x}, & \text{if } x \ge 0, \\ 0 & \text{if } x < 0, \end{cases}$$

then we get a function $\vartheta_\xi^+ \in \mathcal{L}_+^1(\mathbb{R}, \mathbb{C})$. Note that

$$\theta_0^+ = i\vartheta_{-i}^+ \qquad \text{and} \qquad \theta_0^- = i\vartheta_i^-, \tag{8.7.25}$$

where $\theta_0^\pm$ are the functions introduced in Section 8.7.8.

8.7.11 Lemma. (i) *We have*

$$\frac{1}{\lambda - \xi} = \begin{cases} \widehat{\vartheta_\xi^-}(\lambda) & \text{if } \xi \in \mathbb{H}_+, \\ \widehat{\vartheta_\xi^+}(\lambda) & \text{if } \xi \in \mathbb{H}_-, \end{cases} \qquad \lambda \in \mathbb{R}.$$

(ii) *For all* $n \in \mathbb{N}^*$, *we have*

$$\begin{aligned}\left(\frac{\lambda - i}{\lambda + i} - 1\right)^n &= \frac{(-2)^n}{n!}\widehat{\theta}_{n-1}^+(\lambda), \qquad \lambda \in \mathbb{R}, \\ \left(\frac{\lambda + i}{\lambda - i} - 1\right)^n &= \frac{2^n}{n!}\widehat{\theta}_{n-1}^-(\lambda), \qquad \lambda \in \mathbb{R}.\end{aligned}$$

Proof. If $\xi \in \mathbb{H}_+$, then

$$\widehat{\vartheta}_\xi^-(\lambda) = \int_{-\infty}^{\infty} e^{i\lambda x}\vartheta_\xi^-(x)dx = i\int_{-\infty}^{0} e^{(i\lambda - i\xi)x}dx = i\frac{e^{(i\lambda-i\xi)x}}{i\lambda - i\xi}\bigg|_{x=-\infty}^{x=0} = \frac{1}{\lambda-\xi},$$

and if $\xi \in \mathbb{H}_-$, then

$$\widehat{\vartheta}_\xi^+(\lambda) = \int_{-\infty}^{\infty} e^{i\lambda x}\vartheta_\xi^+(x)dx = -i\int_{0}^{\infty} e^{(i\lambda - i\xi)x}dx = -i\frac{e^{(i\lambda-i\xi)x}}{i\lambda - i\xi}\bigg|_{x=0}^{x=\infty} = \frac{1}{\lambda-\xi}.$$

This proves part (i). Setting $\xi = \pm i$ in part (i), we get

$$\frac{i}{\lambda+i} = \widehat{\theta}_0^+(\lambda) \qquad \text{and} \qquad \frac{i}{\lambda-i} = \widehat{\theta}_0^-(\lambda).$$

Hence

$$\frac{\lambda-i}{\lambda+i} - 1 = -\frac{2i}{\lambda+i} = -2\widehat{\theta}_0^+(\lambda) \quad \text{and} \quad \frac{\lambda+i}{\lambda-i} - 1 = \frac{2i}{\lambda-i} = 2\widehat{\theta}_0^-(\lambda).$$

By Lemma 8.7.9 this yields part (ii). □

8.8 The Fourier isometry U of $\mathcal{L}^2(\mathbb{R}, H)$

In this section H is a separable Hilbert space with the scalar product $\langle\cdot,\cdot\rangle = \langle\cdot,\cdot\rangle_H$.

8.8.1. The space $\mathcal{L}^2(\mathbb{R}, H)$ introduced in Section 8.7.1, then will be considered a Hilbert space with the scalar product $\langle\cdot,\cdot\rangle_{\mathcal{L}^2(\mathbb{R},H)}$ defined as follows: For functions $f, g \in \mathcal{C}_0^0(\mathbb{R}, H)$, we set

$$\langle f, g\rangle_{\mathcal{L}^2(\mathbb{R},H)} = \int_{-\infty}^{\infty} \langle f(x), g(x)\rangle_H dx.$$

If f and g are two arbitrary elements of $\mathcal{L}^2(\Gamma, H)$, then a function $\langle f, g\rangle_H \in \mathcal{L}^1(\mathbb{R}, \mathbb{C})$ is well defined. Indeed, let $\{f_n\}_{n\in\mathbb{N}}$ and $\{g_n\}_{n\in\mathbb{N}}$ be two Cauchy sequences of functions from $\mathcal{C}^0(\mathbb{R}, H)$ which represent f and g, respectively. Then, by the Cauchy-Schwarz inequality in H, pointwise on $\mathbb{R}$ we have

$$\left|\langle f_n, g_n\rangle_H - \langle f_m, g_m\rangle_H\right| \le \|f_n - f_m\|_H \|g_n\|_H + \|f_m\|_H \|g_n - g_m\|_H,$$

which implies, by the Cauchy-Schwarz inequality in $L^2(\mathbb{R}, \mathbb{C})$, that

$$\begin{aligned}&\left\|\langle f_n, g_n\rangle_H - \langle f_m, g_m\rangle_H\right\|_{\mathcal{L}^1(\Gamma,\mathbb{C})} \\ &\quad\le \|f_n - f_m\|_{\mathcal{L}^2(\Gamma,H)} \|g_n\|_{\mathcal{L}^2(\Gamma,H)} + \|f_m\|_{\mathcal{L}^2(\Gamma,H)} \|g_n - g_m\|_{\mathcal{L}^2(\Gamma,H)}.\end{aligned}$$

Hence $\big(\langle f_n, g_n\rangle_H\big)_{n\in\mathbb{N}}$ is a Cauchy sequence in $\mathcal{L}^1(\Gamma,\mathbb{C})$, and the function $\langle f,g\rangle_H$ can be defined to be the limit of this sequence in $\mathcal{L}^1(\Gamma,\mathbb{C})$. We now define the scalar product in $\mathcal{L}^2(\mathbb{R},H)$ by

$$\langle f,g\rangle_{\mathcal{L}^2(\Gamma,H)} = \int_{-\infty}^{\infty} \langle f,g\rangle_H \,dx\,, \qquad f,g\in\mathcal{L}^2(\Gamma,H).$$

Finally we note that, for each $f\in\mathcal{L}^2(\Gamma,H)$, the function

$$\|f\|_H := \sqrt{\langle f,f\rangle_H}$$

belongs to $\mathcal{L}^2(\Gamma,\mathbb{C})$ and that, for the norm corresponding to the scalar product $\langle f,g\rangle_{\mathcal{L}^2(\Gamma,H)}$, we have

$$\|f\|^2_{\mathcal{L}^2(\mathbb{R},H)} = \int_{-\infty}^{\infty} \|f\|^2_H \,dx.$$

8.8.2. Recall also the subspaces $\mathcal{L}^2_+(\mathbb{R},H)$ and $\mathcal{L}^2_-(\mathbb{R},H)$ introduced in Section 8.7.3, and note that (obviously) now the direct sum (8.7.3) is even orthogonal, i.e., we have the orthogonal decomposition

$$\mathcal{L}^2(\mathbb{R},H) = \mathcal{L}^2_+(\mathbb{R},H)\oplus\mathcal{L}^2_-(\mathbb{R},H).$$

8.8.3. Let $(e_n)_{n=1}^{\infty}$ be an orthonormal basis of H, and let H_n be the subspace of H spanned by e_n. Then $\mathcal{L}^2(\mathbb{R},H)$ splits into the orthogonal sum

$$\mathcal{L}^2(\mathbb{R},H) = \bigoplus_{n=1}^{\infty}\mathcal{L}^2(\mathbb{R},H_n). \tag{8.8.1}$$

Therefore each function $f\in\mathcal{L}^2(\mathbb{R},H)$ has a uniquely determined representation in the form

$$f=\sum_{n=1}^{\infty} f_n e_n \qquad \text{with } f_n\in\mathcal{L}^2(\mathbb{R},\mathbb{C}) \tag{8.8.2}$$

and

$$\|f\|^2_{\mathcal{L}^2(\mathbb{R},H)} = \sum_{n=1}^{\infty}\|f_n\|^2_{\mathcal{L}^2(\mathbb{R},\mathbb{C})}. \tag{8.8.3}$$

8.8.4 Proposition. *Let $f\in C_0^0(\mathbb{R},H)$. Then the series* (8.8.1) *converges uniformly.*

Proof. We denote by P_n the orthogonal projector from H onto $\bigoplus_{j=1}^{n} H_j$. Let $\varepsilon>0$. Then, for each fixed $x\in\mathbb{R}$, there exists $n_\varepsilon(x)\in\mathbb{N}$ with

$$\|f(x)-P_nf(x)\|\le\frac{\varepsilon}{2} \qquad \text{for } n\ge n_\varepsilon(x).$$

As f is continuous and $\|P_n\| = 1$, this implies that each $x \in \mathbb{R}$ has a neighborhood $U(x)$ such that, for $y \in U(x)$ and $n \geq n_\varepsilon(x)$,

$$\begin{aligned}\|f(y) - P_n f(y)\| &\leq \|f(y) - f(x)\| + \|f(x) - P_n f(x)\| + \|P_n f(x) - P_n f(y)\| \\ &\leq 2\|f(y) - f(x)\| + \|f(x) - P_n f(x)\| \leq \varepsilon.\end{aligned}$$

Since the support of f is compact, this completes the proof. □

8.8.5. The Fourier transform on $\mathcal{L}^2(\mathbb{R}, H)$. As, by the preceding proposition, the series (8.8.1) converges uniformly if $f \in \mathcal{C}_0^0(\mathbb{R}, H)$, for the Fourier transform (Section 8.7.4) $\widehat{f}$ of a function $f \in \mathcal{C}_0^0(\mathbb{R}, H)$, it follows from (8.8.1) that

$$\widehat{f}(\lambda) = \int_{-\infty}^{\infty} e^{i\lambda x} dx = \sum_{n=1}^{\infty} \left(\int_{-\infty}^{\infty} e^{i\lambda x} f_n dx \right) e_n = \sum_{n=1}^{\infty} \widehat{f}_n(\lambda)\, e_n, \qquad \lambda \in \mathbb{R}.$$

By (8.8.3) this implies that

$$\|\widehat{f}\|_{\mathcal{L}^2(\mathbb{R},\mathbb{C})} = \sum_{n=1}^{\infty} \|\widehat{f}_n\|_{\mathcal{L}^2(\mathbb{R},H)} \qquad \text{for all } f \in \mathcal{C}_0^0(\mathbb{R}, \mathbb{C}).$$

As, by the Plancherel theorem ([Ru], theorem 9.13), $\|\widehat{f}_n\|_{\mathcal{L}^2(\mathbb{R},\mathbb{C})} = \|f_n\|_{\mathcal{L}^2(\mathbb{R},\mathbb{C})}$ for all n, this further implies (again using (8.8.3))

$$\|\widehat{f}\|_{\mathcal{L}^2(\mathbb{R},H)} = \|f\|_{\mathcal{L}^2(\mathbb{R},H)} \qquad \text{for all } f \in \mathcal{C}_0^0(\mathbb{R}, H).$$

Since, by definition, $\mathcal{C}_0^0(\mathbb{R}, H)$ is dense in $\mathcal{L}^2(\mathbb{R}, H)$, it follows that the Fourier transformation extends to an isometry of $\mathcal{L}^2(\mathbb{R}, H)$, which we denote by **U**. For $f \in \mathcal{L}^2(\mathbb{R}, H)$ we write again $\widehat{f} = \mathbf{U}f$.

8.8.6. Convolution between $\mathcal{L}^1(\mathbb{R}, L(H))$ and $\mathcal{L}^2(\mathbb{R}, H)$.
First let $K \in \mathcal{C}_0^0(\mathbb{R}, L(H))$ and $f \in \mathcal{C}_0^0(\mathbb{R}, H)$. Then we define (as usual)

$$(K * f)(y) = \int_{-\infty}^{\infty} K(x) f(y - x) dx = \int_{-\infty}^{\infty} K(y - x) f(x) dx, \qquad y \in \mathbb{R},$$

and it follows from Fubini's theorem that

$$\begin{aligned}(\widehat{K * f})(\lambda) &= \int_{-\infty}^{\infty} e^{i\lambda y} \left(\int_{-\infty}^{\infty} K(y - x) f(x) dx \right) dy \\ &= \int_{-\infty}^{\infty} \left(\int_{-\infty}^{\infty} e^{i\lambda(y-x)} K(y - x)\, dy \right) e^{i\lambda x} f(x) dx \\ &= \int_{-\infty}^{\infty} \left(\int_{-\infty}^{\infty} e^{i\lambda(y)} K(y)\, dy \right) e^{i\lambda x} f(x) dx = \widehat{K}(\lambda) \widehat{f}(\lambda), \quad \lambda \in \mathbb{C}.\end{aligned} \tag{8.8.4}$$

Since the Fourier transformation is an isometry of $\mathcal{L}^2(\mathbb{R},H)$, this implies

$$\|K*f\|_{\mathcal{L}^2(\mathbb{R},H)} = \|\widehat{K}\widehat{f}\|_{\mathcal{L}^2(\mathbb{R},H)} \le \sup_{\lambda\in\mathbb{R}} \|\widehat{K}(\lambda)\|\|\widehat{f}\|_{\mathcal{L}^2(\mathbb{R},H)}$$

and further, by (8.7.9) and (8.7.8),

$$\|K*f\|_{\mathcal{L}^2(\mathbb{R},H)} \le \|K\|_{\mathcal{L}^1(\mathbb{R},L(H))}\|f\|_{\mathcal{L}^2(\mathbb{R},H)}.$$

Since, by definition, $\mathcal{C}_0^0(\mathbb{R},H)$ is dense in $\mathcal{L}^2(\mathbb{R},H)$ and $\mathcal{C}_0^0\big(\mathbb{R},L(H)\big)$ is dense in $\mathcal{L}^1\big(\mathbb{R},L(H)\big)$, this implies that the convolution extends to a continuous bilinear map

$$*: \mathcal{L}^1\big(\mathbb{R},L(H)\big)\times\mathcal{L}^2(\mathbb{R},H)\longrightarrow\mathcal{L}^2(\mathbb{R},H)$$

(called **convolution**), where, for all $K\in\mathcal{L}^1\big(\mathbb{R},L(H)\big)$ and $f\in\mathcal{L}^2(\mathbb{R},H)$,

$$\|K*f\|_{\mathcal{L}^2(\mathbb{R},H)} \le \|K\|_{\mathcal{L}^1(\mathbb{R},L(H))}\|f\|_{\mathcal{L}^2(\mathbb{R},H)} \tag{8.8.5}$$

and, by (8.8.4),

$$\widehat{K*f} = \widehat{K}\widehat{f}. \tag{8.8.6}$$

.

This convolution is associative:

8.8.7 Proposition. *If $K,L\in\mathcal{L}^1\big(\mathbb{R},L(H)\big)$ and $f\in\mathcal{L}^2(\mathbb{R},H)$, then*

$$(K*L)*f = K*(L*f).$$

Proof. If K,L,f are continuous and with compact support, then this follows by Fubini's theorem:

$$\begin{aligned}\Big((K*L)*f\Big)(y) &= \int_{-\infty}^{\infty}(K*L)(y-x)f(x)\,dx\\ &= \int_{-\infty}^{\infty}\left(\int_{-\infty}^{\infty}K(t)L(y-x-t)\,dt\right)f(x)\,dx\\ &= \int_{-\infty}^{\infty}K(t)\left(\int_{-\infty}^{\infty}L(y-x-t)f(x)\,dx\right)dt\\ &= \int_{-\infty}^{\infty}K(t)(L*f)(y-t)\,dt = \Big(K*(L*f)\Big)(y), \qquad y\in\mathbb{R}.\end{aligned}$$

By continuity (see estimate (8.8.5)), this implies that the assertion holds also in the general case. □

8.8.8 Proposition. *Let $\mathcal{P}_+$ be the orthogonal projector from $\mathcal{L}^2(\mathbb{R},H)$ onto $\mathcal{L}^2_+(\mathbb{R},H)$. Then, for all $f\in\mathcal{L}^2(\mathbb{R},H)$, $K_+\in\mathcal{L}^1_+\big(\mathbb{R},L(H)\big)$ and $K_-\in\mathcal{L}^1_-\big(\mathbb{R},L(H)\big)$,*

$$K_+*(\mathcal{P}_+f) = \mathcal{P}_+\Big(K_+*(\mathcal{P}_+f)\Big), \tag{8.8.7}$$

$$\mathcal{P}_+\Big(K_-*f\Big) = \mathcal{P}_+\Big(K_-*(\mathcal{P}_+f)\Big). \tag{8.8.8}$$

Proof. In view of (8.8.5), we may assume that the functions f, K_+ and K_- are piecewise continuous. Then

$$\Big(K_+ * (\mathcal{P}_+ f)\Big)(y) = \int_{-\infty}^{\infty} K_+(y-x)(\mathcal{P}_+ f)(x)\,dx = \int_0^y K_+(y-x)f(x)\,dx$$

for all $y \in \mathbb{R}$. This implies that $K_+ * (\mathcal{P}_+ f) \in \mathcal{L}^2_+(\mathbb{R}, H)$, i.e., we have (8.8.7). Moreover

$$\mathcal{P}_+\Big(K_- * f\Big)(y) = \int_{-\infty}^{\infty} K_-(y-x)f(x)\,dx = \int_y^{\infty} K_-(y-x)f(x)\,dx$$

for all $y \in \mathbb{R}$. For $y \geq 0$, this implies that

$$\mathcal{P}_+\Big(K_- * f\Big)(y) = \int_y^{\infty} K_-(y-x)\big(\mathcal{P}_+ f\big)(x)\,dx = \mathcal{P}_+\Big(K_- * (\mathcal{P}_+ f)\Big)(y),$$

i.e., we have (8.8.8). □

8.9 The isometry V from $\mathcal{L}^2(\mathbb{T}, H)$ onto $\mathcal{L}^2(\mathbb{R}, H)$

In this section, H is again a separable Hilbert space, and $\mathcal{L}^2(\mathbb{R}, H)$, $\mathcal{L}^2_+(\mathbb{R}, H)$, $\mathcal{L}^2_-(\mathbb{R}, H)$ are the Hilbert spaces introduced in Section 8.8.1. Recall the (obvious) orthogonal decomposition

$$\mathcal{L}^2(\mathbb{R}, H) = \mathcal{L}^2_+(\mathbb{R}, H) \oplus \mathcal{L}^2_-(\mathbb{R}, H). \tag{8.9.1}$$

Further, in this section, $\mathbb{T}$ is the unit circle, and $\mathcal{L}^2(\mathbb{T}, H)$, $\mathcal{L}^2_+(\mathbb{T}, H)$, $\mathcal{L}^2_-(\mathbb{T}, H)$ are the Hilbert spaces introduced in Section 8.3.1 and in Theorem and Definition 8.3.10. Recall that, by part (iii) of Theorem and Definition 8.3.10, we have the orthogonal decomposition

$$\mathcal{L}^2(\mathbb{T}, H) = \mathcal{L}^2_+(\mathbb{T}, H) \oplus \mathcal{L}^2_-(\mathbb{T}, H). \tag{8.9.2}$$

8.9.1. Let Φ be the Möbius transform defined by

$$\Phi(z) = i\frac{1+z}{1-z} \quad \text{if} \quad \lambda \in \mathbb{C} \setminus \{1\}, \quad \Phi(\infty) = -i \quad \text{and} \quad \Phi(1) = \infty.$$

Note that then

$$\Phi^{-1}(\lambda) = \frac{\lambda - i}{\lambda + i} \quad \text{if} \quad \lambda \in \mathbb{C} \setminus \{-i\}, \quad \Phi^{-1}(-i) = \infty \quad \text{and} \quad \Phi^{-1}(\infty) = 1.$$

Further, let $\mathcal{C}^0\big(\mathbb{T} \setminus \{1\}, H\big)$ and $\mathcal{C}^0(\mathbb{R}, H)$ be the spaces of continuous H-valued functions defined on $\mathbb{T} \setminus \{1\}$ and $\mathbb{R}$, respectively. Set

$$(\mathbf{V}f)(\lambda) = \frac{\Phi^{-1}(\lambda) - 1}{\sqrt{2}} f\big(\Phi^{-1}(\lambda)\big) \qquad \text{for } f \in \mathcal{C}^0\big(\mathbb{T} \setminus \{\infty\}\big) \text{ and } \lambda \in \mathbb{R}. \tag{8.9.3}$$

As Φ is a diffeomorphism from $\mathbb{T} \setminus \{1\}$ onto $\mathbb{R}$, in this way a linear isomorphism $\mathbf{V}$ from $\mathcal{C}^0\big(\mathbb{T} \setminus \{1\}, H\big)$ onto $\mathcal{C}^0(\mathbb{R}, H)$ is defined.

It is the aim of this section to study this linear isomorphism $\mathbf{V}$. We begin with the following

8.9.2 Theorem and Definition. For all $f \in \mathcal{C}^0\big(\mathbb{T} \setminus \{1\}, H\big)$,

$$\int_{\mathbb{T}} \|f\|^2 |dz| = \int_{-\infty}^{\infty} \|(\mathbf{V}f)(\lambda)\|^2 d\lambda. \tag{8.9.4}$$

Since (by our definitions of $\mathcal{L}^2(\mathbb{T}, H)$ and $\mathcal{L}^2(\mathbb{R}, H)$), the space $\mathcal{L}^2(\mathbb{T}, H) \cap \mathcal{C}^0(\mathbb{T}, H)$ is dense in $\mathcal{L}^2(\mathbb{T}, H)$, and the space $\mathcal{L}^2(\mathbb{R}, H) \cap \mathcal{C}^0(\mathbb{R}, H)$ is dense in $\mathcal{L}^2(\mathbb{R}, H)$, it follows from (8.9.4) that the restriction of $\mathbf{V}$ to $\mathcal{L}^2(\mathbb{T}, H) \cap \mathcal{C}^0\big(\mathbb{T} \setminus \{1\}, H\big)$ extends to a linear isometry from $\mathcal{L}^2(\mathbb{T}, H)$ onto $\mathcal{L}^2(\mathbb{R}, H)$.

We denote this isometry also by $\mathbf{V}$.

Moreover, also the isometry from $\mathcal{L}^2(\mathbb{T}, \mathbb{C})$ onto $\mathcal{L}^2(\mathbb{R}, \mathbb{C})$ (obtained for $H = \mathbb{C}$) will be denoted by $\mathbf{V}$.

Proof. Set

$$\varphi(t) = \Phi\big(e^{it}\big) = i\frac{1 + e^{it}}{1 - e^{it}} \qquad \text{for } 0 < t < 2\pi.$$

As Φ is a $\mathcal{C}^\infty$-diffeomorphism from $\mathbb{T} \setminus \{1\}$ onto $\mathbb{R}$ and the map $t \to e^{it}$ is a $\mathcal{C}^\infty$ diffeomorphism from $]0, 2\pi[$ onto $\mathbb{T} \setminus \{1\}$, this is a $\mathcal{C}^\infty$-diffeomorphism from $]0, 2\pi[$ onto $\mathbb{R}$. We have

$$\varphi'(t) = i\frac{ie^{it}(1 - e^{it}) + (1 + e^{it})ie^{it}}{(1 - e^{it})^2} = -2\frac{e^{it}}{(1 - e^{it})^2}, \qquad t \in]0, 2\pi[. \tag{8.9.5}$$

Set $\psi = \varphi^{-1}$. Then

$$\lambda = \varphi\big(\psi(\lambda)\big) = \Phi\Big(e^{i\psi(\lambda)}\Big), \qquad \lambda \in \mathbb{R},$$

and therefore

$$\Phi^{-1}(\lambda) = e^{i\psi(\lambda)}, \qquad \lambda \in \mathbb{R}. \tag{8.9.6}$$

Further, by (8.9.5),

$$\psi'(\lambda) = -\frac{1}{\varphi'\big(\psi(\lambda)\big)} = -2\frac{\big(1 - e^{i\psi(\lambda)}\big)^2}{e^{i\psi(\lambda)}}, \qquad \lambda \in \mathbb{R}.$$

Together with (8.9.6) this implies that

$$\psi'(\lambda) = -\frac{\big(1 - \Phi^{-1}(\lambda)\big)^2}{2\Phi^{-1}(\lambda)}, \qquad \lambda \in \mathbb{R}. \tag{8.9.7}$$

Now let a function $f \in \mathcal{C}^0\big(\mathbb{T} \setminus \{1\}, H\big)$ be given. Then, by definition of $|dz|$,

$$\int_{\mathbb{T}} \|f(z)\|^2 |dz| = \int_0^{2\pi} \|f\big(e^{it}\big)\|^2 dt.$$

Since ψ is a diffeomorphism from $\mathbb{R}$ onto $]0, 2\pi[$ and by (8.9.6) and (8.9.7), this yields

$$\begin{aligned}\int_{\mathbb{T}} \|f(z)\|^2 |dz| &= \int_{-\infty}^{\infty} \left\|f\left(e^{i\psi(\lambda)}\right)\right\|^2 |\psi'(\lambda)|\, d\lambda \\ &= \int_{-\infty}^{\infty} \left\|f\left(\Phi^{-1}(\lambda)\right)\right\|^2 \left|\frac{\left(1-\Phi^{-1}(\lambda)\right)^2}{2\Phi^{-1}(\lambda)}\right| d\lambda.\end{aligned}$$

As $\left|\Phi^{-1}(\lambda)\right| = 1$ for all $\lambda \in \mathbb{R}$, this further implies that

$$\int_{\mathbb{T}} \|f(z)\|^2 |dz| = \int_{-\infty}^{\infty} \left\|\frac{1-\Phi^{-1}(\lambda)}{\sqrt{2}} f\left(\Phi^{-1}(\lambda)\right)\right\|^2 d\lambda.$$

By definition of $\mathbf{V}$ this is (8.9.4). □

8.9.3. In what follows the functions $\omega_n^+ \in \mathcal{L}_+^2(\mathbb{T}, \mathbb{C})$, $n \in \mathbb{N}$, and $\omega_n^- \in \mathcal{L}_-^2(\mathbb{T}, \mathbb{C})$, $n \in \mathbb{N}$, defined by

$$\omega_n^+(z) = (z-1)^n \qquad \text{and} \qquad \omega_n^-(z) = -\frac{1}{z}\left(\frac{1}{z}-1\right)^n,$$

play a special role.

Recall that, by part (iii) of Theorem and Definition 8.3.10, the linear space spanned by the functions z^n, $n \in \mathbb{N}$, is a dense subspace of $\mathcal{L}_+^2(\mathbb{T}, H)$, and the linear space spanned by the functions $1/z^n$, $n \in \mathbb{N}^*$, is a dense subspace of $\mathcal{L}_-^2(\mathbb{T}, H)$. As each of the functions z^n, $n \in \mathbb{N}$, is a linear combination of some of the functions ω_n^+, $n \in \mathbb{N}$, and each of the functions $1/z^n$, $n \in \mathbb{N}^*$, is a linear combination of some of the functions ω_n^-, $n \in \mathbb{N}$, this implies:

The linear space spanned by the functions ω_n^+, $n \in \mathbb{N}$, is a dense subspace of $\mathcal{L}_+^2(\mathbb{T}, \mathbb{C})$, and the linear space spanned by the functions ω_n^-, $n \in \mathbb{N}^*$, is a dense subspace of $\mathcal{L}_-^2(\mathbb{T}, \mathbb{C})$.

In view of the orthogonal decomposition (8.9.2), this further implies:

Then linear space spanned by the functions ω_n^+ and ω_n^-, $n \in \mathbb{N}$, is a dense subspace of $\mathcal{L}^2(\mathbb{T}, H)$.

Further note that, by definition of $\mathbf{V}$, for all $n \in \mathbb{N}$,

$$\begin{aligned}(\mathbf{V}\omega_n^+)(\lambda) &= \frac{1}{\sqrt{2}}\left(\frac{\lambda-i}{\lambda+i}-1\right)^{n+1}, \qquad \lambda \in \mathbb{R}, \\ (\mathbf{V}\omega_n^-)(\lambda) &= \frac{1}{\sqrt{2}}\left(\frac{\lambda+i}{\lambda-i}-1\right)^{n+1}, \qquad \lambda \in \mathbb{R},\end{aligned} \tag{8.9.8}$$

and therefore, by lemma (8.7.11) (ii),

$$\mathbf{V}\omega_n^+ = \frac{(-2)^{n+1}}{\sqrt{2}(n+1)!}\,\widehat{\theta}_n^+ \quad \text{and} \quad \mathbf{V}\omega_n^- = \frac{2^{n+1}}{\sqrt{2}(n+1)!}\,\widehat{\theta}_n^-. \tag{8.9.9}$$

8.9.4 Theorem. *Let $\theta_n^+ \in \mathcal{L}^2_+(\mathbb{R},\mathbb{C})$ and $\theta_n^- \in \mathcal{L}^2_-(\mathbb{R},\mathbb{C})$, $n \in \mathbb{N}$, be the functions introduced in Section 8.7.8. Then the linear space spanned by the functions θ_n^+, $n \in \mathbb{N}$, is dense in $\mathcal{L}^2_+(\mathbb{R},\mathbb{C})$, and the linear space spanned by the functions θ_n^-, $n \in \mathbb{N}$, is dense in $\mathcal{L}^2_-(\mathbb{R},\mathbb{C})$.*

Proof. Since the linear space spanned by the functions ω_n^+ and ω_n^-, $n \in \mathbb{N}$, is dense in $\mathcal{L}^2(\mathbb{T},\mathbb{C})$ (Section 8.9.3), it follows from Theorem and Definition 8.9.2 that the linear space spanned by the functions $\mathbf{V}\omega_n^+$ and $\mathbf{V}\omega_n^-$, $n \in \mathbb{N}$, is dense in $\mathcal{L}^2(\mathbb{R},\mathbb{C})$. By (8.9.9) this implies that the linear space spanned by the functions $\widehat{\theta}_n^+$ and $\widehat{\theta}_n^-$, $n \in \mathbb{N}$, is dense in $\mathcal{L}^2(\mathbb{R},\mathbb{C})$. As the Fourier transformation is an isometry of $\mathcal{L}^2(\mathbb{R},\mathbb{C})$, this further implies that the linear space spanned by the functions θ_n^+ and θ_n^-, $n \in \mathbb{N}$, is dense in $\mathcal{L}^2(\mathbb{R},\mathbb{C})$. Taking into account the orthogonal decomposition (8.9.1) and the fact that $\theta_n^+ \in \mathcal{L}^2_+(\mathbb{R},\mathbb{C})$ and $\theta_n^- \in \mathcal{L}^2_-(\mathbb{R},\mathbb{C})$, this completes the proof. □

8.9.5 Theorem. *Let $\mathcal{P}^{\mathbb{T}}$ be the orthogonal projector from $\mathcal{L}^2(\mathbb{T},H)$ onto $\mathcal{L}^2_+(\mathbb{T},H)$, let $\mathcal{P}^{\mathbb{R}}$ be the orthogonal projector from $\mathcal{L}^2(\mathbb{R},H)$ onto $\mathcal{L}^2_+(\mathbb{R},H)$ and let $\mathbf{U}$ be the Fourier isometry of $\mathcal{L}^2(\mathbb{R},H)$ (Section 8.8.5). Then*

$$\mathcal{P}^{\mathbb{R}} = \mathbf{U}^{-1}\mathbf{V}\mathcal{P}^{\mathbb{T}}\mathbf{V}^{-1}\mathbf{U}. \tag{8.9.10}$$

Proof. It follows from (8.9.9) that, for all $n \in \mathbb{N}$,

$$\mathbf{U}^{-1}\mathbf{V}\omega_n^+ = \frac{(-2)^{n+1}}{\sqrt{2}(n+1)!}\theta_n^+ \qquad \text{and} \qquad \mathbf{U}^{-1}\mathbf{V}\omega_n^- = \frac{2^{n+1}}{\sqrt{2}(n+1)!}\theta_n^-. \tag{8.9.11}$$

As the decomposition (8.9.2) is orthogonal, we have $\mathcal{P}^{\mathbb{T}}\omega_n^+ = \omega_n^+$ and $\mathcal{P}^{\mathbb{T}}\omega_n^- = 0$ for all $n \in \mathbb{N}$. Therefore it follows from (8.9.11) that, for all $n \in \mathbb{N}$,

$$\mathbf{U}^{-1}\mathbf{V}\mathcal{P}^{\mathbb{T}}\mathbf{V}^{-1}\mathbf{U}\theta_n^+ = \theta_n^+ \qquad \text{and} \qquad \mathbf{U}^{-1}\mathbf{V}\mathcal{P}^{\mathbb{T}}\mathbf{V}^{-1}\mathbf{U}\theta_n^- = 0.$$

Taking into account Theorem 8.9.4 and the fact that also the decomposition (8.9.1) is orthogonal, this implies (8.9.10). □

8.10 The algebra of operator functions $L(H) \oplus \mathcal{L}^1\big(\mathbb{R}, L(H)\big)$

In this section, H is again a Hilbert space.

8.10.1. Let $\widehat{\mathcal{L}}^1\big(\mathbb{R},L(H)\big)$, $\widehat{\mathcal{L}}^1_+\big(\mathbb{R},L(H)\big)$ and $\widehat{\mathcal{L}}^1_-\big(\mathbb{R},L(H)\big)$ be the Banach algebras introduced in Section 8.7.5. Then we denote by $L(H) \oplus \widehat{\mathcal{L}}^1\big(\mathbb{R},L(H)\big)$ the algebra of functions $W : \mathbb{R} \to L(H)$ of the form

$$W(\lambda) = A + \widehat{K}(\lambda), \qquad \lambda \in \mathbb{R}, \tag{8.10.1}$$

where $K \in \mathcal{L}^1(\mathbb{R}, L(H))$ and $A \in L(H)$ is a constant operator. Since $\widehat{K}(\infty) = 0$ for all $K \in \mathcal{L}^1(\mathbb{R}, L(H))$ (Theorem 8.7.7 (iii)), the representation of a function $W \in L(H) \oplus \widehat{\mathcal{L}}^1(\mathbb{R}, L(H))$ in the form (8.10.1) is uniquely determined. Therefore we can introduce a norm $\|\cdot\|_{L(H)\oplus\widehat{\mathcal{L}}^1(\mathbb{R},L(H))}$ in $L(H) \oplus \widehat{\mathcal{L}}^1(\mathbb{R}, L(H))$, setting

$$\|A + \widehat{K}\|_{L(H)\oplus\widehat{\mathcal{L}}^1(\mathbb{R},L(H))} = \|A\|_{L(H)} + \|\widehat{K}\|_{\widehat{\mathcal{L}}^1(\mathbb{R},L(H))}$$

for $A \in L(H)$ and $K \in \mathcal{L}^1(\mathbb{R}, L(H))$. Note that then, by definition of the norm in $\widehat{\mathcal{L}}^1(\mathbb{R}, L(H))$ (see (8.7.8)),

$$\|A + \widehat{K}\|_{L(H)\oplus\widehat{\mathcal{L}}^1(\mathbb{R},L(H))} = \|A\|_{L(H)} + \|K\|_{\mathcal{L}^1(\mathbb{R},L(H))}$$

for all $A \in L(H)$ and $K \in \mathcal{L}^1(\mathbb{R}, L(H))$. It is easy to see that in this way, $L(H) \oplus \widehat{\mathcal{L}}^1(\mathbb{R}, L(H))$ becomes a Banach space and that, for all $V, W \in L(H) \oplus \widehat{\mathcal{L}}^1(\mathbb{R}, L(H))$,

$$\|VW\|_{L(H)\oplus\widehat{\mathcal{L}}^1(\mathbb{R},L(H))} \le \|V\|_{L(H)\oplus\widehat{\mathcal{L}}^1(\mathbb{R},L(H))}\|W\|_{L(H)\oplus\widehat{\mathcal{L}}^1(\mathbb{R},L(H))}. \tag{8.10.2}$$

Hence, with this norm, $L(H) \oplus \widehat{\mathcal{L}}^1(\mathbb{R}, L(H))$ is a Banach algebra. As $\widehat{\mathcal{L}}^1_+(\mathbb{R}, L(H))$ and $\widehat{\mathcal{L}}^1_-(\mathbb{R}, L(H))$ are subalgebras of $\widehat{\mathcal{L}}^1(\mathbb{R}, L(H))$, it follows that $L(H) \oplus \widehat{\mathcal{L}}^1_+(\mathbb{R}, L(H))$ and $L(H) \oplus \widehat{\mathcal{L}}^1_{(}\mathbb{R}, L(H))$ are subalgebras of this Banach algebra.

We identify $L(H)$ with the subalgebra of $L(H) \oplus \widehat{\mathcal{L}}^1(\mathbb{R}, L(H))$ which consists of the constant $L(H)$-valued functions. As $\widehat{\mathcal{L}}^1(\mathbb{R}, L(H))$ splits into the direct sum $\widehat{\mathcal{L}}^1(\mathbb{R}, L(H)) = \widehat{\mathcal{L}}^1_-(\mathbb{R}, L(H)) \oplus \widehat{\mathcal{L}}^1_+(\mathbb{R}, L(H))$, then $L(H) \oplus \widehat{\mathcal{L}}^1(\mathbb{R}, L(H))$ splits into the direct sum

$$L(H) \oplus \widehat{\mathcal{L}}^1(\mathbb{R}, L(H)) = L(H) \oplus \widehat{\mathcal{L}}^1_-(\mathbb{R}, L(H)) \oplus \widehat{\mathcal{L}}^1_+(\mathbb{R}, L(H)). \tag{8.10.3}$$

From Theorem 8.7.7 we immediately get the following

8.10.2 Theorem. (i) *A function* $W \in L(H) \oplus \widehat{\mathcal{L}}^1(\mathbb{R}, L(H))$ *belongs to* $L(H) \oplus \widehat{\mathcal{L}}^1_+(\mathbb{R}, L(H))$*, if and only if it admits a continuous extension to* $\overline{\mathbb{H}}_+ \cup \{\infty\}$ *which is holomorphic in* $\mathbb{H}_+$*, and then*

$$\max_{\lambda\in\overline{\mathbb{H}}_+\cup\{\infty\}} \|W(\lambda)\| \le \|W\|_{L(H)\oplus\widehat{\mathcal{L}}^1(\mathbb{R},L(H))}.$$

(ii) *A function* $W \in L(H) \oplus \widehat{\mathcal{L}}^1(\mathbb{R}, L(H))$ *belongs to* $L(H) \oplus \widehat{\mathcal{L}}^1_-(\mathbb{R}, L(H))$*, if and only if it admits a continuous extension to* $\overline{\mathbb{H}}_- \cup \{\infty\}$ *which is holomorphic in* $\mathbb{H}_-$*, and then*

$$\max_{\lambda\in\overline{\mathbb{H}}_-\cup\{\infty\}} \|W(\lambda)\| \le \|W\|_{L(H)\oplus\widehat{\mathcal{L}}^1(\mathbb{R},L(H))}.$$

(iii) *Each function* $W \in L(H) \oplus \widehat{\mathcal{L}}^1\big(\mathbb{R}, L(H)\big)$ *is continuous on* $\mathbb{R} \cup \{\infty\}$. *Moreover, for each* $W \in L(H) \oplus \widehat{\mathcal{L}}^1\big(\mathbb{R}, L(H)\big)$,

$$\max_{\lambda \in \mathbb{R} \cup \{\infty\}} \|W(\lambda)\| \le \|W\|_{L(H) \oplus \widehat{\mathcal{L}}^1(\mathbb{R}, L(H))}.$$

The main result of the present section is the following

8.10.3 Theorem. *The space of* $L(H)$*-valued rational functions without poles on* $\mathbb{R} \cup \{\infty\}$ *is contained in* $L(H) \oplus \widehat{\mathcal{L}}^1(\mathbb{R}, L(H))$ *as a dense subset.*

To prove this, we first deduce from Theorem 8.9.4 the following

8.10.4 Lemma. *Let* $\theta_n^{\pm}$, $n \in \mathbb{N}$, *be the functions introduced in Section* 8.7.8. *Then the linear space spanned by the functions* θ_n^+, $n \in \mathbb{N}$, *is dense in* $\mathcal{L}_+^1(\mathbb{R}, \mathbb{C})$, *and the linear space spanned by the functions* θ_n^-, $n \in \mathbb{N}$, *is dense in* $\mathcal{L}_-^1(\mathbb{R}, \mathbb{C})$.

Proof. It is sufficient to prove that the functions θ_n^+, $n \in \mathbb{N}$, span a dense subspace of $\mathcal{L}_+^1(\mathbb{R}, \mathbb{C})$, because the statement with respect to the functions θ_n^- then follows by the substitution $x \to -x$.

Let $f : [0, \infty[\to \mathbb{C}$ be an arbitrary bounded measurable function with

$$\int_0^\infty x^n e^{-x} f(x)\, dx = 0 \qquad \text{for all } n \in \mathbb{N}. \tag{8.10.4}$$

By the Hahn-Banach theorem it is sufficient to prove that then $f \equiv 0$. Set

$$h(x) = e^{-x} f(2x), \qquad x \ge 0.$$

Then h belongs to $\mathcal{L}_+^2(\mathbb{R}, \mathbb{C})$, and, with the substituion $x \to x/2$, it follows from (8.10.4) that, for all $n \in \mathbb{N}$,

$$\int_0^\infty x^n e^{-x} h(x) dx = \int_0^\infty x^n e^{-2x} f(2x) dx = \frac{1}{2^{n+1}} \int_0^\infty x^n e^{-x} f(x) dx = 0.$$

Since, by Theorem 8.9.4, the space spanned by the functions θ_n^+ is dense in $\mathcal{L}_+^2(\mathbb{R}, \mathbb{C})$, this implies that $h \equiv 0$. Hence $f \equiv 0$. □

8.10.5. Proof of Theorem 8.10.3. It is sufficient to prove this for $H = L(H) = \mathbb{C}$. By Lemma 8.7.11 (i), for each $\xi \in \mathbb{C} \setminus \mathbb{R}$, the function

$$\frac{1}{\lambda - \xi}$$

belongs to $\widehat{\mathcal{L}}^1(\mathbb{R}, \mathbb{C})$. Since $\mathbb{C} \oplus \widehat{\mathcal{L}}^1(\mathbb{R}, \mathbb{C})$ is an algebra and since also the constant functions belong to this algebra, this implies that $\mathbb{C} \oplus \widehat{\mathcal{L}}^1(\mathbb{R}, \mathbb{C})$ contains all rational functions without poles on $\mathbb{R} \cup \{\infty\}$.

It remains to prove the density. By Lemma 8.10.4, the space spanned by the functions θ_n^- and θ_n^+, $n \in \mathbb{N}$, is dense in $\mathcal{L}^1(\mathbb{R}, \mathbb{C})$. Hence (by definition of the norm in $\widehat{\mathcal{L}}^1(\mathbb{R},\mathbb{C})$) the space spanned by the functions $\widehat{\theta}_n^-$ and $\widehat{\theta}_n^+$, $n \in \mathbb{N}$, is dense in $\widehat{\mathcal{L}}^1(\mathbb{R},\mathbb{C})$. By part (ii) of Lemma 8.7.11, this implies that the space spanned by the functions

$$\left(\frac{\lambda - i}{\lambda + i} - 1\right)^n \quad \text{and} \quad \left(\frac{\lambda + i}{\lambda - i} - 1\right)^n, \qquad n \in \mathbb{N}^*, \tag{8.10.5}$$

is dense in $\widehat{\mathcal{L}}^1(\mathbb{R},\mathbb{C})$. Hence the space spanned by the functions (8.10.5) and the constant functions is dense in $\mathbb{C} \oplus \widehat{\mathcal{L}}^1(\mathbb{R},\mathbb{C})$. Since all these functions are rational and without poles on $\mathbb{R} \cup \{\infty\}$, this completes the proof. □

Using again the Möbius transformation $(\lambda - i)/(\lambda + i)$ (Section 8.9.1), we now obtain the following version of Theorem 8.10.1:

8.10.6 Theorem. *Let $W \in L(H) \oplus \widehat{\mathcal{L}}^1(\mathbb{R}, L(H))$ such that $W(\lambda) \in GL(H)$ for all $\lambda \in \mathbb{R} \cup \{\infty\}$. Then:*

(i) *The pointwise defined function W^{-1} again belongs to $L(H) \oplus \widehat{\mathcal{L}}^1(\mathbb{R}, L(H))$.*

(ii) *The function W can be written in the form*

$$W = V_- V V_+,$$

where $V_- : \overline{\mathbb{H}}_- \cup \{\infty\} \to GL(H)$ is continuous on $\overline{\mathbb{H}}_- \cup \{\infty\}$ and holomorphic in $\mathbb{H}_- \cup \{\infty\}$, $V_+ : \overline{\mathbb{H}}_+ \to GL(H)$ is continuous on $\overline{\mathbb{H}}_+$ and holomorphic in $\mathbb{H}_+$, V is an $L(H)$-valued rational function without poles on $\mathbb{R} \cup \{\infty\}$, $V(\lambda) \in GL(H)$ for all $\lambda \in \mathbb{R} \cup \{\infty\}$, and the functions V_-, V_-^{-1}, V_+ and V_+^{-1} belong to $L(H) \oplus \widehat{\mathcal{L}}^1(\mathbb{R}, L(H))$.

(iii) *Let $\widetilde{W} \in L(H) \oplus \widehat{\mathcal{L}}^1(\mathbb{R}, L(H))$ be a second function with $W(\lambda) \in GL(H)$ for all $\lambda \in \mathbb{R} \cup \{\infty\}$, and assume that*

$$W = W_- \widetilde{W} W_+ \qquad \text{on } \mathbb{R} \cup \{\infty\},$$

where $W_- : \overline{\mathbb{H}}_- \cup \{\infty\} \to GL(H)$ and $W_+ : \overline{\mathbb{H}}_+ \to GA$ are continuous functions which are holomorphic in $\mathbb{H}_- \cup \{\infty\}$ and $\mathbb{H}_+$, respectively. Then

$$W_-, W_-^{-1} \in L(H) \oplus \widehat{\mathcal{L}}^1_-(\mathbb{R}, L(H)) \quad \text{and} \quad W_+, W_+^{-1} \in L(H) \oplus \widehat{\mathcal{L}}^1_-(\mathbb{R}, L(H)).$$

Proof. Let Φ be the Möbius transform introduced in Section 8.9.1, let $\mathbb{T}$ be the unit disc, let D_+ be the unit circle and let $D_- := \mathbb{C} \setminus \overline{D}_+$. We denote by $\mathcal{R}$ the Banach algebra of all operator functions $A : \mathbb{T} \to L(H)$ of the form $A = W \circ \Phi$ with $W \in L(H) \oplus \widehat{\mathcal{L}}^1(\mathbb{R}, L(H))$, endowed with the norm

$$\|A\|_{\mathcal{R}} := \|W\|_{L(H) \oplus \widehat{\mathcal{L}}^1(\mathbb{R}, L(H))}.$$

Since Φ maps $\mathbb{T}$ onto $\mathbb{T} \cup \{\infty\}$, D_+ onto $\mathbb{H}_+$ and D_- onto $\mathbb{H}_-$, then it follows from Theorem 8.10.2 (iii), Theorem 8.10.3 and the decomposition (8.10.3) that $\mathcal{R}$ satisfies conditions (A), (B), (C) from Section 7.2.4. Therefore, we can apply Theorem 7.2.5 to $\mathcal{R}$, and we obtain:

(i′) $W^{-1} \circ \Phi \in L(H) \oplus \widehat{\mathcal{L}}^1\big(\mathbb{R}, L(H)\big)$.

(ii′) The function $W \circ \Phi$ can be written in the form

$$W \circ \Phi = V_-^{\mathbb{T}} V^{\mathbb{T}} V_+^{\mathbb{T}},$$

where $V_-^{\mathbb{T}} : \overline{D}_- \cup \{\infty\} \to GL(H)$ is continuous on $\overline{D}_- \cup \{\infty\}$ and holomorphic in $D_- \cup \{\infty\}$, $V_+^{\mathbb{T}} : \overline{D}_+ \to GL(H)$ is continuous on $\overline{D}_+$ and holomorphic in D_+, $V^{\mathbb{T}}$ is an $L(H)$-valued rational function without poles on $\mathbb{T}$, $V^{\mathbb{T}}(z) \in GL(H)$ for all $z \in \mathbb{T}$, and the functions $V_-^{\mathbb{T}}$, $(V_-^{\mathbb{T}})^{-1}$, $V_+^{\mathbb{T}}$ and $(V_+^{\mathbb{T}})^{-1}$ belong to $\mathcal{R}$.

(iii′) The functions $W_- \circ \Phi$, $W_-^{-1} \circ \Phi$, $W_+ \circ \Phi$ and $W_+^{-1} \circ \Phi$ belong to $\mathcal{R}$.

Then (i) and (iii) follow from (i′) and (iii′) by definition of $\mathcal{R}$, and (ii) follows from (ii′) setting $V_\pm := V_\pm^{\mathbb{T}} \circ \Phi^{-1}$ and $V := V^{\mathbb{T}} \circ \Phi^{-1}$. □

8.11 Factorization with respect to the real line

Throughout this section, E is a Banach space.

Here we introduce the notion of factorization with respect to the real line, and we show that this is equivalent to the notion of factorization with respect to the unit circle.

8.11.1 Definition. Let G be one of the groups $GL(E)$, $\mathcal{G}^\infty(E)$ or $\mathcal{G}^\omega(E)$ (Def. 5.12.1), and let $A : \mathbb{R} \cup \{\infty\} \to G$ be a continuous function (cf. Section 8.7.6). A representation of A in the form

$$A = A_- \Delta A_+ \tag{8.11.1}$$

will be called a **factorization** of A relative to $\mathbb{R}$ and G if the factors A_-, A_+, Δ have the following properties:

- Either $\Delta \equiv I$ or Δ is of the form

$$\Delta(\lambda) = P_0 + \sum_{j=1}^{n} \left(\frac{\lambda - i}{\lambda + i}\right)^{\kappa_j} P_j\,, \qquad \lambda \in \mathbb{R}, \tag{8.11.2}$$

  where $n \in \mathbb{N}^*$, $\kappa_1 > \ldots > \kappa_n$ are non-zero integers, $P_1, \ldots, P_n$ are mutually disjoint finite dimensional projectors in E, and $P_0 = I - P_1 - \ldots - P_n$;

- A_+ is a continuous $GL(E)$-valued function on $\overline{\mathbb{H}}_+ \cup \{\infty\}$, which is holomorphic in $\mathbb{H}_+$;

- A_- is a continuous $GL(E)$-valued function on $\overline{\mathbb{H}}_- \cup \{\infty\}$, which is holomorphic in $\mathbb{H}_-$.

If $\Delta \equiv I$, then this factorization will be called **canonical** .

8.11.2. Let $\Phi : \mathbb{C} \cup \{\infty\} \to \mathbb{C} \cup \{\infty\}$ be the Möbius transform defined by

$$\Phi(z) = i\frac{1+z}{1-z}$$

(cf. Section (8.9.1), let $\mathbb{T}$ be the unit circle, let D_+ be the open unit disc, and let $D_- := \mathbb{C} \setminus D_+$. Since $\Phi|_{\overline{D}_+}$ is a homeomorphism from $\overline{D}_+$ onto $\overline{\mathbb{H}}_+ \cup \{\infty\}$, which is biholomorphic from D_+ onto $\mathbb{H}_+$, and $\Phi|_{\overline{D}_- \cup \{\infty\}}$ is a homeomorphism from $\overline{D}_- \cup \{\infty\}$ onto $\overline{\mathbb{H}}_- \cup \{\infty\}$, which is biholomorphic from $D_- \cup \{\infty\}$ onto $\mathbb{H}_-$, we get the following simple but important

8.11.3 Proposition. *Let G be one of the groups $GL(E)$, $\mathcal{G}^\infty(E)$ or $\mathcal{G}^\omega(E)$ (Def. 5.12.1), and let $A : \mathbb{R} \cup \{\infty\} \to G$ be a continuous function (cf. Section 8.7.6). Then*

$$A = A_-\Delta A_+$$

is a factorization of A relative to $\mathbb{R}$ and G, if and only if,

$$A \circ \Phi = \big(A_- \circ \Phi\big)\big(\Delta \circ \Phi\big)\big(A_+ \circ \Phi\big)$$

is a factorization of $A \circ \Phi$ relative to $\mathbb{T}$ and G.

8.11.4 Definition. In view of the corresponding fact for factorizations with respect to $\mathbb{T}$ (see Section 7.1.2), this proposition implies that the integers $\kappa_1, \ldots, \kappa_n$ and the dimensions of the projectors $P_1, \ldots, P_n$ in Definition 8.11.1 are uniquely determined by A. The integers $\kappa_1, \ldots, \kappa_n$ will be called the **non-zero partial indices** of A, and the number $\dim P_j$ will be called the **multiplicity** of κ_j.

Moreover, Proposition 8.11.3 shows that, for each factorization result relative to $\mathbb{T}$, there is a corresponding factorization result with respect to the real line. In the following sections we use this fact to study the Wiener-Hopf integral equation on the half line.

8.12 Wiener-Hopf integral operators in $\mathcal{L}^2\big([0,\infty[,H\big)$

Let H be a separable Hilbert space. In this section we use without further reference the notations introduced in sections 8.7.1, 8.8.1, and 8.8.6.

Here we study the Wiener-Hopf integral equation

$$u(y) - \int_{-\infty}^{\infty} K(y-x)f(x)\,dx = f(y), \qquad y \geq 0, \tag{8.12.1}$$

where $K \in \mathcal{L}^1\big(\mathbb{R}, L(H)\big)$ and $f \in \mathcal{L}^2_+(\mathbb{R}, H)$ are given, and $u \in \mathcal{L}^2(\mathbb{R}, H)$ is sought.

Throughout this section we moreover use without further reference the notations and facts given in the following Section 8.12.1.

8.12.1. Let $\mathcal{P}^{\mathbb{R}}$ be the orthogonal projector from $\mathcal{L}^2(\mathbb{R}, H)$ onto $\mathcal{L}^2_+(\mathbb{R}, H)$. Then $\operatorname{Ker} \mathcal{P}^{\mathbb{R}} = \mathcal{L}^2_-(\mathbb{R}, H)$, because of the (obvious) orthogonal decomposition

$$\mathcal{L}^2(\mathbb{R}, H) = \mathcal{L}^2_+(\mathbb{R}, H) \oplus \mathcal{L}^2_-(\mathbb{R}, H). \tag{8.12.2}$$

If $K \in \mathcal{L}^1\big(\mathbb{R}, L(H)\big)$, then the operator T acting in $\mathcal{L}^2_+(\mathbb{R}, H)$ by

$$Tu = u - \mathcal{P}^{\mathbb{R}}(K * u), \qquad u \in \mathcal{L}^2_+(\mathbb{R}, H),$$

is called the **Wiener-Hopf operator** with **kernel function** K. With a Wiener-Hopf operator with kernel function K we moreover associate the **symbol** of this Wiener-Hopf operator, which is defined to be the operator function

$$I - \widehat{K}. \tag{8.12.3}$$

Then we speak also about the **symbol** $I - \widehat{K}$ of the **kernel function** K and the **kernel function** K of the **symbol** $I - \widehat{K}$

The set of all symbols with kernel function in $\mathcal{L}^1\big(\mathbb{R}, L(H)\big)$, i.e., the set of all functions of the form (8.12.3) with $K \in \mathcal{L}^1\big(\mathbb{R}, L(H)\big)$ will be denoted by $\mathcal{S}\big(\mathbb{R}, L(H)\big)$. If $L(H) \oplus \widehat{\mathcal{L}}^1\big(\mathbb{R}, L(H)\big)$ is the Banach algebra from Section 8.10, then

$$\mathcal{S}\big(\mathbb{R}, L(H)\big) = \Big\{ W \in L(H) \oplus \widehat{\mathcal{L}}^1\big(\mathbb{R}, L(H)\big) \,\Big|\, W(\infty) = I \Big\}. \tag{8.12.4}$$

The Wiener-Hopf operator with symbol $W \in \mathcal{S}\big(\mathbb{R}, L(H)\big)$ will be denoted by T_W. So, if $K \in \mathcal{L}^1(\mathbb{R}, H)$ and W is the symbol of K, then equation (8.12.1) takes the form $T_W u = f$.

We denote by $\mathcal{S}_-\big(\mathbb{R}, L(H)\big)$ and $\mathcal{S}_+\big(\mathbb{R}, L(H)\big)$ the subsets of $\mathcal{S}\big(\mathbb{R}, L(H)\big)$ which consist of the symbols with kernel function in $\mathcal{L}^1_-\big(\mathbb{R}, L(H)\big)$ and $\mathcal{L}^1_+\big(\mathbb{R}, L(H)\big)$, respectively, i.e.,

$$\begin{aligned} \mathcal{S}_+\big(\mathbb{R}, L(H)\big) &= \Big\{ W \in L(H) \oplus \widehat{\mathcal{L}}^1_+\big(\mathbb{R}, L(H)\big) \,\Big|\, W(\infty) = I \Big\}, \\ \mathcal{S}_-\big(\mathbb{R}, L(H)\big) &= \Big\{ W \in L(H) \oplus \widehat{\mathcal{L}}^1_-\big(\mathbb{R}, L(H)\big) \,\Big|\, W(\infty) = I \Big\}. \end{aligned} \tag{8.12.5}$$

From (8.12.5), we get the following simple

8.12.2 Proposition. *Let $W \in \mathcal{S}\big(\mathbb{R}, L(H)\big)$ admit a factorization $W = W_- \Delta W_+$ relative to $\mathbb{R}$ and $GL(H)$ (Def. 8.11.1). Then this factorization can be chosen with*

$$W_-, W_-^{-1} \in \mathcal{S}_-\big(\mathbb{R}, L(H)\big) \quad \text{and} \quad W_+, W_+^{-1} \in \mathcal{S}_+\big(\mathbb{R}, L(H)\big). \tag{8.12.6}$$

If W admits a canonical factorization, then the canonical factorization with (8.12.6) is uniquely determined.

Proof. Let

$$W(\lambda) = \widetilde{W}_-(\lambda)\left(\widetilde{P}_0 + \sum_{j=1}^{n}\left(\frac{\lambda-i}{\lambda+i}\right)^{\kappa_j}\widetilde{P}_j\right)\widetilde{W}_+(\lambda), \qquad \lambda\in\mathbb{R}, \tag{8.12.7}$$

be an arbitrary factorization of W. As $W(\infty)=I$, then

$$I = W(\infty) = \widetilde{W}_-(\infty)\widetilde{W}_+(\infty).$$

Set $A=\widetilde{W}_-(\infty)$, $W_-=\widetilde{W}_-A^{-1}$, $W_+=A\widetilde{W}_+$, and $P_j=A\widetilde{P}_jA^{-1}$ for $0\le j\le n$. Then $W_-(\infty)=W_+(\infty)=I$. Therefore, by (8.12.5), we have (8.12.6). Moreover, then we get a factorization

$$\begin{aligned}
W(\lambda) &= \widetilde{W}_-(\lambda)\left(\widetilde{P}_0 + \sum_{j=1}^{n}\left(\frac{\lambda-i}{\lambda+i}\right)^{\kappa_j}\widetilde{P}_j\right)\widetilde{W}_+(\lambda)\\
&= \widetilde{W}_-(\lambda)A^{-1}\left(A\widetilde{P}_0A^{-1} + \sum_{j=1}^{n}\left(\frac{\lambda-i}{\lambda+i}\right)^{\kappa_j}A\widetilde{P}_jA^{-1}\right)A\widetilde{W}_+(\lambda)\\
&= W_-(\lambda)\left(P_0 + \sum_{j=1}^{n}\left(\frac{\lambda-i}{\lambda+i}\right)^{\kappa_j}P_j\right)W_+(\lambda), \qquad \lambda\in\mathbb{R},
\end{aligned}$$

of the required form.

The assertion of uniqueness in the case of canonical factorizations follows from the fact that holomorphic functions on $\mathbb{C}\cup\{\infty\}$ are constant (using first Theorem 1.5.4). □

Moreover, in view of (8.12.4) and (8.12.5), we immediately obtain the following corollaries of theorems 8.10.2 and 8.10.6:

8.12.3 Corollary. (i) *A symbol $W\in\mathcal{S}(\mathbb{R},L(H))$ belongs to $\mathcal{S}_+(\mathbb{R},L(H))$, if and only if it admits a continuous extension to $\overline{\mathbb{H}}_+\cup\{\infty\}$ which is holomorphic in $\mathbb{H}_+$, and then*

$$\max_{\lambda\in\overline{\mathbb{H}}_+\cup\{\infty\}}\|W(\lambda)\| \le \|W\|_{L(H)\oplus\widehat{\mathcal{L}}^1(\mathbb{R},L(H))}.$$

(ii) *A symbol $W\in\mathcal{S}(\mathbb{R},L(H))$ belongs to $\mathcal{S}_-(\mathbb{R},L(H))$, if and only if it admits a continuous extension to $\overline{\mathbb{H}}_-\cup\{\infty\}$ which is holomorphic in $\mathbb{H}_-$, and then*

$$\max_{\lambda\in\overline{\mathbb{H}}_-\cup\{\infty\}}\|W(\lambda)\| \le \|W\|_{L(H)\oplus\widehat{\mathcal{L}}^1(\mathbb{R},L(H))}.$$

(iii) *Each symbol* $W \in \mathcal{S}(\mathbb{R}, L(H))$ *is continuous on* $\mathbb{R} \cup \{\infty\}$. *Moreover, for each* $W \in \mathcal{S}(\mathbb{R}, L(H))$,

$$\max_{\lambda \in \mathbb{R} \cup \{\infty\}} \|W(\lambda)\| \le \|W\|_{L(H) \oplus \widehat{\mathcal{L}}^1(\mathbb{R}, L(H))}.$$

8.12.4 Corollary. *Let* $W \in \mathcal{S}(\mathbb{R}, L(H))$ *such that* $W(\lambda) \in GL(H)$ *for all* $\lambda \in \mathbb{R}$.[9] *Then:*

(i) *The pointwise defined function* W^{-1} *again belongs to* $\mathcal{S}(\mathbb{R}, L(H))$.

(ii) *The function* W *can be written in the form*

$$W = V_- V V_+,$$

where

$$\begin{aligned} &V_-(\lambda) \in GL(H), \quad \lambda \in \overline{\mathbb{H}}_- \cup \{\infty\}, \quad V_-, V_-^{-1} \in \mathcal{S}_-(\mathbb{R}, L(H)), \\ &V_+(\lambda) \in GL(H), \quad \lambda \in \overline{\mathbb{H}}_+ \cup \{\infty\}, \quad V_+, V_+^{-1} \in \mathcal{S}_+(\mathbb{R}, L(H)). \end{aligned}$$

(iii) *Let* $\widetilde{W} \in \mathcal{S}(\mathbb{R}, L(H))$ *be a second symbol with* $\widetilde{W}(\lambda) \in GL(H)$ *for all* $\lambda \in \mathbb{R}$, *and assume that*

$$W = W_- \widetilde{W} W_+ \qquad \text{on } \mathbb{R},$$

where $W_- : \overline{\mathbb{H}}_- \cup \{\infty\} \to GL(H)$ *and* $W_+ : \overline{\mathbb{H}}_+ \cup \{\infty\} \to GA$ *are continuous functions which are holomorphic in* $\mathbb{H}_-$ *and* $\overline{\mathbb{H}}_+$, *respectively, and*

$$W_-(\infty) = W_+(\infty) = I.$$

Then $W_-, W_-^{-1} \in \mathcal{S}_-(\mathbb{R}, L(H))$ *and* $W_+, W_+^{-1} \in \mathcal{S}_+(\mathbb{R}, L(H))$.

In this section we prove the following three theorems:

8.12.5 Theorem. *Let* $W \in \mathcal{S}(\mathbb{R}, L(H))$. *Then the following two conditions are equivalent:*

(i) $W(\lambda) \neq 0$ *for all* $\lambda \in \mathbb{R}$, *and* W *admits a canonical factorization relative to* $\mathbb{R}$ *and* $GL(H)$ (*Def.* 8.11.1).

(ii) *The Wiener-Hopf operator* T_W *with symbol* W *is invertible.*

In that case we have: If $W = W_- W_+$ *is the canonical factorization of* W *with*

$$W_-, W_-^{-1} \in \mathcal{S}_-(\mathbb{R}, L(H)) \qquad \text{and} \qquad W_+, W_+^{-1} \in \mathcal{S}_+(\mathbb{R}, L(H)), \tag{8.12.8}$$

(*see Proposition* 8.12.2), *then the inverse of* T_W *is obtained in the following way: If* $K_\pm \in \mathcal{L}^1_\pm(\mathbb{R}, L(H))$ *are the kernel functions of* $W_\pm^{-1}$, *i.e.,*

$$W_-^{-1}(\lambda) = I - \int_{-\infty}^0 e^{i\lambda x} K_-(x)\,dx \quad \text{and} \quad W_+^{-1}(\lambda) = I - \int_0^\infty e^{i\lambda x} K_+(x)\,dx,$$

[9] Note that then $W(\lambda) \in GL(H)$ for all $\lambda \in \mathbb{R} \cup \{\infty\}$, as always $W(\infty) = I$.

then, for all $u \in \mathcal{L}^2_+(\mathbb{R},H)$,

$$(T_W^{-1}u)(y) = u(y) + \int_0^\infty K(y,x)u(x)\,dx, \qquad y \geq 0, \tag{8.12.9}$$

where

$$K(y,x) = \begin{cases} -K_+(y-x) + \int\limits_0^x K_+(y-t)K_-(t-x)\,dt & \text{if } 0 \leq x \leq y < \infty, \\ -K_-(y-x) + \int\limits_0^y K_+(y-t)K_-(t-x)\,dt & \text{if } 0 \leq y \leq x < \infty. \end{cases}$$

8.12.6 Theorem. *Let* $W \in \mathcal{S}(\mathbb{R}, L(H))$. *Then the following two conditions are equivalent:*

(i) $W(\lambda) \neq 0$ *for all* $\lambda \in \mathbb{R}$, *and* W *admits a factorization relative to* $\mathbb{R}$ *and* $GL(H)$ (*Def.* 8.11.1).

(ii) *The Wiener-Hopf operator* T_W *with symbol* W *is a Fredholm operator.*

In that case we have: If

$$W(\lambda) = W_-(\lambda)\left(P_0 + \sum_{j=1}^n \left(\frac{\lambda-i}{\lambda+i}\right)^{\kappa_j} P_j\right)W_+(\lambda), \qquad \lambda \in \mathbb{R},$$

is a factorization of W *with respect to* $\mathbb{R}$ *and* $GL(H)$, *and if* r *is the index with* $\kappa_1 > \ldots > \kappa_r > 0 > \kappa_{r+1} > \ldots > \kappa_n$, *then*

$$\dim \operatorname{Ker} T_W = -\sum_{j=1}^r \kappa_j \dim P_j \quad \textit{and} \quad \dim \operatorname{Coker} T_W = \sum_{j=r+1}^n \kappa_j \dim P_j. \tag{8.12.10}$$

8.12.7 Theorem. *Let* $W \in \mathcal{S}(\mathbb{R}, L(H))$ *be a symbol satisfying the equivalent conditions* (i) *and* (ii) *in Theorem* 8.12.6, *let*

$$W(\lambda) = W_-(\lambda)\left(P_0 + \sum_{j=1}^n \left(\frac{\lambda-i}{\lambda+i}\right)^{\kappa_j} P_j\right) W_+(\lambda), \qquad \lambda \in \mathbb{R},$$

be a factorization of W *relative to* $\mathbb{R}$ *and* $GL(H)$ *with*

$$W_-, W_-^{-1} \in \mathcal{S}_-(\mathbb{R}, L(H)) \quad \textit{and} \quad W_+, W_+^{-1} \in \mathcal{S}_+(\mathbb{R}, L(H)) \tag{8.12.11}$$

(*which then exists by Proposition* 8.12.2), *let* r *be the index with* $\kappa_1 > \ldots > \kappa_r > 0 > \kappa_{r+1} > \ldots > \kappa_n$, *and let*

$$\Delta(\lambda) := P_0 + \sum_{j=1}^n \left(\frac{\lambda-i}{\lambda+i}\right)^{\kappa_j} P_j, \qquad \lambda \in \mathbb{R}.$$

Then: $\Delta, \Delta^{-1} \in \mathcal{S}\big(\mathbb{R}, L(H)\big)$ *(this holds by* (8.12.4)) *and (assertion of the theorem)*

$$T_{W_+^{-1}} T_{\Delta^{-1}} T_{W_-^{-1}}$$

is a generalized inverse (Section 6.10.2) *of* T_W*, and the operator* $T_{\Delta^{-1}}$ *is given by*

$$T_{\Delta^{-1}} = I + \sum_{j=1}^{n} T_j P_j, \tag{8.12.12}$$

where, for all $u \in \mathcal{L}^2_+(\mathbb{R}, H)$ *and* $y \geq 0$,

$$(T_j u)(y) = (P_j u)(y) - \sum_{\nu=1}^{\kappa_j} \binom{\kappa_j}{\nu} \frac{2^\nu}{\nu!} \int_y^\infty (y-x)^{\nu-1} e^{y-x} (P_j u)(x)\,dx \quad \text{if } r+1 \leq j \leq n, \tag{8.12.13}$$

and

$$(T_j u)(y) = (P_j u)(y) + \sum_{\nu=1}^{-\kappa_j} \binom{-\kappa_j}{\nu} \frac{(-2)^\nu}{\nu!} \int_0^y (y-x)^{\nu-1} e^{x-y} (P_j) u(x)\,dx \quad \text{if } r+1 \leq j \leq n. \tag{8.12.14}$$

The remainder of this section is devoted to the proof of these three theorems.

8.12.8. Let $\Phi : \mathbb{C} \cup \{\infty\} \to \mathbb{C} \cup \{\infty\}$ be the Möbius transform defined by

$$\Phi(z) = i\frac{1+z}{1-z}, \qquad z \in \mathbb{C},$$

(cf. Section (8.9.1)), let $\mathbb{T}$ be the unit circle, and let $\mathcal{L}^2(\mathbb{T}, H)$, $\mathcal{L}^2_+(\mathbb{T}, H)$, $\mathcal{L}^2_-(\mathbb{T}, H)$ be the Hilbert spaces introduced in Section 8.3.1 and in Theorem and Definition 8.3.10. Recall that, by part (iii) of Theorem and Definition 8.3.10, then we have the orthogonal decomposition

$$\mathcal{L}^2(\mathbb{T}, H) = \mathcal{L}^2_+(\mathbb{T}, H) \oplus \mathcal{L}^2_-(\mathbb{T}, H). \tag{8.12.15}$$

Let $\mathcal{P}^{\mathbb{T}}$ be the orthogonal projector from $\mathcal{L}^2(\mathbb{T}, H)$ onto $\mathcal{L}^2_+(\mathbb{T}, H)$. Note that then $\operatorname{Ker} \mathcal{P}^{\mathbb{T}} = \mathcal{L}^2_-(\mathbb{T}, H)$, as it follows from the orthogonal decomposition (8.12.15).

For each symbol $S \in \mathcal{S}\big(\mathbb{R}, L(H)\big)$, together with the Wiener-Hopf operator T_S acting in $\mathcal{L}^2(\mathbb{R}, H)$, we also consider the Wiener-Hopf operator $\mathcal{W}_{S\circ\Phi}$ acting in $\mathcal{L}^2_+(\mathbb{T}, H)$ by

$$\mathcal{W}_{S\circ\Phi} f = \mathcal{P}^{\mathbb{T}}\Big((S \circ \phi) f\Big), \qquad f \in \mathcal{L}^2_+(\mathbb{T}, H),$$

which was studied already in Section 8.4.

8.12.9 Proposition. *Let* $\mathbf{V}$ *be the isometry introduced in Theorem and Definition* 8.9.2, *and let* $\mathbf{U}$ *be the Fourier isometry of* $\mathcal{L}^2(\mathbb{R},H)$ (*Section* 8.8.5). *Then*

$$\mathcal{W}_{S\circ\Phi} = \mathbf{V}^{-1}\mathbf{U}T_S\mathbf{U}^{-1}\mathbf{V} \qquad \textit{for all } S\in\mathcal{S}\big(\mathbb{R},L(H)\big). \tag{8.12.16}$$

Proof. Let $S\in\mathcal{S}\big(\mathbb{R},L(H)\big)$ be given. First note that

$$(S\circ\phi)f = \mathbf{V}^{-1}\Big(S(\mathbf{V}f)\Big) \qquad \text{for all } f\in\mathcal{L}^2(\mathbb{T},H). \tag{8.12.17}$$

Indeed, if f belongs to the linear space $\mathcal{C}^0(\mathbb{T},H)\cap\mathcal{L}^2(\mathbb{T},H)$, this follows from the definition (8.9.3) of $\mathbf{V}$. Since this space is dense in $\mathcal{L}^2(\mathbb{T},H)$ (by our definition of $\mathcal{L}^2(\mathbb{T},H)$, it follows for all $f\in\mathcal{L}^2(\mathbb{T},H)$.

Now let K be the kernel function of S. Then, for all $u\in\mathcal{L}^2_+(\mathbb{R},H)$, it follows from (8.8.6) and (8.12.17) that

$$\begin{aligned}T_Su &= \mathcal{P}^{\mathbb{R}}(u-K*u) = \mathcal{P}^{\mathbb{R}}\mathbf{U}^{-1}\big(\mathbf{U}u-\mathbf{U}(K*u)\big) = \mathcal{P}^{\mathbb{R}}\mathbf{U}^{-1}\big(\mathbf{U}u-\widehat{K}\mathbf{U}u\big)\\ &= \mathcal{P}^{\mathbb{R}}\mathbf{U}^{-1}\big((I-\widehat{K})\mathbf{U}u\big) = \mathcal{P}^{\mathbb{R}}\mathbf{U}^{-1}\Big(S(\mathbf{U}u)\Big)\end{aligned}$$

for all $u\in\mathcal{L}^2_+(\mathbb{R},H)$.

Now let $f\in\mathcal{L}^2(\mathbb{T},H)$ be given. Then this implies that

$$\begin{aligned}\mathbf{V}^{-1}\mathbf{U}T_S\mathbf{U}^{-1}\mathbf{V}f &= \mathbf{V}^{-1}\mathbf{U}\mathcal{P}^{\mathbb{R}}\mathbf{U}^{-1}\Big(S(\mathbf{U}\mathbf{U}^{-1}\mathbf{V}f)\Big) = \mathbf{V}^{-1}\mathbf{U}\mathcal{P}^{\mathbb{R}}\mathbf{U}^{-1}\Big(S(\mathbf{V}f)\Big)\\ &= \mathbf{V}^{-1}\mathbf{U}\mathcal{P}^{\mathbb{R}}\mathbf{U}^{-1}\mathbf{V}\Big(\mathbf{V}^{-1}\big(S(\mathbf{V}f)\big)\Big).\end{aligned}$$

As, by Theorem 8.9.5, $\mathbf{V}^{-1}\mathbf{U}\mathcal{P}^{\mathbb{R}}\mathbf{U}^{-1}\mathbf{V} = \mathcal{P}^{\mathbb{T}}$ and, by (8.12.17), $\mathbf{V}^{-1}\big(S(\mathbf{V}f)\big) = (S\circ\phi)f$, this further implies

$$\mathbf{V}^{-1}\mathbf{U}T_S\mathbf{U}^{-1}\mathbf{V}f = \mathcal{P}^{\mathbb{T}}\Big((S\circ\phi)f\Big),$$

i.e.,

$$\mathbf{V}^{-1}\mathbf{U}T_S\mathbf{U}^{-1}\mathbf{V}f = \mathcal{W}_{S\circ\Phi}f.$$

□

8.12.10 Lemma. *Let* $W\in\mathcal{S}\big(\mathbb{R},L(H)\big)$ *such that* T_W *is a Fredholm operator. Then* $W(\lambda)$ *is invertible for all* $\lambda\in\mathbb{R}$.

Proof. Let Φ be the Möbius transformation introduced in Section 8.9.1. Then, by Proposition 8.12.9, also the Wiener-Hopf operator $\mathcal{W}_{W\circ\Phi}$ is a Fredholm operator. Therefore the assertion follows from Proposition 8.6.6. □

8.12.11. Proof of Theorem 8.12.5. By Lemma 8.12.10 we may already assume that

$$W(\lambda)\in GL(H) \qquad \text{for all } \lambda\in\mathbb{R}. \tag{8.12.18}$$

Let

$$\Phi(z) := z\frac{1+z}{1-z}.$$

As $W(\infty) = I$, then it follows from (8.12.18) that $(W \circ \Phi)(z)$ is invertible for all $z \in \mathbb{T}$, where $\mathbb{T}$ is the unit circle. Moreover, part (ii) of Corollary 8.12.4 in particular implies that $W \circ \Phi$ admits local factorizations with respect to $\mathbb{T}$ and $GL(H)$ (Def. 7.1.3). Therefore, by Theorem 8.4.2, the Wiener-Hopf operator $\mathcal{W}_{W\circ\Phi}$ is invertible, if and only if, $W \circ \Phi$ admits a canonical factorization relative to $\mathbb{T}$ and $GL(H)$. As, by Proposition 8.12.9,

$$\mathcal{W}_{W\circ\Phi} = \mathbf{V}^{-1}\mathbf{U}T_W\mathbf{U}^{-1}\mathbf{V},$$

this implies that T_W is invertible, if and only if, $W \circ \Phi$ admits a canonical factorization relative to $\mathbb{T}$ and $GL(H)$. Finally, by Proposition 8.11.3, this yields that T_W is invertible, if and only if, W admits a canonical factorization relative to $\mathbb{R}$ and $GL(H)$. So the equivalence of the conditions (i) and (ii) in Theorem 8.12.5 is proved.

Now let $W = W_-W_+$ be the canonical factorization of W with (8.12.8). As the constant function with value I belongs to $\mathcal{S}\big(\mathbb{R}, L(H)\big)$, then part (iii) of Corollary 8.12.4 implies (8.12.8). Hence, by parts (ii) and (iii) of Proposition 8.12.14, the operators T_{W_-} and T_{W_+} are invertible, where

$$T_{W_-}^{-1} = T_{W_-^{-1}} \qquad \text{and} \qquad T_{W_+}^{-1} = T_{W_+^{-1}},$$

and, by part (i) of this proposition,

$$T_W = T_{W_-}T_{W_+}.$$

Together this implies that the inverse of T_W is given by

$$T_W^{-1} = T_{W_+^{-1}}T_{W_-^{-1}}. \tag{8.12.19}$$

Let K_+ and K_- be the kernels of W_+^{-1} and W_-^{-1}, respectively, and let $u \in \mathcal{L}_+^2(\mathbb{R}, H)$. Then, by (8.12.19),

$$\begin{aligned}
&(T_W^{-1}u)(y) = \Big(T_{W_+^{-1}}\big(T_{W_-^{-1}}u\big)\Big)(y) \\
&= \big(T_{W_-^{-1}}u\big)(y) - \int_{-\infty}^{\infty} K_+(y-x)\big(T_{W_-^{-1}}u\big)(x)\,dx \\
&= u(y) - \int_{-\infty}^{\infty} \Big((K_-(y-x) + K_+(y-x)\Big)u(x)dx \\
&\quad + \int_{-\infty}^{\infty}\left(\int_{-\infty}^{\infty} K_+(y-x)K_-(x-t)\,u(t)\,dt\right)dx, \qquad y \geq 0.
\end{aligned} \tag{8.12.20}$$

First applying Fubini's theorem and then changing the role of x and t, we get

$$\begin{aligned}\int_{-\infty}^{\infty}\left(\int_{-\infty}^{\infty} K_+(y-x)K_-(x-t)\,u(t)\,dt\right)dx \\ &= \int_{-\infty}^{\infty}\left(\int_{-\infty}^{\infty} K_+(y-x)K_-(x-t)\,u(t)\,dx\right)dt \\ &= \int_{-\infty}^{\infty}\left(\int_{-\infty}^{\infty} K_+(y-t)K_-(t-x)\,dt\right)u(x)\,dx, \quad y\geq 0.\end{aligned}$$

Therefore it follows from (8.12.20) that

$$\left(T_W^{-1}u\right)(y) = u(y) + \int_{-\infty}^{\infty} L(y,x)u(x)dx, \qquad y \geq 0, \tag{8.12.21}$$

where

$$L(y,x) := -K_-(x-y) - K_+(x-y) + \int_{-\infty}^{\infty} K_+(y-t)K_-(t-x)\,dt, \quad y,x\geq 0.$$

Since $K_+(s) = 0$ if $s \leq 0$ and $K_-(s) = 0$ if $s \geq 0$, we see that $L(y,x) = K(y,x)$ for all $x, y \geq 0$. This completes the proof of Theorem 8.12.5. □

8.12.12. Proof of Theorem 8.12.6. By Lemma 8.12.10 we may already assume that

$$W(\lambda) \in GL(H) \qquad \text{for all } \lambda \in \mathbb{R}. \tag{8.12.22}$$

Let

$$\Phi(z) := z\frac{1+z}{1-z}.$$

As $W(\infty) = I$, then it follows from (8.12.22) that $(W \circ \Phi)(z)$ is invertible for all $z \in \mathbb{T}$, where $\mathbb{T}$ is the unit circle. Moreover, part (ii) of Corollary 8.12.4 in particular implies that $W \circ \Phi$ admits local factorizations with respect to $\mathbb{T}$ and $GL(H)$ (Def. 7.1.3).

Therefore, by Theorem 8.4.2, the Wiener-Hopf operator $\mathcal{W}_{W\circ\Phi}$ is a Fredholm operator, if and only if, $W \circ \Phi$ admits a factorization relative to $\mathbb{T}$ and $GL(H)$. As, by Proposition 8.12.9,

$$\mathcal{W}_{W\circ\Phi} = \mathbf{V}^{-1}\mathbf{U}T_W\mathbf{U}^{-1}\mathbf{V}, \tag{8.12.23}$$

this implies that T_W is a Fredholm operator, if and only if $W \circ \Phi$ admits a factorization relative to $\mathbb{T}$ and $GL(H)$. Finally, by Proposition 8.11.3, this yields that T_W is a Fredholm operator, if and only if, W admits a canonical factorization relative to $\mathbb{R}$ and $GL(H)$. So the equivalence of the conditions (i) and (ii) in Theorem 8.12.6 is proved.

Now we assume that these two conditions are satisfied, that

$$W(\lambda) = W_-(\lambda)\Big(P_0 + \sum_{j=1}^{n}\Big(\frac{\lambda - i}{\lambda + i}\Big)^{\kappa_j} P_j\Big)W_+(\lambda), \qquad \lambda \in \mathbb{R},$$

is a factorization of W relative to $\mathbb{R}$ and $GL(H)$, and that r is the index with $\kappa_1 > \ldots > \kappa_r > 0 > \kappa_{r+1} > \ldots > \kappa_n$.

By Proposition 8.11.3, then

$$(W \circ \Phi)(z) = (W_- \circ \Phi)(z)\Big(P_0 + \sum_{j=1}^{n} z^{\kappa_j} P_j\Big)(W_+ \circ \Phi)(z), \qquad z \in \mathbb{T},$$

is a factorization of $W \circ \Phi$ relative to $\mathbb{T}$ and $GL(H)$. By Theorem 8.4.2, this implies that

$$\dim \operatorname{Ker} \mathcal{W}_{W\circ\Phi} = -\sum_{j=r+1}^{n} \kappa_j P_j \quad \text{and} \quad \dim \operatorname{Coker} \mathcal{W}_{W\circ\Phi} = \sum_{j=1}^{r} \kappa_j P_j.$$

By (8.12.23), this further implies (8.12.10). This completes the proof of Theorem 8.12.6. □

8.12.13 Proposition. *Let $V, W \in \mathcal{S}(\mathbb{R}, L(H))$. Then $V, W \in \mathcal{S}(\mathbb{R}, L(H))$ and if K_V and K_W are the kernel functions of V and W, respectively, then $K_V + K_W - K_V * K_W$ is the kernel function of VW.*

Proof. As $\widehat{K}_V \widehat{K}_W = \widehat{K_V * K_W}$ (see (8.7.12)), we have

$$VW = (I - \widehat{K}_V)(I - \widehat{K}_W) = I - \widehat{K}_V - \widehat{K}_W + \widehat{K}_V\widehat{K}_W = I - \widehat{K}_V - \widehat{K}_W + \widehat{K_V * K_W}.$$

□

8.12.14 Proposition. *Let $W_- \in \mathcal{S}_-(\mathbb{R}, L(H))$, $W_+ \in \mathcal{S}_+(\mathbb{R}, L(H))$ and $W \in \mathcal{S}(\mathbb{R}, L(H))$. Then*

(i) $T_{W_-W} = T_{W_-}T_W$ *and* $T_{WW_+} = T_W T_{W_+}$.

(ii) *Suppose $W_-(\lambda)$ is invertible for all $\lambda \in \overline{\mathbb{H}}_- \cup \{\infty\}$, and, hence (Corollaries 8.12.3 and 8.12.4), also $W_-^{-1} \in \mathcal{S}_-(\mathbb{R}, L(H))$. Then T_{W_-} is invertible and*

$$T_{W_-}^{-1} = T_{W_-^{-1}}.$$

(iii) *Suppose $W_+(\lambda)$ is invertible for all $\lambda \in \overline{\mathbb{H}}_+ \cup \{\infty\}$, and, hence (Corollaries 8.12.3 and 8.12.4), also $W_+^{-1} \in \mathcal{S}_+(\mathbb{R}, L(H))$. Then T_{W_+} is invertible and*

$$T_{W_+}^{-1} = T_{W_+^{-1}}.$$

We only have to prove part (i), because (ii) and (iii) follow from (i).

Let $K_\pm$ and K be the kernel functions of $W_\pm$ and W, respectively. Then, for all $f \in \mathcal{L}^2_+(\mathbb{R},H)$,

$$\begin{aligned} T_{W_-}T_W f &= T_{W_-}\Big(f - \mathcal{P}^{\mathbb{R}}(K*f)\Big) \\ &= f - \mathcal{P}^{\mathbb{R}}(K*f) - \mathcal{P}^{\mathbb{R}}\Big(K_- * \Big(f - \mathcal{P}^{\mathbb{R}}(K*f)\Big)\Big) \\ &= f - \mathcal{P}^{\mathbb{R}}\Big((K+K_-)*f\Big) + \mathcal{P}^{\mathbb{R}}\Big(K_- * \mathcal{P}^{\mathbb{R}}(K*f)\Big) \end{aligned}$$

and

$$\begin{aligned} T_W T_{W_+} f &= T_W\Big(f - \mathcal{P}^{\mathbb{R}}(K_+*f)\Big) \\ &= f - \mathcal{P}^{\mathbb{R}}(K_+*f) - \mathcal{P}^{\mathbb{R}}\Big(K * \Big(f - \mathcal{P}^{\mathbb{R}}(K_+*f)\Big)\Big) \\ &= f - \mathcal{P}^{\mathbb{R}}\Big((K_++K)*f\Big) + \mathcal{P}^{\mathbb{R}}\Big(K * \mathcal{P}^{\mathbb{R}}(K_+*f)\Big). \end{aligned}$$

Since, by proposition (8.8.8),

$$\mathcal{P}^{\mathbb{R}}\Big(K_- * \mathcal{P}^{\mathbb{R}}(K*f)\Big) = \mathcal{P}^{\mathbb{R}}\Big(K_- * (K*f)\Big)$$

and

$$\mathcal{P}^{\mathbb{R}}\Big(K * \mathcal{P}^{\mathbb{R}}(K_+*f)\Big) = \mathcal{P}^{\mathbb{R}}\Big(K * (K_+*f)\Big),$$

this implies that

$$T_{W_-}T_W f = f - \mathcal{P}^{\mathbb{R}}\Big((K+K_-)*f\Big) + \mathcal{P}^{\mathbb{R}}\Big(K_- * (K*f)\Big)$$

and

$$T_W T_{W_+} f = f - \mathcal{P}^{\mathbb{R}}\Big((K_++K)*f\Big) + \mathcal{P}^{\mathbb{R}}\Big(K * (K_+*f)\Big).$$

As the convolution is associative (Proposition 8.8.7), this further implies

$$T_{W_-}T_W f = f - \mathcal{P}^{\mathbb{R}}\Big((K+K_- - K_- * K)*f\Big)$$

and

$$T_W T_{W_+} f = f - \mathcal{P}^{\mathbb{R}}\Big((K_+ + K - K*K_+)*f\Big).$$

Since, by Proposition 8.12.13, $K + K_- - K_- * K$ is the kernel function of T_{W_-W} and $K_+ + K - K * K_+$ is the kernel function of T_{WW_+}, this proves part (i) of the proposition.

8.12.15. Proof of Theorem 8.12.6. We introduce the abbreviations

$$\Psi(\lambda) = \frac{\lambda - i}{\lambda + i} \qquad \text{and} \qquad \Theta(\lambda) = \frac{\lambda + i}{\lambda - i}, \qquad \lambda \in \mathbb{C}.$$

Then, by definition of Δ,

$$\begin{aligned}\Delta &= P_0 + \sum_{j=1}^{r} \Psi^{\kappa_j} P_j + \sum_{j=r+1}^{n} \Theta^{-\kappa_j} P_j,\\ \Delta^{-1} &= P_0 + \sum_{j=1}^{r} \Theta^{\kappa_j} P_j + \sum_{j=r+1}^{n} \Psi^{-\kappa_j} P_j,\end{aligned} \tag{8.12.24}$$

and, by (8.12.5),

$$\Psi^{\kappa} \in \mathcal{S}_+\big(\mathbb{R}, \mathbb{C}\big) \quad \text{and} \quad \Theta^{\kappa} \in \mathcal{S}_-\big(\mathbb{R}, \mathbb{C}\big), \qquad \kappa \in \mathbb{N}^*.$$

By Proposition 8.12.14, the latter relation implies that

$$T_{\Theta^{\kappa}} T_{\Psi^{\kappa}} = I, \qquad \kappa \in \mathbb{N}^*. \tag{8.12.25}$$

From (8.12.24) it follows that

$$\begin{aligned}T_{\Delta} &= P_0 + \sum_{j=1}^{r} T_{\Psi^{\kappa_j}} P_j + \sum_{j=r+1}^{r} T_{\Theta^{-\kappa_j}} P_j,\\ T_{\Delta^{-1}} &= P_0 + \sum_{j=1}^{r} T_{\Theta^{\kappa_j}} P_j + \sum_{j=r+1}^{n} T_{\Psi^{-\kappa_j}} P_j,\end{aligned} \tag{8.12.26}$$

and, further, by (8.12.25),

$$\begin{aligned}T_{\Delta^{-1}} T_{\Delta}^{-1} T_{\Delta^{-1}} &= P_0 + \sum_{j=1}^{r} T_{\Theta^{\kappa_j}} T_{\Psi^{\kappa_j}} T_{\Theta^{\kappa_j}} P_j + \sum_{j=r+1}^{r} T_{\Psi^{-\kappa_j}} T_{\Theta^{-\kappa_j}} T_{\Psi^{-\kappa_j}} P_j\\ &= T_{\Delta^{-1}},\\ T_{\Delta} T_{\Delta^{-1}}^{-1} T_{\Delta} &= P_0 + \sum_{j=1}^{r} T_{\Psi^{\kappa_j}} T_{\Theta^{\kappa_j}} T_{\Psi^{\kappa_j}} P_j + \sum_{j=r+1}^{n} T_{\Theta^{-\kappa_j}} T_{\Psi^{-\kappa_j}} T_{\Theta^{-\kappa_j}} P_j\\ &= T_{\Delta}.\end{aligned}$$

Moreover, again by Proposition 8.12.14,

$$\begin{aligned}&T_{W_-^{-1}} T_{W_-} = T_{W_-} T_{W_-^{-1}} = I,\\ &T_{W_+^{-1}} T_{W_+} = T_{W_+} T_{W_+^{-1}} = I,\\ &T_W = T_{W_-} T_{\Delta} T_{W_+}.\end{aligned}$$

Together this yields

$$\begin{aligned}
&T_W\left(T_{W_+^{-1}}T_{\Delta^{-1}}T_{W_-^{-1}}\right)T_W\\
&\qquad = \left(T_{W_-}T_\Delta T_{W_+}\right)\left(T_{W_+^{-1}}T_{\Delta^{-1}}T_{W_-^{-1}}\right)\left(T_{W_-}T_\Delta T_{W_+}\right)\\
&\qquad = T_{W_-^{-1}}T_{\Delta^{-1}}T_\Delta T_{\Delta^{-1}}T_{W_+^{-1}}\\
&\qquad = T_{W_-^{-1}}T_{\Delta^{-1}}T_{W_+^{-1}}\\
&\qquad = T_W
\end{aligned}$$

and

$$\begin{aligned}
&\left(T_{W_+^{-1}}T_{\Delta^{-1}}T_{W_-^{-1}}\right)T_W\left(T_{W_+^{-1}}T_{\Delta^{-1}}T_{W_-^{-1}}\right)\\
&\qquad = \left(T_{W_+^{-1}}T_{\Delta^{-1}}T_{W_-^{-1}}\right)\left(T_{W_-}T_\Delta T_{W_+}\right)\left(T_{W_+^{-1}}T_{\Delta^{-1}}T_{W_-^{-1}}\right)\\
&\qquad = T_{W_+^{-1}}T_{\Delta^{-1}}T_\Delta T_{\Delta^{-1}}T_{W_-^{-1}}\\
&\qquad = T_{W_+^{-1}}T_{\Delta^{-1}}T_{W_-^{-1}},
\end{aligned}$$

which shows that $T_{W_+^{-1}}T_{\Delta^{-1}}T_{W_-^{-1}}$ is a generalized inverse of T_W.

It remains to prove (8.12.12).

It follows from Proposition 8.7.11 that

$$(\Psi-1)^\nu = \frac{(-2)^\nu}{\nu!}\widehat{\theta}^+_{\nu-1} \quad\text{and}\quad (\Theta-1)^\nu = \frac{2^\nu}{\nu!}\widehat{\theta}^-_{\nu-1}, \qquad \nu\in\mathbb{N}^*,$$

and further

$$\begin{aligned}
\Psi^\kappa - 1 = (\Psi-1+1)^\kappa - 1 &= \sum_{\nu=1}^{\kappa}\binom{\kappa}{\nu}(\Psi-1)^\nu\\
&= \sum_{\nu=1}^{\kappa}\binom{\kappa}{\nu}\frac{(-2)^\nu}{\nu!}\widehat{\theta}^+_{\nu-1}, \quad \kappa\in\mathbb{N}^*,\\
\Theta^\kappa - 1 &= \sum_{\nu=1}^{\kappa}\binom{\kappa}{\nu}\frac{2^\nu}{\nu!}\widehat{\theta}^-_{\nu-1}, \qquad \kappa\in\mathbb{N}^*.
\end{aligned}$$

This implies that, for all $u\in\mathcal{L}^2_+(\mathbb{R},H)$,

$$\begin{aligned}
(T_{\Psi^\kappa}u)(y) &= u(y) + \sum_{\nu=1}^{\kappa}\binom{\kappa}{\nu}\frac{(-2)^\nu}{\nu!}\int_{-\infty}^{\infty}\widehat{\theta}^+_{\nu-1}(x)u(x)dx, \quad y\geq 0,\\
(T_{\Theta^\kappa}u)(y) &= u(y) + \sum_{\nu=1}^{\kappa}\binom{\kappa}{\nu}\frac{2^\nu}{\nu!}\int_{-\infty}^{\infty}\widehat{\theta}^-_{\nu-1}(x)u(x)dx, \quad y\geq 0.
\end{aligned}$$

Since, by definition of the functions $\theta^{\pm}_{\nu-1}$ (Section 8.7.8), for all $\nu \in \mathbb{N}^*$,

$$\theta^{+}_{\nu-1}(y-x) = \begin{cases} (y-x)^{\nu-1}e^{x-y} & \text{if } y-x \geq 0, \\ 0 & \text{if } y-x < 0, \end{cases}$$

and

$$\theta^{-}_{\nu-1}(y-x) = \begin{cases} -(y-x)^{\nu-1}e^{y-x} & \text{if } y-x \leq 0, \\ 0 & \text{if } y-x > 0, \end{cases}$$

and since $u(x) = 0$ for $x < 0$ if $u \in \mathcal{L}^2_+(\mathbb{R}, \mathbb{C})$, this further implies that, for all $\nu \in \mathbb{N}^*$,

$$\begin{aligned} (T_{\Psi^\kappa}u)(y) &= u(y) + \sum_{\nu=1}^{\kappa} \binom{\kappa}{\nu} \frac{(-2)^\nu}{\nu!} \int_0^y (y-x)^{\nu-1} e^{x-y} u(x)dx, \quad y \geq 0, \\ \big(T_{\Theta^\kappa}u\big)(y) &= u(y) - \sum_{\nu=1}^{\kappa} \binom{\kappa}{\nu} \frac{2^\nu}{\nu!} \int_y^\infty (y-x)^{\nu-1} e^{y-x} u(x)dx, \quad y \geq 0. \end{aligned} \tag{8.12.27}$$

Set

$$T_j = \begin{cases} T_{\Theta^{\kappa_j}} & \text{for } 1 \leq j \leq r, \\ T_{\Psi^{-\kappa_j}} & \text{for } r+1 \leq n. \end{cases}$$

Then (8.12.12) follows from (8.12.26) and (8.12.27). □

8.13 An example

In this section, $D_+ \subseteq \mathbb{C}$ is a bounded connected open set with piecewise $\mathcal{C}^1$-boundary Γ such that $0 \in D_+$.

If E is a Banach space, and $A \in L(E)$ is a Fredholm operator, then (see, e.g., [GGK2]) there exists $\varepsilon > 0$ such that the following holds: If $B \in L(E)$ with $\|A - B\| < \varepsilon$, then also B is a Fredholm operator, where $\operatorname{ind} A = \operatorname{ind} B$, and if A is invertible, then also B is invertible.

In view of the connection with Wiener-Hopf operators, this implies different stability statements for the factorization problem. For example, Theorem 8.4.2 immediately implies:

8.13.1 Corollary (to Theorem 8.4.2). *Let H be a separable Hilbert space, and let $A : \Gamma \to GL(H)$ be a continuous function which admits a factorization with respect to Γ (Def. 7.1.1). Then there exists $\varepsilon > 0$ such that the following holds:*

Let $B : \Gamma \to GL(H)$ be a continuous function which admits local factorizations with respect to Γ (Def. 7.1.3) and which satisfies the estimate

$$\max_{z \in \Gamma} \|A(z) - B(z)\|_{L(H)} \leq \varepsilon\,.$$

Then also B admits a factorization with respect to Γ and, moreover:

(i) *If* $\kappa_1(A), \dots, \kappa_n(A)$ *and* $\kappa_1(B), \dots, \kappa_m(B)$ *are the non-zero partial indices of* A *and* B*, respectively (Def. 7.9.6), if* $d_j(A)$ *is the multiplicity of* $\kappa_j(A)$ *as a partial index of* A *(Def. 7.9.8), and if* $d_j(B)$ *is the multiplicity of* $\kappa_j(B)$ *as a partial index of* B*, then*

$$\sum_{j=1}^{n} \kappa_j(A) d_j(A) = \sum_{j=1}^{m} \kappa_j(B) d_j(B) . \tag{8.13.1}$$

(ii) *If* A *admits a canonical factorization with respect to* Γ*, then also* B *admits a canonical factorization with respect to* Γ.

However, each partial index alone is not stable with respect to small pertubations. Consider, for $\varepsilon \in \mathbb{C}$, the following example of a 2×2-matrix:

$$A_\varepsilon(z) := \begin{pmatrix} z^{-1} & 0 \\ \varepsilon & z \end{pmatrix} , \qquad z \in \mathbb{C}^* . \tag{8.13.2}$$

For $\varepsilon = 0$, the partial indices with respect to Γ of this function are -1 and 1. However, for all $\varepsilon \neq 0$, A_ε admits a canonical factorization with respect to Γ, namely:

$$\begin{pmatrix} z^{-1} & 0 \\ \varepsilon & z \end{pmatrix} = \begin{pmatrix} z^{-1} & -\varepsilon^{-1} \\ \varepsilon & 0 \end{pmatrix} \begin{pmatrix} 1 & \varepsilon^{-1} z \\ 0 & 1 \end{pmatrix} . \tag{8.13.3}$$

This example is good to explain also some other questions.

Factorizations with respect to Γ as considered in this book, are also called **right** factorizations with respect to Γ. The notion of a **left** factorization with respect to Γ one obtains by interchanging the roles of A_- and A_+ in Definition 7.1.1, i.e., a left factorization of a continuous function $A : \Gamma \to GL(E)$ (E being a Banach space) with respect to Γ is a representation of the form $A = A_+ \Delta A_-$, where A_-, A_+ and Δ are as in Definition 7.1.1. We restrict ourselves to right factorizations with respect to Γ, because the theories of right and left factorizations with respect to Γ are equivalent.

But we point out that the partial indices (Def. 7.9.6) need not be the same for the left and the right factorization with respect to Γ of the same function (if both exist). An example is given by the function A_ε above if $\varepsilon \neq 0$. Then (8.13.3) shows that zero is the only "right" partial index of A_ε with respect to Γ, whereas the representation

$$\begin{pmatrix} z^{-1} & 0 \\ \varepsilon & z \end{pmatrix} = \begin{pmatrix} 1 & 0 \\ \varepsilon z & 1 \end{pmatrix} \begin{pmatrix} z^{-1} & 0 \\ 0 & z \end{pmatrix} . \tag{8.13.4}$$

shows that the "left" partial indices of A_ε with respect to Γ are -1 and 1.

Finally, we discuss the following generalization.

Let E be a Banach space, and let $A : \Gamma \to GL(E)$ be a continuous function.

A representation of the form $A = A_- \Delta A_+$ will be called a **generalized factorization of** A **with respect to** Γ if everything is as in Definition 7.1.1, with one exception: We do not require that the projectors $P_1, \dots, P_n$ are finite dimensional.

It is easy to see that in terms of the filtration of A with respect to Γ (Def. 7.9.8), this can be characterized as follows: A generalized factorization of A with respect to Γ exists, if and only if A admits local factorizations with respect to Γ and:

If $k_1 > \ldots > k_n$ are the partial indices of A with respect to Γ (Def. 7.9.6), then, for $1 \leq j \leq n-1$, the spaces $M_-(z, k_j, \Gamma, A)$, $z \in \overline{D}_- \cup \{\infty\}$, and $M_+(z, k_j, \Gamma, A)$, $z \in \overline{D}_+$, are topologically closed and complemented in $M_-(z, k_{j+1}, \Gamma, A)$ and $M_+(z, k_{j+1}, \Gamma, A)$, respectively, and the families of subspaces

$$\Big\{M_-(z, k_j, \Gamma, A)\Big\}_{z \in \overline{D}_- \cup \{\infty\}} \quad \text{and} \quad \Big\{M_+(z, k_j, \Gamma, A)\Big\}_{z \in \overline{D}_+}$$

are continuous (Def. 6.2.1) and holomorphic (Def. 6.4.1) over $D_- \cup \{\infty\}$ and D_+, respectively.

It turns out that also a generalized factorization of A with respect to Γ does not always exist, even if A is a polynomial in z and z^{-1}. To show this, consider the following infinite dimensional version of example (8.13.3): Let H be an infinite dimensional separable Hilbert space and let $V \in L(H)$ be an operator such that $\operatorname{Ker} V = \{0\}$ and $\operatorname{Im} V$ is not topologically closed in H, and let $A : \mathbb{C}^* \to L(H \oplus H)$ be defined by the block matrix

$$A(z) = \begin{pmatrix} z^{-1}I & 0 \\ V & zI \end{pmatrix}, \qquad z \in \mathbb{C}^*, \tag{8.13.5}$$

where I is the unit operator on H. It is easy to see that the only $(\Gamma, -1)$ section of A (Def. 7.9.1) is the zero section, and that the space of $(\Gamma, 0)$-sections of A consists of all pairs (φ^-, φ^+) of the form

$$\varphi^+(z) = \begin{pmatrix} a + zb \\ -Vb \end{pmatrix} \quad \text{and} \quad \varphi^-(z) = \begin{pmatrix} z^{-1}a + b \\ Va \end{pmatrix} \qquad \text{with } a, b \in H\,.$$

Hence, 0 is a partial index of A with respect to Γ, and $M_+(0, 0, \Gamma, A) = H \oplus \operatorname{Im} V$, which is not topologically closed in $H \oplus H$ (as $\operatorname{Im} V$ is not topologically closed in H). Therefore, A does not admit a generalized factorization with respect to Γ.

8.14 Comments

The material of this chapter for matrix-valued functions was first presented in [GK] (see also [Go4, CG]). It consists of an infinite dimensional generalization for a wider set of functions. Some operator generalization of the Wiener-Hopf equation one can find in the paper of Feldman [F]. The presented operator-valued generalizations are used for the solution of the linear transport equation. The transport theory concerns the mathematical analysis of equations that describe transport phenomena in matter, e.g. a flow of electrons through a metal strip or radiated transport bin stellar atmosphere. This phenomenon concerns the migration of particles in a medium. For the finite dimensional case, see [BGK], chapter 6. For the linear transport theory, see [KLH].

Chapter 9

Multiplicative cocycles with restrictions ($\mathcal{F}$-cocycles)

Here we present a generalization of the theory of cocycles developed in Chapter 5. We study multiplicative cocycles with *restrictions*. To formulate this, it is convenient to use the language of sheaves.

9.1 $\mathcal{F}$-cocycles

In this section, A is a Banach algebra with unit 1, and G is an open subgroup of the group of invertible elements of A. By $\mathbb{P}^1$ we denote the Riemann sphere (see the beginning of Section 5.10). If $U \subseteq \mathbb{P}^1$ is a non-empty open set, then we denote by $\mathcal{O}^G(U)$ the group of G-valued holomorphic functions defined on U, and we set $\mathcal{O}^G(\emptyset) = \{1\}$.

9.1.1 Definition ($\mathcal{O}^G$-sheaf). Let $D \subseteq \mathbb{P}^1$ be an open set.

A map $\mathcal{F}$, which assigns to each open set $U \subseteq D$ a subgroup $\mathcal{F}(U)$ of $\mathcal{O}^G(U)$, will be called an **$\mathcal{O}^G$-sheaf** over D if, for each open $U \subseteq D$, the following two conditions are satisfied:

(i) If $f \in \mathcal{F}(U)$, then $f\big|_V \in \mathcal{F}(V)$ for each open $V \subseteq U$. Here $f|_\emptyset := 1$.

(ii) Suppose $f \in \mathcal{O}^G(U)$ such that, for each $w \in U$ there exists a neighborhood $W \subseteq U$ of w with $f\big|_W \in \mathcal{F}(W)$. Then $f \in \mathcal{F}(U)$.

If $\mathcal{F}$ is an $\mathcal{O}^G$-sheaf over D and $U \subseteq D$ is open, then the functions from $\mathcal{F}(U)$ are called **sections** of $\mathcal{F}$ over U.

If $\mathcal{F}$ is an $\mathcal{O}^G$-sheaf over D and $Y \subseteq D$ is open, then we denote by $\mathcal{F}\big|_Y$ the restriction of $\mathcal{F}$ to the open subsets of Y.

9.1.2 Definition (Sheaves defined by a data of zeros). A pair (Z, m) is called a **data of zeros** if $Z \subseteq \mathbb{P}^1$ and $m = \{m_w\}_{w\in Z}$ is a family of numbers $m_w \in \mathbb{N}$.

Let (Z, m) be a data of zeros. Then, for each open set $U \subseteq \mathbb{P}^1$, we denote by $\mathcal{O}^G_{Z,m}(U)$ the group of all $f \in \mathcal{O}^G(U)$ such that, for each $w \in U \cap Z$, the function $f - 1$ has a zero of order $\geq m_w$ at w.[1] The map $\mathcal{O}^G_{Z,m}$ is an $\mathcal{O}^G$-sheaf over $\mathbb{P}^1$. The restriction of $\mathcal{O}^G_{Z,m}$ to the open subsets of an open set $D \subseteq \mathbb{P}^1$ will be denoted by $\mathcal{O}^G_{D,Z,m}$.

9.1.3 Definition ($\mathcal{O}^G$-sheaves of finite order). Let $D \subseteq \mathbb{P}^1$ be an open set. An $\mathcal{O}^G$-sheaf $\mathcal{F}$ over D will be called of **finite order** if there exists a data of zeros (Z, m) such that $Z \cap D$ is discrete and closed in D and, for each open set $U \subseteq D$,

$$\mathcal{F}(U \setminus Z) = \mathcal{O}^G(U \setminus Z)\,, \tag{9.1.1}$$

$$\mathcal{F}(U) \supseteq \mathcal{O}^G_{Z,m}(U)\,. \tag{9.1.2}$$

In particular, if (Z, m) is a data of zeros and $D \subseteq \mathbb{P}^1$ is an open set such that $Z \cap D$ is discrete and closed in D, then $\mathcal{O}^G_{D,Z,m}$ is of finite order.

9.1.4 Definition ($(\mathcal{U}, \mathcal{F})$-cocycles). Let $D \subseteq \mathbb{P}^1$ be an open set, let $\mathcal{F}$ be an $\mathcal{O}^G$-sheaf over D, and let $\mathcal{U} = \{U_j\}_{j\in I}$ be an open covering of D.

Then we denote by $Z^1(\mathcal{U}, \mathcal{F})$ the set of all $f \in Z^1(\mathcal{U}, \mathcal{O}^G)$ (Def. 5.6.1) with $f_{jk} \in \mathcal{F}(U_j \cap U_k)$, $j, k \in I$. The elements of $Z^1(\mathcal{U}, \mathcal{F})$ will be called **$(\mathcal{U}, \mathcal{F})$-cocycles.** Two cocycles $f, g \in Z^1(\mathcal{U}, \mathcal{F})$ will be called **$\mathcal{F}$-equivalent** if there exists a family $h = \{h_j\}_{j\in I}$ of functions $h_j \in \mathcal{F}(U_j)$ such that

$$h_j f_{jk} h_k^{-1} = g_{jk} \qquad \text{on } U_j \cap U_k \tag{9.1.3}$$

for all $j, k \in I$ with $U_j \cap U_k \neq \emptyset$.

A cocycle $f \in Z^1(\mathcal{U}, \mathcal{F})$ will be called **$\mathcal{F}$-trivial** if it is equivalent to the cocycle $\{e_{jk}\}_{jk\in I}$ defined by $e_{jk} \equiv 1$ on $U_j \cap U_k$ for all $j, k \in I$ with $U_j \cap U_k \neq \emptyset$.

More directly, a cocycle $f \in Z^1(\mathcal{U}, \mathcal{F})$ will be called **$\mathcal{F}$-trivial** if there exists a family $\{h_j\}_{j\in I}$ of sections $f_j \in \mathcal{F}(U_j)$ such that

$$f_{jk} = h_j^{-1} h_k \qquad \text{on } U_j \cap U_k \tag{9.1.4}$$

for all $j, k \in I$ with $U_j \cap U_k \neq \emptyset$. The family $\{h_j\}_{j\in I}$ then will be called a **splitting** of f.

9.1.5 Definition (Passing to refinements). Let $\mathcal{F}$ be an $\mathcal{O}^G$-sheaf over an open set $D \subseteq \mathbb{P}^1$, and let $\mathcal{U} = \{U_j\}_{j\in I}$ and $\mathcal{V} = \{V_j\}_{j\in J}$ be two open coverings of D such that $\mathcal{V}$ is a refinement of $\mathcal{U}$.

If $\tau : J \to I$ is a map with $V_j \subseteq U_{\tau(j)}$ for all $j \in J$ (by definition of a refinement at least one such map exists), then, for each $f \in \mathcal{Z}^1(\mathcal{U}, \mathcal{F})$, we define

[1] Since $f^{-1} = \big(1-(1-f)\big)^{-1} = 1 + (1-f)\sum_{n=0}^{\infty}(1-f)^n$ in some neighborhood of w if $f - 1$ has a zero at w, this is indeed a group.

a cocycle $\tau^* f \in \mathcal{Z}^1(\mathcal{V}, \mathcal{F})$, by setting

$$(\tau^* f)_{jk} = f_{\tau(j)\tau(k)}\big|_{V_j \cap V_k}, \qquad j, k \in J.$$

Here again $f_{\tau(j)\tau(k)}\big|_{\emptyset} := 1$. We shall say that $g \in Z^1(\mathcal{V}, \mathcal{F})$ is **induced** by $f \in \mathcal{Z}^1(\mathcal{U}, \mathcal{F})$ if there exists a map $\tau : J \to I$ with $V_j \subseteq U_{\tau(j)}$, $j \in J$, and $g = \tau^* f$.

There is the following generalization of Proposition 5.7.2:

9.1.6 Proposition. *Let $\mathcal{F}$ be an $\mathcal{O}^G$-sheaf over an open set $D \subseteq \mathbb{P}^1$, and let $\mathcal{U} = \{U_j\}_{j \in I}$ and $\mathcal{V} = \{V_j\}_{j \in J}$ be two open coverings of D such that $\mathcal{V}$ is a refinement of $\mathcal{U}$.*

If $f, g \in Z^1(\mathcal{U}, \mathcal{F})$ and $\widetilde{f}, \widetilde{g} \in Z^1(\mathcal{V}, \mathcal{F})$ such that $\widetilde{f}$ is induced by f and $\widetilde{g}$ is induced by g, then the following are equivalent:

(i) *f and g are $\mathcal{F}$-equivalent.*

(ii) *$\widetilde{f}$ and $\widetilde{g}$ are $\mathcal{F}$-equivalent.*

In particular, the following are equivalent:

(i′) *f is $\mathcal{F}$-trivial.*

(ii′) *$\widetilde{f}$ is $\mathcal{F}$-trivial.*

Proof. Repetition of the proof of Proposition 5.7.2. □

9.1.7 Definition ($\mathcal{F}$-cocycles). Let $D \subseteq \mathbb{P}^1$ be an open set, and let $\mathcal{F}$ be an $\mathcal{O}^G$-sheaf over D.

(i) By an **$\mathcal{F}$-cocycle** we mean a $(\mathcal{U}, \mathcal{F})$-cocycle such that $\mathcal{U}$ is an open covering of D. The covering $\mathcal{U}$ then is called **the covering of this cocycle**.

(ii) Let f and g be two $\mathcal{F}$-cocycles over D (possibly with different coverings). The cocycles f and g will be called **$\mathcal{F}$-equivalent** if the following two equivalent (by Proposition 9.1.6) conditions are satisfied:

1) There exists an open covering $\mathcal{W}$ of D, which is a refinement both of the covering of f and of the covering of g, such that at least one of the $(\mathcal{W}, \mathcal{F})$-cocycles induced by f is $\mathcal{F}$-equivalent to at least one of the $(\mathcal{W}, \mathcal{F})$-cocycles induced by g.

2) For each open covering $\mathcal{W}$ of D, which is a refinement both of the covering of f and of the covering of g, each $(\mathcal{W}, \mathcal{F})$-cocycle induced by f is $\mathcal{F}$-equivalent to each $(\mathcal{W}, \mathcal{F})$-cocycle induced by g.

9.1.8 Definition (Restriction to subsets). Let $D \subseteq \mathbb{C}$ be an open set, let $\mathcal{U} = \{U_j\}_{j \in I}$ be an open covering of D, and let Y be an open subset of D. Set

$$\mathcal{U} \cap Y = \left\{ U_j \cap Y \,\middle|\, j \in I \right\}.$$

Let $\mathcal{F}$ be an $\mathcal{O}^G$-sheaf over D. Then we define:

(i) Let f be an $\mathcal{F}$-cocycle over D with the covering $\mathcal{U}$. Then we denote by $f\big|_Y$ the $\mathcal{F}\big|_Y$-cocycle with the covering $\mathcal{U} \cap Y$ defined by

$$(f|_Y)_{jk} = f_{jk}\big|_{U_j \cap U_k \cap Y}$$

for $j, k \in I$ with $U_j \cap U_k \cap Y \neq \emptyset$. This cocycle $f\big|_Y$ will be called the **restriction** of f to Y. We shall say that f is $\mathcal{F}$-trivial **over** Y if $f\big|_Y$ is $\mathcal{F}$-trivial.

(ii) Let f, g be two $\mathcal{F}$-cocycles over D. Then we shall say that f and g are $\mathcal{F}$-equivalent **over** Y if $f\big|_Y$ and $g\big|_Y$ are $\mathcal{F}\big|_Y$-equivalent.

9.1.9 Proposition. *Let $D \subseteq \mathbb{C}$ be an open set, let $U = \{U_j\}_{j \in I}$ be an open covering of D, let $\mathcal{F}$ be an $\mathcal{O}^G$-sheaf over D, and let f be an $\mathcal{F}$-cocycle over D, which is $\mathcal{F}$-trivial over each U_j.*

Then f is $\mathcal{F}$-equivalent to some $(\mathcal{U}, \mathcal{F})$-cocycle.

Proof. Repetition of the proof of Proposition 5.7.6 □

9.2 The main results on cocycles with restrictions. Formulation and reduction to $\mathcal{O}_{D,Z,m}$

In this section we use the notations and definitions introduced in the preceding section. The main results for cocycles obtained in this chapter can be stated as follows:

9.2.1 Theorem. *Let $\mathcal{F}$ be an $\mathcal{O}^G$-sheaf of finite order over an open set $D \subseteq \mathbb{C}$, and let f be an $\mathcal{F}$-cocycle. Assume that at least one of the following three conditions is satisfied:*

(i) *The cocycle f is $\mathcal{C}^G$-trivial over D (Definition* 5.6.1).

(ii) *The group G is connected.*

(iii) *D is simply connected.*

Then f is $\mathcal{F}$-trivial.

Note also the following corollary:

9.2.2 Corollary. *Let $\mathcal{F}$ be an $\mathcal{O}^G$-sheaf of finite order over an open set $D \subseteq \mathbb{C}$, and let $D_1, D_2 \subseteq D$ be two open sets with $D = D_1 \cup D_2$. Further let $f \in \mathcal{F}(D_1 \cap D_2)$ and assume that at least one of the following three conditions is satisfied:*

(i) *There exist continuous functions $c_j : D_j \to G$, $j = 1, 2$, such that $f = c_1^{-1} c_2$ on $D_1 \cap D_2$.*

(ii) *The group G is connected.*

(iii) *$D_1 \cup D_2$ is simply connected.*

Then there exist functions $f_j \in \mathcal{F}(D_j)$, $j = 1, 2$, *such that* $f = f_1^{-1} f_2$ *on* $D_1 \cap D_2$.

9.2.3 Remark. It is sufficient to prove Theorem 9.2.1 for the case when $\mathcal{F}$ is of the form $\mathcal{O}^G_{D,Z,m}$, where (Z, m) is a data of zeros such that $Z \cap D$ is discrete and closed in D (Def. 9.1.2).

Indeed, assume this is done, and let $\mathcal{F}$ be an arbitrary $\mathcal{O}^G$-sheaf of finite order over an open set $D \subseteq \mathbb{C}$. Further, let f be an $\mathcal{F}$-cocycle such that at least one of the following three conditions (i)–(iii) in Theorem 9.2.1 is satisfied.

Let $\mathcal{U}$ be the open covering of D associated to f, and let (Z, m) be a data of zeros as in Definition 9.1.3. Since $Z \cap D$ is discrete and closed in D, then, for each point $w \in D$, we can find a neighborhood $V_w \subseteq D$ of w such that V_w is contained in at least one of the sets of the covering $\mathcal{U}$ and

$$V_w \cap Z = \begin{cases} \{w\} & \text{if } w \in Z\,, \\ \emptyset & \text{if } w \notin Z\,. \end{cases}$$

Then $\mathcal{V} := \{V_w\}_{w \in D}$ is an open covering of D and a refinement of $\mathcal{U}$. Let f^* be a $(\mathcal{V}, \mathcal{F})$-cocycle induced by f. If w, z are two different points in D, then $Z \cap V_w \cap V_z = \emptyset$. By (9.1.1) this implies that

$$\mathcal{F}(V_w \cap V_z) = \mathcal{O}^G(V_w \cap V_z) = \mathcal{O}^G_{Z,m}(V_w \cap V_z) \qquad \text{if } w, z \in D \text{ with } w \neq z\,.$$

Therefore f^* can be interpreted as a $(\mathcal{V}, \mathcal{O}^G_{D,Z,m})$-cocycle.

If one of the conditions (ii) or (iii) is satisfied for $\mathcal{F}$, then it is clear that this condition is satisfied for $\mathcal{O}^G_{D,Z,m}$ (as it depends only on D resp. G). If condition (i) is satisfied for the cocycle f, then it follows from Proposition 9.1.6 that this condition is satisfied also for f^*.

Therefore, as Theorem 9.2.1 is already proved for the sheaf $\mathcal{O}^G_{D,Z,m}$ (by our hypothesis), it follows that f^* is $\mathcal{O}^G_{D,Z,m}$-trivial. By (9.1.2) this implies that f^* is $\mathcal{F}$-trivial, which means by Proposition 9.1.6 that f is $\mathcal{F}$-trivial.

The following Sections 9.3–9.8 are devoted to the proof of Theorem 9.2.1 in the case when $\mathcal{F}$ is of the form $\mathcal{O}^G_{D,Z,m}$, where (Z, m) is a data of zeros such that $Z \cap D$ is discrete and closed in D. Then the assertion of Theorem 9.2.1 is contained in Theorems 9.6.1 (if D is simply connected) and 9.8.1 (if f is $\mathcal{C}^G$-trivial or G is connected).

9.3 The Cartan lemma with restrictions

In this section, A is a Banach algebra with unit 1, G is an open subgroup of the group of invertible elements of A, G_1A is the connected component of 1 in G, and (Z, m) is a data of zeros (Def. 9.1.2).

Here we prove a version of the Cartan lemma for sections of the sheaf $\mathcal{O}^G_{Z,m}$. To linearize the problem, we need the following definition:

9.3.1 Definition. For each open set $U \subseteq \mathbb{P}^1$, we denote by $\mathcal{O}^A_{Z,m}(U)$ the set of all $f \in \mathcal{O}^A(U)$ such that, for each point $w \in U \cap Z$, the function f has a zero of order $\geq m_w$ at w.

9.3.2 Lemma. (*For the definitions of* log *and* exp, *see Section* 5.4.3). *For each open set $U \subseteq \mathbb{P}^1$, we have:*

(i) *If $f \in \mathcal{O}^A_{Z,m}(U)$, then $\exp f \in \mathcal{O}^{G_1 A}_{Z,m}(U)$.*

(ii) *If $f \in \mathcal{O}^{G_1 A}_{Z,m}(U)$ and $\|f(z) - 1\| < 1$, $z \in U$, then $\log f \in \mathcal{O}^A_{Z,m}(U)$.*

Proof. (i) Let $f \in \mathcal{O}^A_{Z,m}(U)$ and $w \in Z$. Then

$$\exp f - 1 = \sum_{n=1}^{\infty} \frac{f^n}{n!} .$$

As f has a zero of order $\geq m_w$ at w and therefore each f^n with $n \geq 1$ has a zero of order $\geq m_w$ at w, this implies that $\exp f - 1$ has a zero of order $\geq m_w$ at w.

(ii) Let $f \in \mathcal{O}^{G_1 A}_{Z,m}(U)$ with $\|f(z) - 1\| < 1$ for all $z \in U$, and let $w \in Z$. Then

$$\log f = \sum_{n=1}^{\infty} \frac{(-1)^{n+1}}{n} (f-1)^n .$$

As $f - 1$ has a zero of order $\geq m_w$ at w and therefore each $(f-1)^n$ with $n \geq 1$ has a zero of order $\geq m_w$ at w, this implies that $\log f$ has a zero of order $\geq m_w$ at w. □

9.3.3 Definition. Let Γ be an arbitrary subset of $\mathbb{P}^1$, and let Γ^0 be the set of interior points of Γ.

(i) We denote by $\overline{\mathcal{O}}^G_{Z,m}(\Gamma)$ the group of continuous functions $f : \Gamma \to G$ such that $f|_{\Gamma^0} \in \mathcal{O}^G_{Z,m}(\Gamma^0)$.[2] If Γ is compact, then $\overline{\mathcal{O}}^G_{Z,m}(\Gamma)$ will be considered as a topological group with respect to uniform convergence on Γ. It is easy to see that this is a closed subgroup of $\mathcal{C}^G(\Gamma)$ endowed with the same topology.

(ii) We denote by $\overline{\mathcal{O}}^A_{Z,m}(\Gamma)$ the algebra of continuous functions $f : \Gamma \to A$ such that $f|_{\Gamma^0} \in \mathcal{O}^A_{Z,m}(\Gamma^0)$. If Γ is compact, then $\overline{\mathcal{O}}^A_{Z,m}(\Gamma)$ will be considered as the Banach algebra endowed with the max-norm. It is easy to see that this is indeed a *Banach* algebra, i.e., a closed subalgebra of the Banach algebra $\mathcal{C}^A(\Gamma)$ endowed with the same norm.

[2] Here it is allowed that points of Z lie on $\Gamma \setminus \Gamma^0$, but for $f \in \overline{\mathcal{O}}^G_{Z,m}(\Gamma)$, the function $f - 1$ need not have zeros at such points.

9.3.4 Proposition. *Let $D \subseteq \mathbb{C}$ be a bounded open set with piecewise C^1-boundary, and let $\Omega \supseteq \overline{D}$ be an open set such that each bounded connected component of $\mathbb{C}\setminus\overline{D}$ contains at least one point of $\mathbb{C} \setminus \Omega$. Assume that $Z \cap \Omega$ is discrete and closed in Ω and $Z \cap \partial D = \emptyset$.*

Then each $f \in \overline{\mathcal{O}}^A_{Z,m}(\overline{D})$ can be approximated uniformly on $\overline{D}$ by functions from $\mathcal{O}^A_{Z,m}(\Omega)$.

Proof. Let $U_1, \dots, U_n$ be the bounded connected components of $\mathbb{C} \setminus \overline{D}$. By hypothesis, we can find points $a_j \in U_j \cap (\mathbb{C} \setminus \Omega)$, $1 \le j \le n$. Since $Z \cap \Omega$ is discrete and closed in Ω, by the Weierstrass product Theorem 2.5.7, we can find a scalar holomorphic function $\varphi : \Omega \to \mathbb{C}$ such that $\varphi(z) \neq 0$ for $z \in \Omega \setminus Z$, and, for each $w \in Z$, φ has a zero of order m_w at w. Since $Z \cap \partial D = \emptyset$, then

$$\widetilde{f} := \frac{f}{\varphi}$$

is continuous on $\overline{D}$ and holomorphic in D. Therefore, by the Runge approximation Theorem 2.2.2 (ii), we can find a sequence $(\widetilde{f}_\nu)_{\nu\in\mathbb{N}}$ of functions $\widetilde{f}_\nu \in \mathcal{O}^A(\mathbb{C} \setminus \{a_1, \dots, a_n\})$ which converges to $\widetilde{f}$ uniformly on $\overline{D}$. Setting $f_\nu = \varphi\widetilde{f}_\nu$ on Ω, we get a sequence $(f_\nu)_{\nu\in\mathbb{N}}$ of functions $f_\nu \in \mathcal{O}^A_{Z,m}(\Omega)$. Since $\overline{D}$ is compact and, therefore, φ is bounded on $\overline{D}$, this sequence converges to $f = \varphi\widetilde{f}$ uniformly on $\overline{D}$. □

9.3.5 Lemma. *Let $\Omega \subseteq \mathbb{C}$ be a bounded open set with piecewise C^1-boundary such that $\mathbb{C} \setminus \Omega$ is connected. Suppose that $Z \cap \mathbb{C}$ is discrete and closed in $\mathbb{C}$ and*

$$Z \cap \overline{\Omega} = \emptyset. \text{ [3]}$$

Then each $f \in \overline{\mathcal{O}}^{G_1A}(\overline{\Omega})$ can be approximated uniformly on $\overline{\Omega}$ by maps from $\mathcal{O}^{G_1A}_{Z,m}(\mathbb{C})$.

Proof. Let $f \in \overline{\mathcal{O}}^{G_1A}(\overline{\Omega})$ be given. By Lemma 5.4.7, $\overline{\mathcal{O}}^{G_1A}(\overline{\Omega})$ is the connected component of the unit in the group of invertible elements of the Banach algebra $\overline{\mathcal{O}}^A(\overline{\Omega})$. Therefore, by Proposition 5.4.1, f can be written as a finite product

$$f = g_1 \cdot \ldots \cdot g_n$$

where $g_j \in \overline{\mathcal{O}}^{G_1A}(\overline{\Omega})$ and $\|1 - g_j\|_{\overline{\mathcal{O}}^A(\overline{\Omega})} < 1$ for $1 \le j \le n$. By Lemma 5.4.4 (ii), $e^{\log g_j} = g_j$. Hence

$$f = e^{\log g_1} \cdot \ldots \cdot e^{\log g_n}.$$

Since $Z \cap \overline{\Omega} = \emptyset$, it is trivial that $\log g_j \in \overline{\mathcal{O}}^A_{Z,m}(\overline{\Omega})$. Since $\mathbb{C} \setminus \Omega$ is connected, by Proposition 9.3.4 (i), for each j, we can find a sequence $(\varphi_{j\nu})_{\nu\in\mathbb{N}}$ of functions $\varphi_{j\nu} \in \mathcal{O}^A_{Z,m}(\mathbb{C})$ which converges to $\log g_j$ uniformly on $\overline{\Omega}$. Setting

$$f_\nu = e^{\varphi_{j\nu}} \ldots e^{\varphi_{j\nu}}$$

[3] We shall see below (Theorem 9.5.3) that this condition can be replaced by $Z \cap \partial\Omega = \emptyset$.

then we get a sequence $(f_\nu)_{\nu\in\mathbb{N}}$ of holomorphic functions $f_\nu : \mathbb{C} \to A$ which converges to f uniformly on $\overline{D}$. By Lemma 9.3.2 (i) each f_ν belongs to $\mathcal{O}^{G_1 A}_{Z,m}(\mathbb{C})$. □

A modification of the proof of Lemma 5.5.3 gives the following generalization of this lemma:

9.3.6 Lemma. *Let (D_1, D_2) be a Cartan pair such that*

$$Z \cap \partial D_1 = Z \cap \partial D_2 = \emptyset \quad \textit{and} \quad Z \cap (D_1 \cup D_2) \textit{ is finite.}$$

Let $\varepsilon > 0$. Then there exists $\delta > 0$ such that, for all $g \in \overline{\mathcal{O}}^A_{Z,m}(\overline{D}_1 \cap \overline{D}_2)$ with

$$\max_{z\in\overline{D}_1\cap\overline{D}_2} \|g(z)\| < \delta,$$

there exist $g_j \in \overline{\mathcal{O}}^A_{Z,m}(\overline{D}_j)$ such that $\max\limits_{z\in\overline{D}_j} \|g_j(z)\| < \varepsilon$, $j = 1, 2$, and

$$1 + g = (1 + g_1)(1 + g_2) \qquad \textit{on } \overline{D}_1 \cap \overline{D}_2. \tag{9.3.1}$$

Proof. We consider $\overline{\mathcal{O}}^A_{Z,m}(\overline{D}_1 \cap \overline{D}_2)$, $\overline{\mathcal{O}}^A_{Z,m}(\overline{D}_1)$ and $\overline{\mathcal{O}}^A_{Z,m}(\overline{D}_2)$ as Banach algebras endowed with the maximum norm. It is sufficient to prove that each $f \in \overline{\mathcal{O}}^A_{Z,m}(\overline{D}_1 \cap \overline{D}_2)$ can be written in the form $f = f_1 + f_2$ with $f_j \in \overline{\mathcal{O}}^A_{Z,m}(\overline{D}_j)$, because then the assertion follows from Lemma 5.2.1.

If $Z \cap (D_1 \cup D_2) = \emptyset$, this follows from Corollary 5.3.5.

If $Z \cap (D_1 \cup D_2) = \{w_1, \ldots, w_n\}$, $n \in \mathbb{N}^*$, then we set

$$p(z) = (z - z_1)^{m_{w_1}} \cdot \ldots \cdot (z - z_n)^{m_{w_n}}.$$

Now let an arbitrary $f \in \overline{\mathcal{O}}^A_{Z,m}(\overline{D}_1 \cap \overline{D}_2)$ be given. Since Z does not meet the boundaries of D_1 and D_2, then $f/p \in \overline{\mathcal{O}}^A(\overline{D}_1 \cap \overline{D}_2)$, and, again by Corollary 5.3.5, we can find functions $h_j \in \overline{\mathcal{O}}^A(\overline{D}_j)$ with $f/p = h_1 + h_2$. Setting $f_j = ph_j$, we get the required functions $f_j \in \overline{\mathcal{O}}^A_{Z,m}(\overline{D}_j)$ with $f = f_1 + f_2$. □

Now we can prove the following generalization of the Cartan Lemma 5.5.2:

9.3.7 Lemma. *Let (D_1, D_2) be a Cartan pair such that*

$$Z \cap \partial(D_1 \cup D_2) = \emptyset \quad \textit{and} \quad Z \cap (D_1 \cup D_2) \textit{ is finite.}$$

Let $f \in \overline{\mathcal{O}}^G_{Z,m}(\overline{D}_1 \cap \overline{D}_2)$ such that all values of f belong to the same connected component of G. (This is always the case if $D_1 \cap D_2$ is connected.) Then there exist $f_j \in \overline{\mathcal{O}}^G_{Z,m}(\overline{D}_j)$, $j = 1, 2$, such that

$$f = f_1 f_2 \qquad \textit{on} \quad \overline{D}_1 \cap \overline{D}_2.\ [4] \tag{9.3.2}$$

[4] From the proof of this lemma it will be clear that also the more precise statements corresponding to assertions (i) and (ii) of Lemma 5.5.2 are true. We omit these finer points, because we will not use them.

Proof. Since all values of f belong to the same connected component of G, after multiplying by a constant, we may assume that all values of f belong to G_1A. The main problem compared to the proof of Lemma 5.5.2 is that, possibly, $Z \cap \overline{D}_1 \cap \overline{D}_2 \neq \emptyset$. We avoid this problem by the following trick:

Since $Z \cap (D_1 \cup D_2)$ is finite, we can find a "smaller" Cartan pair (X_1, X_2) such that $X_j \subseteq D_j$, $X_1 \cup X_2 = D_1 \cup D_2$ and

$$Z \cap \overline{X}_1 \cap \overline{X}_2 = \emptyset.$$

Now it is sufficient to find $f_j \in \overline{\mathcal{O}}^{G_1A}_{Z,m}(\overline{X}_j)$, $j = 1, 2$, such that

$$f = f_1 f_2 \qquad \text{on} \quad \overline{X}_1 \cap \overline{X}_2. \tag{9.3.3}$$

Indeed, then it follows from the equations

$$f_1 = f f_2^{-1} \qquad \text{and} \qquad f_2 = f_1^{-1} f$$

that the maps f_j extend to maps $f_j \in \overline{\mathcal{O}}^{G_1A}_{Z,m}(\overline{D}_j)$, and (9.3.2) follows from (9.3.3) by uniqueness of holomorphic functions.

Now we continue as in the proof of Lemma 5.5.2:

Since $Z \cap \overline{X}_1 \cap \overline{X}_2 = \emptyset$, it follows from Lemma 9.3.5 that, for each $\delta > 0$, we can find $f_\delta \in \mathcal{O}^{G_1A}_{Z,m}(\mathbb{C})$ with

$$\max_{z \in \overline{U}_1 \cup \ldots \cup \overline{U}_n} \|f(z) f_\delta^{-1}(z) - 1\| < \delta \quad \text{and} \quad \max_{z \in \overline{U}_1 \cup \ldots \cup \overline{U}_n} \|f_\delta^{-1}(z) f(z) - 1\| < \delta.$$

Therefore, by the preceding Lemma 9.3.6, for sufficiently small δ, we can find $g_j \in \overline{\mathcal{O}}^A(\overline{D}_j)$ such that

$$\max_{z \in \overline{D}_j} \|g_j(z) < 1$$

and

$$f f_\delta^{-1} = (1 + g_1)(1 + g_2)$$

on $\overline{D_1 \cap D_2}$. Setting $f_1 = 1 + g_1$ and $f_2 = (1 + g_2) f_\delta$, we conclude the proof. □

9.3.8 Corollary. *Let $D \subseteq \mathbb{C}$ be an open set such that $Z \cap D$ is discrete and closed in D. Let (D_1, D_2) be a Cartan pair such that*

$$\overline{D}_1 \cup \overline{D}_2 \subseteq D.$$

Let $U \subseteq D$ be a neighborhood of $\overline{D}_1 \cap \overline{D}_2$ and $f \in \mathcal{O}^G_{Z,m}(U)$ such that all values of f belong to the same connected component of G. (This is always the case if U is connected.) Then there exist neighborhoods $U_j \subseteq D$ of $\overline{D}_j$ and functions $f_j \in \mathcal{O}^G_{Z,m}(U_j)$, $j = 1, 2$, such that

$$f = f_1 f_2 \qquad \text{on} \quad U_1 \cap U_2. \tag{9.3.4}$$

Proof. Since $Z \cap D$ is discrete and closed in D, we can find a slightly larger Cartan pair (U_1, U_2) such that $D_j \subseteq U_j$, $\overline{U}_j \subseteq D$, $\overline{U}_1 \cap \overline{U}_2 \subseteq U$ and

$$Z \cap \partial(U_1 \cup U_2) = \emptyset.$$

Then $Z \cap (U_1 \cup U_1)$ is finite, $f|_{\overline{U}_1 \cap \overline{U}_2} \in \overline{\mathcal{O}}^G_{Z,m}(\overline{U}_1 \cap \overline{U}_2)$ and we can apply Lemma 9.3.7 to the Cartan pair (U_1, U_2). □

9.4 Splitting over simply connected open sets after shrinking

In this section A is a Banach algebra with unit 1, G is an open subgroup of the group of invertible elements of A, and (Z, m) is a data of zeros (Def. 9.1.2). By $G_1 A$ we denote again the connected component of 1 in the group of all invertible elements of A.

Here we want to generalize the splitting Theorem 5.6.3 to $\mathcal{O}^G_{Z,m}$-cocycles. The problem is that (at the moment) we do not have a generalization to $\mathcal{O}^G_{Z,m}$ of part (i) of the Runge approximation Theorem 5.0.1, which we used in the proof of Lemma 5.9.1 and then again in the proof of Theorem 5.6.3. In the next section we will prove this generalization, but first we have to prove the following splitting statement "with shrinking":

9.4.1 Lemma. *Let $D \subseteq \mathbb{C}$ be a simply connected open set such that $Z \cap D$ is discrete and closed in D, let Ω be a relatively compact open subset of D, and let f be an $\mathcal{O}^G_{Z,m}$-cocycle over D. Then the restriction $f|_\Omega$ is $\mathcal{O}^G_{Z,m}$-trivial* (*Def.* 9.1.4).

Proof. We proceed similar to the proof of Lemma 5.9.2. But instead of the precise factorization statement of the Cartan Lemma 5.9.1, here we can use only the factorization statement "with shrinking" of Corollary 9.3.8. Therefore, in each of the induction steps, we have to shrink the covering which makes the arguments more technical. We give now the details.

Set

$$K_t = \Big\{ z \in \mathbb{C} \;\Big|\; -t < \operatorname{Re} z < 1 + t \text{ and } -t < \operatorname{Im} z < 1 + t \Big\}, \quad t > 0.$$

Since D is simply connected and Ω is relatively compact in D, by the Riemann mapping theorem, we may assume that, for some $\varepsilon > 0$, $D = K_\varepsilon$ and $\Omega \subseteq K_0$.

Now let an $\mathcal{O}^G_{Z,m}$-cocycle f over $D = K_\varepsilon$ be given.

We choose $n \in \mathbb{N}^*$ sufficiently large (will be specified some lines below) and denote by U^t_{jk}, $j, k = 1, \ldots, n$, $0 < t < 1$, the open rectangle of all $z \in \mathbb{C}$ with

$$\left(k - 1 - \frac{t}{3}\right)\frac{1}{n} < \operatorname{Re} z < \left(k + \frac{t}{3}\right)\frac{1}{n}$$

and

$$\left(j-1-\frac{t}{3}\right)\frac{1}{n} < \operatorname{Im} z < \left(j+\frac{t}{3}\right)\frac{1}{n}.$$

Then, for all $0 < t < 1$, $\mathcal{U}^t := \{U^t_{jk}\}_{1\le j,k\le n}$ is an open covering of $K_{t/3n}$. We choose n so large that

$$K_{t/3n} \subseteq K_{\varepsilon/2} \qquad \text{for all } 0 < t < 1.$$

Let $\mathcal{V} = \{V_\nu\}_{\nu\in I}$ be the covering associated to f. Then

$$\mathcal{V}\cap K_{t/3n} := \left\{V_\nu \cap K_{t/3n}\right\}_{\nu\in I}$$

is a second open covering of $K_{t/3n}$. Since $\overline{K}_{\varepsilon/2}$ is a compact subset of $D = K_\varepsilon$, the covering $\mathcal{V}$ contains a finite subcovering which already covers $\overline{K}_{\varepsilon/2} \supseteq K_{t/3n}$. Therefore, we can n further enlarge (and now we fix it) so that each U^t_{jk} is contained in at least one of the sets V_ν. Now, for all $0 < t < 1$, the covering $\mathcal{U}^t$ is a refinement of $\mathcal{V}\cap K_{t/3n}$.

We give the family $\mathcal{U}^t$ an order saying that $U^t_{jk} < U^t_{j'k'}$, if and only if, either $j < j'$ or $j = j'$ and $k < k'$. Let $U^t_1, \dots U^t_{n^2}$ be the family $\mathcal{U}^t$ numbered in this way. For $\mu = 1,\dots,n^2$, we consider the statement

$S(\mu)$: There exists $0 < t < 1$ such that $f|_{U^t_1\cup\ldots\cup U^t_\mu}$ is $\mathcal{O}^G_{Z,m}$-trivial.

Since $\Omega \subseteq K_0 \subseteq K_t = U^t_1 \cup \ldots \cup U^t_{n^2}$ for all $0 < t < 1$, it is sufficient to prove $S(n^2)$. Since U^t is a refinement of $\mathcal{V}\cap K_{t/3n}$ and $\mathcal{V}$ is associated to f, it is trivial that $f|_{U^t_\mu}$ is $\mathcal{O}^G_{Z,m}$-trivial for all $1 \le \mu \le n^2$ and $0 < t < 1$. In particular, $S(1)$ is true.

Therefore it is sufficient to prove that $S(\mu) \Rightarrow S(\mu+1)$ for all $1 \le \mu \le n^2-1$.

Let $1 \le \mu \le n^2-1$ be given such that $S(\mu)$ is true. Set $W^t_1 = U^t_1 \cup \ldots \cup U^t_\mu$ and $W^t_2 = U^t_{\mu+1}$ for $0 < t < 1$. Since $S(\mu)$ is true, we can fix $0 < t < 1$ such that $f|_{W^t_1}$ is $\mathcal{O}^G_{W^t_1,Z,m}$-trivial. Since, trivially, also $f|_{W^t_2}$ is $\mathcal{O}^G_{Z,m}$-trivial, it follows from Proposition 9.1.9 that $f|_{W^t_1\cup W^t_2}$ is $\mathcal{O}^G_{Z,m}$-equivalent to some $(\{W^t_1, W^t_2\}, \mathcal{O}^G_{Z,m})$-cocycle g.

Now we choose some $0 < t' < t$. Then $(W^{t'}_1, W^{t'}_2)$ is a Cartan pair with $\overline{W}^{t'}_1 \cup \overline{W}^{t'}_2 \subseteq D$ and $W^t_1 \cap W^t_2$ is a connected neighborhood of $\overline{W}^{t'}_1 \cup \overline{W}^{t'}_2$. Therefore, from Corollary 9.3.8 we get functions $g_j \in \mathcal{O}^G_{Z,m}(W'_j)$, $j = 1, 2$, such that

$$g_{12} = g_1 g_1 \qquad \text{on } W'_1 \cap W'_2,$$

which means that the restricted cocycle $g|_{W'_1\cup W'_2}$ is $\mathcal{O}^G_{Z,m}$-trivial. Since $W'_1 \cup W'_2 = U^{t'}_1 \cup \ldots \cup U^{t'}_{\mu+1}$, it follows that $S(\mu+1)$ is true. □

9.5 Runge approximation on simply connected open sets

In this section A is a Banach algebra with unit 1, G is an open subgroup of the group of invertible elements of A, G_1A is the connected component of 1 in the group of invertible elements of A, and (Z, m) is a data of zeros (Def. 9.1.2). Here we prove part (i) of the Runge approximation Theorem 5.0.1 for sections of the sheaf $\mathcal{O}^G_{Z,m}$. The first step is the following Mergelyan approximation theorem:

9.5.1 Theorem. *Let $D \subseteq \mathbb{C}$ be a bounded open set with piecewise $\mathcal{C}^1$-boundary such that $Z \cap \overline{D}$ is finite and $Z \cap \partial D = \emptyset$. Then, for each $f \in \overline{\mathcal{O}}^G_{Z,m}(\overline{D})$ and all $\varepsilon > 0$, there exists a neighborhood U of $\overline{D}$ and a function $\widetilde{f} \in \mathcal{O}^G_{Z,m}(U)$ such that*

$$\max_{z\in\overline{D}} \left\| f(z) - \widetilde{f}(z) \right\| < \varepsilon.$$

Proof. Since the assertion of the theorem does not change if we replace Z by $Z \setminus (\mathbb{C} \setminus D)$, we may assume that Z is a finite subset of D. Then we can find an open set $\Omega \subseteq D$ with piecewise $\mathcal{C}^1$-boundary such that $\overline{\Omega} \subseteq D$ and $Z \subseteq \Omega$. (For example, we can surround the points of Z by small discs.) Further, let $\triangle$ be an open disc such that $\overline{D} \subseteq \triangle$.

Now let $f \in \overline{\mathcal{O}}^G_{Z,m}(\overline{D})$ and $\varepsilon > 0$ be given. Since G is an open subset of A, by the Mergelyan approximation Theorem 2.2.1, we can find a sequence $(U_n)_{n\in\mathbb{N}}$ of neighborhoods $U_n \subseteq \triangle$ of $\overline{D}$ and a sequence $(h_n)_{n\in\mathbb{N}}$ of functions $h_n \in \mathcal{O}^G(U_n)$ such that

$$\lim_{n\to\infty} \max_{z\in\overline{D}} \left\| f(z)h_n^{-1}(z) - 1 \right\| = 0.$$

Then, in particular

$$\lim_{n\to\infty} \max_{z\in\overline{D}\cap(\overline{\triangle}\setminus\Omega)} \left\| f(z)h_n^{-1}(z) - 1 \right\| = 0.$$

Therefore, by Lemma 9.3.6, we can find $n_0 \in \mathbb{N}$, a sequence $\left(g_n^{(1)}\right)_{n\geq n_0}$ of functions $g_n^{(1)} \in \overline{\mathcal{O}}^A_{Z,m}(\overline{D})$ and a sequence $\left(g_n^{(2)}\right)_{n\geq n_0}$ of functions $g_n^{(2)} \in \overline{\mathcal{O}}^A_{Z,m}(\overline{\triangle} \setminus \Omega) = \overline{\mathcal{O}}^G(\overline{\triangle} \setminus \Omega)$ such that

$$f h_n^{-1} = \left(1 + g_n^{(1)}\right)\left(1 + g_n^{(2)}\right) \qquad \text{on } \overline{D} \cap \left(\overline{\triangle} \setminus \Omega\right),\ n \geq n_0, \tag{9.5.1}$$

and

$$\lim_{n\to\infty} \max_{z\in\overline{D}} \left\| g_n^{(1)}(z) \right\| = 0 \tag{9.5.2}$$

and

$$\lim_{n\to\infty} \max_{z\in\overline{\triangle}\setminus\Omega} \left\| g_n^{(2)}(z) \right\| = 0.$$

Moreover we can assume that

$$1 + g_n^{(1)} \in \overline{\mathcal{O}}^G_{Z,m}(\overline{D}) \quad \text{and} \quad 1 + g_n^{(2)} \in \overline{\mathcal{O}}^G(\overline{\triangle} \setminus \Omega), \quad n \geq n_0.$$

Set

$$\widetilde{f}_n := \left(1 + g_n^{(1)}\right)^{-1} f \quad \text{on } \overline{D}.$$

Then it follows from (9.5.1) that

$$\widetilde{f}_n := \left(1 + g_n^{(2)}\right) h_n \qquad \text{on } \overline{D} \cap (\overline{\triangle} \setminus \Omega), \ n \geq n_0,$$

which shows that $\widetilde{f}_n \in \mathcal{O}^G_{Z,m}(U_n)$ for all $n \geq n_0$, and from (9.5.2) it follows that

$$\lim_{n\to\infty} \max_{z\in\overline{D}} \left\| f(z) - \widetilde{f}_n(z) \right\| = 0.$$

It remains to choose $n_1 \geq n_0$ so large that

$$\max_{z\in\overline{D}} \left\| f(z) - \widetilde{f}_{n_1}(z) \right\| < \varepsilon$$

and to set $U = U_{n_1}$ and $\widetilde{f} = f_{n_1}$. □

Now we can prove Lemma 5.4.7 for the sheaf $\mathcal{O}^G_{Z,m}$:

9.5.2 Lemma. *Let $D \subseteq \mathbb{C}$ be a bounded simply connected open set with piecewise $\mathcal{C}^1$-boundary. Suppose that $Z \cap D$ is finite and $Z \cap \partial D = \emptyset$. Then the group $\overline{\mathcal{O}}^{G_1A}_{Z,m}(\overline{D})$ is connected.*

Proof. Let $f \in \mathcal{O}^{G_1A}_{Z,m}(\overline{D})$ be given. Since G_1A is open and by the Mergelyan Theorem 9.5.1, we may assume that f is defined and holomorphic, and with values in G_1A in some neighborhood Ω of $\overline{D}$. Since D is a bounded simply connected open set with piecewise $\mathcal{C}^1$-boundary, we may assume that also Ω is simply connected. Choose an open covering $\mathcal{U} = \{U_j\}_{j\in I}$ of Ω by open discs U_j with center z_j such that $z_j \neq z_k$ if $j \neq k$ and, for each $j \in I$, either $U_j \cap Z = \emptyset$ or $U_j \cap Z = \{z_j\}$. Then $U_j \cap U_k \cap Z = \emptyset$ for $j \neq k$.

Since G_1A is connected, for each $j \in I$, we can find a continuous curve $\gamma_j : [-1, 0] \to G_1A$ with $\gamma(-1) = 1$ and $\gamma(0) = f(z_j)$. Moreover, if $U_j \cap Z = \{z_j\}$ and hence $f(z_j) = 1$, then we may assume that $\gamma_j \equiv 1$. Now, for all $j \in I$ and $z \in U_j$, we define

$$\phi_j(z,t) = \begin{cases} \gamma(t) & \text{if } -1 \leq t \leq 0, \\ f\big(z_j + t(z - z_j)\big) & \text{if } 0 \leq t \leq 1. \end{cases}$$

So we get continuous G_1A-valued functions on $U_j \times [-1, 1]$, holomorphic for fixed t, such that

$$\phi_j(z, -1) = 1 \quad \text{and} \quad \phi_j(z, 1) = f(z), \qquad z \in U_j.$$

Moreover

$$\phi_j(\cdot,t)\in \mathcal{O}^{G_1A}_{Z,m}(U_j) \qquad \text{for each fixed } t\in[-1,1] \text{ and } j\in I. \tag{9.5.3}$$

Indeed, if $U_j\cap Z=\emptyset$, this is trivial. If $U_j\cap Z=\{z_j\}$ and $-1\le t\le 0$, this is also trivial (because then $\gamma_j\equiv 1$). Now let $U_j\cap Z=\{z_j\}$ and $0\le t\le 1$. Then f is of the form

$$f(z)=1+\sum_{\nu=m_{z_j}}^{\infty} f_{j\nu}(z-z_j)^{\nu}\,, \qquad z\in U_j\,,$$

and therefore $\phi_j(\cdot,t)$ is of the form

$$\phi_j(z,t)=1+\sum_{\nu=m_{z_j}}^{\infty}\big(f_{j\nu}t^{\nu}\big)(z-z_j)^{\nu}\,, \qquad z\in U_j\,.$$

This implies (9.5.3). Set

$$\psi_{jk}(z,t)=\phi_j^{-1}(z,t)\phi_k(z,t) \qquad \text{for } z\in U_j\cap U_k\,,\quad j,k\in I\,. \tag{9.5.4}$$

Now let $\widetilde{A}$ be the algebra of all continuous maps $\varphi:[-1,1]\to A$ such that, for some $\lambda\in\mathbb{C}$,

$$\varphi(-1)=\varphi(1)=\lambda\cdot 1\,.$$

Endowed with the maximum norm, this is a Banach algebra with unit. Let $G\widetilde{A}$ be the group of invertible elements of $\widetilde{A}$. It consists of all continuous functions $\varphi:[-1,1]\to G_1A$ such that, for some $\lambda\in\mathbb{C}\setminus\{0\}$,

$$\varphi(-1)=\varphi(1)=\lambda\cdot 1\,. \tag{9.5.5}$$

With this notation, $\psi:=\{\psi_{jk}(z,t)\}_{j,k\in I}$ can be considered as a $(\mathcal{U},\mathcal{O}^{G\widetilde{A}})$-cocycle. Moreover, since $U_j\cap U_k\cap Z=\emptyset$ if $j\neq k$, it can be considered as a $(\mathcal{U},\mathcal{O}^{G\widetilde{A}}_{Z,m})$-cocycle.

Let W be a neighborhood of $\overline{D}$ with $\overline{W}\subseteq\Omega$. Since Ω is simply connected, then it follows from Lemma 9.4.1 that $\psi|_W$ is $\mathcal{O}^{G\widetilde{A}}_{Z,m}$-trivial. Hence we have functions $\psi_j\in\mathcal{O}^{G\widetilde{A}}_{Z,m}(W\cap U_j)$ such that, if we interpret them as functions $\psi_j: W\cap U_j\times[-1,1]\to G_1A$,

$$\psi_{jk}(z,t)=\psi_j^{-1}(z,t)\psi_k(z,t)\,, \qquad z\in W\cap U_j\cap U_k\,,\ j,k\in I\,.$$

Together with (9.5.4) this implies that

$$\phi_j(z,t)\psi_j^{-1}(z,t)=\phi_k(z,t)\psi_k^{-1}(z,t)\,, \qquad z\in W\cap U_j\cap U_k\,,\ j,k\in I\,.$$

Therefore, we have a global continuous function $F(z,t): W\times[-1,1]\to G_1A$ holomorphic for fixed t, such that, for each $j\in I$,

$$F(z,t)=\phi_j(z,t)\psi_j^{-1}(z,t)\,, \qquad z\in U_j\,,\ t\in[-1,1]. \tag{9.5.6}$$

That ψ_j belongs to $\mathcal{O}_{Z,m}^{G\widetilde{A}}(W \cap U_j)$, implies that

$$\psi_j(\cdot,t) \in \mathcal{O}_{Z,m}^{G_1 A}(W \cap U_j) \qquad \text{for all } t \in [-1,1] .$$

Together with (9.5.3) this implies that

$$\phi_j(\cdot,t)\psi^{-1}(\cdot,t) \in \mathcal{O}_{Z,m}^{G_1 A}(W \cap U_j) \qquad \text{for all } t \in [-1,1] .$$

Hence, by (9.5.6),

$$F(\cdot,t) \in \mathcal{O}_{Z,m}^{G_1 A}(W) \qquad \text{for all } t \in [-1,1] .$$

Therefore

$$[-1,1] \ni t \longrightarrow F(\cdot,t)\big|_{\overline{D}} \tag{9.5.7}$$

is a continuous curve in $\overline{\mathcal{O}}_{Z,m}^{G_1 A}(\overline{D})$.

Let us look for the beginning and the end of this curve. Since $\psi_j \in \mathcal{O}_{Z,m}^{G\widetilde{A}}(W \cap U_j)$ and by (9.5.5), we have non-vanishing holomorphic functions $\lambda_j : W \cap U_j \to \mathbb{C}$ with

$$\psi_j(z,-1) = \psi_j(z,1) = \lambda_j(z) \cdot 1 , \qquad z \in W \cap U_j , \quad j \in I .$$

Since $\phi_j(z,-1) = 1$ and $\phi_j(z,1) = f(z)$, this shows that

$$\phi_j(z,-1)\psi_j^{-1}(z,-1) = \lambda_j^{-1}(z) \cdot 1 , \qquad z \in W \cap U_j , \quad j \in I ,$$

and

$$\phi_j(z,1)\psi_j^{-1}(z,-1) = \lambda_j^{-1}(z) f(z) , \qquad z \in W \cap U_j , \quad j \in I .$$

Together with (9.5.6) this gives

$$F(z,-1) = \lambda_j^{-1}(z) \quad \text{and} \quad F(z,1) = \lambda_j^{-1}(z) f(z) , \quad z \in W \cap U_j , \quad j \in I .$$

This implies that there is a non-vanishing scalar holomorphic function λ globally defined on W with

$$F(z,-1) = \lambda^{-1}(z) \cdot 1 \quad \text{and} \quad F(z,1) = \lambda^{-1}(z) f(z) , \qquad z \in W ,$$

and the curve (9.5.7) connects the functions $\lambda^{-1} f$ and $\lambda^{-1} \cdot 1$. Hence

$$[-1,1] \ni t \longrightarrow \lambda F(\cdot,t)\big|_{\overline{D}}$$

is a continuous curve in $\overline{\mathcal{O}}_{Z,m}^{G_1 A}(\overline{D})$ which connects f and the unit element of $\overline{\mathcal{O}}_{Z,m}^{G_1 A}(\overline{D})$. □

9.5.3 Theorem. *Let $D \subseteq \mathbb{C}$ be a bounded open set with piecewise $\mathcal{C}^1$-boundary such that $\mathbb{C} \setminus \overline{D}$ is connected. Assume $Z \cap \mathbb{C}$ is discrete and closed in $\mathbb{C}$ and $Z \cap \partial D = \emptyset$. Let $f \in \overline{\mathcal{O}}_{Z,m}^{G}(\overline{D})$ such that all values of f belong to the same connected component of G (which is always the case if D is connected). Then f can be approximated uniformly on $\overline{D}$ by functions from $\mathcal{O}_{Z,m}^{G}(\mathbb{C})$.*

Proof. Since all values of f belong to the same connected component of G, after multiplying by a constant element of G, we may assume that f belongs to $\overline{\mathcal{O}}^{G_1A}_{Z,m}(\overline{D})$. By Lemma 9.5.2, $\overline{\mathcal{O}}^{G_1A}_{Z,m}(\overline{D})$ is connected. Therefore, by Proposition 5.4.1, f can be written in the form

$$f = g_1 \cdot \ldots \cdot g_n$$

where $g_j \in \overline{\mathcal{O}}^{G_1A}_{Z,m}(\overline{D})$ and

$$\max_{z \in \overline{D}} \|1 - g_j(z)\| < 1 \qquad \text{for } 1 \le j \le n.$$

By Lemma 5.4.4 (ii), $e^{\log g_j} = g_j$. Hence

$$f = e^{\log g_1} \cdot \ldots \cdot e^{\log g_n}.$$

By Proposition 9.3.2, $\log g_j \in \overline{\mathcal{O}}^{A}_{Z,m}(\overline{D})$. Therefore and since $\mathbb{C} \setminus \overline{D}$ is connected, we can apply Proposition 9.3.4 and get, for each j, a sequence $(\varphi_{j\nu})_{\nu \in \mathbb{N}}$ of functions $\varphi_{j\nu} \in \mathcal{O}^{A}_{Z,m}(\mathbb{C})$ which converges to $\log g_j$ uniformly on $\overline{D}$. Setting

$$f_\nu = e^{\varphi_{j\nu}} \cdot \ldots \cdot e^{\varphi_{j\nu}}$$

then we get a sequence $(f_\nu)_{\nu \in \mathbb{N}}$ of holomorphic functions $f_\nu : \mathbb{C} \to A$ which converges to f uniformly on $\overline{D}$. By Proposition 9.3.2, each f_ν belongs to $\mathcal{O}^{G_1A}_{Z,m}(\mathbb{C})$. □

9.6 Splitting over simply connected open sets without shrinking

In this section A is a Banach algebra with unit 1, G is an open subgroup of the group of invertible elements of A, G_1A is the connected component of 1 in the group of invertible elements of A, and (Z, m) is a data of zeros (Def. 9.1.2). It is the aim of this section to prove the following theorem:

9.6.1 Theorem. *Let $D \subseteq \mathbb{C}$ be a simply connected open set such that $Z \cap D$ is discrete and closed in D. Then any $\mathcal{O}^G_{Z,m}$-cocycle over D is $\mathcal{O}^G_{Z,m}$-trivial (Def. 9.1.4).*

We will deduce this from Lemma 9.4.1. For that we need the following generalization of the technical Lemma 5.8.1 to the sheaf $\mathcal{O}^G_{Z,m}$:

9.6.2 Lemma. *Let $D \subseteq \mathbb{C}$ be an open set such that $Z \cap D$ is discrete and closed in D, and let $(D_n)_{n \in \mathbb{N}}$ be a sequence of bounded open sets such that*

(0) *$Z \cap \partial D_n = \emptyset$ for all $n \in \mathbb{N}$;*

(1) *$\overline{D}_n \subseteq D_{n+1}$ for all $n \in \mathbb{N}$;*

(2) $\bigcup_{n\in\mathbb{N}} D_n = D$

(3) *for each n, any function from* $\mathcal{O}^G_{Z,m}(D_{n+1})$ *can be approximated uniformly on* $\overline{D}_n$ *by functions from* $\mathcal{O}^G_{Z,m}(D)$.

Further let f be an $\mathcal{O}^G_{Z,m}$*-cocycle over D such that, for each* $n \in \mathbb{N}$*, the restriction* $f|_{D_n}$ *is* $\mathcal{O}^G_{Z,m}$*-trivial (Def. 9.1.4). Then f itself is* $\mathcal{O}^G_{Z,m}$*-trivial .*

Proof. We denote by $\|\cdot\|$ the norm of A and set

$$\operatorname{dist}(a, A\setminus G) = \inf_{b\in A\setminus G} \|a-b\| \qquad \text{for } a \in G.$$

Let $\mathcal{U} = \{U_j\}_{j\in I}$ be the covering associated to f. By Proposition 9.1.6, after passing to a refinement, we may assume that each U_j is a relatively compact open disc in D and $f_{jk} \in \overline{\mathcal{O}}^G_{Z.m}(\overline{U}_j \cap \overline{U}_k)$ for all $j,k \in I$. Note that then, for each $j \in I$, there exists $n_j \in \mathbb{N}$ with

$$\overline{U}_j \subseteq D_n \qquad \text{if } n \geq n_j. \tag{9.6.1}$$

Moreover we may assume that

$$\begin{cases} \text{for each compact set } K \subseteq D \text{ there exists only a} \\ \text{finite number of indices } j \in I \text{ with } U_j \cap K \neq \emptyset. \end{cases} \tag{9.6.2}$$

To prove the lemma now it is sufficient to find a sequence $(f_n)_{n\in\mathbb{N}}$ of families $f_n = \{f_{n,j}\}_{j\in I}$ of functions $f_{n,j} \in \overline{\mathcal{O}}^G_{Z,m}(D_{n+1} \cap \overline{U}_j)$ as well as a sequence $(\varepsilon_n)_{n\in\mathbb{N}}$ of positive numbers, such that, for all $n \in \mathbb{N}$,

$$f_{jk} = f^{-1}_{n,j} f_{n,k} \qquad \text{on } D_{n+1} \cap \overline{U}_j \cap \overline{U}_k\,, \quad j,k \in I, \tag{9.6.3}$$

$$\varepsilon_n < \frac{1}{4} \min_{z\in\overline{D}_n\cap\overline{U}_j} \operatorname{dist}\Big(f_{n,j}(z), A\setminus G\Big)\,, \quad j \in I, \tag{9.6.4}$$

$$\max_{z\in\overline{D}_n\cap\overline{U}_j} \|f_{n,j}(z) - f_{n-1,j}(z)\| < \varepsilon_{n-1} \text{ if } n \geq 1\,, \quad j \in I, \quad \text{and} \tag{9.6.5}$$

$$\varepsilon_n < \frac{\varepsilon_{n-1}}{2} \text{ if } n \geq 1. \tag{9.6.6}$$

Indeed, then it follows from (9.6.1), (9.6.5) and (9.6.6) that, for all $n, p \in \mathbb{N}$ with $n_j \leq n < p$,

$$\max_{z\in\overline{U}_j} \|f_{p,j}(z) - f_{n,j}(z)\| < \varepsilon_n + \frac{\varepsilon_n}{2} + \ldots + \frac{\varepsilon_n}{2^{p-n-1}} < 2\varepsilon_n,$$

which implies, by (9.6.6), that for each $j \in I$ the sequence $(f_{n,j})_{n\geq n_j}$ converges uniformly on $\overline{U}_j$ to some function $f_j \in \overline{\mathcal{O}}^A(\overline{U}_j)$ where

$$\max_{z\in\overline{U}_j} \|f_j(z) - f_{n,j}(z)\| \leq 2\varepsilon_n \qquad \text{for } n \geq n_j.$$

By (9.6.4), this inequality implies that

$$\max_{z\in\overline{U}_j} \|f_j(z) - f_{n,j}(z)\| < \frac{1}{2} \inf_{z\in\overline{U}_j} \operatorname{dist}\Big(f_{n,j}(z), G\setminus A\Big) \qquad \text{for } n \geq n_j.$$

Hence $f_j \in \overline{\mathcal{O}}^G(\overline{U}_j)$, $j\in I$. Moreover, since, for each $j\in I$, the functions $f_{n,j}$ with $n\geq n_j$ belong to $\overline{\mathcal{O}}^G_{Z,m}(\overline{U}_j)$ and the sequence $\big(f_{n,j}\big)_{n\geq n_j}$ converges uniformly on $\overline{U}_j$ to f_j, it follows that even $f_j\in\overline{\mathcal{O}}^G_{Z,m}(\overline{U}_j)$, $j\in I$. It remains to observe that now we can pass to the limit for $n\to\infty$ in (5.8.3), which gives $f_{jk} = f_j^{-1}f_k$ on $\overline{U_j\cap U_k}$ for all $j,k\in I$. Hence f is $\mathcal{O}^G_{Z,m}$-trivial.

To prove the existence of such sequences, we first recall that, by hypothesis of the lemma, each $f|_{D_{n+2}}$ is $\mathcal{O}^G_{Z,m}$-trivial. Therefore we can find a sequence $\big(\widetilde{f}_n\big)_{n\in\mathbb{N}}$ of families $\widetilde{f}_n = \{\widetilde{f}_{n,j}\}_{j\in I}$ of functions $\widetilde{f}_{n,j}\in\mathcal{O}^G_{Z,m}(D_{n+2}\cap U_j)$ such that

$$f_{jk} = \widetilde{f}_{n,j}^{-1}\widetilde{f}_{n,k} \qquad \text{on } D_{n+2}\cap U_j\cap U_k \tag{9.6.7}$$

for all $n\in\mathbb{N}$ and $j,k\in I$. We claim that even

$$\widetilde{f}_{n,j}\in\overline{\mathcal{O}}^G_{Z.m}(D_{n+2}\cap\overline{U}_j) \qquad \text{for all } j\in I \text{ and } n\in\mathbb{N}. \tag{9.6.8}$$

Indeed, let $(z_\nu)_{\nu\in\mathbb{N}}$ be a sequence in $D_{n+2}\cap U_j$ which converges to some point $z\in D_{n+2}\cap\overline{U}_j$. Since $\mathcal{U}$ covers D, we can find $k\in I$ with $z\in U_k$. Since U_k is open, then $z_\nu\in D_{n+2}\cap\overline{U}_j\cap U_k$ for sufficiently large ν, where, by (9.6.7),

$$\widetilde{f}_{n,j}(z_\nu) = \widetilde{f}_{n,k}(z_\nu)\,f_{jk}^{-1}(z_\nu).$$

Since both $\widetilde{f}_{n,k}$ and f_{jk} are continuous on $D_{n+2}\cap\overline{U}_j\cap U_k$ and since $z\in D_{n+2}\cap\overline{U}_j\cap U_k$, this implies that

$$\lim_{\nu\to\infty}\widetilde{f}_{n,j}(z_\nu) = \widetilde{f}_{n,k}(z)\,f_{jk}^{-1}(z)\,.$$

Now we proceed by induction.

Beginning of the induction: Since $\widetilde{f}_{0,j}\in\overline{\mathcal{O}}^G_{Z,m}(D_2\cap\overline{U}_j)$ and $D_1\subseteq D_2$, we can define $f_{0,j} = \widetilde{f}_{0,j}\big|_{D_1\cap\overline{U}_j}$. It follows from condition (9.6.2) that $\bigcup_{j\in I} f_{0,j}(\overline{D}_0\cap\overline{U}_j)$ is a compact subset of G. Hence we can find $\varepsilon_0>0$ such that

$$\varepsilon_0 < \frac{1}{4}\min_{z\in\overline{D_0\cap U_j}}\operatorname{dist}\Big(f_{0,j}(z), G\setminus A\Big) \qquad \text{for all } j\in I.$$

With this choice of the family $\{f_{0,j}\}_{j\in I}$ and the number ε_0, conditions (9.6.3)–(9.6.6) are satisfied for $n=0$.

Hypothesis of induction: Assume, for some $p\in\mathbb{N}$, we already have families $f_0 = \{f_{0,j}\}_{j\in I},\dots,f_p=\{f_{p,j}\}_{j\in I}$ of functions

$$f_{0,j}\in\overline{\mathcal{O}}^G_{Z,m}(D_0\cap\overline{U}_j),\dots,f_{p,j}\in\overline{\mathcal{O}}^G_{Z,m}(D_p\cap\overline{U}_j)$$

as well as positive numbers $\varepsilon_0, \dots, \varepsilon_p$ such that (9.6.3)–(9.6.6) hold for $n = 0, \dots, p$.

Step of induction: Since the compact set $\overline{D}_p \cap \overline{U}_j$ is contained in $D_{p+1} \cap \overline{U}_j$ and $f_{p,j}$ is continuous on $D_{p+1} \cap \overline{U}_j$, the function $f_{p,j}$ is bounded on $\overline{D}_p \cap \overline{U}_j$. By condition (9.6.2), this implies that

$$\max_{j \in I} \max_{z \in \overline{D}_p \cap \overline{U}_j} \|f_{p,j}(z)\| < \infty. \tag{9.6.9}$$

By (9.6.7),

$$f_{jk} = \widetilde{f}_{p,j}^{-1} \widetilde{f}_{p,k} \qquad \text{on } D_{p+2} \cap U_j \cap U_k.$$

Moreover, by hypothesis of induction,

$$f_{jk} = f_{p,j}^{-1} f_{p,k} \qquad \text{on } D_{p+1} \cap U_j \cap U_k.$$

Since $D_{p+1} \subseteq D_{p+2}$, this yields

$$\widetilde{f}_{p,k} f_{p,k}^{-1} = \widetilde{f}_{p,j} f_{p,j}^{-1} \qquad \text{on } D_{p+1} \cap U_j \cap U_k.$$

Hence, there is a well-defined function $\Phi \in \mathcal{O}^G_{Z,m}(D_{p+1})$ with

$$\Phi = \widetilde{f}_{p,j} f_{p,j}^{-1} \tag{9.6.10}$$

on $D_{p+1} \cap U_j$ for all $j \in I$. Note that, since $f_{p,j}$ is continuous on $D_{p+1} \cap \overline{U}_j$ and, by (9.6.8), $\widetilde{f}_{p,j}$ is continuous on $D_{p+2} \cap \overline{U}_j$, (9.6.10) even holds on $D_{p+1} \cap \overline{U}_j$, $j \in I$. By hypothesis of the lemma, Φ can be approximated uniformly on $\overline{D}_p$ by functions from $\mathcal{O}^G_{Z,m}(D)$. Therefore and by (9.6.9), we can find $\Psi \in \mathcal{O}^G_{Z,m}(D)$ such that

$$\max_{\overline{D}_p} \|\Psi\Phi - 1\| < \frac{\varepsilon_p}{\max\limits_{\overline{D}_p \cap \overline{U}_j} \|f_{p,j}\|} \qquad \text{for all } j \in I.$$

Since (9.6.10) holds over $D_{p+1} \cap \overline{U}_j$ and $\overline{D}_p \cap \overline{U}_j \subseteq D_{p+1} \cap \overline{U}_j$, this implies that

$$\max_{\overline{D}_p \cap \overline{U}_j} \|\Psi \widetilde{f}_{p,j} f_{p,j}^{-1} - 1\| < \frac{\varepsilon_p}{\max\limits_{\overline{D}_p \cap \overline{U}_j} \|f_{p,j}\|} \qquad \text{for all } j \in I. \tag{9.6.11}$$

Setting

$$f_{p+1,j} = \Psi \widetilde{f}_{p,j} \qquad \text{on } D_{p+2} \cap \overline{U}_j,$$

now we obtain a family $f_{p+1} = \{f_{p+1,j}\}_{j \in I}$ of functions $f_{p+1,j} \in \overline{\mathcal{O}}^G_{Z,m}(D_{p+2} \cap \overline{U}_j)$. Further, it follows from condition (9.6.2) that $\bigcup_{j \in I} f_{p+1,j}(\overline{D}_{p+1} \cap \overline{U}_j)$ is a compact subset of G. Hence we can find $\varepsilon_{p+1} > 0$ so small that condition (9.6.4) is satisfied for $n = p+1$. As $\varepsilon_p > 0$, we may moreover assume that (9.6.6) holds for $n = p+1$. From (9.6.7) we get

$$f_{p+1,j}^{-1} f_{p+1,k} = \widetilde{f}_{p,j}^{-1} \Psi^{-1} \Psi \widetilde{f}_{p,k} = \widetilde{f}_{p,j}^{-1} \widetilde{f}_{p,k} = f_{jk}$$

on $D_{p+2} \cap \overline{U}_j \cap \overline{U}_k$, i.e., (9.6.3) holds for $n = p+1$. From (9.6.11) it follows that

$$\max_{\overline{D}_p \cap \overline{U}_j} \left\| f_{p+1,j} - f_{p,j} \right\| = \max_{\overline{D}_p \cap \overline{U}_j} \left\| \left(\Psi \widetilde{f}_{p,j} f_{p,j}^{-1} - 1 \right) f_{p,j} \right\| < \varepsilon_p.$$

Hence also (9.6.5) holds for $n = p+1$. □

Proof of Theorem 9.6.1. Since D is simply connected, by the Riemann mapping theorem we may assume that either $D = \mathbb{C}$ or D is the open unit disc. In both cases, there exists a sequence $(D_n)_{n \in \mathbb{N}}$ of open discs $D_n \subseteq \mathbb{C}$ such that

- $\overline{D}_n \subseteq D$,
- $\overline{D}_n \subseteq D_{n+1}$ for all $n \in \mathbb{N}$,
- $\bigcup_{n \in \mathbb{N}} D_n = D$,
- by the Runge approximation Theorem 9.5.3, for all $n \in \mathbb{N}$, any $f \in \mathcal{O}^G_{Z,m}(D_{n+1})$ can be approximated uniformly on $\overline{D}_n$ by functions from $\mathcal{O}^G_{Z,m}(D)$.

Hence, by Lemma 9.6.2, it is sufficient to prove that for any $\mathcal{O}^G_{Z,m}$-cocycle f over D, the restrictions $f|_{D_n}$, $n \in \mathbb{N}$, are $\mathcal{O}^G_{Z,m}$-trivial, which is indeed the case by Lemma 9.4.1. □

Finally we point out the following special case of Theorem 9.6.1:

9.6.3 Corollary. *Let $D_1, D_2 \subseteq \mathbb{C}$ be two open sets such that $D_1 \cup D_2$ is simply connected and $Z \cap (D_1 \cup D_2)$ is discrete and closed in $D_1 \cup D_2$. Then, for each $f \in \mathcal{O}^G_{Z,m}(D_1 \cap D_2)$, there exist functions $f_j \in \mathcal{O}^G_{Z,m}(D_j)$, $j = 1, 2$, such that $f = f_1^{-1} f_2$ on $D_1 \cap D_2$.*

9.7 Runge approximation. The general case

In this section A is a Banach algebra with unit 1, G is an open subgroup of the group of invertible elements of A, $G_1 A$ is the connected component of 1 in the group of invertible elements of A, and (Z, m) is a data of zeros (Def. 9.1.2).

Here we generalize part (ii) of the Runge approximation Theorem 5.0.1 to sections of $\mathcal{O}^G_{Z,m}$. For that we pass again to the Riemann sphere $\mathbb{P}^1 = \mathbb{C} \cup \{\infty\}$ and use the notation concerning the Riemann sphere introduced in the beginning of Section 5.10.

Since, for each $a \in \mathbb{P}^1$, T_a is biholomorphic between $\mathbb{P}^1 \setminus \{a\}$ and the complex plane, from the Runge approximation Theorem 9.5.3 we immediately get:

9.7.1 Proposition. *Let $a \in \mathbb{P}^1$ be a fixed point. Let D be a relatively compact open subset of $\mathbb{P}^1 \setminus \{a\}$ with piecewise $\mathcal{C}^1$-boundary such that $\mathbb{P}^1 \setminus \overline{D}$ is connected. Assume $Z \cap \left(\mathbb{P}^1 \setminus \{a\}\right)$ is discrete and closed in $\mathbb{P}^1 \setminus \{a\}$ and*

$$Z \cap \partial D = \emptyset.$$

Let $f \in \overline{\mathcal{O}}^G_{Z,m}(\overline{D})$ such that all values of f belong to the same connected component of G (which is always the case if D is connected). Then f can be approximated uniformly on $\overline{D}$ by functions from $\mathcal{O}^G_{Z,m}(\mathbb{P}^1 \setminus \{a\})$.

Moreover, from Corollary 5.9.3 we immediately get:

9.7.2 Proposition. *Let $a \in \mathbb{P}^1$ and let $D_1, D_2 \subseteq \mathbb{P}^1 \setminus \{a\}$ be two open sets such that $D_1 \cup D_2$ is simply connected and $Z \cap (D_1 \cup D_2)$ is discrete and closed in $D_1 \cup D_1$. Then, for each $f \in \mathcal{O}^G_{Z,m}(D_1 \cap D_2)$, there exist functions $f_j \in \mathcal{O}^G_{Z,m}(D_j)$, $j = 1, 2$, such that $f = f_1^{-1} f_2$ on $D_1 \cap D_2$.*

9.7.3 Lemma. *Let $D \subseteq \mathbb{P}^1$ be an open set with piecewise $\mathcal{C}^1$-boundary and let $U_1, \dots, U_n$ be the connected components of $\mathbb{P}^1 \setminus \overline{D}$. Let $n \geq 2$ and let some points $a_j \in U_j$, $1 \leq j \leq n$, be fixed such that $Z \cap \big(\mathbb{P}^1 \setminus \{a_1, \dots, a_n\}\big)$ is discrete and closed in $\mathbb{P}^1 \setminus \{a_1, \dots, a_n\}$. Then, for each $f \in \overline{\mathcal{O}}^G_{Z,m}(\overline{D})$, there exist functions $f_j \in \overline{\mathcal{O}}^G_{Z,m}(\mathbb{P}^1 \setminus U_j)$, $1 \leq j \leq n$, and a function $h \in \mathcal{O}^G_{Z,m}\big(\mathbb{P}^1 \setminus \{a_1, \dots, a_n\}\big)$ such that $f = h f_n \cdots f_1$ on $\overline{D}$.*

Proof. This proof is similar to the proof of Lemma 5.10.4. For $1 \leq k \leq n$, we consider the following statement:

$A(k)$: There exist functions $f_j \in \overline{\mathcal{O}}^G_{Z,m}(\mathbb{P}^1 \setminus U_j)$, $1 \leq j \leq k$, and a function $h_k \in \overline{\mathcal{O}}^G_{Z,m}\Big(\overline{D} \cup \big(U_1 \setminus \{a_1\}\big) \cup \ldots \cup \big(U_k \setminus \{a_k\}\big)\Big)$ such that $f = h_k f_k \cdots f_1$ on $\overline{D}$.

Since

$$\overline{D} \cup \big(U_1 \setminus \{a_1\}\big) \cup \ldots \cup \big(U_n \setminus \{a_n\}\big) = \mathbb{P}^1 \setminus \{a_1, \dots, a_n\},$$

then $A(n)$ is the assertion of the lemma. Therefore it is sufficient to prove $A(1)$ and the conclusions $A(k) \Rightarrow A(k+1)$, $1 \leq k \leq n-1$.

Proof of $A(1)$: Since

$$\big(\mathbb{P}^1 \setminus \overline{U}_1\big) \cup \big(D \cup (\overline{U}_1 \setminus \{a_1\})\big) = \mathbb{P}^1 \setminus \{a_1\}$$

and

$$\big(\mathbb{P}^1 \setminus \overline{U}_1\big) \cap \big(D \cup (\overline{U}_1 \setminus \{a_1\})\big) = D,$$

from Proposition 9.7.2 we get functions $f_1 \in \mathcal{O}^G_{Z,m}\big(\mathbb{P}^1 \setminus \overline{U}_1\big)$ and $h_1 \in \mathcal{O}^G_{Z,m}\big(D \cup (\overline{U}_1 \setminus \{a_1\}\big)$ such that

$$f = h_1 f_1 \tag{9.7.1}$$

on D. Since f is continuous and with values in G on $\overline{D}$, since h_1 is continuous and with values in G on $D \cup \partial U_1$, since f_1 is continuous and with values in G on $\overline{D} \setminus \partial U_1$ and since (9.7.1) holds in D, it follows that $f_1 \in \overline{\mathcal{O}}^G_{Z,m}(\mathbb{P}^1 \setminus U_1)$, $h_1 \in \overline{\mathcal{O}}^G_{Z,m}\big(\overline{D} \cup (U_1 \setminus \{a_1\})\big)$ and (9.7.1) holds on $\overline{D}$, i.e., assertion $A(1)$ is valid.

Proof of $A(k) \Rightarrow A(k+1)$: Let $1 \le k \le n-1$ be given, assume that statement $A(k)$ is valid, and let $f_1, \dots, f_k$ and h_k be as in this statement. Since

$$\left(\mathbb{P}^1 \setminus \overline{U}_{k+1}\right) \cup \left(D \cup \left(\overline{U}_1 \setminus \{a_1\}\right) \cup \ldots \cup \left(\overline{U}_{k+1} \setminus \{a_{k+1}\}\right)\right) = \mathbb{P}^1 \setminus \{a_{k+1}\}$$

and

$$\left(\mathbb{P}^1 \setminus \overline{U}_{k+1}\right) \cap \left(D \cup \left(\overline{U}_1 \setminus \{a_1\}\right) \cup \ldots \cup \left(\overline{U}_{k+1} \setminus \{a_{k+1}\}\right)\right)$$
$$= D \cup \left(\overline{U}_1 \setminus \{a_1\}\right) \cup \ldots \cup \left(\overline{U}_k \setminus \{a_k\}\right),$$

from Proposition 9.7.2 we get functions

$$f_{k+1} \in \mathcal{O}^G_{Z,m}\left(\mathbb{P}^1 \setminus \overline{U}_{k+1}\right)$$

and

$$h_{k+1} \in \mathcal{O}^G_{Z,m}\left(D \cup \left(\overline{U}_1 \setminus \{a_1\}\right) \cup \ldots \cup \left(\overline{U}_{k+1} \setminus \{a_{k+1}\}\right)\right)$$

such that

$$h_k = h_{k+1} f_{k+1} \tag{9.7.2}$$

in $D \cup \left(\overline{U}_1 \setminus \{a_1\}\right) \cup \ldots \cup \left(\overline{U}_k \setminus \{a_k\}\right)$. Since h_k is continuous and with values in G on $\overline{D}$, since h_{k+1} is continuous and with values in G on $D \cup \partial U_{k+1}$, since f_{k+1} is continuous and with values in G on $\overline{D} \setminus \partial U_{k+1}$ and since (9.7.2) holds in D, it follows that

$$f_{k+1} \in \overline{\mathcal{O}}^G_{Z,m}\left(\mathbb{P}^1 \setminus U_{k+1}\right),$$

$$h_{k+1} \in \overline{\mathcal{O}}^G_{Z,m}\left(\overline{D} \cup \left(U_1 \setminus \{a_1\}\right) \cup \ldots \cup \left(U_{k+1} \setminus \{a_{k+1}\}\right)\right)$$

and (9.7.2) holds on $\overline{D}$. Since $f = h_k f_k \cdots f_1$ on $\overline{D}$, this implies that

$$f = h_{k+1} f_{k+1} f_k \cdots f_1$$

on $\overline{D}$, i.e., assertion $A(k+1)$ is valid. □

9.7.4 Theorem. *Let $D \subseteq \mathbb{P}^1$ be an open set with piecewise $\mathcal{C}^1$-boundary, and let $U_1, \dots, U_n$ be the connected components of $\mathbb{P}^1 \setminus \overline{D}$. Let $n \ge 2$ and let some points $a_j \in U_j$, $1 \le j \le n$, be fixed such that $Z \cap \left(\mathbb{P}^1 \setminus \{a_1, \dots, a_n\}\right)$ is discrete and closed in $\mathbb{P}^1 \setminus \{a_1, \dots, a_n\}$ and*

$$Z \cap \partial D = \emptyset\,.$$

Then the functions from $\overline{\mathcal{O}}^G_{Z,m}(\overline{D})$ can be approximated uniformly on $\overline{D}$ by functions from $\mathcal{O}_{Z,m}\left(\mathbb{P}^1 \setminus \{a_1, \dots, a_n\}, G\right)$.

Proof. This proof is similar to the proof of Theorem 5.10.5. If $n = 1$, the assertion of the theorem is that of Proposition 9.7.1. If $n \geq 2$, then, by Lemma 9.7.3, each $f \in \overline{\mathcal{O}}^G_{Z,m}(\overline{D})$ can be written in the form

$$f = hf_n \dots f_1 \qquad \text{on } \overline{D}, \tag{9.7.3}$$

where $f_j \in \overline{\mathcal{O}}^G_{Z,m}(\mathbb{P}^1 \setminus U_j)$, $1 \leq j \leq n$, and $h \in \mathcal{O}^G_{Z,m}(\mathbb{P}^1 \setminus \{a_1, \dots, a_n\})$. Let V be the interior of $\mathbb{P}^1 \setminus U_j$. Since the boundary of U_j is piecewise $\mathcal{C}^1$ (as a part of the boundary of D), also the boundary of V is piecewise $\mathcal{C}^1$ and $\overline{V} = \mathbb{P}^1 \setminus U_j$. Since U_j is connected, $\mathbb{P}^1 \setminus \overline{V} = U_j$ is connected. Therefore, Proposition 9.7.1 can be applied to each V_j. Hence, each f_j can be approximated uniformly on $\overline{V} = \mathbb{P}^1 \setminus U_j$ by functions from $\mathcal{O}^G_{Z,m}(\mathbb{P}^1 \setminus \{a_j\})$. Since $\mathcal{O}^G(\mathbb{P}^1 \setminus \{a_j\}) \subseteq \mathcal{O}^G_{Z,m}(\mathbb{P}^1 \setminus \{p_1, \dots, p_n\})$ and $\overline{D} \subseteq \mathbb{P}^1 \setminus U_j$), this means in particular that each f_j can be approximated uniformly on $\overline{D}$ by functions from $\mathcal{O}^G_{Z,m}(\mathbb{P}^1 \setminus \{p_1, \dots, p_n\})$. Since h belongs to $\mathcal{O}^G_{Z,m}(\mathbb{P}^1 \setminus \{p_1, \dots, p_n\})$ and by (9.7.3), this implies the assertion of the theorem. □

9.8 The Oka-Grauert principle

In this section A is a Banach algebra with unit 1, G is an open subgroup of the group of invertible elements of A, G_1A is the group of invertible elements of A, and (Z, m) is a data of zeros (Def. 9.1.2). Here we prove the following theorem:

9.8.1 Theorem. *Let $D \subseteq \mathbb{C}$ be an open set such that $Z \cap D$ is discrete and closed in D, and let f be an $\mathcal{O}^G_{Z,m}$-cocycle. Then:*

(i) *If f is $\mathcal{C}^G$-trivial*[5] *over D, then f is $\mathcal{O}^G_{Z,m}$-trivial.*

(ii) *If G is connected, then f is $\mathcal{O}^G_{Z,m}$-trivial.*

First we prove the following generalization of Lemma 5.11.3:

9.8.2 Lemma. *Let $D \subseteq \mathbb{C}$ be an open set with piecewise $\mathcal{C}^1$-boundary such that $Z \cap D$ is finite and $Z \cap \partial D = \emptyset$. Further, let U be a neighborhood of $\overline{D}$ and f an $\mathcal{O}^G_{Z,m}$-cocycle over U. Suppose that at least one of the following two conditions is fulfilled:*

(i) *f is $\mathcal{C}^G$-trivial over U.*

(ii) *G is connected.*

Then the restriction $f|_D$ is $\mathcal{O}^G_{Z,m}$-trivial.

[5]This is not a misprint. We really mean "if f is $\mathcal{C}^G$-trivial" and not "if f is $\mathcal{C}^G_{Z,m}$-trivial". The latter notion we even did not define, because we do not use it.

Proof. We may assume that D is connected. We proceed by induction over the number of connected components of $\mathbb{C} \setminus \overline{D}$.

Beginning of induction: Suppose this number is 1, i.e., $\mathbb{C} \setminus \overline{D}$ is connected. As the boundary of D is piecewise $\mathcal{C}^1$, then also $\mathbb{C} \setminus D$ is connected, which means (cf., e.g., theorem 13.11 in Rudin's book [Ru]) that D is simply connected. Therefore the assertion of the lemma follows from Theorem 9.6.1. (Even if none of the conditions (i) or (ii) is satisfied.)

Hypothesis of induction: Assume, for some $n \in \mathbb{N}$ with $n \geq 2$, the assertion of the lemma is already proved if the number of connected components of $\mathbb{C} \setminus \overline{D}$ is $n-1$.

Step of induction: Assume that the number of connected components of $\mathbb{C} \setminus \overline{D}$ is equal to n. Then, by Lemma 5.11.2, we can find a Cartan pair (D_1, D_2) with $D = D_1 \cup D_2$ satisfying conditions (1), (2), (3) (of this lemma). Since the boundaries of D_1, D_2, $D_1 \cap D_2$ and D are piecewise $\mathcal{C}^1$, we can find a Cartan pair (D_1', D_2') satisfying the same conditions (1), (2), (3) such that $\overline{D}_j \subseteq D_j'$ and $\overline{D}_1' \cup \overline{D}_2' \subseteq U$.

Then, again by Theorem 9.6.1, the cocycle $f|_{D_1'}$ is $\mathcal{O}^G_{Z,m}$-trivial. Moreover, since the number of connected components of $\mathbb{C} \setminus \overline{D}_2'$ is equal to $n-1$, it follows from the hypothesis of induction that also $f|_{D_2'}$ is $\mathcal{O}^G$-trivial. Hence both restrictions $f|_{D_1'}$ and $f|_{D_2'}$ are $\mathcal{O}^G$-trivial. By Proposition 9.1.9, this implies that $f\big|_{D_1' \cup D_2'}$ is $\mathcal{O}^G_{Z,m}$-equivalent to certain $\big(\{D_1', D_2'\}, \mathcal{O}^G_{Z,m}\big)$-cocycle f'. Since $\overline{D}_1 \cap \overline{D}_2 \subseteq D_1' \cap D_2'$, setting

$$F = f_{12}'|_{\overline{D}_1 \cap \overline{D}_2},$$

we get a function $F \in \overline{\mathcal{O}}^G_{Z,m}(\overline{D}_1 \cap \overline{D}_2)$. We claim that all values of F belong to the same connected component of G.

If $G = G_1 A$, this is trivial. If not, then condition (i) in the lemma under proof is satisfied, i.e., f is $\mathcal{C}^G$-trivial over U. As $D_1' \cup D_2' \subseteq U$, then also the restriction $f|_{D_1' \cup D_1'}$ is $\mathcal{C}^G$-trivial over $D_1' \cup D_1'$. Since $f|_{D_1' \cup D_1'}$ is $\mathcal{C}^G$-equivalent to f' over $D_1' \cup D_2'$ (it is even $\mathcal{O}^G_{Z,m}$-equivalent to f'), this implies that also f' is $\mathcal{C}^G$-trivial, i.e., we can find $C_j \in \mathcal{C}^G(D_j')$, $j = 1, 2$, with

$$f_{12}' = C_1^{-1} C_2 \qquad \text{on } D_1' \cap D_2'.$$

Hence condition (iii) in Lemma 5.11.1 is satisfied, and it follows (from this lemma) that all values of F belong to the same connected component of G.

Since all values of F belong to the same connected component of G, it follows from the Cartan Lemma 9.3.7 that there exist functions $F_j \in \overline{\mathcal{O}}^G_{Z,m}(\overline{D}_j)$, $j = 1, 2$, with

$$F = F_1^{-1} F_2 \qquad \text{on } \overline{D}_j \cap \overline{D}_j.$$

Since $F|_{D_1 \cap D_2} = f_{12}'|_{D_1 \cap D_2}$, this means in particular that $f'\big|_D$ is $\mathcal{O}^G_{Z,m}$-trivial. Finally, as $f\big|_{D_1' \cup D_2'}$ and f' are $\mathcal{O}^G_{Z,m}$-equivalent and therefore $f|_D$ and $f'|_D$ are $\mathcal{O}^G_{Z,m}$-equivalent, it follows that also $f|_D$ is $\mathcal{O}^G_{Z,m}$-trivial. □

Proof of Theorem 9.8.1. Take a sequence $(\Omega_n)_{n\in\mathbb{N}}$ of bounded open sets with $\mathcal{C}^1$-boundaries such that $\overline{\Omega}_n \subseteq \Omega_{n+1}$ and $Z\cap\partial\Omega_n = \emptyset$ for all $n \in \mathbb{N}$ and $\bigcup_{n\in\mathbb{N}} \Omega_n = D$. Let U_n be the union of all bounded connected components of $\mathbb{C} \setminus \overline{\Omega}_n$ which are subsets of D (if there is any – otherwise $U_n := \emptyset$), and set

$$D_n = \Omega_n \cup \overline{U}_n.$$

Then also $(D_n)_{n\in\mathbb{N}}$ is a sequence of bounded open sets with $\mathcal{C}^1$-boundaries such that $Z \cap \partial D_n = \emptyset$, $\overline{D}_n \subseteq D_{n+1}$ for all $n \in \mathbb{N}$ and $\bigcup_{n\in\mathbb{N}} D_n = D$. Moreover this sequence has the important property that each bounded connected component of $\mathbb{C} \setminus \overline{D}_n$ (if there is any) contains at least one point of $\mathbb{C} \setminus D$. By the Runge approximation Theorem 9.7.4, this implies that, for each n, the functions from $\overline{\mathcal{O}}^G_{Z,m}(\overline{D}_n)$ can be approximated uniformly on $\overline{D}_n$ by functions from $\mathcal{O}^G_{Z,m}(D)$. In particular we see that the sequence $(D_n)_{n\in\mathbb{N}}$ has the properties (0)–(2) of Lemma 9.6.2. Therefore, by this lemma, f is $\mathcal{O}^G_{Z,m}$-trivial if each $f|_{D_n}$ is $\mathcal{O}^G_{Z,m}$-trivial. That each $f|_{D_n}$ is $\mathcal{O}^G_{Z,m}$-trivial, follows from Lemma 9.8.2. □

9.9 Comments

The results of this chapter are practically new, and here they are published for the first time, although they could be viewed as special cases of a much more general theory (see [FoRa] for finite dimensional groups, and [Le2, Le7] for infinite dimensional groups). However this is far not obvious. It is simpler to prove them again. Note also that some elements of the theory of cocycles with restrictions can be pointed out in the proofs in the papers [GR1, GR2]. The results of this chapter are used in the consequent chapters only.

Chapter 10

Generalized interpolation problems

Here we prove further generalizations of the Weierstrass product theorem.

10.1 Weierstrass theorems

10.1.1 Theorem. *Let A be a Banach algebra with unit 1, let G be an open subgroup of the group of invertible elements of A, let $D \subseteq \mathbb{C}$ be an open set, and let Z be a discrete and closed subset of D. Suppose, for each $w \in Z$, there is given a neighborhood U_w of w with $U_w \cap Z = \{w\}$ and a holomorphic function $f_w : U_w \setminus \{w\} \to G$. Moreover, we assume that at least one of the following conditions is fulfilled:*

(i) *G is connected.*

(ii) *D is simply connected.*

Then there exist holomorphic functions $h_w : U_w \to G$, $w \in Z$, and a holomorphic function $h : D \setminus Z \to G$ such that

$$h_w f_w = h \qquad \text{on } U_w \setminus \{w\}. \tag{10.1.1}$$

Moreover, for any given family of positive integers m_w, $w \in Z$, the functions h_w can be chosen so that, for each $w \in Z$, the functions $h_w - 1$ and $h_w^{-1} - 1$ have a zero of order $\geq m_w$ at w.

10.1.2. In this theorem, it would be sufficient to claim that one of the functions $h_w - 1$ *or* $h_w^{-1} - 1$ has a zero of order $\geq m_w$ at w. For the other one this follows automatically. Indeed, assume, for example, that this is the case for $h_w - 1$. Then, in a neighborhood of w,

$$h_w^{-1} = 1 - (1 - h_w^{-1}) = \sum_{j=0}^{\infty} (1 - h_w)^j = 1 + (1 - h_w) \sum_{j=0}^{\infty} (1 - h_w)^j ,$$

which shows that also $h_w^{-1} - 1$ has a zero of order $\geq m_w$ at w.

The topological conditions (i) and (ii) in Theorem 10.1.1 can be replaced by the more general condition that *the problem can be solved continuously*, i.e., there is the following Oka-Grauert principle:

10.1.3 Theorem. *Let A be a Banach algebra with unit* 1, *let G be an open subgroup of the group of invertible elements of A, let $D \subseteq \mathbb{C}$ be an open set, and let Z be a discrete and closed subset of D. Suppose, for each $w \in Z$, there is given a neighborhood U_w of w with $U_w \cap Z = \{w\}$ and a holomorphic function $f_w : U_w \setminus \{w\} \to G$. Assume that:*

(iii) *There exist continuous functions $c_w : U_w \to G$ and a continuous function $c : D \setminus Z \to G$ such that $c_w f_w = c$ on $U_w \setminus \{w\}$, $w \in Z$.*

Then there exist holomorphic functions $h_w : U_w \to G$, $w \in Z$, and a holomorphic function $h : D \setminus Z \to G$ such that

$$h_w f_w = h \qquad \text{on } U_w \setminus \{w\} . \tag{10.1.2}$$

Moreover, for any given family of positive integers m_w, $w \in Z$, the functions h_w can be chosen so that, for each $w \in Z$, the functions $h_w - 1$ and $h_w^{-1} - 1$ have a zero of order $\geq m_w$ at w.

The first parts of theorems 10.1.1 and 10.1.3 coincide with theorems 5.13.1 and 5.13.2, respectively. The additional information at the end about the zeros of $h_w - 1$ and $(h_w - 1)^{-1}$ has important consequences. For example, if the functions f_w are meromorphic at w and, hence, h is meromorphic at each w (which is already clear by the first part of the theorem), then, for any given orders $m_w \in \mathbb{N}^*$, the family of functions h_w can be chosen so that $h - f_w$ has a zero of order m_w, i.e., it can be achieved that arbitrarily prescribed finite pieces of the Laurent expansions of h and f_w concide. The corresponding is true if the functions f_w^{-1} are meromorphic. This is contained in the following immediate corollary of theorems 10.1.1 and 10.1.3:

10.1.4 Corollary. *Assume, under the hypotheses of Theorem* 10.1.1 *or under the hypotheses of Theorem* 10.1.3, *two subsets Z_+ and Z_- of Z are given, where the cases $Z_+ = \emptyset$, $Z_- = \emptyset$ and $Z_+ \cap Z_- \neq \emptyset$ are possible. Moreover assume that:*

- *If $w \in Z_+$, then f_w is meromorphic at w.*
- *If $w \in Z_-$, then f_w^{-1} is meromorphic at w.*

Then, for any given family of positive integers m_w, $w \in Z$, in the claims of these theorems, it can be shown that:

- *If $w \in Z_+$, then $f - f_w$ has a zero of order m_w at w.*
- *If $w \in Z_-$, then $f^{-1} - f_w^{-1}$ has a zero of order m_w at w.*

Proof of Theorems 10.1.1 *and* 10.1.3. Let a family $m = \{m_w\}_{w \in Z}$ of positive integers be given. Since Z is discrete and closed in D, then (Z, m) is a data of zeros in the sense of Definition 9.1.2. In terms of this definition, we have to find $h \in \mathcal{O}_{Z,m}(D \setminus Z)$ and $h_w \in \mathcal{O}_{Z,m}(U_w)$, $w \in Z$, with $h_w f_w = h$ on $U_w \setminus \{w\}$.

Choose neighborhoods $V_w \subseteq U_w$ so small that $V_w \cap V_v = \emptyset$ if $w \neq v$.

It is sufficient to find $h \in \mathcal{O}_{Z,m}(D \setminus Z)$ and $h_w \in \mathcal{O}_{Z,m}(V_w)$, $w \in Z$, such that

$$h_w f_w = h \tag{10.1.3}$$

on $V_w \setminus \{w\}$. Indeed, since $V_w \cap Z = U_w \cap Z = \{w\}$, then, by (10.1.3), each h_w admits an extension to a function from $\mathcal{O}^G_{Z,m}(U_w)$ such that (10.1.3) holds also for this extension.

Set $D_1 = \bigcup_{w \in Z} V_w$ and $D_2 = D \setminus Z$. Since the sets V_w are pairwise disjoint and $V_w \cap Z = \{w\}$, the family of functions f_w can be interpreted as a single holomorphic function $f \in \mathcal{O}^G(D_1 \setminus Z)$. Since $Z \cap D_1 \cap D_2 = \emptyset$, the restriction $f\big|_{D_1 \cap D_2}$ belongs to $\mathcal{O}^G_{Z,m}(D_1 \cap D_2)$. Now, by Corollary 9.2.2, there exist $h_1 \in \mathcal{O}^G_{Z,m}(D_j)$ and $h_2 \in \mathcal{O}^G_{Z,m}(D_2)$ with $f = h_1^{-1} h_2$ on $D_1 \cap D_2$. Setting $h_w = h_1\big|_{V_w}$ and $h = h_2$, we complete the proof. □

Since, in theorems 10.1.1 and 10.1.3, the multiplication by the functions h_w is carried out from the left, we call these theorems *left-sided* Weierstrass theorems. There are also *right-* and *two-sided* versions.

10.2 Right- and two-sided Weierstrass theorems

If A is a Banach algebra with the multiplication "$\cdot$", then we can pass to the Banach Algebra $\widetilde{A}$ which consists of the same additive group A but with the multiplication "$\tilde{\cdot}$" defined by $a \,\tilde{\cdot}\, b = b \cdot a$. In this way, from theorems 10.1.1 and 10.1.3 we get the following *right-sided* Weierstrass theorems:

10.2.1 Theorem. *Let A be a Banach algebra with unit 1, let G be an open subgroup of the group of invertible elements of A, let $D \subseteq \mathbb{C}$ be an open set, and let Z be a discrete and closed subset of D. Suppose, for each $w \in Z$, there is given a neighborhood U_w of w with $U_w \cap Z = \{w\}$ and a holomorphic function $f_w : U_w \setminus \{w\} \to G$. Moreover, we assume that at least one of the following conditions is fulfilled:*

(i) *G is connected.*

(ii) *D is simply connected.*

Then there exist holomorphic functions $h_w : U_w \to G$, $w \in Z$, and a holomorphic function $h : D \setminus Z \to G$ such that

$$f_w h_w = h \qquad \text{on } U_w \setminus \{w\}\,. \tag{10.2.1}$$

Moreover, for any given family of positive integers m_w, $w \in Z$, the functions h_w can be chosen so that, for each $w \in Z$, the functions $h_w - 1$ and $h_w^{-1} - 1$ have a zero of order $\geq m_w$ at w.

10.2.2 Theorem. *Let A be a Banach algebra with unit 1, let G be an open subgroup of the group of invertible elements of A, let $D \subseteq \mathbb{C}$ be an open set, and let Z be a discrete and closed subset of D. Suppose, for each $w \in Z$, there is given a neighborhood U_w of w with $U_w \cap Z = \{w\}$ and a holomorphic function $f_w : U_w \setminus \{w\} \to G$. Moreover, we assume that:*

(iii) *There exist continuous functions $c_w : U_w \to G$ and a continuous function $c : D \setminus Z \to G$ such that $f_w c_w = c$ on $U_w \setminus \{w\}$, $w \in Z$.*

Then there exist holomorphic functions $h_w : U_w \to G$, $w \in Z$, and a holomorphic function $h : D \setminus Z \to G$ such that

$$f_w h_w = h \qquad \text{on } U_w \setminus \{w\}\,. \tag{10.2.2}$$

Moreover, for any given family of positive integers m_w, $w \in Z$, the functions h_w can be chosen so that, for each $w \in Z$, the functions $h_w - 1$ and $h_w^{-1} - 1$ have a zero of order $\geq m_w$ at w.

Finally, we present a *two-sided* Weierstrass theorem:

10.2.3 Theorem. *Let A be a Banach algebra with unit 1, let G be an open subgroup of the group of invertible elements of A, let $D \subseteq \mathbb{C}$ be an open set, and let Z be a discrete and closed subset of D. Suppose, for each $w \in Z$, there are given a neighborhood U_w of w with $U_w \cap Z = \{w\}$ and two holomorphic functions $f_w, g_w : U_w \setminus \{w\} \to G$. Moreover, we assume that at least one of the following conditions is fulfilled:*

(i) *G is connected.*

(ii) *D is simply connected.*

Then there exist holomorphic functions $h_w : U_w \to G$, $w \in Z$, and a holomorphic function $h : D \setminus Z \to G$ such that

$$f_w h_w g_w = h \qquad \text{on } U_w \setminus \{w\}\,. \tag{10.2.3}$$

Moreover, for any given family of positive integers m_w, $w \in Z$, the functions f_w and g_w can be chosen so that, for each $w \in Z$, the functions $f_w - 1$, $f_w^{-1} - 1$, $g_w - 1$ and $g_w^{-1} - 1$ have a zero of order $\geq m_w$ at w.

Proof. Let a family of positive integers m_w, $w \in Z$, be given. Then from the left-sided Weierstrass Theorem 10.1.1 we get holomorphic functions $h^l_w : U_w \to G$, $w \in Z$, and a holomorphic function $h^l : D \setminus Z \to G$ such that

$$h^l_w g_w = h^l \qquad \text{on } U_w \setminus \{w\} \tag{10.2.4}$$

and the functions $h^l_w - 1$ have a zero of order $\geq m_w$ at w. From the right-sided Weierstrass Theorem 10.2.1 we get holomorphic functions $h^r_w : U_w \to G$, $w \in Z$, and a holomorphic function $h^r : D \setminus Z \to G$ such that

$$f_w h^r_w = h^r \qquad \text{on } U_w \setminus \{w\} \tag{10.2.5}$$

and the functions $h^r_w - 1$ have a zero of order $\geq m_w$ at w. Set $h = h^l h^r$ and $h_w = h^r_w h^l_w$, $w \in Z$. Then $h \in \mathcal{O}^G(D \setminus Z)$, $h_w \in \mathcal{O}^G(U_w)$ and

$$f_w h_w g_w = f_w h^r_w h^l_w g_w = h^r h^l = h \qquad \text{on } U_w \setminus \{w\}$$

and the functions $f_w - 1$, $f_w^{-1} - 1$, $g_w - 1$ and $g_w^{-1} - 1$ have a zero of order $\geq m_w$ at w. □

10.2.4 Remark. Instead of conditions (i) or (ii) in Theorem 10.2.3 also the following condition would be sufficient (Oka-Grauert principle):

(iii) There exist continuous functions $c_w : U_w \to G$, $w \in Z$, and a continuous function $c : D \setminus Z \to G$ such that $f_w c_w g_w = c$ on $U_w \setminus \{w\}$, $w \in Z$.

But to prove this we would need a further generalization of the theory of multiplicative cocycles where the group G in Definition 9.1.2 is replaced by a *fiber bundle* of groups with characteristic fiber G. Also this generalization of Grauert's theory is known in Complex analysis of several variables (see [FoRa] for finite dimensional groups, and [Le2, Le7] for infinite dimensional groups). But the direct proof of this result for the case of one complex variable would require a further chapter larger than Chapter 9. To keep the book shorter we omit this extension of the theory of cocycles.

By the same arguments as in the case of Corollary 10.1.4, the above three theorems lead to the following two corollaries:

10.2.5 Corollary. *Assume, under the hypotheses of one of the Theorems* 10.2.1 *or* 10.2.2, *two subsets Z_+ and Z_- of Z are given, where the cases $Z_+ = \emptyset$, $Z_- = \emptyset$ and $Z_+ \cap Z_- \neq \emptyset$ are possible. Moreover assume that:*

- *If $w \in Z_+$, then f_w is meromorphic at w.*
- *If $w \in Z_-$, then f_w^{-1} is meromorphic at w.*

Then, for any given family of positive integers m_w, $w \in Z$, in the claims of these theorems, it can be shown that:

- *If $w \in Z_+$, then $h - f_w$ has a zero of order m_w at w.*

- *If $w \in Z_-$, then $h^{-1} - f_w^{-1}$ has a zero of order m_w at w.*

10.2.6 Corollary. *Assume, under the hypotheses of Theorem 10.3.3, two subsets Z_+ and Z_- of Z are given, where the cases $Z_+ = \emptyset$, $Z_- = \emptyset$ and $Z_+ \cap Z_- \neq \emptyset$ are possible. Moreover assume that:*

- *If $w \in Z_+$, then the functions f_w and g_w are meromorphic at w.*
- *If $w \in Z_-$, then the functions f_w^{-1} and g_w^{-1} are meromorphic at w.*

Then, for any given family of positive integers m_w, $w \in Z$, in the claim of this theorem, it can be shown that:

- *If $w \in Z_+$, then the function $h - f_w g_w$ has a zero of order m_w at w.*
- *If $w \in Z_-$, then the function $h^{-1} - g_w^{-1} f_w^{-1}$ has a zero of order m_w at w.*

10.3 Weierstrass theorems for $\mathcal{G}^\infty(E)$- and $\mathcal{G}^\omega(E)$-valued functions

Let E be a Banach space, let $\mathcal{F}^\infty(E)$ be the ideal in $L(E)$ of operators which can be approximated by finite dimensional operators, and let $\mathcal{F}^\omega(E)$ be the ideal of compact operators in $L(E)$. Throughout this section, $\aleph$ stands for one of the symbols ∞ or ω, $\mathcal{G}^\aleph(E)$ is the group of invertible operators in E which are of the form $I + K$, where $K \in \mathcal{F}^\aleph(E)$.

We first prove the following strengthening of the left-sided Weierstrass Theorem 5.14.1:

10.3.1 Theorem. *Let $D \subseteq \mathbb{C}$ be an open set, and let Z be a discrete and closed subset of D. Suppose, for each $w \in Z$, there are given a neighborhood U_w of w with $U_w \cap Z = \{w\}$ and a holomorphic function $A_w : U_w \setminus \{w\} \to \mathcal{G}^\aleph(E)$. Then there exist holomorphic functions $H_w : U_w \to \mathcal{G}^\aleph(E)$, $w \in Z$, and a holomorphic function $H : D \setminus Z \to \mathcal{G}^\aleph(E)$ such that*

$$H_w A_w = H \qquad \text{on } U_w \setminus \{w\}. \tag{10.3.1}$$

Moreover, for any given family of positive integers m_w, $w \in Z$, the functions H_w can be chosen so that, for each $w \in Z$, the functions $H_w - I$ and $H_w^{-1} - I$ have a zero of order $\geq m_w$ at w.

Proof. Let a family of positive integers m_w, $w \in Z$, be given. Since $\mathcal{G}^\aleph(E) \subseteq G\mathcal{F}_I^\aleph(E)$, the functions A_w can be interpreted as functions with values in $G\mathcal{F}_I^\aleph(E)$. Since the latter group is the group of invertible elements of a Banach algebra and since this group is connected, we can apply Theorem 10.1.1 to it and obtain holomorphic functions $\widetilde{H}_w : U_w \to G\mathcal{F}_I^\aleph(E)$, $w \in Z$, and a holomorphic function $\widetilde{H} : D \setminus Z \to G\mathcal{F}_I^\aleph(E)$ such that

$$\widetilde{H}_w A_w = \widetilde{H} \qquad \text{on } U_w \setminus \{w\} \tag{10.3.2}$$

and , for each $w \in Z$, the functions $\widetilde{H}_w - I$ and $\widetilde{H}_w^{-1} - I$ have a zero of order $\geq m_w$ at w. If $\dim E < \infty$ and therefore $\mathcal{G}^{\aleph}(E) = GL(E) = G\mathcal{F}_I^{\aleph}(E)$, this completes the proof.

Let $\dim E = \infty$, and let $\lambda_w : U_w \to \mathbb{C}$, $\lambda : D \setminus Z \to \mathbb{C}$, $K_w : U_w \to \mathcal{F}^{\aleph}$ and $K : D \setminus Z \to \mathcal{F}^{\aleph}$ be the holomorphic functions with $\widetilde{H}_w = \lambda_w I + K_w$ and $\widetilde{H} = \lambda I + K$. Then, passing to the factor algebra $\mathcal{F}_I(E)/\mathcal{F}^{\aleph}(E) \cong \mathbb{C}$, we see:

- Since $\widetilde{H}$ and $\widetilde{H}_w$ are invertible, the functions λ and λ_w have no zeros.
- It follows from (10.3.2) that $\lambda_w = \lambda$ on U_w.
- Since $\widetilde{H}_w - I$ and $\widetilde{H}_w^{-1} - I$ have a zero of order $\geq m_w$ at w, the functions $\lambda_w - 1$ and $\lambda_w^{-1} - 1$ have a zero of order $\geq m_w$ at w.

Hence the functions $H_w := \widetilde{H}_w/\lambda$ and $H := \widetilde{H}/\lambda$ have the required properties. □

Precisely in the same way, replacing the left-sided Theorem 10.1.1 by the right-sided Theorem 10.2.1, we get the corresponding left-sided result:

10.3.2 Theorem. *Let $D \subseteq \mathbb{C}$ be an open set, and let Z be a discrete and closed subset of D. Suppose, for each $w \in Z$, there are given a neighborhood U_w of w with $U_w \cap Z = \{w\}$ and a holomorphic function $A_w : U_w \setminus \{w\} \to \mathcal{G}^{\aleph}(E)$. Then there exist a holomorphic function $H : D \setminus Z \to \mathcal{G}^{\aleph}(E)$ and holomorphic functions $H_w : U_w \to \mathcal{G}^{\aleph}(E)$, $w \in Z$ such that*

$$A_w H_w = H \qquad \text{on } U_w \setminus \{w\}. \tag{10.3.3}$$

Moreover, for any given family of positive integers m_w, $w \in Z$, the functions H_w can be chosen so that, for each $w \in Z$, the functions $H_w - I$ and $H_w^{-1} - I$ have a zero of order $\geq m_w$ at w.

Both theorems together again give a two-sided version:

10.3.3 Theorem. *Let $D \subseteq \mathbb{C}$ be an open set, and let Z be a discrete and closed subset of D. Suppose, for each $w \in Z$, there are given a neighborhood U_w of w with $U_w \cap Z = \{w\}$ and two holomorphic functions $F_w, G_w : U_w \setminus \{w\} \to \mathcal{G}^{\aleph}(E)$.*

Then there exist a holomorphic function $H : D \setminus Z \to \mathcal{G}^{\aleph}(E)$ and holomorphic functions $H_w : U_w \to \mathcal{G}^{\aleph}(E)$, $w \in Z$, such that

$$F_w H_w G_w = H \qquad \text{on } U_w \setminus \{w\}. \tag{10.3.4}$$

Moreover, for any given family of positive integers m_w, $w \in Z$, the functions F_w and G_w can be chosen so that, for each $w \in Z$, the functions $F_w - 1$, $F_w^{-1} - 1$, $G_w - 1$ and $G_w^{-1} - 1$ have a zero of order $\geq m_w$ at w.

Proof. Let a family of positive integers m_w, $w \in Z$, be given. Then from the left-sided Theorem 10.3.1 we get holomorphic functions $H_w^l : U_w \to \mathcal{G}^{\aleph}$, $w \in Z$, and a holomorphic function $H^l : D \setminus Z \to \mathcal{G}^{\aleph}(E)$ such that

$$H_w^l G_w = H^l \qquad \text{on } U_w \setminus \{w\}$$

and the functions $H_w^l - 1$ have a zero of order $\geq m_w$ at w. From the right-sided Theorem 10.3.2 we get holomorphic functions $H_w^r : U_w \to \mathcal{G}^{\aleph}(E)$, $w \in Z$, and a holomorphic function $H^r : D \setminus Z \to \mathcal{G}^{\infty}(E)$ such that

$$F_w H_w^r = H^r \qquad \text{on } U_w \setminus \{w\}$$

and the functions $H_w^r - 1$ have a zero of order $\geq m_w$ at w. Setting $H = H^l H^r$ and $H_w = H_w^r H_w^l$, $w \in Z$, we get holomorphic functions $H : D \setminus Z \to \mathcal{G}^{\aleph}(E))$, $H_w : U_w \to \mathcal{G}^{\aleph}(E)$ such that

$$F_w H_w G_w = F_w H_w^r H_w^l G_w = H^r H^l = H \qquad \text{on } U_w \setminus \{w\},$$

and the functions $F_w - 1$, $F_w^{-1} - 1$, $G_w - 1$ and $G_w^{-1} - 1$ have a zero of order $\geq m_w$ at w. □

By the same arguments as in the case of Corollary 10.1.4, the above three theorems lead to the following two corollaries:

10.3.4 Corollary. *Assume, under the hypotheses of one of the theorems* 10.3.1 *or* 10.3.2, *two subsets Z_+ and Z_- of Z are given, where the cases $Z_+ = \emptyset$, $Z_- = \emptyset$ and $Z_+ \cap Z_- \neq \emptyset$ are possible. Moreover assume that:*

- *If $w \in Z_+$, then A_w is meromorphic at w.*
- *If $w \in Z_-$, then A_w^{-1} is meromorphic at w.*

Then, for any given family of positive integers m_w, $w \in Z$, in the claims of these theorems, it can be shown that:

- *If $w \in Z_+$, then $H - A_w$ has a zero of order m_w at w.*
- *If $w \in Z_-$, then $H^{-1} - A_w^{-1}$ has a zero of order m_w at w.*

10.3.5 Corollary. *Assume, under the hypotheses of Theorem* 10.3.3, *two subsets Z_+ and Z_- of Z are given, where the cases $Z_+ = \emptyset$, $Z_- = \emptyset$ and $Z_+ \cap Z_- \neq \emptyset$ are possible. Moreover assume that:*

- *If $w \in Z_+$, then the functions F_w and G_w are meromorphic at w.*
- *If $w \in Z_-$, then the functions F_w^{-1} and G_w^{-1} are meromorphic at w.*

Then, for any given family of positive integers m_w, $w \in Z$, in the claim of this theorem, it can be shown that:

- *If $w \in Z_+$, then the function $H - F_w G_w$ has a zero of order m_w at w.*
- *If $w \in Z_-$, then the function $H^{-1} - G_w^{-1} F_w^{-1}$ has a zero of order m_w at w.*

10.3.6 Theorem. *Let $D \subseteq \mathbb{C}$ be an open set, and let Z be a discrete and closed subset of D. Suppose, for each $w \in Z$, there is given a neighborhood U_w of w with $U_w \cap Z = \{w\}$ and a holomorphic function $A_w : U_w \setminus \{w\} \to \mathcal{G}^{\aleph}(E)$ which is finite meromorphic at w. Then there exist a holomorphic function $H : D \setminus Z \to \mathcal{G}^{\aleph}(E)$,* *which is finite meromorphic and Fredholm (Def. 4.1.1) at the points of* Z, *and holomorphic functions* $H_w : U_w \to \mathcal{G}^{\aleph}(E)$, $w \in Z$, *which are finite meromorphic and Fredholm at* w, *such that*

$$H_w A_w = H \qquad \text{on } U_w \setminus \{w\}. \tag{10.3.5}$$

Moreover, for any given family of positive integers m_w, $w \in Z$, *the functions* H_w *can be chosen so that, for each* $w \in Z$, *the functions* $H - A_w$ *and* $H^{-1} - A_w^{-1}$ *have a zero of order* $\geq m_w$ *at* w.

Proof. By hypothesis, the functions A_w are finite meromorphic and Fredholm, and, by Corollary 4.1.3, also the functions A_w^{-1} are finite meromorphic and Fredholm. In particular, these functions are meromorphic. Therefore, the assertion follows from Corollary 10.2.5 with $Z_+ = Z_- = Z$. □

10.4 Holomorphic $\mathcal{G}^\infty(E)$-valued functions with given principal parts of the inverse

In this section E is a Banach space, $\mathcal{F}^\infty(E)$ is the ideal in $L(E)$ of the operators which can be approximated by finite dimensional operators, and $\mathcal{G}^\infty(E)$ is the group of invertible operators in E which are of the form $I + K$, where $K \in \mathcal{F}^\infty(E)$ (Def. 5.12.1). We first prove the following generalization of Corollary 4.3.3 to the Smith factorization theorem:

10.4.1 Lemma. *Let* $w \in \mathbb{C}$, *and let* $K : \mathbb{C} \setminus L(E)$ *be a rational function of the form*

$$K(z) = \sum_{n=-m}^{-1} (z - w)^n K_n\,, \tag{10.4.1}$$

where $K_{-m}, \ldots, K_{-1}$, $1 \leq m < \infty$, *are finite dimensional operators. Then there exists a neighborhood* U *of* w *and a holomorphic operator function* $V : U \to \mathcal{F}^\infty(E)$ *such that* $I + V(z)$ *is invertible for all* $z \in U \setminus \{w\}$ *and* K *is the principal part of the Laurent expansion of* $(I + V)^{-1}$ *at* w.

Proof. Since the operators $K_{-m}, \ldots, K_{-1}$ are finite dimensional, we can find a finite dimensional projector P in E such that $K = PKP$. Let $\widetilde{K} : \mathbb{C} \setminus \{w\} \to L(\operatorname{Im} P)$ be the function defined by

$$\widetilde{K}(z) = K(z)\big|_{\operatorname{Im} P} \qquad \text{for } z \in \mathbb{C} \setminus \{w\}.$$

Then $K = P\widetilde{K}P$ and, by Corollary 4.3.3, there exists a neighborhood U of w and a holomorphic function $\widetilde{A} : U \to L(\operatorname{Im} P)$ such that $\widetilde{A}(z)$ is invertible for all $z \in U \setminus \{w\}$ and $\widetilde{K}$ is the principal part of the Laurent expansion of $\widetilde{A}^{-1}$ at w. Then $A := I - P + P\widetilde{A}P$ is the required function. □

10.4.2 Theorem. *Let $D \subseteq \mathbb{C}$ be an open set, and let Z be a discrete and closed subset of D. Suppose, for each $w \in Z$, there is given a rational function $K : \mathbb{C} \to L(E)$ of the form*

$$K_w(z) = \sum_{n=-m_w}^{-1} (z-w)^n K_n^w \,, \tag{10.4.2}$$

where $K^w_{-m_w}, \ldots, K^w_{-1}$, $1 \le m_w < \infty$, are finite dimensional operators. Then there exists a holomorphic function $V : D \to \mathcal{F}^\infty$ such that $I + V(z)$ is invertible for all $z \in D \setminus Z$ and, for all $w \in Z$, the given function K_w is the principal part of the Laurent expansion of $(I+V)^{-1}$ at w.

Proof. By Lemma 10.4.1, for each $w \in Z$, there exists a neighborhood U_w of w and a holomorphic operator function $V_w : U_w \to \mathcal{F}^\infty(E)$ such that $I + V_w(z)$ is invertible for all $z \in U_w \setminus \{w\}$ and K_w is the principal part of the Laurent expansion of $(I+V_w)^{-1}$ at w.

By Corollary 10.2.5, now we can find holomorphic functions $H_w : U_w \to \mathcal{G}^\infty(E)$, $w \in Z$, and a holomorphic function $H : D \setminus Z \to \mathcal{G}^\infty(E)$, which is finite meromorphic at the points of Z, such that, for each $w \in Z$,

$$H_w(I+V_w) = H \qquad \text{on } U_w \setminus \{w\},$$

and $H^{-1} - (I+V_w)^{-1}$ has a zero of order m_w at w. This implies, as K^w is the principal part of the Laurent expansion of $(I+V_w)^{-1}$ at w, that K^w is also the principal part of the Laurent expansion of H^{-1} at w. It remains to set $V = H - I$. □

It is impossible to replace in Theorem 10.4.2 the prescribed function (10.4.2) by an arbitrary function of the form

$$K(z) = \sum_{n=-m}^{0} (z-w)^n K_n \,,$$

where the operators $K_{-m}, \ldots, K_{-1}$ are finite dimensional and $K_0 \neq 0$. This follows from counterexample 4.3.4.

10.5 Comments

In such a generality the material of this chapter is published here for the first time. Less general versions of the interpolation theorems for matrix and operator functions were published earlier in [GR1, GR2].

Chapter 11

Holomorphic equivalence, linearization and diagonalization

11.1 Introductory remarks

11.1.1 Definition. Let X_1, X_2, Y_1, Y_2 be Banach spaces such that X_1 is isomorphic to X_2 and Y_1 is isomorphic to Y_2. Let $D \subseteq \mathbb{C}$ be an open set, and let $S : D \to L(X_1, Y_1)$, $T : D \to L(X_2, Y_2)$ be two holomorphic functions. The functions T and S are called **holomorphically equivalent** over D if there exist holomorphic functions $E : D \to L(X_2, X_1)$ and $F : D \to L(Y_1, Y_2)$ with invertible values such that

$$T = FSE\,, \qquad \text{on } D\,. \tag{11.1.1}$$

We give an example. Let X be a Banach space and

$$P(z) = z^n I + z^{n-1} A_{n-1} + \ldots + A_0$$

a polynomial where I is the identity operator in X and $A_0, \ldots, A_{n-1}$ are arbitrary operators from $L(X)$, $n \in \mathbb{N}^*$. Set

$$X^n := \underbrace{X \oplus \ldots \oplus X}_{n \text{ times}}\,.$$

Then the $L(X^n)$-valued function

$$\widetilde{P}(z) := \begin{pmatrix} P(z) & 0 & \cdots & 0 \\ 0 & I & \cdots & 0 \\ \vdots & \vdots & \ddots & \vdots \\ 0 & 0 & \cdots & I \end{pmatrix}$$

is equivalent to the linear $L(X^n)$-valued function $zI_{X^n} - C$ where I_{X^n} is the identity operator in X^n and

$$C := \begin{pmatrix} 0 & I & 0 & \cdots & 0 \\ 0 & 0 & I & \cdots & 0 \\ \vdots & \vdots & \vdots & \ddots & \vdots \\ 0 & 0 & 0 & \cdots & I \\ -A_0 & -A_1 & -A_2 & \cdots & -A_{n-1} \end{pmatrix}.$$

Indeed, set

$$E(z) := \begin{pmatrix} I & 0 & \cdots & 0 \\ zI & I & \cdots & 0 \\ \vdots & \vdots & \ddots & \vdots \\ z^{n-1}I & z^{n-2}I & \cdots & I \end{pmatrix}$$

and

$$F(z) := \begin{pmatrix} B_{n-1}(z) & B_{n-2}(z) & \cdots & B_1(z) & B_0(z) \\ -I & 0 & \cdots & 0 & 0 \\ 0 & -I & \cdots & 0 & 0 \\ \vdots & \vdots & \ddots & \vdots & \vdots \\ 0 & 0 & \cdots & -I & 0 \end{pmatrix}$$

where

$$B_0(z) := I \quad \text{and} \quad B_{k+1}(z) = zB_k(z) + A_{n-1-k} \quad \text{for } k = 0, 1, \dots, n-2.$$

Then

$$\widetilde{P}(z) = F(z)\big(zI_{X^n} - C\big)E(z) \qquad , z \in D.$$

11.1.2 Definition. Let $D \subseteq \mathbb{C}$ be an open set, and X, Y, Z Banach spaces. Given an operator function $T : D \to L(X, Y)$, we call the operator function

$$\begin{pmatrix} T & 0 \\ 0 & I_Z \end{pmatrix} : D \longrightarrow L(X \oplus Z, Y \oplus Z)$$

the Z-extension of T. Here I_Z is the identity operator of Z.

According to the example given above a suitable extension of an operator polynomial is equivalent on $\mathbb{C}$ to a linear function. In the next section, a more general example is given for linearization of analytic operator functions by extension and equivalence.

11.2 Linearization by extension and equivalence

11.2.1 Theorem. *Let $D \subseteq \mathbb{C}$ be a bounded open set with piecewise C^1-boundary and such that $0 \in D$. Let X be a Banach space, and*

$$T : \overline{D} \longrightarrow L(X)$$

a continuous operator function which is holomorphic in D. Denote by $\mathcal{C}(\partial D, X)$ the Banach space of all X-valued continuous functions on ∂D endowed with the maximum norm. Let

$$Z := \left\{ f \in \mathcal{C}(\partial D, X) \;\middle|\; \int_{\partial D} \frac{f(z)}{z}\, dz = 0 \right\}$$

and define an operator A on $\mathcal{C}(\partial D, X)$ by setting

$$(Af)(z) = zf(z) - \frac{1}{2\pi i} \int_{\partial D} \Big(I - T(\zeta)\Big) f(\zeta)\, d\zeta\,, \qquad z \in \partial D. \tag{11.2.1}$$

Then $X \oplus Z$ and $\mathcal{C}(\partial D, X)$ are isomorphic and the Z-extension of T is holomorphically equivalent on D to the linear operator function $\lambda - A$, $\lambda \in D$, i.e., there are holomorphic operator functions $E : D \to L\big(X \oplus Z, \mathcal{C}(\partial D, X)\big)$ and $F : D \to L\big(X \oplus Z, \mathcal{C}(\partial D, X)\big)$ with invertible values such that

$$E(\lambda)(\lambda - A)F(\lambda) = \begin{pmatrix} T(\lambda) & 0 \\ 0 & I_Z \end{pmatrix}, \qquad \lambda \in D. \tag{11.2.2}$$

Proof. Let $\tau : X \to \mathcal{C}(\partial D, X)$ be the canonical embedding, i.e., $(\tau x)(z) = x$ for all $x \in X$ and $z \in \partial D$. Furthermore, define an operator $\omega : \mathcal{C}(\partial D, X) \to X$ by

$$\omega f = \frac{1}{2\pi i} \int_{\partial D} \frac{f(z)}{z}\, dz\,.$$

Since $0 \in D$, then $\omega\tau = I_X$ and $P := \tau\omega$ is the projector in $\mathcal{C}(\partial D, X)$ with $\operatorname{Im} P = \tau(X)$ and $\operatorname{Ker} P = Z$. Let $J : X \oplus Z \to \mathcal{C}(\partial D, X)$ be given by $J(x, g) = \tau x + g$. Then J is invertible and $J^{-1} f = \big(\omega f, (I - P) f\big)$.

Next, consider on $\mathcal{C}(\partial D, X)$ the following auxiliary operator

$$(Vf)(z) = zf(z)\,, \qquad z \in \partial D.$$

Then D is in the resolvent set of V, where

$$\Big((\lambda - V)^{-1} f\Big)(z) = \frac{f(z)}{\lambda - z}\,, \qquad \lambda \in D, z \in \partial D. \tag{11.2.3}$$

Now we define holomorphic operator functions $E : D \to L\big(X \oplus Z, C(\partial D, X)\big)$ and $F : D \to L\big(X \oplus Z, C(\partial D, X)\big)$, setting for $\lambda \in D$:

$$E(\lambda) = (\lambda - V)^{-1} J \quad \text{and} \quad F(\lambda) = J^{-1}\big(I - PB(\lambda)(I - P)\big)\,.$$

It is clear that all values of E and F are invertible (as J is invertible, $\lambda I - V$ is invertible for $\lambda \in D$ and P is a projector). Moreover we introduce a holomorphic operator function $B : D \to L\big(C(\partial D, X)\big)$ defined by

$$B(\lambda) = I + PV(\lambda - V)^{-1} - PV(\lambda - V)^{-1} T\,.$$

(Here T means the operator of multiplication by the operator function T.) Note that $A = V - PV + PVT$. Therefore

$$F(\lambda)(\lambda - A)E(\lambda) = F(\lambda)\Big(I + PV(\lambda I - V)^{-1} - PVT(\lambda I - V)^{-1}\Big)J\,.$$

Since $(\lambda I - V)^{-1}$ is defined by multiplication by a scalar function, $(\lambda - V)^{-1}$ commutes with the multiplication by T. Therefore it follows that

$$F(\lambda)(\lambda I - A)E(\lambda) = F(\lambda)B(\lambda)J = J^{-1}\Big(I - PB(\lambda)(I-P)\Big)B(\lambda)$$

and further, since $(I - P)B(\lambda) = I - P$,

$$\begin{aligned} F(\lambda)(\lambda - A)E(\lambda) &= J^{-1}\Big(B(\lambda) - PB(\lambda)(I-P)\Big)J \\ &= J^{-1}\Big(I - P + B(\lambda)P\Big)J \\ &= I - J^{-1}PJ + \Big(J^{-1}B(\lambda)J\Big)J^{-1}PJ\,. \end{aligned}$$

Since $J^{-1}PJ$ is the projector in $X \oplus Z$ with image X and kernel Z, it remains to prove that

$$J^{-1}B(\lambda)Jx = T(\lambda)x$$

for all $x \in X$ and $\lambda \in D$. Let such x and λ be fixed. Then, by (11.2.3),

$$\Big(V(\lambda - V)^{-1}\tau x\Big)(z) = \frac{zx}{\lambda - z}\,, \qquad z \in \partial D,$$

and

$$\Big(V(\lambda - V)^{-1}Tx\Big)(z) = \frac{zT(z)x}{\lambda - z}\,, \qquad z \in \partial D\,.$$

By the Cauchy formula this implies that

$$\omega V(\lambda - V)^{-1}\tau x = \frac{1}{2\pi i}\int_{\partial D} \frac{x}{\lambda - z}\,dz = -x\,,$$

$$\omega V(\lambda - V)^{-1}Tx = \frac{1}{2\pi i}\int_{\partial D} \frac{T(z)x}{\lambda - z}\,dz = -T(\lambda)x\,,$$

and therefore

$$PV(\lambda - V)^{-1}\tau x = -\tau x \qquad \text{and} \qquad PV(\lambda - V)^{-1}Tx = -\tau T(\lambda)x\,.$$

Hence

$$\begin{aligned} J^{-1}B(\lambda)Jx = J^{-1}B(\lambda)\tau x &= J^{-1}\Big(\tau x + PV(\lambda - V)^{-1}\tau x - PV(\lambda - V)^{-1}Tx\Big) \\ &= J^{-1}\tau T(\lambda)x = T(\lambda)x. \end{aligned}$$

□

It can be shown (see [GKL]) that the spectrum $\sigma(A)$ of the operator A defined by (11.2.1) is given by

$$\sigma(A) = \left\{\lambda \in D \;\middle|\; T(\lambda) \text{ not invertible}\right\} \cup \partial D\,.$$

The next theorem shows that for an operator function of the form $\lambda - A$ the procedure of linearization by extension and equivalence does not simplify further the operator A and leads to operators that are similar to A.

11.2.2 Theorem. *Let A_1 and A_2 be operators acting on the Banach spaces X_1 and X_2, respectively, and suppose that for some Banach spaces Z_1 and Z_2 the extensions $(\lambda - A_1) \oplus I_{Z_1}$ and $(\lambda - A_2) \oplus I_{Z_2}$ are holomorphically equivalent over some open set D containing $\sigma(A_1) \cup \sigma(A_2)$ (here $\sigma(A_j)$ denotes the spectrum of A_j). Then A_1 and A_2 are similar. More precisely: Let the holomorphic equivalence be given by*

$$\begin{pmatrix} \lambda - A_1 & 0 \\ 0 & I_{Z_1} \end{pmatrix} = F(\lambda) \begin{pmatrix} \lambda - A_2 & 0 \\ 0 & I_{Z_2} \end{pmatrix} E(\lambda)\,, \qquad \lambda \in D, \tag{11.2.4}$$

let U be an open neighborhood of $\sigma(A_1) \cup \sigma(A_2)$ with piecewise C^1-boundary such that $\overline{U} \subseteq D$, and let $\pi_j : X_j \oplus Z_j \to X_j$, $\tau_j : X_j \to X_j \oplus Z_j$ be the canonical projectors and embeddings, respectively. Then

$$S := \frac{1}{2\pi i} \int\limits_{\partial U} (\lambda - A_2)^{-1} \pi_2 F(\lambda)^{-1} \tau_1 \, d\lambda \tag{11.2.5}$$

is a well-defined operator $S : X_1 \to X_2$ (as ∂U is contained in the resolvent set of A_2). Moreover, this operator is invertible, where

$$S^{-1} = \frac{1}{2\pi i} \int\limits_{\partial U} \pi_1 E(\lambda)^{-1} \tau_2 (\lambda - A_2)^{-1} \, d\lambda\,, \tag{11.2.6}$$

and

$$S A_1 S^{-1} = A_2.$$

Proof. From the equivalence (11.2.4) it follows that the integrands in (11.2.5) and (11.2.6) satisfy the following identities:

$$(\lambda - A_2)^{-1} \pi_2 F(\lambda)^{-1} \tau_1 = \pi_2 E(\lambda) \tau_1 (\lambda - A_1)^{-1}\,, \qquad \lambda \in \partial U, \tag{11.2.7}$$

$$\pi_1 E(\lambda)^{-1} \tau_2 (\lambda - A_2)^{-1} = (\lambda - A_1)^{-1} \pi_1 F(\lambda) \tau_2\,, \qquad \lambda \in \partial U. \tag{11.2.8}$$

Since, by Cauchy's theorem,

$$\int\limits_{\partial U} \pi_2 F(\lambda)^{-1} \tau_1 \, d\lambda = \int\limits_{\partial U} \pi_2 E(\lambda)^{-1} \tau_1 \, d\lambda = 0,$$

we get

$$A_2S = \frac{1}{2\pi i}\int_{\partial U}(A_2 - \lambda + \lambda)(\lambda - A_2)^{-1}\pi_2F(\lambda)^{-1}\tau_1\,d\lambda$$
$$= \frac{1}{2\pi i}\int_{\partial U}\lambda(\lambda - A_2)^{-1}\pi_2F(\lambda)^{-1}\tau_1\,d\lambda$$

and therefore, by (11.2.7),

$$A_2S = \frac{1}{2\pi i}\int_{\partial U}\lambda\pi_2E(\lambda)\tau_1(\lambda - A_1)^{-1}\,d\lambda$$
$$= \frac{1}{2\pi i}\int_{\partial U}\pi_2E(\lambda)\tau_1(\lambda - A_1 + A_1)(\lambda - A_1)^{-1}\,d\lambda$$
$$= \left(\frac{1}{2\pi i}\int_{\partial U}\pi_2E(\lambda)\tau_1(\lambda - A_1)^{-1}d\lambda\right)A_1\,.$$

Again by (11.2.7), this gives $A_2S = SA_1$.

It remains to prove that S is invertible. To do this, we first note that, for $\lambda \in D \setminus \big(\sigma(A_1) \cup \sigma(A_2)\big)$, we have the identities

$$\pi_1E(\lambda)^{-1}\tau_2(\lambda - A_2)^{-1}\pi_2F(\lambda)^{-1}\tau_1 - (\lambda - A_1)^{-1} = -\pi_1E(\lambda)^{-1}\begin{pmatrix}0 & 0\\ 0 & I_{Z_2}\end{pmatrix}F(\lambda)^{-1}\tau_1 \quad (11.2.9)$$

and

$$\pi_2E(\lambda)\tau_1(\lambda - A_1)^{-1}\pi_1F(\lambda)\tau_2 - (\lambda - A_2)^{-1} = -\pi_2E(\lambda)\begin{pmatrix}0 & 0\\ 0 & I_{Z_1}\end{pmatrix}F(\lambda)\tau_2\,. \quad (11.2.10)$$

Indeed, from the equivalence (11.2.4) we get

$$\begin{pmatrix}(\lambda - A_1)^{-1} & 0\\ 0 & I_{Z_1}\end{pmatrix} = E(\lambda)^{-1}\begin{pmatrix}(\lambda - A_2)^{-1} & 0\\ 0 & I_{Z_2}\end{pmatrix}F(\lambda)^{-1}$$

and

$$E(\lambda)\begin{pmatrix}(\lambda - A_1)^{-1} & 0\\ 0 & I_{Z_1}\end{pmatrix}F(\lambda) = \begin{pmatrix}(\lambda - A_2)^{-1} & 0\\ 0 & I_{Z_2}\end{pmatrix}.$$

Multiplying the first equation from left by π_1 and from the right by τ_1 this gives (11.2.9). Multiplying the second equation from left by π_2 and from the right by τ_2 this gives (11.2.10).

Now we denote by T the operator defined by the right-hand side of (11.2.6), and we choose two neighborhoods U_1 and U_2 of $\sigma(A_1) \cup \sigma(A_2)$ with piecewise $\mathcal{C}^1$-boundary such that $\overline{U}_1 \subseteq U_2$ and $\overline{U}_2 \subseteq U$. Then, by Cauchy's theorem, S and T can be written also in the form

$$S = \frac{1}{2\pi i} \int\limits_{\partial U_1} (\lambda - A_2)^{-1} \pi_2 F(\lambda)^{-1} \tau_1 \, d\lambda$$

and

$$T = \frac{1}{2\pi i} \int\limits_{\partial U_2} \pi_1 E(\mu)^{-1} \tau_2 (\mu - A_2)^{-1} \, d\mu \, .$$

By (11.2.7) and (11.2.8), this implies that

$$S = \frac{1}{2\pi i} \int\limits_{\partial U_1} \pi_2 E(\lambda) \tau_1 (\lambda - A_1)^{-1} \, d\lambda$$

and

$$T = \frac{1}{2\pi i} \int\limits_{\partial U_2} (\mu - A_1)^{-1} \pi_1 F(\mu) \tau_2 \, d\mu \, .$$

Hence

$$ST = \left(\frac{1}{2\pi i}\right)^2 \int\limits_{\partial U_1} \int\limits_{\partial U_2} \pi_2 E(\lambda) \tau_1 (\lambda - A_1)^{-1} (\mu - A_1)^{-1} \pi_1 F(\mu) \tau_2 \, d\mu d\lambda.$$

For $\lambda \in \partial U_1$ and $\mu \in \partial U_2$, we have the so-called resolvent equation

$$(\lambda - A_1)^{-1}(\mu - A_1)^{-1} = \frac{(\lambda - A_1)^{-1} - (\mu - A_1)^{-1}}{\mu - \lambda}.$$

(For the proof just multiply it from the left by $\lambda - A_1$ and from the right by $\mu - A_1$.) Therefore it follows that

$$ST = A - B,$$

where

$$A = \left(\frac{1}{2\pi i}\right)^2 \int\limits_{\partial U_1} \int\limits_{\partial U_2} \frac{\pi_2 E(\lambda) \tau_1 (\lambda - A_1)^{-1} \pi_1 F(\mu) \tau_2}{\mu - \lambda} \, d\mu d\lambda$$

and

$$B = \left(\frac{1}{2\pi i}\right)^2 \int\limits_{\partial U_1} \int\limits_{\partial U_2} \frac{\pi_2 E(\lambda) \tau_1 (\mu - A_1)^{-1} \pi_1 F(\mu) \tau_2}{\mu - \lambda} \, d\mu d\lambda.$$

We have

$$\begin{aligned}A =&\frac{1}{2\pi i}\int\limits_{\partial U_1}\pi_2E(\lambda)\tau_1(\lambda-A_1)^{-1}\left(\frac{1}{2\pi i}\int\limits_{\partial U_2}\frac{\pi_1F(\mu)\tau_2}{\mu-\lambda}\,d\mu\right)d\lambda\\ =&\frac{1}{2\pi i}\int\limits_{\partial U_1}\pi_2E(\lambda)\tau_1(\lambda-A_1)^{-1}\pi_1F(\lambda)\tau_2 d\lambda.\end{aligned}$$

Since the right-hand side of (11.2.10) is holomorphic in D and $\sigma(A_2)\subseteq U_1\subseteq \overline{U}_1\subseteq D$, this implies

$$A=\frac{1}{2\pi i}\int\limits_{\partial U_1}(\lambda-A_2)^{-1}\pi_1 d\lambda = I_{X_2}.$$

Moreover,

$$B=\frac{1}{2\pi i}\int\limits_{\partial U_2}\left(\frac{1}{2\pi i}\int\limits_{\partial U_1}\frac{\pi_2E(\lambda)\tau_1}{\mu-\lambda}\,d\lambda\right)(\mu-A_1)^{-1}\pi_1F(\mu)\tau_2 d\mu\,.$$

Since $\partial U_2\cap\overline{U}_1=\emptyset$, this implies that $B=0$. We have now proved that $ST=I_{X_2}$. In a similar way, using (11.2.9) instead of (11.2.10), one obtains $TS=I_{X_1}$. □

For linear functions $\lambda-A_1$ and $\lambda-A_2$ global holomorphic equivalence on $\mathbb{C}$ means just that A_1 and A_2 are similar. This follows from the next corollary.

11.2.3 Corollary. *Two operators A_1 and A_2 are similar if and only if some extensions of $\lambda-A_1$ and $\lambda-A_2$ are holomorphically equivalent on some open set containing $\sigma(A_1)$ and $\sigma(A_2)$.*

Proof. If A_1 and A_2 are similar, then, obviously, $\lambda-A_1$ and $\lambda-A_2$ are equivalent on $\mathbb{C}$. Theorem 11.2.2 gives the reverse implication. □

11.3 Local equivalence

11.3.1 Definition. Let X, Y be Banach spaces, let $D\subseteq\mathbb{C}$ be an open set, and let $S,T: D\to L(X,Y)$ be meromorphic (Section 1.10.6).

(i) Let $w\in D$. The functions T and S are called **holomorphically equivalent at** w if there exist a neighborhood $U_w\subseteq D$ of w and holomorphic functions $E: U_w\to GL(X)$, $F: U_w\to GL(Y)$ such that

$$T=FSE\qquad\text{on } U_w\setminus\{w\}\,. \tag{11.3.1}$$

(ii) The functions T and S are called **locally holomorphically equivalent over** D if they are holomorphically equivalent at each point of D.

(iii) The functions T and S are called **holomorphically equivalent over** D if there exist holomorphic functions $E : D \to GL(X)$, $F : D \to GL(Y)$ such that

$$T(z) = F(z)S(z)E(z) \tag{11.3.2}$$

for all $z \in D$ which are not singular for T and S. To point out the difference from local holomorphic equivalence over D, then we speak also about **global** holomorphic equivalence over D.

If T and S have no singular points, part (iii) of this definition coincides with Definition 11.1.1 above.

11.3.2 Definition. (i) Let X be a Banach space, let $D \subseteq \mathbb{C}$ be an open set, and let $f : D \to B$ be meromorphic. Let $w \in D$. If f identically vanishes in a neighborhood of w, then we set $\operatorname{ord}_w f = \infty$. If not, then we denote by $\operatorname{ord}_w f$ the uniquely determined integer such the Laurent expansion of f at w is of the form

$$f(z) = \sum_{n=\operatorname{ord}_w f}^{\infty} f_n(z-w)^n \qquad \text{with } f_{\operatorname{ord}_w f} \neq 0 .$$

We call $\operatorname{ord}_w f$ the **order of f at w**.

(ii) Let X, Y be Banach spaces, let $D \subseteq \mathbb{C}$ be an open set, and let $A : D \to L(X, Y)$ be meromorphic.

Let $w \in D$. Then, for $k \in \mathbb{Z}$, we denote by $X_{A,w}(k)$ the set of all $x \in X$ such that there exists a neighborhood U of w and a holomorphic function $f : U \to X$ with

$$f(w) = x \qquad \text{and} \qquad \operatorname{ord}_w(Af) \geq k.$$

Obviously, $X_{A,w}(k)$ is a linear subspace of X for all $k \in \mathbb{Z}$. The family $\{X_{A,w}(k)\}_{k\in\mathbb{Z}}$ will be called the **characteristic filtration** of A at w. We set

$$X_{A,w}(\infty) = \bigcap_{k\in\mathbb{Z}} X_{A,w}(k) .$$

Note that, obviously,

$$X_{A,w}(k+1) \subseteq X_{A,w}(k) \qquad \text{for all } k \in \mathbb{Z} \tag{11.3.3}$$

and

$$X = \bigcup_{k\in\mathbb{Z}} X_{A,w}(k) . \tag{11.3.4}$$

11.3.3 Proposition. *Let X, Y be Banach spaces, let $D \subseteq \mathbb{C}$ be an open set, let $A, B : D \to L(X, Y)$ be meromorphic, and let $w \in D$. If A and B are holomorphically equivalent at w, then the characteristic filtrations of A and B at w are "isomorphic" in the following sense: There exists $M \in GL(X)$ with*

$$X_{A,w}(k) = MX_{B,w}(k) \qquad \textit{for all } k \in \mathbb{Z} . \tag{11.3.5}$$

Moreover, if $U \subseteq D$ is a neighborhood of w and $E : U \to GL(Y)$, $F : U \to GL(X)$ are holomorphic functions with $EAF = B$ on $U \setminus \{w\}$, then (11.3.5) is valid for $M = F(w)$.

Proof. Let $k \in \mathbb{Z} \cup \{\infty\}$ and $x \in X_{B,w}(k)$ be given. By definition of $X_{B,w}(k)$, then, after shrinking U if necessary, we have a holomorphic function $f : U \to X$ with $f(w) = x$ and $\operatorname{ord}_w(Bf) \geq k$. Set $g = Ff$. Then g is a holomorphic function on U with $g(w) = F(w)f(w) = F(w)x$. Moreover, since E is holomorphic and invertible, it follows that $\operatorname{ord}_w(Ag) = \operatorname{ord}_w(EAg) = \operatorname{ord}_w(EAFf) = \operatorname{ord}_w(Bf) \geq k$. Hence $F(w)x \in X_{A,w}(k)$. This proves "$\supseteq$" in (11.3.5). In the same way we prove that

$$F^{-1}(w)X_{A,w}(k) \subseteq X_{B,w}(k) \qquad \text{for all } k \in \mathbb{Z}\,,$$

i.e., "$\subseteq$"in (11.3.5). □

11.3.4 Definition. Let X, Y be Banach spaces, let $D \subseteq \mathbb{C}$ be an open set, and let $A : D \to L(X, Y)$ be meromorphic.

(i) (For $X = Y$, this is Definition 4.1.1) Let $w \in D$. The function A is called **finite meromorphic at** w if the Laurent expansion of A at w is of the form

$$A(z) = \sum_{n=m}^{\infty} (z - w)^n A_n\,, \tag{11.3.6}$$

where (if $m < 0$) the operators $A_m, \ldots, A_{-1}$ are finite dimensional. If, moreover, $m \leq 0$ and A_0 is a Fredholm operator, then A is called **finite meromorphic and Fredholm** at w.

(ii) The function A is called a **finite meromorphic Fredholm function** on D if it is finite meromorphic and Fredholm at each point of D.

(iii) If D is connected, and A is a **finite meromorphic Fredholm function** on D, then it follows from the stability properties of Fredholm operators (see, e.g., [GGK2]) that the index of A_0 in (11.3.6) does not depend on w. We call it the **index** of A and denote it by $\operatorname{ind} A$.

Note that, obviously, this notion is invariant with respect to local holomorphic equivalence (Def. 11.3.1), i.e., if two meromorphic operator functions A and B are locally holomorphically equivalent, and A is a finite meromorphic Fredholm function, then B is a finite meromorphic Fredholm function.

For finite meromorphic Fredholm functions the characteristic filtrations (Def. 11.3.2) are especially simple:

11.3.5 Proposition. *Let X, Y be Banach spaces, let $D \subseteq \mathbb{C}$ be an open set, let $A : D \to L(X, Y)$ be a finite meromorphic Fredholm function, and let $w \in D$. Then:*

(i) *There exists $m \in \mathbb{Z}$ with $X_{A,w}(k) = X$ for $k \leq m$.*

(ii) *For all $k \leq 0$, the space $X_{A,w}(k)$ is closed and of finite codimension in X.*

(iii) $\dim X_{A,w}(k) < \infty$ *for all $k \geq 1$.*

Proof. We may assume that $A \neq 0$. Since A is finite meromorphic and Fredholm at w, then the Laurent expansion of A at w is of the form

$$A(z) = \sum_{n=m}^{\infty} (z-w)^n A_n \,,$$

where $m \leq 0$ is finite, A_0 is a Fredholm operator, and (if $m < 0$) the operators $A_m, \ldots, A_{-1}$ are finite dimensional. Then, it is clear that $\operatorname{ord}_w(Af) \geq m$ for each X-valued holomorphic f in a neighborhood of w. Hence $X_{A,w}(k) = X$ if $k \leq m$. This proves part (i).

Since the operators $A_m, \ldots, A_{-1}$ are finite dimensional (if $m < 0$), the space

$$L := \bigcap_{m \leq \nu \leq -1} \operatorname{Ker} A_\nu$$

is closed and of finite codimension in X. Then, for each vector $x \in L$, the function $f_x := Ax$ admits a holomorphic extension $\widetilde{f_x}$ to w, where $\widetilde{f_x}(w) = A_0 x$. Hence $L \subseteq X_{A,w}(0)$. By (11.3.3) this proves (ii).

Moreover, this implies that $x \notin X_{A,w}(1)$ if $x \in L$ and $x \notin \operatorname{Ker} A_0$. As A_0 is a Fredholm operator, it follows that $X_{A,w}(1) \cap L \subseteq \operatorname{Ker} A_0$ and therefore

$$\dim X_{A,w}(1) \leq \dim \operatorname{Ker} A_0 < \infty \,.$$

By (11.3.3) this proves (iii). □

11.3.6 Definition. Let X, Y be Banach spaces, let $D \subseteq \mathbb{C}$ be an open set, let $A : D \to L(X,Y)$ be a finite meromorphic Fredholm function, let $w \in D$, and let

$$A(z) = \sum_{n=m}^{\infty} (z-w)^n A_n$$

be the Laurent expansion of A at w.

(i) By Proposition 11.3.5,

$$\alpha := X_{A,w}(\infty) < \infty \,.$$

The number α will be called the **generic kernel dimension of A at w.**

(ii) As A_0 is a Fredholm operator, we have the index $\operatorname{ind} A_0 = \dim \operatorname{Ker} A_0 - \dim(Y/\operatorname{Im} A_0)$. The number

$$\beta := \alpha - \operatorname{ind} A_0$$

will be called the **generic cokernel dimension of A at w.**

(iii) If $X_{A,w}(k) = X_{A,w}(0)$ for $k \le 0$, then we say that A **has no negative powers.**

Assume this is not the case. Then, by Proposition 11.3.5, there are uniquely determined negative integers $0 > s_1 > \ldots > s_p$ such that

$$X_{A,w}(0) \subsetneq X_{A,w}(s_1) \subsetneq X_{A,w}(s_2) \subsetneq \ldots \subsetneq X_{A,w}(s_p) = X\,,$$

$$\begin{aligned} X_{A,w}(\nu) &= X_{A,w}(0) && \text{if } 0 > \nu > s_1\,, \\ X_{A,w}(\nu) &= X_{A,w}(s_{j-1}) && \text{if } s_{j-1} \ge \nu > s_j\,,\ 2 \le j \le p\,. \end{aligned}$$

Again by Proposition 11.3.5 the dimensions

$$\begin{aligned} d_1 &:= \dim\Big(X_{A,w}(s_1)\Big/X_{A,w}(0)\Big)\,, \\ d_j &:= \dim\Big(X_{A,w}(s_j)\Big/X_{A,w}(s_{j-1})\Big)\,, \qquad 2 \le j \le p \end{aligned}$$

are finite. Set $d = d_1 + \ldots + d_p$ and denote by $\kappa_1 \ge \ldots \ge \kappa_d$ the integers such that, for each $1 \le j \le p$, precisely d_j components of the vector $(\kappa_1, \ldots, \kappa_d)$ are equal to s_j. These numbers $\kappa_1 \ge \ldots \ge \kappa_d$ will be called the **negative powers of** A **at** w.

(iv) If $X_{A,w}(k) = X_{A,w}(\infty)$ for $k \ge 1$, then we say that A **has no positive powers.**

Assume this is not the case. Then, by Proposition 11.3.5, there are uniquely determined positive integers $s_1 > \ldots > s_p > 0$ such that

$$X_{A,w}(\infty) \subsetneq X_{A,w}(s_1) \subsetneq X_{A,w}(s_2) \subsetneq \ldots \subsetneq X_{A,w}(s_p) \subsetneq X_{A,w}(0)\,,$$

$$\begin{aligned} X_{A,w}(\nu) &= X_{A,w}(\infty) && \text{if } \nu > s_1\,, \\ X_{A,w}(\nu) &= X_{A,w}(s_{j-1}) && \text{if } s_{j-1} \ge \nu > s_j\,,\ 2 \le j \le p-1\,, \\ X_{A,w}(\nu) &= X_{A,w}(s_p) && \text{if } s_p \ge \nu \ge 1\,. \end{aligned}$$

Again by Proposition 11.3.5 the dimensions

$$\begin{aligned} d_1 &:= \dim\Big(X_{A,w}(s_1)\Big/X_{A,w}(\infty)\Big)\,, \\ d_j &:= \dim\Big(X_{A,w}(s_j)\Big/X_{A,w}(s_{j-1})\Big)\,, \qquad 2 \le j \le p\,, \end{aligned}$$

are finite. Set $d = d_1 + \ldots + d_p$, and denote by $\kappa_1 \ge \ldots \ge \kappa_d$ the integers such that, for each $1 \le j \le p$, precisely d_j components of the vector $(\kappa_1, \ldots, \kappa_d)$ are equal to s_j. These numbers $\kappa_1 \ge \ldots \ge \kappa_d$ will be called the **positive powers of** A **at** w.

(v) Let α be the generic kernel dimension of A at w, and let β be the generic cokernel dimension of A at w.

If A has no negative and no positive powers at w, then the pair (α, β) (defined in (i) and (ii)) will be called the **numerical characteristic of A at w.**

If A has powers at w (negative or positive or both), and if $\kappa_1 > \ldots > \kappa_n$ are all powers of A, then the vector $(\kappa_1, \ldots, \kappa_n, \alpha, \beta)$ will be called the **numerical characteristic of A at w.**

11.3.7 Proposition. *Let X, Y be Banach spaces, let $D \subseteq \mathbb{C}$ an open set, let $A, A' : D \to L(X,Y)$ be two finite meromorphic Fredholm functions, and let $w \in D$. Assume that A and A' are holomorphically equivalent at w. Then A and A' have the same numerical characteristic at w.*

Proof. Let

$$\left(\kappa_1, \ldots, \kappa_d, \alpha, \alpha\right)$$

and

$$\left(\kappa'_1, \ldots, \kappa'_{d'}, \alpha', \beta'\right)$$

be the numerical characteristics of A and A', respectively.

By definition, the numbers $\kappa_1 \geq \ldots \geq \kappa_d$ and α depend only on the sequence

$$\dim \left(X_{A,w}(k+1) \big/ X_{A,w}(k) \right), \qquad k \in \mathbb{Z}.$$

By Proposition 11.3.3 this sequence is invariant with respect to holomorphic equivalence at w. As A and A' are holomorphically equivalent at w, this implies that

$$\alpha' = \alpha, \qquad d' = d \qquad \text{and} \qquad \kappa'_j = \kappa_j \text{ for } 1 \leq j \leq d.$$

It remains to prove that $\beta' = \beta$. Let

$$A(z) = \sum_{n=m'}^{\infty} (z-w)^n A_n \qquad \text{and} \qquad A'(z) = \sum_{n=m'}^{\infty} (z-w)^n A'_{n'}$$

be the Laurent expansions of A and A' at w. By hypothesis, we have a neighborhood $U \subseteq D$ of w and holomorphic functions $E : U \to GL(Y)$, $F : U \to GL(X)$ such that

$$A'(z) = E(z)A(z)F(z) \qquad \text{for all } z \in U \setminus \{w\}. \tag{11.3.7}$$

From the stability properties of Fredholm operators it follows that

$$\operatorname{ind} A_0 = \operatorname{ind} A(z) \quad \text{and} \quad \operatorname{ind} A'_0 = \operatorname{ind} A'(z) \qquad \text{for all } z \in U \setminus \{w\}.$$

By (11.3.7) this implies that $\operatorname{ind} A'_0 = \operatorname{ind} A_0$. Hence

$$\beta' = \alpha' - \operatorname{ind} A'_0 = \alpha - \operatorname{ind} A_0 = \beta. \qquad \square$$

11.4 A theorem on local and global equivalence

In this section we prove that, under certain conditions, local holomorphical equivalence of meromorphic operator functions implies global holomorphical equivalence. We will state this result in a more general setting for Banach algebras.

Throughout this section, A is a Banach algebra with unit 1, and GA is the group of invertible elements of A.

11.4.1 Definition. Let $D \subseteq \mathbb{C}$ be an open set, and let $f : D \to A$ be a meromorphic operator function (Section 1.10.6). Let Z be the set of all $w \in D$ such that either f has a pole at w or f is holomorphic at w and $f(w) \notin GA$. The function f will be called **meromorphically invertible** if Z is a discrete and closed subset of D, and f^{-1} (which is well defined and holomorphic on $D \setminus Z$) is meromorphic on D. The set Z then will be called the **spectrum** of f.

If $D \subseteq \mathbb{C}$ is open and connected and A is the algebra of complex $n \times n$-matrices, then it is clear that any meromorphic function $f : D \to A$, which is invertible in at least one point, is meromorphically invertible. By Proposition 4.1.4 the same is true for all finite meromorphic Fredholm functions.

11.4.2 Theorem. *Let $D \subseteq \mathbb{C}$ be an open set, let $f, g : D \to A$ be two meromorphically invertible meromorphic functions, and let G be an open subgroup of GA. Suppose f and g have the same spectrum Z, and the following condition (of local equivalence) is satisfied:*

$$\begin{aligned}&\text{For each } w \in D \text{ there exist a neighborhood } U \subseteq D \text{ and}\\ &\text{holomorphic functions } a_w, b_w : U \to G \text{ with } a_w f b_w = g \text{ on } U \setminus Z\,.\end{aligned} \tag{11.4.1}$$

Then there exist holomorphic functions $a, b : D \to G$ with $afb = g$ on $D \setminus Z$.

Proof. By hypothesis we have an open covering $\{U_j\}_{j \in I}$ of D and families $\{a_j\}_{j \in I}$, $\{b_j\}_{j \in I}$ of holomorphic functions $a_j, b_j : U_j \to G$ such that

$$a_j f b_j = g \qquad \text{on } U_j \setminus Z\,, \quad j \in I. \tag{11.4.2}$$

Then

$$f^{-1} a_i^{-1} a_j f = b_i b_j^{-1} \qquad \text{on } U_i \cap U_j \setminus Z\,, \quad i, j \in I. \tag{11.4.3}$$

Since both f and f^{-1} are meromorphic on D, for each $w \in Z$, we can choose $n_w \in \mathbb{N}$ such that both functions

$$(z - w)^{n_w} f(z) \qquad \text{and} \qquad (z - w)^{n_w} f^{-1}(z) \tag{11.4.4}$$

are holomorphic at w. Setting $m_w = 2n_w + 1$, $w \in Z$, we introduce a data of zeros as well as the corresponding sheaf $\mathcal{O}^g_{Z,m}$ (cf. Definition 9.1.2).

Now, for each open set $U \subseteq D$, we denote by $\mathcal{F}(U)$ the set of all holomorphic functions $h \in \mathcal{O}^G(U)$ such that the function

$$f^{-1} h f,$$

which is a well-defined G-valued holomorphic function on $U \setminus Z$, admits a G-valued holomorphic extension to U. It is easy to see that in this way a $\mathcal{O}^G$-sheaf $\mathcal{F}$ over D is defined (Def. 9.1.1).

We now prove that $\mathcal{F}$ is of finite order. It is clear that

$$\mathcal{F}(U \setminus Z) = \mathcal{O}^G(U \setminus Z)$$

for each open set $U \subseteq D$. Since Z is a discrete and closed subset of D, therefore it is sufficient to prove that

$$\mathcal{O}^G_{Z,m}(U) \subseteq \mathcal{F}(U)$$

for all open sets $U \subseteq D$. Let an open set $U \subseteq D$ and $h \in \mathcal{O}^G_{Z,m}(U)$ be given. Since $m_w = 2n_w + 1$, then, by definition of $\mathcal{O}^G_{Z,m}$, for each $w \in Z$, the function

$$\frac{h(z) - 1}{(z-w)^{2n_w+1}} \tag{11.4.5}$$

is holomorphic at w. Hence, for each $w \in Z$ and $z \in U \setminus Z$, the function $f^{-1}hf$ can be written in the form

$$\begin{aligned} f^{-1}(z)h(z)f(z) &= 1 + f^{-1}(z)\Big(h(z) - 1\Big)f(z) \\ &= 1 + (z-w)\left((z-w)^{n_w} f^{-1}(z)\left(\frac{h(z)-1}{(z-w)^{2n_w}+1}\right)(z-w)^{n_w} f(z)\right). \end{aligned}$$

Since the functions (6.8.3) and (11.4.5) are holomorphic at w, for all $w \in Z$, this implies that $h \in \mathcal{F}(U)$. Hence it is proved that $\mathcal{F}$ is of finite order.

Therefore, Theorem 9.2.1 applies to $\mathcal{F}$. As $a_j \in \mathcal{O}^G(U_j)$, $j \in I$, the family $\{a_i^{-1}a_j\}_{i,j\in I}$ is an $\mathcal{O}^G$-trivial $\mathcal{O}^G$-cocycle. Since also $b_j \in \mathcal{O}^G(U_j)$, $j \in I$, it follows from (11.4.3), that $\{a_i^{-1}a_j\}_{i,j\in I}$ is even an $\mathcal{F}$-cocycle over D. As it is $\mathcal{O}^G$-trivial, condition (i) in Theorem 9.2.1 is satisfied. It follows that $\{a_i^{-1}a_j\}_{i,j\in I}$ is $\mathcal{F}$-trivial, i.e., there are functions $\widetilde{a}_j \in \mathcal{F}(U_j)$, $j \in I$ with

$$a_i^{-1}a_j = \widetilde{a}_i\widetilde{a}_j^{-1} \qquad \text{on } U_i \cap U_j\,, \quad i,j \in I\,.$$

Therefore, setting

$$a = a_j\widetilde{a}_j \qquad \text{on } U_j\,, \quad j \in I\,,$$

we get a global holomorphic function $a : D \to G$.

Moreover, we set $b = f^{-1}a^{-1}g$ on $D \setminus Z$. Then, obviously,

$$afb = aff^{-1}a^{-1}g = g \qquad \text{on } D \setminus Z\,, \tag{11.4.6}$$

and, on each U_j, we have

$$b = f^{-1}\widetilde{a}_j^{-1}a_j^{-1}g.$$

By (11.4.2), this implies that

$$b = f^{-1}\widetilde{a}_j^{-1} f b_j^{-1} \qquad \text{on } U_j\,, \quad j \in U\,. \tag{11.4.7}$$

As $\widetilde{a}_j^{-1} \in \mathcal{F}(U_j)$, the function $f^{-1}\widetilde{a}_j^{-1}f$ extends to a G-valued holomorphic function on U_j. Since also $b_j^{-1} \in \mathcal{O}^G(U_j)$, this implies together with (11.4.7) that b extends to a G-valued holomorphic function on D. In view of (11.4.6), this completes the proof. □

11.5 The finite dimensional case

11.5.1 Proposition. *Let $D \subseteq \mathbb{C}$ be a connected open set, let A be an $n \times m$ matrix of scalar meromorphic functions on D, and let P be the set of poles of A. Denote by* rank $A(z)$, $z \in D \setminus P$, *the rank of $A(z)$, and set*

$$r := \max_{z \in D\setminus P} \operatorname{rank} A(z)$$

Then

$$N := \Big\{ z \in D \setminus P \;\Big|\; \operatorname{rank} A(z) < r \Big\}$$

is a discrete and closed subset of D.

Clearly, this follows from the Smith factorization Lemma 4.3.1, but it can be seen also more directly:

Proof of Proposition 11.5.1. Take a point $z_0 \in D \setminus P$ with

$$r = \operatorname{rank} A(z_0)\,.$$

Then there is an $r \times r$ submatrix B of A such that $\det B(z_0) \neq 0$. Since D is connected and $\det B$ is meromorphic on D without poles in $D \setminus P$, then the set

$$M := \Big\{ z \in D \setminus P \;\Big|\; \det B(z) = 0 \Big\}$$

is discrete and closed in D. As $N \subseteq M$, it follows that N is discrete and closed in D. □

11.5.2 Definition. We use the notation from the preceding Proposition 11.5.1, and we set $Z = P \cup N$. Then the points in $D \setminus Z$ will be called the **generic points** of A, and the number r will be called the **generic rank** of A. The points in Z will be called the **non-generic points** of A.

Obviously, a quadratic matrix of scalar meromorphic functions has maximal generic rank if and only if it is meromorphically invertible in the sense of Definition 11.4.1. Therefore, the following lemma is a special case of Theorem 11.4.2:

11.5.3 Lemma. *Let $D \subseteq \mathbb{C}$ be a connected open set, and let A, B be two $r \times r$-matrices of scalar meromorphic functions of generic rank r on D which are locally holomorphically equivalent on D. Then A and B are globally holomorphically equivalent on D.*

Since any meromorphic matrix function is a finite meromorphic Fredholm function in the sense of Definition 11.3.4, we have the notion of the numerical characteristic (Def. 11.3.6) of a meromorphic matrix function. We now explain the relation of this to the Smith factorization Lemma 4.3.1.

Let $D \subseteq \mathbb{C}$ be an open set, let A be an $n \times m$ matrix of scalar meromorphic functions on D, and let $w \in D$ be a point such that A is not identically zero in a punctured neighborhood of w.

Using the notion of equivalence (Def. 11.3.1), now the Smith factorization lemma can be stated as follows:

There are uniquely determined integers $\kappa_1 \geq \ldots \geq \kappa_r$ such that, at w, the matrix A is holomorphically equivalent to the $(n \times m)$-matrix

$$\begin{pmatrix} \Delta & 0 \\ 0 & 0 \end{pmatrix} \tag{11.5.1}$$

where Δ is the $r \times r$ diagonal matrix with the diagonal

$$(z-w)^{\kappa_1}, \ldots, (z-w)^{\kappa_r} \,.$$

Obviously, r is the rank of the matrix (11.5.1) in a punctured neighborhood of w. Therefore

$$r = \operatorname{rank} A(z) \tag{11.5.2}$$

for all z in a punctured neighborhood of w (which again proves Proposition 11.5.2).

Let p be the number of zero components of the vector $(\kappa_1, \ldots, \kappa_r)$, and let $(\widetilde{\kappa}_1, \ldots, \widetilde{\kappa}_{r-p})$ be the vector obtained from $(\kappa_1, \ldots, \kappa_r)$ omitting the zero components. Then it is easy to see that

$$(\widetilde{\kappa}_1, \ldots, \widetilde{\kappa}_{r-p}, m-r, n-r) \tag{11.5.3}$$

is the numerical characteristic (Def. 11.3.6) of the matrix (11.5.1) at w. Since the numerical characteristic is invariant with respect to holomorphic equivalence (Proposition 11.3.7), this implies that (11.5.3) is also the numerical characteristic of A at w.

Conversely, this also shows that from the numerical characteristic of A at w one can obtain the powers (Def. 4.3.2) $\kappa_1 \geq \ldots \geq \kappa_r$ in the Smith factorization theorem. We summarize:

11.5.4 Proposition. *Let $D \subseteq \mathbb{C}$ be an open set, let A, B be two $n \times m$ matrices of scalar meromorphic functions on D and let $w \in D$ be a point such that A and B are not identially zero in a punctured neighborhood of w. Then A and B have the same numerical characteristic at w if and only if they have the same powers at w.*

11.5.5 Lemma. *Let $D \subseteq \mathbb{C}$ be a connected open set, and let A be an $n \times m$-matrix of scalar meromorphic functions on D. Let r be the generic rank of A, and assume that $r > 0$. Then there exist holomorphic matrix functions $E : D \to GL(n, \mathbb{C})$ and $F : D \to GL(m, \mathbb{C})$ such that EAF is a block matrix of the form*

$$EAF := \begin{pmatrix} B & 0 \\ 0 & 0 \end{pmatrix}, \tag{11.5.4}$$

where B is an $r \times r$-matrix of maximal generic rank of scalar meromorphic functions on D.

Proof. By the Smith factorization Theorem 4.3.1, for each $w \in D$, we have a neighborhood $U_w \subseteq D$ of w and holomorphic functions $E_w : U_w \to GL(n, \mathbb{C})$, $F_w : U_w \to GL(m, \mathbb{C})$ such that

$$E_w(z)A(z)F_w(z) = \begin{pmatrix} \Delta(z) & 0 \\ 0 & 0 \end{pmatrix}, \qquad z \in U_w \setminus \{w\},$$

where Δ is the $r \times r$ diagonal matrix with the diagonal $(z-w)^{\kappa_1}, \dots, (z-w)^{\kappa_r}$. Setting

$$R_0 = \left\{ z \in \mathbb{C}^n \;\middle|\; z_{r+1} = \ldots = z_n = 0 \right\}$$

and

$$K_0 = \left\{ z \in \mathbb{C}^m \;\middle|\; z_1 = \ldots = z_r = 0 \right\},$$

this implies that, for all $w \in D$ and $z \in U_w \setminus \{w\}$,

$$\operatorname{Im} A(z) = E_w^{-1}(z) R_0 \qquad \text{and} \qquad \operatorname{Ker} A(z) = F_w(z) K_0 \,. \tag{11.5.5}$$

Let P be the set of poles of A and let N be the set of all $z \in D \setminus P$ such that $\operatorname{rank} A(z) < r$. If $w \in D \setminus (P \cup N)$, then (11.5.5) holds also for $z = w$. Therefore, we have a uniquely determined holomorphic family $\{R(z)\}_{z \in D}$ of subspaces (see Def. 6.4.1) of $\mathbb{C}^n$ as well as a uniquely determined holomorphic family $\{K(z)\}_{z \in D}$ of subspaces of $\mathbb{C}^m$ such that

$$R(z) = \operatorname{Im} A(z) \quad \text{and} \quad K(z) = \operatorname{Ker} A(z) \qquad \text{for } z \in D \setminus (P \cup N) \,.$$

Then, by Theorem 6.9.1, we can find holomorphic functions $E : D \to GL(n, \mathbb{C})$ and $F : D \to GL(m, \mathbb{C})$ such that

$$E(z)R(z) = R_0 \quad \text{and} \quad F(z)K_0 = K(z) \qquad \text{for all } z \in D \,.$$

Then, for $z \in D \setminus (P \cup N)$,

$$\operatorname{Im} \Big(E(z)A(z)F(z) \Big) = E(z)A(z)F(z)\mathbb{C}^n = E(z)A(z)F(z)\Big(R_0 \oplus K_0 \Big) = R_0$$

and

$$\operatorname{Im} \Big(E(z)A(z)F(z) \Big) = R_0 \quad \text{and} \quad \operatorname{Ker} \Big(E(z)A(z)F(z) \Big) = K_0 \,,$$

i.e.,, over $D \setminus (P \cup N)$, EAF is of the form (11.5.4). Since N is discrete and closed in $D \setminus P$, this implies by continuity that EAF is of the form (11.5.4) also in the points of N. □

11.5.6 Lemma. *Let $D \subseteq \mathbb{C}$ be a connected open set, and let A be an $r \times r$-matrix of scalar meromorphic functions of generic rank r on D. Then there exist not identically vanishing scalar meromorphic functions $\varphi_1, \dots, \varphi_r$ on D, such that the quotients φ_j/φ_{j+1}, $1 \le j \le r-1$, are holomorphic on D, and A is locally holomorphically equivalent over D to the diagonal matrix with the diagonal $\varphi_1, \dots, \varphi_r$.*

Supplement: *If $w \in D$ and $\kappa_1(w) \ge \ldots \ge \kappa_r(w)$ is the vector of powers of A at w (Def. 4.3.2), then the functions*

$$\frac{\varphi_j(z)}{(z-w)^{\kappa_j(w)}}, \qquad 1 \le j \le r,$$

are holomorphic and $\neq 0$ in some neighborhood of w.

Proof. Since the generic rank of A is maximal, by the Smith factorization Lemma 4.3.1, for each $w \in D$, there exist integers $\kappa_1(w) \ge \ldots \ge \kappa_r(w)$ such that A is holomorphically equivalent at w to the diagonal matrix function with the diagonal

$$(z-w)^{\kappa_1(w)}, \dots, (z-w)^{\kappa_r(w)}. \tag{11.5.6}$$

Let Z be the set of points $w \in D$ such that at least one of the numbers $\kappa_1(w), \dots, \kappa_r(w)$ is different from zero. Since Z is discrete and closed in D, then by the Weierstraß product theorem we can find scalar meromorphic functions $\varphi_1, \dots, \varphi_r$ on D which are holomorphic and $\neq 0$ on $D \setminus Z$ and such that, for each $w \in Z$, there is a neighborhood $U_w \subseteq D$ of w such that the functions

$$h_j^w(z) := \frac{\varphi_j(z)}{(z-w)^{\kappa_j(w)}}, \qquad 1 \le j \le r,$$

are holomorphic and $\neq 0$ on U_w. Since $\kappa_1(w) \ge \ldots \ge \kappa_r(w)$, then the quotients φ_j/φ_{j+1}, $1 \le j \le r-1$, are holomorphic on D. Denote by Δ the diagonal matrix with the diagonal

$$\varphi_1, \dots, \varphi_k .$$

Further, for $w \in Z$, we denote by H_w the diagonal matrix with the diagonal

$$h_1^w, \dots, h_r^w .$$

Then each H_w is a holomorphic and invertible matrix function on U_w such that $H_w^{-1}\Delta$ is the diagonal matrix with the diagonal (11.5.6). Therefore A is holomorphically equivalent to Δ at each point $w \in Z$. Moreover, it is clear that A is holomorphically equivalent to Δ at each point $w \in D \setminus Z$, because, on $D \setminus Z$ both A and Δ are holomorphic and of maximal rank r. Hence A and Δ are locally holomorphically equivalent over D. □

Now we can prove the following global version of the Smith factorization lemma:

11.5.7 Theorem. *Let $D \subseteq \mathbb{C}$ be a connected open set, and let A be an $n \times m$ matrix of scalar meromorphic functions on D. Let r be the generic rank of A, and assume that $r > 0$. Then there exist not identically vanishing scalar meromorphic functions $\varphi_1, \dots, \varphi_r$ on D such that the quotients φ_j/φ_{j+1}, $1 \le j \le r-1$, are holomorphic on D, as well as holomorphic matrix functions $E : D \to GL(n, \mathbb{C})$ and $F : D \to GL(m, \mathbb{C})$ such that*

$$EAF = \begin{pmatrix} \Delta & 0 \\ 0 & 0 \end{pmatrix} \qquad (11.5.7)$$

where Δ is the $r \times r$ diagonal matrix with the diagonal $\varphi_1, \dots, \varphi_r$.

Supplement: *If $w \in D$ and $\kappa_1(w) \ge \dots \ge \kappa_r(w)$ is the vector of powers of A at w (Def. 4.3.2), then the functions*

$$\frac{\varphi_j(z)}{(z-w)^{\kappa_j(w)}}, \qquad 1 \le j \le r, \qquad (11.5.8)$$

are holomorphic and $\neq 0$ in some neighborhood of w.

Proof. By Lemma 11.5.5 we may assume that $r = n = m$. By Lemma 11.5.6, we can find not identically vanishing scalar meromorphic functions $\varphi_1, \dots, \varphi_r$ on D, such that, for each $w \in D$, the functions (11.5.8) are holomorphic and $\neq 0$ in some neighborhood of w (and, hence, the quotients φ_j/φ_{j+1}, $1 \le j \le r-1$, are holomorphic on D), and A is locally holomorphically equivalent over D to the diagonal matrix Δ with the diagonal $\varphi_1, \dots, \varphi_r$. Now the assertion follows from Lemma 11.5.3. □

Taking into account also Proposition 11.5 from this theorem we get the following corollary:

11.5.8 Corollary. *Let $D \subseteq \mathbb{C}$ be a connected open set, and let A_1, A_2 be two $n \times m$-matrices of scalar meromorphic functions on D. Then the following are equivalent:*

(i) *The matrix functions A_1 and A_2 are globally holomorphically equivalent over D.*

(ii) *The matrix functions A_1 and A_2 are locally holomorphically equivalent over D.*

(iii) *The matrix functions A_1 and A_2 have the same vectors of powers at each point in D (Def. 4.3.2).*

(iv) *The matrix functions A_1 and A_2 have the same numerical characteristics at each point in D.*

11.6 Local and global equivalence for finite meromorphic Fredholm functions

First we generalize the Smith factorization Theorem 4.3.1 to finite meromorphic Fredholm functions. For that we introduce the notion of a **local diagonal power function**:

11.6.1 Definition. Let X be a Banach space.

By a **projector** in X we always mean a *continuous linear* projector in X, i.e., an operator $P \in L(X)$ with $P^2 = P$. A family $\{P_j\}_{j\in I}$ of projectors in X will be called **mutually disjoint** if $P_jP_k = 0$ for all $j, k \in I$ with $j \neq k$.

Let X, Y be Banach spaces, and let $w \in \mathbb{C}$. A function $\Delta : \mathbb{C}\setminus\{w\} \to L(X,Y)$ will be called a **local diagonal power function** at w if either $\Delta \in L(X,Y)$ is a constant Fredholm operator or, for some $n \in \mathbb{N}^*$, Δ is of the form

$$\Delta(z) = Q_0B_0P_0 + \sum_{j=1}^{n}(z-w)^{\kappa_j}Q_jB_jP_j\,, \qquad z \in \mathbb{C}\setminus\{w\} \tag{11.6.1}$$

where $\kappa_1 \geq \ldots \geq \kappa_n$ are integers $\neq 0$, $P_0, \ldots, P_n$ are non-zero mutually disjoint projectors in X, $Q_0, \ldots, Q_n$ are non-zero mutually disjoint projectors in Y such that

$$\dim \operatorname{Ker} P_0 < \infty \quad \text{and} \quad \dim \operatorname{Ker} Q_0 < \infty\,, \tag{11.6.2}$$

$$\dim \operatorname{Im} P_j = \dim \operatorname{Im} Q_j = 1 \qquad \text{if } 1 \leq j \leq n\,, \tag{11.6.3}$$

and B_j is an invertible operator from $\operatorname{Im} P_j$ onto $\operatorname{Im} Q_j$, $0 \leq j \leq n$.

11.6.2 Remark. It is clear that any local diagonal power function Δ at $w \in \mathbb{C}$ is a finite meromorphic Fredholm function on $\mathbb{C}$. Moreover it is easy to see that if Δ is written in the form (11.6.1), then

$$\Big(\kappa_1, \ldots, \kappa_n, \dim \operatorname{Ker} P_0, \dim \operatorname{Ker} Q_0\Big)$$

is the numerical characteristic of Δ at w (Def. 11.3.6).

11.6.3 Proposition. *Let X, Y be Banach spaces, let $w \in \mathbb{C}$, and let $\Delta, \widetilde{\Delta} : \mathbb{C} \setminus \{w\} \to L(X,Y)$ be two local diagonal power functions at w. Then the following are equivalent:*

(i) *The functions Δ and $\widetilde{\Delta}$ have the same numerical characteristic at w.*

(ii) *There exist operators $E \in GL(Y)$ and $F \in GL(X)$ with $\widetilde{\Delta} = E\Delta F$. In particular, $\widetilde{\Delta}$ and Δ are globally holomorphically equivalent over $\mathbb{C}$ in the sense of Definition* 11.3.1 (iii).

Proof. The implication (ii)⇒(i) is a special case of Proposition 11.3.7.

We prove (i)⇒(ii). Let Δ be written in the form (11.6.1), and let

$$\widetilde{\Delta}(z) = \widetilde{Q}_0 \widetilde{B}_0 \widetilde{P}_0 + \sum_{j=1}^{\widetilde{n}} (z-w)^{\widetilde{\kappa}_j} \widetilde{Q}_j \widetilde{B}_j \widetilde{P}_j\,, \qquad z \in \mathbb{C} \setminus \{w\}$$

be the corresponding representation of $\widetilde{\Delta}$. Since $\widetilde{\Delta}$ has the same numerical characteristic at w, then $\widetilde{n} = n$, $\widetilde{\kappa}_j = \kappa_j$,

$$\dim \operatorname{Ker} \widetilde{P}_0 = \dim \operatorname{Ker} P_0 \quad \text{and} \quad \dim \operatorname{Ker} \widetilde{Q}_0 = \dim \operatorname{Ker} Q_0\,.$$

Then we can find $G \in GL(Y)$ and $H \in GL(X)$ with

$$\widetilde{P}_j = H^{-1} P_j H \quad \text{and} \quad \widetilde{Q}_j = G Q_j G^{-1} \qquad \text{for } 0 \le j \le n,$$

and we obtain

$$\widetilde{\Delta}(z) = G \left(Q_0 G^{-1} \widetilde{B}_0 H^{-1} P_0 + \sum_{j=1}^{n} (z-w)^{\kappa_j} Q_j G^{-1} \widetilde{B}_j H^{-1} P_j \right) H\,.$$

Since the operators $G^{-1}\widetilde{B}_j H^{-1}$ are invertible from $\operatorname{Im} P_j$ to $\operatorname{Im} Q_j$, further we can find operators $T_j \in GL(\operatorname{Im} Q_j$ and $S_j \in GL(\operatorname{Im} P_j)$ such that

$$T_j G^{-1} \widetilde{B}_j H^{-1} S_j = B_j \qquad \text{for } 0 \le j \le n\,.$$

Set

$$T = (I - Q_0 - \ldots - Q_n) + \sum_{j=0}^{n} Q_j T_j Q_j$$

and

$$S = (I - P_0 - \ldots - P_n) + \sum_{j=0}^{n} P_j S_j P_j\,.$$

Then $T \in GL(Y)$, $S \in GL(X)$ and

$$T \left(Q_0 G^{-1} \widetilde{B}_0 H^{-1} P_0 + \sum_{j=1}^{n} (z-w)^{\kappa_j} Q_j G^{-1} \widetilde{B}_j H^{-1} P_j \right) S$$

$$= Q_0 B_0 P_0 + \sum_{j=1}^{n} (z-w)^{\kappa_j} Q_j B_j P_j = \Delta(z)\,.$$

It remains to set $E = GT$ and $F = SH$. □

If X and Y are finite dimensional Banach spaces, then the Smith factorization Theorem 4.3.1 says that an $L(X,Y)$-valued meromorphic operator function, at any point, is holomorphically equivalent to a local diagonal power function. This can be generalized to arbitrary finite meromorhic Fredholm functions (Def. 11.3.4).

11.6.4 Theorem. *Let X and Y be Banach spaces, let $D \subseteq \mathbb{C}$ be an open set and let $A : D \to L(X,Y)$ be a finite meromorphic Fredholm function. Then, at each point $w \in D$, A is holomorphically equivalent to a local diagonal power function at w.*

Proof. Let $w \in D$ be given, and let

$$A(z) = \sum_{n=m}^{\infty} (z-w)^n A_n$$

be the Laurent expansion of A at w. Since the operators $A_m, \dots, A_{-1}$ are finite dimensional and A_0 is a Fredholm operator, then there exists a projector P_X in X such that

$$\dim\left(X/\operatorname{Im} P_X\right) < \infty,$$
$$A_j P_X = 0 \qquad \text{if } m \le j \le -1,$$

and $A_0\big|_{\operatorname{Im} P_X}$ is an isomorphism between $\operatorname{Im} P_X$ and $A_0 \operatorname{Im} P_X$. Since $A_0 \operatorname{Im} P_X$ is of finite codimension in Y, we can find a projector P_Y in Y with $\operatorname{Im} P_Y = A_0 \operatorname{Im} P_X$. Then there is a neighborhood U of w such that the function

$$A' : U \setminus \{w\} \longrightarrow L(\operatorname{Im} P_X, Y),$$

defined by

$$A'(z) = A(z)\big|_{\operatorname{Im} P_X}, \qquad z \in U \setminus \{w\},$$

admits a holomorphic extension to w (which we also denote by A') where $A'(w) = A_0\big|_{\operatorname{Im} P_X}$. Let

$$A'' : U \longrightarrow L(\operatorname{Im} P_X, \operatorname{Im} P_Y)$$

be the function defined by $A'' = P_Y A'$. Then $A''(w) = A_0\big|_{\operatorname{Im} P_X}$ is invertible and, after shrinking U, we may assume that each $A''(z)$, $z \in U$, is invertible. Let $Q_X := I_X - P_X$ and $Q_Y := I_Y - P_Y$. Then, setting

$$E(z) = Q_Y + A'(z)\big(A''(w)\big)^{-1} P_Y, \qquad z \in U,$$

we get a holomorphic function $E : U \to L(Y)$ with $E(w) = I_Y$. After a further shrinking of U we may assume that $E(z) \in GL(Y)$ for all $z \in U$. Then $E^{-1}A$ is holomorphically equivalent to A over U. Setting

$$B(z) = E^{-1}(z) A'(z)\big|_{\operatorname{Im} P_X}, \qquad z \in U,$$

we get a holomorphic function

$$B : U \longrightarrow L(\operatorname{Im} P_X, \operatorname{Im} P_Y),$$

such that

$$B(z) = E^{-1}(z)A(z)\big|_{\operatorname{Im} P_X} \qquad \text{for } z \in U \setminus \{w\}.$$

Since

$$B(w) = A_0\big|_{\operatorname{Im} P_X}$$

is an isomorphism between $\operatorname{Im} P_X$ and $\operatorname{Im} P_Y$, after a further shrinking of U, we may assume that $B'(z)$ is an isomorphism between $\operatorname{Im} P_X$ and $\operatorname{Im} P_Y$ for all $z \in U$. Moreover,

$$E(z)\big(\operatorname{Im} P_Y\big) = A'(z)\Big(\big(A''(w)\big)^{-1}\big(\operatorname{Im} P_Y\big)\Big) = \operatorname{Im} A'(z), \quad z \in U,$$

and therefore

$$\operatorname{Im} P_Y = E^{-1}(z)\big(\operatorname{Im} A'(z)\big) = B(z)\big(\operatorname{Im} P_X\big). \qquad z \in U.$$

Hence

$$E^{-1}(z)A(z) = P_Y B(z) P_X + P_Y E^{-1}(z)A(z)Q_X + Q_Y E^{-1}(z)A(z)Q_X$$

for $z \in U \setminus \{w\}$ and, setting

$$F(z) = P_X - P_X(B(z))^{-1}P_Y E^{-1}(z)A(z)Q_X + Q_X,$$

for all $z \in U$, we get a holomorphic function $F : U \to GL(X)$ such that

$$E^{-1}(z)A(z)F(z) = P_Y B(z)P_X + Q_Y E^{-1}(z)A(z)Q_X, \qquad z \in U.$$

Since each $B(z)$, $z \in U$, is invertible from $\operatorname{Im} P_X$ to $\operatorname{Im} P_Y$, and since the spaces $\operatorname{Im} Q_X$ and $\operatorname{Im} Q_Y$ are finite dimensional, now the assertion of the lemma follows from the Smith factorization Theorem 4.3.1 applied to the operator function $Q_Y E^{-1}(z)A(z)\big|_{\operatorname{Im} Q_X}$. □

As a first consequence of Theorem 11.6.4 we obtain:

11.6.5 Corollary. *Let X and Y be Banach spaces, let $D \subseteq \mathbb{C}$ be an open set, and let $A, B : D \to L(X,Y)$ be two finite meromorphic Fredholm functions. Then, for each point $w \in D$, the following are equivalent:*

(i) *The functions A and B have the same numerical characteristics at w.*

(ii) *The functions A and B are holomorphically equivalent at w.*

Proof. That (ii) implies (i) is the statement of Proposition 11.3.7. To prove (i) ⇒ (ii), we consider a point $w \in D$ and assume that A and B have the same numerical characteristics at w. By Theorem 11.6.4, then we have local diagonal power functions $\Delta_{w,A}$ and $\Delta_{w,B}$ at w which are holomorphically equivalent at w to A and B, respectively. By Proposition 11.3.7, then, at w, $\Delta_{w,A}$ has the same numerical characteristic as A, and $\Delta_{w,B}$ has the same numerical characteristic as B. Since A and B have the same numerical characteristics at w, it follows that also $\Delta_{w,A}$ and $\Delta_{w,B}$ have the same numerical characteristics at w. Hence, by Proposition 11.6.3, the functions $\Delta_{w,A}$ and $\Delta_{w,B}$ are globally holomorphically equivalent over $\mathbb{C}$. Since, at $w \in D$, the function A is holomorphically equivalent to $\Delta_{w,A}$, and B is holomorphically equivalent to $\Delta_{w,B}$, this implies that A and B are holomorphically equivalent at w. □

The following corollary is an immediate consequence of Theorem 11.6.4[1] :

11.6.6 Corollary. *Let X, Y be Banach spaces, let $D \subseteq \mathbb{C}$ be a connected open set, let $A : D \to L(X, Y)$ be a finite meromorphic Fredholm function, and let P be the set of poles of A. Then there exist numbers $n, m \in \mathbb{N}$ and a discrete and closed subset Z of D with $P \subseteq Z$ such that*

$$\dim \operatorname{Ker} A(z) = n \quad \textit{and} \quad \dim \left(Y \big/ \operatorname{Im} A(z)\right) = m \qquad \textit{if } z \in D \setminus Z$$

and

$$\dim \operatorname{Ker} A(z) > n \quad \textit{and} \quad \dim \Big(Y / \operatorname{Im} A(z)\Big) < m \qquad \textit{if } z \in Z \setminus P.$$

In particular: If $A(z_0)$ is invertible for at least one point $z_0 \in D \setminus P$, then there exists a discrete and closed subset Z of D with $P \subseteq Z$ such that $A(z)$ is invertible for all $z \in D \setminus Z$.

11.6.7 Definition. With the notation from the preceding corollary we define: The points in $D \setminus Z$ will be called the **generic points** of A and the points in Z will be called the **non-generic points** of A.

Also from Theorem 11.6.4 we get the following

11.6.8 Theorem. *Let X, Y be Banach spaces, let $D \subseteq \mathbb{C}$ be a connected open set, and let $A : D \to L(X, Y)$ be a finite meromorphic Fredholm function. Let P be the set of poles of A, and let Z be the set of all non-generic points of A (Def. 11.6.7). Then there exist a holomorphic family $\{K(z)\}_{z \in D}$ of finite dimensional subspaces of X (Def. 6.4.1) and a holomorphic family $\{R(z)\}_{z \in D}$ of finite codimensional subspaces of Y such that*

$$\operatorname{Ker} A(z) = K(z) \quad \textit{and} \quad \operatorname{Im} A(z) = R(z) \qquad \textit{if } z \in D \setminus Z, \tag{11.6.4}$$

and

$$\operatorname{Ker} A(z) \underset{\neq}{\supset} K(z) \quad \textit{and} \quad \operatorname{Im} A(z) \underset{\neq}{\subset} R(z) \qquad \textit{if } z \in Z \setminus P. \tag{11.6.5}$$

[1]Note that this consequence can be obtained also in a more direct way without using the Smith factorization Theorem 4.3.1 which is contained in Theorem 11.6.4. Namely, by the same arguments as we deduced Theorem 11.6.4 from the Smith factorization lemma, one can deduce it from the simpler Proposition 11.5.1.

Proof. By Corollary 6.4.2 the families $\{\operatorname{Ker} A(z)\}_{z\in D\setminus Z}$ and $\{\operatorname{Im} A(z)\}_{z\in D\setminus Z}$ are holomorphic families of subspaces of X and Y, respectively. Therefore, over $D\setminus Z$, we can (and have to) define the required families K and R by (11.6.4).

Now consider a point $w\in Z$. Then, by Theorem 11.6.4, there exist a neighborhood $U\subseteq D$ of w, holomorphic functions $E: U\to GL(Y)$, $F: U\to GL(X)$ and a local diagonal power function

$$\Delta(z) = Q_0B_0P_0 + \sum_{j=1}^{n}(z-w)^{\kappa_j}Q_jB_jP_j\,, \qquad z\in\mathbb{C}\setminus\{w\}\,,$$

(with the properties as in Def. 11.6.1) such that

$$A = E\Delta F \qquad \text{on } U\setminus\{w\}\,.$$

By (11.6.4), then

$$K(z) = F^{-1}(z)\operatorname{Ker}\big(P_0+\ldots+P_n\big) \quad\text{and}\quad R(z) = E(z)\operatorname{Im}\big(Q_0+\ldots+Q_n\big)$$

for $z\in D\setminus Z$. Therefore we can (and have to) define

$$K(w) = F^{-1}(z)\operatorname{Ker}\big(P_0+\ldots+P_n\big) \quad\text{and}\quad R(w) = E(w)\operatorname{Im}\big(Q_0+\ldots+Q_n\big)\,.$$

Doing this for all points in Z, we obtain a holomorphic family $\{K(z)\}_{z\in D}$ of subspaces of X and a holomorphic family $\{R(z)\}_{z\in D}$ of subspaces of Y such that (11.6.4) is satisfied. Moreover, if $w\in Z\setminus P$, then $\kappa_j>0$ for all $1\le j\le n$, which implies (11.6.5). □

11.6.9 Definition. With the notation from the preceding theorem we define: The family $\{K(z)\}_{z\in D}$ will be called the **smoothing of the kernel** of A. The family $\{R(z)\}_{z\in D}$ will be called the **smoothing of the image** of A.

With this definition, as an immediate consequence of Theorem 6.9.1 we obtain:

11.6.10 Theorem. *Let X, Y be Banach spaces, let $D\subseteq\mathbb{C}$ be a connected open set, let $A: D\to L(X,Y)$ be a finite meromorphic Fredholm function, and let Z be the set of non-generic points of A. Let $\{K(z)\}_{z\in D}$ be the smoothing of the kernel of A, and let $\{R(z)\}_{z\in D}$ be the smoothing of the image of A, and let $z_0\in D\setminus Z$. Then there exist holomorphic functions $E: D\to GL(Y)$ and $F: D\to GL(X)$ with*

$$K(z) = F(z)K(z_0) \quad \textit{and} \quad R(z) = E(z)R(z_0) \qquad \textit{for all } z\in D. \tag{11.6.6}$$

11.6.11 Theorem. *Let X, Y be Banach spaces, let $D\subseteq\mathbb{C}$ be a connected open set, let $A: D\to L(X,Y)$ be a finite meromorphic Fredholm function, and let Z be the set of non-generic points of A (Def. 11.6.7). Then there exist*

– *a Banach space M,*

- *a surjective operator* $\Psi \in L(X, M)$ *with* $\dim \operatorname{Ker} \Psi < \infty$,
- *an injective operator* $\Phi \in L(M, Y)$ *with finite codimensional and, hence*[2], *closed image in* Y,
- *a finite meromorphic Fredholm function* $B_M : D \to L(M)$ *which is holomorphic and invertible on* $D \setminus Z$,

such that A *is holomorphically equivalent to* $\Phi B_M \Psi$ *over* D.

Proof. We use the notation of the preceding theorem; for $z \in D \setminus Z$ we put $\widetilde{A}(z) = E^{-1}(z)A(z)F(z)$. By (11.6.6), then

$$K(z_0) = \operatorname{Ker} \widetilde{A}(z) \quad \text{and} \quad R(z_0) = \operatorname{Im} \widetilde{A}(z) \qquad \text{for } z \in D \setminus Z. \tag{11.6.7}$$

Choose the required space M as a direct complement of $K(z_0)$ in X. Then, by the first relation in (11.6.6), $F(z)M$ is a direct complement in E of $K(z) = \operatorname{Ker} A(z)$ for $z \in D \setminus Z$. Hence $A(z)F(z)M = \operatorname{Im} A(z) = R(z)$ for $z \in D \setminus Z$. By the second relation in (11.6.6), this implies that

$$\widetilde{A}(z)M = R(z_0) \qquad \text{for } z \in D \setminus Z. \tag{11.6.8}$$

Moreover, since $K(z_0)$ is the kernel of $A(z_0)$ and $R(z_0)$ is the image of $A(z_0)$, $A(z_0)\big|_M$ is an invertible operator from M onto $R(z_0)$. Set $\Phi = A(z_0)\big|_M$. Then it follows from (11.6.8) that, by setting

$$B_M(z) = \Phi^{-1}\widetilde{A}(z)\big|_M \qquad \text{for } z \in D \setminus Z,$$

we obtain a holomorphic function $B_M : D \setminus Z \to GL(M)$. Since $\widetilde{A}$ is finite meromorphic and Fredholm at the points of Z also as a function with values in $L\big(X, R(z_0)\big)$, and since Φ^{-1} is an invertible operator from $R(z_0)$ to M, it follows that B_M is finite meromorphic and Fredholm at the points of Z. Finally we choose Ψ as the projector from X onto M parallel to $K(z_0)$. Then it is clear from the definition of B_M and (11.6.7) that

$$E^{-1}(z)A(z)F(z)\big|_M = \widetilde{A}(z)\big|_M = \Phi B_M(z)\Psi\big|_M \qquad \text{for all } z \in D \setminus Z.$$

Moreover, by the first relation in (11.6.6),

$$E^{-1}(z)A(z)F(z)K(z_0) = E^{-1}(z)A(z)K(z) = \{0\} \qquad \text{for all } z \in D \setminus Z\,.$$

Hence

$$E^{-1}(z)A(z)F(z) = \Phi B_M(z)\Psi \qquad \text{for all } z \in D \setminus Z\,,$$

i.e., A and $\Phi B_M(z)\Psi$ are holomorphically equivalent on D. □

[2]It follows from the Banach open mapping theorem that $\operatorname{Im}\Phi$ is closed if it is of finite codimension in Y.

11.6.12 Theorem. *Let X, Y be Banach spaces, let $D \subseteq \mathbb{C}$ be a connected open set, and let $A, B : D \to L(X,Y)$ be finite meromorphic Fredholm functions. Then the following are equivalent:*

(i) *The functions A and B have the same numerical characteristics at each point in D.*

(ii) *The functions A and B are locally holomorphically equivalent on D.*

(iii) *The functions A and B are globally holomorphically equivalent over D.*

Proof. The equivalence of (i) and (ii) follows from Corollary 11.6.5. The implication (iii) ⇒(ii) is trivial.

It remains to prove (ii) ⇒ (iii). Assume the functions A and B are locally holomorphically equivalent on D. Then A and B have the same set of non-generic points. We denote this set by Z. By Theorem 11.6.12, we may assume that $X = Y$, and $A(z), B(z) \in GL(X)$ for all $z \in D \setminus Z$. Then, by Corollary 11.6.6, the functions A and B are meromorphically invertible in the sense of Definition 11.4.1. Therefore, it follows from Theorem 11.4.2 that they are globally holomorphically equivalent over D. □

11.7 Global diagonalization of finite meromorphic Fredholm functions

Here we globalize the diagonalization Theorem 11.6.4. First we have to introduce an appropriate notion of a "global diagonal".

11.7.1 Definition. Let M be a Banach space, and let $D \subseteq \mathbb{C}$ be a connected open set. A meromorphic function $\Delta : D \to L(M)$ will be called an **invertible meromorphic diagonal function** on D if, for some $\omega \in \mathbb{N} \cup \{\infty\}$, it is of the form

$$\Delta = I + \sum_{j=1}^{\omega} (\varphi_j - 1) P_j \tag{11.7.1}$$

where:

- $\{P_j\}_{j=1}^{\omega}$ is a family of one-dimensional mutually disjoint projectors in M;
- $\{\varphi_j\}_{j=r+1}^{\omega}$ is a family of not identically vanishing meromorphic functions on D such that the functions φ_j/φ_{j+1}, $r+1 \le j \le \omega - 1$, are holomorphic on D;
- if $\omega = \infty$, then the following condition is satisfied (which ensures the conver-

gence of the infinite sum in (11.7.1)):

$$\begin{gathered}\text{for each compact set } K \subseteq D, \text{ there exists } \omega_K \in \mathbb{N} \text{ such that}\\ \text{the functions } \varphi_j,\ j > \omega_K, \text{ are holomorphic on } K, \text{ and}\\ \sum_{j=\omega_K+1}^{\infty} \|P_j\| \max_{z\in K} \big|\varphi_j(z) - 1\big| < \infty\,.\end{gathered} \tag{11.7.2}$$

11.7.2 Remark. If, in the preceding definition, $\omega < \infty$ and if we set $P := \sum_{j=1}^{\omega} P_j$ and $Q = I - P$, then Δ can be written in the form

$$\Delta = Q + P\left(\sum_{j=1}^{\omega} \varphi_j P_j\right) P\,,$$

which "shows" the diagonal. For $\omega = \infty$ this is impossible, because then the series $\sum_{j=1}^{\infty} \varphi_j P_j$ does not converge, at least not in the operator norm. In the sense of *strong* convergence however, this is sometimes possible.

For example, let M be a separable Hilbert space with the scalar product $(\cdot,\cdot)$ and an orthonormal basis $\{e_j\}_{j\in\mathbb{N}^*}$, and let $P_j(x) = (x, e_j)e_j$ for $x \in H$ and $j \in \mathbb{N}^*$. Then, for each compact set $K \subseteq D$ and sufficiently large $\omega_K \in \mathbb{N}$, the series

$$\sum_{j=\omega_K+1}^{\infty} \varphi_j P_j\, x$$

converges uniformly on K for each vector $x \in H$. This is even the case if instead of condition (11.7.2), we only require the following weaker condition:

For each compact set $K \subseteq D$, there exists $\omega_K \in \mathbb{N}$ such that the functions $\varphi_j,\ j > \omega_K$, are holomorphic on K and

$$\sup_{j>\omega_K\ ,\ z\in K} |\varphi_j(z)| < \infty\,.$$

Therefore, then Δ can be written in the form

$$\Delta = \sum_{j=1}^{\infty} \varphi_j P_j\,.$$

11.7.3 Remark. We use the notation of Definition 11.7.1.

Then Δ is a finite meromorphic Fredholm function on D (Def. 4.1.1), which can be seen as follows:

Let $w \in D$. Take a neighborhood U_w of w which is relatively compact in D. Then, by (11.7.2), we can find $\omega_w \in \mathbb{N}$ such that the functions φ_j, $j > \omega_w$, are holomorphic in a neighborhood of $\overline{U}_w$,

$$\sum_{j=\omega_w+1}^{\infty} \|P_j\| \max_{z\in\overline{U}_w} \big|\varphi_j(z)-1\big| < \frac{1}{2}\,. \tag{11.7.3}$$

Set $P=\sum_{j=1}^{\omega_w} P_j$ and $Q=I-P$. Then Δ can be written in the form

$$\Delta = Q\Big(I+\sum_{j=\omega_w+1}^{\infty}(\varphi_j(z)-1)P_j\Big)Q + P\Big(\sum_{j=1}^{\omega_w}\varphi_j(z)P_j\Big)P\,,$$

where, by (11.7.3),

$$Q\Big(I+\sum_{j=\omega_w+1}^{\infty}(\varphi_j(z)-1)P_j\Big)Q\Big|_{\operatorname{Im} Q}$$

is holomorphic and invertible on U_w as a function with values in $L(\operatorname{Im} Q)$. This shows that Δ is a finite meromorphic Fredholm function on U_w. Moreover, the numerical characteristic of Δ at w can be found as follows: Let $1\le j_1<\ldots<j_n\le\omega_w$ be the indices such that, for all $1\le j\le\omega_w$,

$$\operatorname{ord}_w \varphi_j \begin{cases} \neq 0 & \text{if } j\in\{j_1,\ldots,j_n\},\\ =0 & \text{if } j\notin\{j_1,\ldots,j_n\}.\end{cases}$$

Then

$$\Big(\operatorname{ord}_w \varphi_{j_1},\ldots,\operatorname{ord}_w \varphi_{j_n},0\,,0\Big)$$

is the numerical characteristic of Δ at w.

Now we construct invertible meromorphic diagonal functions with given numerical characteristics.

11.7.4 Lemma. *Let M be an infinite dimensional Banach space, and let $\{P_j\}_{j=1}^{\infty}$ be an infinite sequence of mutually disjoint one-dimensional projectors in M. Let $D\subseteq\mathbb{C}$ be a connected open set, and let Z be a discrete and closed subset of D. Suppose, for each $w\in Z$, a collection of non-zero integers $\kappa_1^w\ge\ldots\ge\kappa_{n_w}^w$ is given, $n_w\in\mathbb{N}^*$. Then there exists an invertible meromorphic diagonal function $\Delta: D\to L(M)$, where the projectors in (11.7.1) can be chosen from the family $\{P_j\}_{j=1}^{\infty}$, such that*

- *Δ is holomorphic and invertible on $D\setminus Z$;*
- *for $w\in Z$, $\big(\kappa_1^w,\ldots,\kappa_{n_w}^w,0,0\big)$ is the numerical characteristic of Δ at w.*

Proof. Let $\{w_\nu\}_{\nu=1}^{\omega}$ be the set Z numbered in some way, where $\omega\in\mathbb{N}^*$ if Z is finite and $\omega=\infty$ if Z is infinite. Set

$$\mathbb{N}_\omega^* = \begin{cases}\{1,2,\ldots,\omega\} & \text{if } \omega<\infty\,,\\ \mathbb{N}^* & \text{if } \omega=\infty\,.\end{cases}$$

By the Weierstrass product theorem (for example, by setting $f_w(z) = (z-w)^{\kappa_j}$ in Theorem 2.7.1), we can find a sequence $\{\phi_j\}_{j=1}^{\infty}$ of scalar meromorphic functions on D, which are holomorphic and $\neq 0$ on $D \setminus \{w_\nu\}_{\nu=1}^{\omega}$, and such that, for all $\nu \in \mathbb{N}_\omega^*$,

$$\operatorname{ord}_{w_\nu} \phi_j = \begin{cases} \kappa_j^{w_\nu} & \text{for } 1 \leq j \leq n_{w_\nu}\,, \\ 0 & \text{for all } j \in \mathbb{N}^* \text{ with } j > n_\nu\,. \end{cases} \tag{11.7.4}$$

As $\kappa_1^{w_\nu} \geq \ldots \geq \kappa_{n_{w_\nu}}^{w_\nu}$, then the quotients ϕ_j/ϕ_{j+1} are holomorphic on D.

First consider the case

$$m := \sup_{\nu \in \mathbb{N}_\omega^*} n_{w_\nu} < \infty\,.$$

Then

$$\Delta := I - \sum_{j=1}^{m} P_j + \sum_{j=1}^{m} \phi_j P_j = I + \sum_{j=1}^{m} (\phi_j - 1) P_j \tag{11.7.5}$$

is a meromorphic diagonal function on D. Since each ϕ_j is holomorphic and $\neq 0$ on $D \setminus \{w_\nu\}_{\nu=1}^{\omega}$, and by (11.7.4), we see that Δ has the same numerical characteristics as A.

Now let

$$\sup_{\nu \in \mathbb{N}_\omega^*} n_{w_\nu} = \infty\,. \tag{11.7.6}$$

Then $\omega = \infty$ and we have to modify the sequence $\{\phi_j\}_{j=1}^{\infty}$ in order to obtain a sequence $\{\varphi_j\}_{j=1}^{\infty}$ of meromorphic functions on D satisfying also condition (11.7.2).

Choose a sequence $\{K_s\}_{s=1}^{\infty}$ of compact subsets of D such that

- $\bigcup_{s=1}^{\infty} K_s = D$,

and, for each $s \in \mathbb{N}^*$,

- K_s is contained in the interior of K_{s+1},
- K_s is simply connected with respect to D (i.e., each connected component of $\mathbb{C} \setminus K_s$ contains at least one point of $\mathbb{C} \setminus D$).

Since the sequence $\{w_\nu\}_{\nu=1}^{\infty}$ is discrete and closed in D, each K_s contains at most a finite number of w_ν's. Therefore

$$m(s) := \sup\left\{n_{w_\nu} \;\middle|\; \nu \in \mathbb{N}^* \text{ and } w_\nu \in K_s\right\} < \infty$$

for all $s \in \mathbb{N}^*$. Since the functions ϕ_j satisfy condition (11.7.4) and, on $D \setminus \{w_\nu\}_{\nu=1}^{\infty}$, they are holomorphic and $\neq 0$, it follows that ϕ_j is holomorphic and $\neq 0$ in a neighborhood of K_s if $j > m(s)$, $s \in \mathbb{N}^*$. Set

$$s(j) = \max\left\{s \in \mathbb{N}^* \;\middle|\; j > m(s)\right\}, \qquad j \in \mathbb{N}^*.$$

Then each ϕ_j is holomorphic and $\neq 0$ in a neighborhood of $K_{s(j)}$. Therefore, by the Runge approximation Theorem 5.0.1 for invertible functions, there exists holomorphic functions $\psi_j : D \to \mathbb{C} \setminus \{0\}$ such that

$$\max_{z \in K_{s(j)}} \left| \frac{\phi_j(z)}{\psi_j(z)} - 1 \right| < \frac{2^{-j}}{\|P_j\|}, \qquad j \in \mathbb{N}^* . \tag{11.7.7}$$

Set

$$\varphi_j = \frac{\phi_j}{\psi_j}, \qquad j \in \mathbb{N}^* .$$

Since the functions ϕ_j satisfy condition (11.7.4) and, on $D \setminus \{w_\nu\}_{\nu=1}^\infty$, they are holomorphic and $\neq 0$, then the same is true for the functions φ_j, i.e.,, for all $\nu \in \mathbb{N}^*$,

$$\operatorname{ord}_{w_\nu} \varphi_j = \begin{cases} \kappa_j^{w_\nu} & \text{for } 1 \leq j \leq n_{w_\nu} , \\ 0 & \text{for all } j \in \mathbb{N}^* \text{ with } j > n_{w_\nu} , \end{cases} \tag{11.7.8}$$

and, on $D \setminus \{w_\nu\}_{\nu=1}^\infty$, each φ_j is holomorphic and $\neq 0$.

Since $\lim_{j\to\infty} s(j) = \infty$, each compact set $K \subseteq D$ is contained in some $K_{s(j)}$. Therefore it follows from (11.7.7) that the sequence $\{\varphi_j\}_{j=1}^\infty$ satisfies condition (11.7.2). Therefore, setting

$$\Delta = I + \sum_{j=1}^{\infty} (\varphi_j - 1) P_j , \tag{11.7.9}$$

we can define a meromorphic diagonal function Δ on D. Since the functions φ_j satisfy condition (11.7.8), and, on $D \setminus \{w_\nu\}_{\nu=1}^\infty$, they are holomorphic and $\neq 0$, we see that Δ is holomorphic and invertible on $D \setminus \{w_\nu\}_{\nu=1}^\infty$ and, at w_ν, $\nu \in \mathbb{N}^*$, Δ has the numerical characteristic (cf. Remark 11.7.3)

$$\left(\kappa_1^{w_\nu}, \ldots, \kappa_{n_{w_\nu}}^{w_\nu}, 0, 0 \right) . \qquad \square$$

11.7.5 Definition. Let X, Y be Banach spaces, and let $D \subseteq \mathbb{C}$ be a connected open set. A meromorphic function $\Delta : D \to L(X,Y)$ will be called a **meromorphic diagonal function** on D if there exists a Banach space M and an invertible meromorphic diagonal function $\Delta_M : D \to L(M)$ such that Δ is of the form

$$\Delta = \Phi \Delta_M \Psi , \tag{11.7.10}$$

where $\Psi \in L(X,M)$ is surjective with $\dim \operatorname{Ker} \Psi < \infty$, and $\Phi \in L(M,Y)$ is injective with finite codimensional and, hence[3], closed image in Y,

[3]It follows from the Banach open mapping theorem that $\operatorname{Im} \Phi$ is closed if it is of finite codimension in Y.

It is easy to see that each meromorphic diagonal function Δ on D (notation as in the preceding definition) is a finite meromorphic Fredholm function on D, where if $(\kappa_1, \ldots, \kappa_n, 0, 0)$ is the numerical characteristic of Δ_M at some point $w \in D$ (cf. Remark 11.7.3), then $\big(\kappa_1, \ldots, \kappa_n, \dim \operatorname{Ker} \Psi, \dim(Y/\operatorname{Im}\Phi)\big)$ is the numerical characteristic of Δ at w.

11.7.6 Theorem. *Let X, Y be Banach spaces, let $D \subseteq \mathbb{C}$ be a connected open set, and let $A : D \to L(X,Y)$ be a finite meromorphic Fredholm function. Then there exists a meromorphic diagonal function $\Delta : D \to L(X,Y)$ such that A and Δ are globally holomorphically equivalent over D.* (Recall that then, by Theorem 11.6.12, the functions A and Δ have the same numerical characteristics.)

Proof. Let Z be the set of non-generic points of A. Now we use the notation from Theorem 11.6.11. Then, by Lemma 11.7.4, there exists an invertible meromorphic diagonal function $\Delta_M : D \to L(M)$ which has the same numerical characteristics as B_M. By therorem 11.6.12, Δ_M is globally holomorphically equivalent to B_M over D, i.e., we have holomorphic functions $T, S : D \to GL(M)$ with $B_M = T\Delta_M S$. Let $\Phi^{(-1]}$ be a left inverse of Φ, let Q be a projector from Y to $\operatorname{Im}\Phi$, let $\Psi^{(-1)}$ be a right inverse of Ψ, and let P be the projector from X to $\operatorname{Im}\Psi^{(-1)}$ parallel to $\operatorname{Ker}\Psi$. Define holomorphic functions $F : D \to GL(X)$ and $D : D \to Gl(Y)$, setting

$$F(z) = (I - P) + P\Psi^{-1}S(z)\Psi P \quad \text{and} \quad E(z) = (I - Q) + Q\Phi T(z)\Phi^{(-1)}Q$$

for $z \in D$. Then

$$\begin{aligned} E(z)\Phi\Delta_M(z)\Psi F(z) &= Q\Phi T(z)\Phi^{(-1)}Q\Phi\Delta_M(z)\Psi P\Psi^{(-1)}S(z)\Psi P \\ &= \Phi T(z)\Delta_M(z)S(z)\Psi = \Phi B_M(z)\Psi \qquad \text{for all } z \in D. \end{aligned}$$

Hence $\Phi B_M \Psi$ and $\Phi\Delta_M\Psi$ are holomorphically equivalent over D, where, by definition, $\Phi\Delta_M\Psi$ is a meromorphic diagonal function. Since A and $\Phi B_M\Psi$ are holomorphically equivalent over D, this completes the proof. □

11.8 Comments

Holomorphic equivalence for polynomial matrix functions was introduced long ago (see [Ge], for holomorphic matrix functions see [BGR]). For operator functions it was considered probably for the first time in [Eni]. Developments concerned with this term can be found in [GKL, GS, Go3, GGK1] and the literature cited there.

The first two sections together with the proofs are borrowed from [GGK1]. The local principle of section 8.3 is new and is published here for the first time. The global diagonalization theorems for finite meromorphic Fredholm functions were obtained in [Le4, Le5]. The corresponding local results were proved in [GS]. The fact that the kernel and the cokernel of such functions can be smoothed, which follows from [GS], was established already in [Go1].

Bibliography

[A] Arlt, D., *Zusammenziehbarkeit der allgemeinen linearen Gruppe des Raumes c_0 der Nullfolgen*, Invent. Math. **1** (1966), 36–44.

[Ban] Banach, S., *Théorie des opérations linéaires,* Chelsea Publishing Company, New York 1955.

[Bar] Bari, N. K., *Trigonometric series* (Russian), Gosudarstvennoe Isdatelstvo Fisiko-Matematičeskoij Literaturoij, Moscou 1961.

[BGR] Ball, J. A., Gohberg, I., Rodman, L., *Interpolaion of rational matrix functions*, Operator Theory: Advances and Applications **45**, Birkhäuser, Basel-Boston, Mass., 1990, 605 pp.

[BGK] Bart, H., Gohberg, I, Kaashoek, M. A., *Minimal factorization of matrix and operator functions,* Operator Theory: Advances and Applications **1**, Birkhäuser, Basel-Boston, Mass., 1979, 228 pp.

[BGK1] Bart, H., Gohberg, I, Kaashoek, M. A., *Wiener-Hopf factorization, inverse Fourier transforms and exponentially dichotomous operators*, J. Functional Analyis, **68**, No. 1 (1986), 1–42.

[BGK2] Bart, H., Gohberg, I, Kaashoek, M. A., *Invariants for Wiener-Hopf equvalence of analytic operator functions,* Operator Theory: Advances and Applications **21**, Birkhäuser, 1986, 317–355.

[BGKR] Bart, H., Gohberg, I, Kaashoek, M. A., Ran, A.C.M. *Factorization of matrix and operator functions: The state space method,* Operator Theory: Advances and Applications **178**, Birkhäuser, 2008.

[Bu] Bungart, L., *On analytic fiber bundles. I. Holomorphic fiber bundles with infinite dimensional fibers*, Topology **7** (1967), 55–68.

[Ca] Cartan, H., *Sur les matrices holomorphes de n variables complexes*, J. Math. Pures Appl. **19** (1940), 1–26.

[CG] Clancey, K. F., Gohberg, I., *Factorization of matrix functions and singular integral operators*, Operator Theory: Advances and Applications **3**, Birkhäuser, Basel-Boston, Mass., 1981, 234 pp.

[CMM] Coifman, R. R., McIntosh, A., Meyer, Y., *L'integrale de Cauchy definit un operateur borne sur L_2 pour les courbes lipschitziennes*, Ann. of Math. (2) **116** (1982), no. 2, 361–387.

[Do1] Douady, A., *Un espace de Banach dont le group linéaire n'est pas connecxe*, Indag. Math. **27** (1965), 787–789.

[Do2] A. Douady, *Le problème de modules pur sous-espaces analytiques compacts d'un espace analyique donné*, Ann. Inst. Fourier, Grenoble **15** (1966), 1, 1–94.

[EMS1] Edelstein, I., Mitjagin, B., Semenov, E., *The linear groups of C and L_1 are contractible,* Bull. Acad. Polon. Sci. Sér. Math. Astronom. Phys. **18** (1970), 27–33.

[EMS2] Edelstein, I., Mitjagin, B., Semenov, E., *Letter to the editors*, Bull. Acad. Polon. Sci. Sér. Math. Astronom. Phys. **18** (1970), 213.

[Eni] Eni, V. M., *Stability of the root-number of a bundle of operators* (Russian), Bul. Akad. Štiince RSS Moldoven, 1966 no. 4, 78–81.

[F] Feldman, I. A., *Wiener-Hopf operator equations and their application to the transport equation,* Integral Equations and Operator Theory **3**, 1 (1980), 43–61.

[Ge] Gelfand, I. M., *Lectures on linear algebra*, with the collaboration of Z. Ya. Shapiro, translated from the second Russian edition by A. Shenitzer, reprint of the 1961 translation, Dover Books on Advanced Mathematics, Dover Publications, Inc., New York, 1989. vi+185 pp.

[FoRa] Forster Ramspott, *Okasche Paare von Garben nicht-abelscher Gruppen*, Invent. math. **1** (1966), 260–286.

[Go1] Gohberg, I. C, *On linear operators depending analytically on a parameter* (Russian), Doklady Akad. Nauk SSSR (N.S.) **78** (1951), 629–632.

[Go2] Gohberg, I. C., *The factorization problem for operator-valued functions*, Izv. Akad. Nauk SSSR, Ser. Mat. **28** (1964) 1055–1082.

[Go3] Gohberg, I. C., *Certain questions of the spectral theory of finitely meromorphic operator functions* (Russian), Izv. Akad. Nauk Armjan. SSR, Ser. Mat. **6** (1971), no. 2–3, 160–181; corrections, ibid. **7** (1972), no. 2, 152.

[Go4] Goherg, I. C., *A factorization problem in normed rings, functions of isometric and symmetric operators, and singular integral equations*, Russian Mathematical Survey, London Mathematical Society, Pitman Press, 19, 1 ,1964, 63–115.

[GGK1] Gohberg, I., Goldberg, S., Kaashoek, M. A., *Classes of linear operators. Vol. I*, Operator Theory: Advances and Applications **49**, Birkhäuser, Basel-Boston, Mass., 1990.

[GGK2] Gohberg, I., Goldberg, S., Kaashoek, M. A., *Basic classes of linear operators*, Birkhäuser, 2003.

[GKL] Gohberg, I. C., Kaashoek, M. A., Lay, D. C., *Equivalence, linearization, and decomposition of holomorphic operator functions*, J. Functional Analysis **28** (1978), no. 1, 102–144.

[GKS] Gohberg, I., Kaashoek, M. A., Spitkovsky, I. M., *An overview of matrix factorization theory and operator applications. Factorization and integrable systems* (Faro, 2000), 1–102, Operator Theory: Advances and Applications **141**, Birkhäuser, Basel-Boston, Mass., 2003.

[GKa] Gohberg, I, Kaashoek, M.A., *Constructive methods of Wiener-Hopf factorization,* Operator Theory: Advances and Applications **21**, Birkhäuser, 1986.

[GK] Gohberg, I. C., Krein, M. G., *Systems of integral equations on a half line with kernels depending on the difference of arguments*, Amer. Math. Soc. Transl. (2) **14** (1960), 217–287.

[GKru] Gohberg, I, Krupnik, N., *One-dimensional linear singular integral equations. Vol. I. Introduction*, translated from the 1979 German translation by Bernd Luderer and Steffen Roch and revised by the authors, Operator Theory: Advances and Applications **53**, Birkhäuser, Basel, 1992, 266 pp.

[GLR] Gohberg, I., Lancaster, P., Rodman, L., *Matrix polynomials. Computer Science and Applied Mathematics*, Academic Press, Inc. [Harcourt Brace Jovanovich, Publishers], New York-London, 1982.

[GL1] Gohberg, I. C., Leiterer, J, *Holomorphic vector-valued functions of one variable. I. Functions on a compact set* (Russian), Mat. Issled. **7** (1972), no. 4(26), 60–84, 252.

[GL2] Gohberg, I. C., Leiterer, J., *Holomorphic vector-valued functions of one variable. II. Functions in a domain* (Russian), Mat. Issled. **8** (1973), no. 1(27), 37–58, 236.

[GL3] Gohberg, I. C., Leiterer, J., *Cocycles, operator-valued functions and families of subspaces* (Russian), Mat. Issled. **8** (1973), no. 2(28), 23–56, 188.

[GL4] Gohberg, I. C., Leiterer, J, *The local principle for the factorization problem of continuous operator functions* (Russian), Hommage au Professeur Miron Nicolescu pour son 70'me anniversaire, Rev. Roumaine Math. Pures Appl. **18** (1973), 1585–1600.

[GL5] Gohberg, I. C., Leiterer, J., *General theorems on the factorization of operator-valued functions with respect to a contour. I. Holomorphic functions* (Russian), Acta Sci. Math. (Szeged) **34** (1973), 103–120, *II. Generalizations* (Russian), Acta Sci. Math. (Szeged) **35** (1973), 39–59.

[GM] Gohberg, I. C., Markus, A. S., *Two theorems on the gap between subspaces of a Banach space* (Russian), Uspehi Mat. Nauk **14** (1959), no. 5 (89), 135–140.

[GR1] Gohberg, I., Rodman, L., *Analytic operator-valued functions with prescribed local data*, Acta. Sci. Math. **45** (1983), 189–199.

[GR2] Gohberg, I., Rodman, L., *Analytic operator-valued functions with prescribed local data*, Acta Sci. Math. (Szeged) **45** (1983), no. 1–4, 189–199.

[GS] Gohberg, I. C., Sigal, E. I., *An operator generalization of the logarithmic residue theorem and the theorem of Rouche*, Math. USSSR Sbornik **13** (1971), 603–625.

[Gr1] Grauert, H., *Approximationssätze für holomorphe Funktionen mit Werten in komplexen Räumen*, Math. Ann. **133** (1957), 139–159.

[Gr2] Grauert, H., *Holomorphe Funktionen mit Werten in komplexen Lieschen Gruppen*, Math. Ann. **133** (1957), 450–472.

[Gr3] Grauert, H., *Analytische Faserungen über holomorph-vollständigen Räumen*, Math. Ann. **135** (1958), 263–273.

[HaGr] Griffiths, P., Harris, J., *Principles of Algebraic Geometry*, Wiley and Sons, 1978.

[Ho] L. Hörmander, *An introduction to complex analysis in several variables,* North-Holland 1990.

[KLH] *Spectral Methods in Linear Transport Theory,* Operator Theory: Advances and Applications **5**, Birkhäuser, Basel-Boston, Mass., 1982.

[K] Krein, M. G., *Integral equations on a half line with a kernel depending upon the difference of the arguments,* Uspekhi Mat. Nauk **13**, 5 (1958), 3–120 (Russian); English Trnsl., Amer. Math. Soc. Transl. (Series 2) **22** (1962), 163–288.

[KKM] Krein, M., Krasnoselski, M., Milman, D., *On the defect numbers of linear operators in Banach space and on some geometric problems* (Russian), Sbornik Trud. Inst. Mat. Akad. Nauk Ukr. SSR **11** (1948), 97–112.

[Ku] Kuiper, N. H., *The homotopy type of the unitary group of Hilbert space*, Topology **3** (1965), 19–30.

[Le1] Leiterer, J., *Über Operatorfunktionen einer komplexen Veränderlichen*, Dissertation B, Friedrich-Schiller-Universität Jena 1973.

[Le2] Leiterer, J., *Banach coherence and Oka's principle*, Contributions to the School on Global Analysis in Ludwigsfelde near Berlin, May 1976, Part 2, Akad. Wiss. DDR, Berlin 1977, VI.2–VI.67.

[Le3] Leiterer, J., *On factorization of matrix and operator functions* (Russian), Soobčenija Akad. Nauk. Grus. SSR. **88** (1977), no. 3, 541–544.

[Le4] Leiterer, J., *Global diagonalization of meromorphic Fredholm operator functions*, Akad. Wiss. DDR, Zentralinstitut für Mathematik und Mechanik, Berlin, September 1977, 1–15.

[Le5] Leiterer, J. *Local and global equivalence of meromorphic operator functions. Part I*, Math. Nachr. **83** (1978), 7–29,*Part II*, Math. Nachr. **84** (1978), 145–170.

[Le6] Leiterer, J., *Banach coherent analytic Fréchet sheaves*, Math. Nachr. **85** (1978), 91–109.

[Le7] Leiterer, J., *Oka's principle for sheaves of infinite-dimensional nonabelian groups* (Russian), in: Complex analysis and its applications (Russian), Nauka, Moscow 1978, 295–317.

[LiTz] Lindenstrauss, J., Tzafiri, L., *Classical Banach spaces. I and II*, Classics in Mathematics, Springer 1996. Reprint of the 1977 and 1979 ed.

[Ma] Markus, A. S., *Introduction to the spectral theory of polynomial operator pencils*, Translations of Mathematical Monographs, vol. **71**, American Mathematical Society.

[Mi] Mitjagin, B. S., *The homotopy structure of a linear group of a Banach space,* Russian Math. Surveys **25** (1070), no. 5 (155), 59–103.

[MiE] Mitjagin, B. S., Edelstein, I. S., *The homotopy type of linear groups of two classes of Banach spaces,* Funkcional. Anal. Appl. **4** (1970), 221–231.

[Mu] Muschelishvili, N. I., *Singular integral equations. Boundary problems of function theory and their application to mathematical physics*, translated from the second (1946) Russian edition and with a preface by J. R. M. Radok, corrected reprint of the 1953 English translation, Dover Publications, Inc., New York, 1992. 447 pp.

[N1] Neubauer, G., *Der Homotopietyp der Automorphismengruppe in den Räumen* l_p *und* c_0, Math. Ann. **174** (1967), 33–40.

[N2] Neubauer, G., Review in Math. Rev.:
MR0341523 (49♯6274a), MR0341524 (49♯6274b),
MR0341525 (49♯6274c), MR0341526 (49♯6274d).

[Ok] Oka, K., *Sur les fonctions analytiques de plusieurs variables. II. Deuxième problème de Cousin*, J. Sci. Hiroshima Univ., Ser. A9, 7–19 (1939).

[Pl] Plemelj, J., *Problems in the sense of Riemann and Klein*, edited and translated by J. R. M. Radok, Interscience Tracts in Pure and Applied Mathematics, no. 16, Interscience Publishers John Wiley & Sons Inc., New York-London-Sydney 1964, vii+175 pp.

[Re] Remmert, R., *Funktionentheorie 2*, Springer-Lehrbuch, 1992.

[Ro] Rodman, L., *An introduction to operator polynomials,* Operator Theory: Advances and Applications **38**, Birkhäuser, 1989.

[Ru] Rudin, W., *Real and complex analysis*, McGraw-Hill 1987.

[Sh1] Shubin, M. A., *On the local principle in the factorization problem*, Mat. Issled. **6** (1971), no. 1(28).

[Sh2] Shubin, M. A., *On holomorphic families of subspaces of a Banach space*, translated from the Russian by K. Clancey, W. Kaballo and G. Ph. A. Thijsse, Integral Equations and Operator Theory **2** (1979), no. 3, 407–420.

[Sz] Szegö, G., *Orthogonal Polynomials,* American Math. Soc., Providence, 1975.

[ViMa] Virozub, A.I., Macaev, V.I., *A theorem on the canonical factorization of operator functions* (Russian), Matem. Issled. Kishinev **8** (1973), no. 3, 145–150.

Index

$\mathcal{C}^1$-contour, 8
$\mathcal{C}^1$-parametrization, 8
$\mathcal{C}^G$-equivalence of cocycles, 135
$\mathcal{C}^k$-contour
closed connected, 72
$\mathcal{C}^k$-parametrization, 72
$\mathcal{F}$-cocycle, 344
trivial, 344
$\mathcal{F}$-equivalence, 344
$\mathcal{G}^\omega(E)$ and $\mathcal{G}^\infty(E)$, 151
$\mathcal{O}^G$-equivalence of cocycles, 135
$\mathcal{O}^G$-sheaf, 343

Abel's lemma, 19
approximation
Mergelyan, 38
Runge, 38
Runge for G-valued functions, 147
Runge for invertible scalar functions, 50

block matrix, 297
block Töplitz operator, 297
block Töplitz operator defined by a continuous operator function, 298

canonical factorization w.r.t. a contour, 220
functions with values in a Banach algebra, 221
canonical factorization w.r.t. the real line, 324
Cartan lemma, 129, 141
Cartan pair, 129
Cauchy formula, 12
Cauchy formula for continuous functions, 31
Cauchy integral, 61
Cauchy integral theorem, 11
homotopy version, 11
Cauchy-Riemann derivative, 29, 30
Cauchy-Riemann equation, 17
class $\mathcal{C}^\alpha$, 33
class $\mathcal{C}^{k+\alpha}$ on $\overline{D}_\pm$, 73
class $\mathcal{C}^k$ on Γ, 73
class $\mathcal{C}^{k+\alpha}$ on Γ, 73
closed $\mathcal{C}^k$-contour, 72
closed contour, 8, 9
cocycle
$(\mathcal{U}, \mathcal{C}^G)$-cocycle, 131
$(\mathcal{U}, \mathcal{O}^G)$-cocycle, 131
$\mathcal{C}^G$-cocycle, 135
$\mathcal{F}$-cocycle, 344
$\mathcal{F}$-trivial, 344
$\mathcal{O}^E$-cocycle, 43
$\mathcal{O}^E_{Z,m}$-cocycle, 52
$\mathcal{O}^G$-cocycle, 135
$\mathcal{O}^{\mathcal{G}^\infty(E)}$-cocycle, 152
$\mathcal{O}^{\mathcal{G}^\infty(E)}$-trivial, 152
$\mathcal{O}^{\mathcal{G}^\omega(E)}$-cocycle, 152
$\mathcal{O}^{\mathcal{G}^\omega(E)}$-trivial, 152
$\overline{\mathcal{O}}^E$-cocycle, 123
additive, 43
induced on a refinement, 134, 344
multiplicative, 131
trivial, 131

with values in a family of subspaces, 186
cocycle condition, 131
complexly differentiable, 1
complexly differentiable function, 1
continuous equivalence
of cocycles, 135
contour
$\mathcal{C}^1$-contour, 8
closed, 9
closed $\mathcal{C}^k$-contour, 72
closed connected $\mathcal{C}^k$-contour, 72
piecewise $\mathcal{C}^1$-contour, 8
convolution, 306
between $\mathcal{L}^1(\mathbb{R}, L(H))$ and $\mathcal{L}^2(\mathbb{R}, H)$, 315
of functions with values in a Banach algebra, 306
Cousin problem
additive, 43
multiplicative, 132

data of zeros, 52
derivative of f w.r.t. Γ, 73
diagonal function
local power, 399
meromorphic, 410
meromorphic invertible, 406

equivalence
of $\mathcal{F}$-cocycles, 344
of cocycles, 131, 135
of holomorphic operator functions, 379
of meromorphic operator functions at a point, 386
of meromorphic operator functions, global, 387
of meromorphic operator functions, local, 386
w.r.t. a contour, 221
essential singularity of f, 26
extension of operator functions, 380

factorization w.r.t. a contour
canonical, 220
global, 84
local, 93, 220
of operator functions, 220
of scalar functions, 84
factorization w.r.t. the real line, 324
canonical, 324
family of subspaces
complemented, 206
continuous, 172
holomorphic, 185
injective, 203
Shubin family, 204
filtration
characteristic of a meromorphic operator function at a point, 387
of an operator function relative to a closed contour, 255
finite meromorphic, 97, 388
finite meromorphic and Fredholm, 97, 388
finite meromorphic Fredholm function, 388
function
complexly differentiable, 1
Hölder continuous with exponent α, 33
Hölder-α continuous, 33
holomorphic, 1
of class $\mathcal{C}^\alpha$, 33
Wiener, 60

gap metric, 159
generalized inverse, 212
generalized invertible, 212
generic cokernel dimension of a finite meromorphic Fredholm function, 389
generic kernel dimension of a finite meromorphic Fredholm function, 389

generic point
of a finite meromorphic Fredholm function, 403
of a meromorphic matrix function, 394
generic rank of a meromorphic matrix function, 394

Hölder continuous, 33
Hölder continuous with exponent α, 33
Hölder-α continuous, 33
Hahn-Banach criterion, 15
holomorphic, 1
holomorphic equivalence
of cocycles, 135
holomorphic function, 1
holomorphic function with isolated singularities, 28
homotopy version of the Cauchy integral theorem, 11

index
of a finite meromorphic Fredholm function, 388
of a function which is finite meromorphic and Fredholm at a point, 97
of a function w.r.t. a contour, 86
w.r.t. a contour of a function which is finite meromorphic Fredholm, 105
Isolated singularities, 25
isolated singularity, 25

kernel function
of a symbol, 326
of a Wiener-Hopf operator, 326

Laurent series, 23
length of a contour, 10
Liouville's theorem, 2
local diagonal power function, 399
local factorization w.r.t. a contour, 220
of scalar functions, 93
local principle
additive, 82
multiplicative, 94
local splitting, additive, 82

maximum pronciple, 3
Mergelyan approximation, 38
meromorphic, 28
meromorphically invertible, 392
Mittag-Leffler theorem, 44, 53, 56
multiplicity
of a partial index of an operator function, relative to $\mathbb{R}$, 325
of a partial index of an operator function, relative to a closed contour, 220, 255

numerical characteristic of a meromorphic Fredholm function at a point, 389

order of the pole z_0, 26
order of the zero z_0, 26
orientation defined by, 11
orientation of a contour, 9
oriented as the boundary of, 11
oriented by, 11

parametrization
$\mathcal{C}^1$-parametrization, 8
$\mathcal{C}^k$-parametrization, 72
piecewise $\mathcal{C}^1$-parametrization, 9
partial index
non-zero of an operator function, relative to a closed contour, 220
of an operator function relative to a closed contour, 254
of an operator function, relative to $\mathbb{R}$, 325
piecewise $\mathcal{C}^1$-boundary, 11
piecewise $\mathcal{C}^1$-contour, 8
piecewise $\mathcal{C}^1$-parametrization, 9

pole, 26
Pompeiju operator, 33
power series
absolutely convergent, 19
convergent, 19
powers of a meromorphic Fredholm function at a point, 389

rational function with values in a Banach space, 223
regular point, 28
regularity of $\overline{\partial}$, 37
removable singularity of f, 25
residuum, 25
resolution of a family of subspaces, 185
globally short, 192
injective, 192
Runge approximation
for G-valued functions, 147
for invertible scalar functions, 50
for scalar functions, 38

section
(Γ, κ)-section of an operator function defined on a closed contour Γ, 251
(Γ, κ)-section of an operator function defined on a closed contour *through* (z, v), 258
of a family of subspaces, continuous, 172
of a family of subspaces, holomorphic, 184
of a sheaf, 343
sheaf
$\mathcal{O}^G$-sheaf, 343
defined by a data of zeros, 344
of finite order, 344
simply connected, 11
singular point, 28
smoothing of the image of a finite meromorphic Fredholm function, 404
smoothing of the kernel of a finite meromorphic Fredholm function, 404
splitting
of holomorphic cocycles, 132
splitting $\mathcal{R}$-spaces and algebras, 224
splitting of multiplicative cocycles, 131
splitting w.r.t. a contour, 60
global, 60
over U, 82
vanishing at infinity, 60
strong maximum principle, 3
symbol
of a Wiener-Hopf operator, 326
of a kernel function, 326

Töplitz operator, 297
trivial cocycle, 131, 152
triviality
of $\mathcal{F}$-cocycles, 344

uniqueness of holomorphic functions, 2

Weierstrass product theorem, 51
generalization, 54–56, 153–157, 369–376
Wiener functions, 60, 222
Wiener-Hopf operator
defined by an operator function on a closed contour, 270, 287
defined by an operator function on the real line, 326
of a kernel function, 326
of a symbol, 326

zero of, 25

Operator Theory: Advances and Applications (OT)

Edited by
Israel Gohberg, Tel Aviv University, Israel

BIRKHÄUSER

This series is devoted to the publication of current research in operator theory, with particular emphasis on applications to classical analysis and the theory of integral equations, as well as to numerical analysis, mathematical physics and mathematical methods in electrical engineering.

OT 192: Gohberg, I. / Leiterer, J., Holomorphic Operator Functions of One Variable and Applications. Methods from Complex Analysis in Several Variables (2009).
ISBN 978-3-0346-0125-2

OT 190/191: Adamyan, V. / Berezansky, Yu.M. / Gohberg, I. / Gorbachuk, M.L. / Gorbachuk, V. / Kochubei, A.N. / Langer, H. / Popov, G. (eds.), Modern Analysis and Applications. The Mark Krein Centenary Conference (2009).
Volume 1 (OT 190): Operator Theory and Related Topics. ISBN 978-3-7643-9918-4
Volume 2 (OT 191): Differential Operators and Mechanics. ISBN 978-3-7643-9920-7
Set Volumes 1/2: ISBN 978-3-7643-9924-5

OT 189: Rodino, L. / Wong, M.W. (eds.), New Developments in Pseudo-Differential Operators (2008). ISBN 978-3-7643-8968-0

OT 188: Behrndt, J. / Förster, K.-H. / Langer, H. / Trunk, C. (eds.), Spectral Theory in Inner Product Spaces and Applications (2008).
ISBN 978-3-7643-8910-9

OT 187: Ando, T. / Curto, R.E. / Jung, I.B. / Lee, W.Y. (eds.), Recent Advances in Operator Theory and Applications (2008).
ISBN 978-3-7643-8892-8

OT 186: Janas, J. / Kurasov, P. / Laptev, A. / Naboko, S. / Stolz, G. (eds.), Methods of Spectral Analysis in Mathematical Physics (2008).
ISBN 978-3-7643-8754-9

OT 185: Vasilevski, N., Commutative Algebras of Toeplitz Operators on the Bergman Space (2008).
ISBN 978-3-7643-8725-9

OT 184: Qin, Y., Nonlinear Parabolic-Hyperbolic Coupled Systems and Their Attractors (2008). Subseries **A**dvances in **P**artial **D**ifferential **E**quations. ISBN 978-3-7643-8813-3

OT 183: Nazaikinskii, V.E.. / Savin, A.Yu. / Sternin, B.Yu., Elliptic Theory and Noncommutative Geometry (2008). Subseries **A**dvances in **P**artial **D**ifferential **E**quations.
ISBN 978-3-7643-8774-7

OT 182: van der Mee, C., Exponentially Dichotomous Operators and Applications (2008). Subseries **L**inear **O**perators and **L**inear **S**ystems. ISBN 978-3-7643-8732-7

OT 181: Bastos, A. / Gohberg, I. / Lebre, A.B. / Speck, F.-O. (eds.), Operator Algebras, Operator Theory and Applications (2008).
ISBN 978-3-7643-8683-2

OT 180: Okada, S. / Ricker, W.J. / Sánchez Pérez, E.A., Optimal Domain and Integral Extension of Operators Acting in Function Spaces (2008). ISBN 978-3-7643-8647-4

OT 179: Ball, J.A. / Eidelman, Y. / Helton, J.W. / Olshevsky, V. / Rovnyak, J. (eds.), Recent Advances in Matrix and Operator Theory (2007). ISBN 978-3-7643-8538-5

OT 178: Bart, H. / Gohberg, I. / Kaashoek, M.A. / Ran, A.C.M., Factorization of Matrix and Operator Functions: The State Space Method (2007).
ISBN 978-3-7643-8267-4

OT 177: López-Gómez, J. / Mora-Corral, C., Algebraic Multiplicity of Eigenvalues of Linear Operators (2007). ISBN 978-3-7643-8400-5

OT 176: Alpay, D. / Vinnikov, V. (Eds.), System Theory, the Schur Algorithm and Multidimensional Analysis (2007).
ISBN 978-3-7643-8136-3

OT 175: Förster, K.-H. / Jonas, P. / Langer, H. / Trunk, C. (Eds.), Operator Theory in Inner Product Spaces (2007).
ISBN 978-3-7643-8269-8

Operator Theory: Advances and Applications (OT)

Edited by
Israel Gohberg, Tel Aviv University, Israel

BIRKHÄUSER

This series is devoted to the publication of current research in operator theory, with particular emphasis on applications to classical analysis and the theory of integral equations, as well as to numerical analysis, mathematical physics and mathematical methods in electrical engineering.

OT 174: Janas, J. / Kurasov, P. / Laptev, A. / Naboko, S. / Stolz, G. (Eds.), Operator Theory, Analysis and Mathematical Physics (2007). ISBN 978-3-7643-8134-9

OT 173: Emel'yanov, E.Yu., Non-spectral Asymptotic Analysis of One-Parameter Operator Semigroups (2006). ISBN 978-3-7643-8095-3

OT 172: Toft, J. / Wong, M.W. / Zhu, H. (Eds.), Modern Trends in Pseudo-Differential Operators (2007). ISBN 978-3-7643-8097-7

OT 171: Dritschel, M. (Ed.), The Extended Field of Operator Theory (2007). ISBN 978-3-7643-7979-7

OT 170: Erusalimsky, Ya.M. / Gohberg, I. / Grudsky, S.M. / Rabinovich, V. / Vasilevski, N. (Eds.), Modern Operator Theory and Applications (2007). ISBN 978-3-7643-7736-6

OT 169: Haase, M., The Functional Calculus for Sectorial Operators (2006). ISBN 978-3-7643-7697-0

OT 168: Koelink, E. / van Neerven, J. / de Pagter, B. / Sweers, G. (Eds.), Partial Differential Equations and Functional Analysis. ISBN 978-3-7643-7600-0

OT 167: Alpay, D. (Ed.), Wavelets, Multiscale Systems and Hypercomplex Analysis (2006). ISBN 978-3-7643-7587-4

OT 166: De Gosson, M., Symplectic Geometry and Quantum Mechanics (2006). Subseries **A**dvances in **P**artial **D**ifferential **E**quations ISBN 978-3-7643-7574-4

OT 165: Alpay, D. / Gohberg, I. (Eds.), Interpolation, Schur Functions and Moment Problems (2006). Subseries **L**inear **O**perators and **L**inear **S**ystems. ISBN 978-3-7643-7546-1

OT 164: Boggiatto, P. / Rodino, L. / Toft, J. / Wong, M.W. (Eds.), Pseudo-Differential Operators and Related Topics (2006). ISBN 978-3-7643-7513-3

OT 163: Langer, M. / Luger, A. / Woracek, H. (Eds.), Operator Theory and Indefinite Inner Product Spaces (2006). ISBN 978-3-7643-7515-7

OT 162: Förster, K.-H. / Jonas, P. / Langer, H. (Eds.), Operator Theory in Krein Spaces and Nonlinear Eigenvalue Problems (2006). ISBN 978-3-7643-7452-5

OT 161: Alpay, D. / Gohberg, I. (Eds.), The State Space Method. Generalizations and Applications (2005). Subseries **L**inear **O**perators and **L**inear **S**ystems. ISBN 978-3-7643-7370-2

OT 160: Kaashoek, M.A. / Seatzu, S. / van der Mee, C. (Eds.), Recent Advances in Operator Theory and its Applications. The Israel Gohberg Anniversary Volume (2005). ISBN 978-3-7643-7290-3

OT 159: Reissig, M. / Schulze, B.-W. (Eds.), New Trends in the Theory of Hyperbolic Functions (2005). Subseries **A**dvances in **P**artial **D**ifferential **E**quations. ISBN 978-3-7643-7283-5

OT 158: Eiderman, V.Ya. / Samokhin, M.V. (Eds.), Selected Topics in Complex Analysis (2005). ISBN 978-3-7643-7251-4

OT 157: Alpay, D. / Vinnikov, V. (Eds.), Operator Theory, Systems Theory and Scattering Theory: Multidimensional Generalizations (2005). ISBN 978-3-7643-7212-5

OT 156: Ebenfelt, P. / Gustafsson, B. / Khavinson, D. / Putinar, M. (Eds.), Quadrature Domains and Their Applications. The Harold S. Shapiro Anniversary Volume (2005). ISBN 978-3-7643-7145-6

MIX
Papier aus verantwortungsvollen Quellen
Paper from responsible sources
FSC® C105338

If you have any concerns about our products,
you can contact us on
ProductSafety@springernature.com

In case Publisher is established outside the EU,
the EU authorized representative is:
Springer Nature Customer Service Center GmbH
Europaplatz 3, 69115 Heidelberg, Germany

Printed by Libri Plureos GmbH
in Hamburg, Germany